Botany: Advances in Plant Biology

Botany: Advances in Plant Biology

Edited by Agatha Wilson

SYRAWOOD
PUBLISHING HOUSE

New York

Published by Syrawood Publishing House,
750 Third Avenue, 9th Floor,
New York, NY 10017, USA
www.syrawoodpublishinghouse.com

Botany: Advances in Plant Biology
Edited by Agatha Wilson

International Standard Book Number: 978-1-68286-646-7 (Hardback)

Cataloging-in-Publication Data

Botany : advances in plant biology / edited by Agatha Wilson.
    p. cm.
Includes bibliographical references and index.
ISBN 978-1-68286-646-7
1. Botany. 2. Plants. 3. Biology. I. Wilson, Agatha.
QK45.2 .B68 2019
580--dc23

TABLE OF CONTENTS

PREFACE

Botany is the study of plants and the life processes related to them. Botany as a scientific field holds relevance across a number of other areas such as plant physiology, medicine, epigenetics, human nutrition, etc. and is crucial for their development. It covers a range of topics such as plant anatomy, plant morphology, plant hormones, evolution and many other aspects related to plants. Research in botany explores the study of plant structure, growth, differentiation, taxonomy, evolution, etc. This book studies, analyzes and upholds the pillars of botany and its utmost significance in modern times. A number of latest researches have been included to keep the readers up-to-date with the global concepts in this area of study. This book includes contributions of experts and scientists which will provide innovative insights into this field. It is appropriate for students seeking detailed information in this area as well as for experts.

This book has been the outcome of endless efforts put in by authors and researchers on various issues and topics within the field. The book is a comprehensive collection of significant researches that are addressed in a variety of chapters. It will surely enhance the knowledge of the field among readers across the globe.

It gives us an immense pleasure to thank our researchers and authors for their efforts to submit their piece of writing before the deadlines. Finally in the end, I would like to thank my family and colleagues who have been a great source of inspiration and support.

Editor

Metabolic and transcriptional elucidation of the carotenoid biosynthesis pathway in peel and flesh tissue of loquat fruit during on-tree development

Margarita Hadjipieri[1], Egli C. Georgiadou[1], Alicia Marin[2], Huertas M. Diaz-Mula[2], Vlasios Goulas[1], Vasileios Fotopoulos[1], Francisco A. Tomás-Barberán[2] and George A. Manganaris[1]*

Abstract

Background: Carotenoids are the main colouring substances found in orange-fleshed loquat fruits. The aim of this study was to unravel the carotenoid biosynthetic pathway of loquat fruit (cv. 'Obusa') in peel and flesh tissue during distinct on-tree developmental stages through a targeted analytical and molecular approach.

Results: Substantial changes regarding colour parameters, both between peel and flesh and among the different developmental stages, were monitored, concomitant with a significant increment in carotenoid content. Key genes and individual compounds that are implicated in the carotenoid biosynthetic pathway were further dissected with the employment of molecular (RT-qPCR) and advanced analytical techniques (LC-MS). Results revealed significant differences in carotenoid composition between peel and flesh. Thirty-two carotenoids were found in the peel, while only eighteen carotenoids were identified in the flesh. *Trans*-lutein and *trans*-β-carotene were the major carotenoids in the peel; the content of the former decreased with the progress of ripening, while the latter registered a 7.2-fold increase. However, carotenoid profiling of loquat flesh indicated *trans*-β-cryptoxanthin, followed by *trans*-β-carotene and 5,8-epoxy-β-carotene to be the most predominant carotenoids. High amounts of *trans*-β-carotene in both tissues were supported by significant induction in a chromoplast-specific *lycopene β-cyclase (CYCB)* transcript levels. *PSY1, ZDS, CYCB* and *BCH* were up-regulated and *CRTISO, LCYE, ECH* and *VDE* were down-regulated in most of the developmental stages compared with the immature stage in both peel and flesh tissue. Overall, differential regulation of expression levels with the progress of on-tree fruit development was more evident in the middle and downstream genes of carotenoid biosynthetic pathway.

Conclusions: Carotenoid composition is greatly affected during on-tree loquat development with striking differences between peel and flesh tissue. A link between gene up- or down-regulation during the developmental stages of the loquat fruit, and how their expression affects carotenoid content per tissue (peel or flesh) was established.

Keywords: *Eriobotrya japonica*, Developmental stages, Maturation, Ripening, β-carotene, β-cryptoxanthin, Lutein, LC-MS, Biosynthetic pathway

* Correspondence: george.manganaris@cut.ac.cy
[1]Department of Agricultural Sciences, Biotechnology and Food Science,
Cyprus University of Technology, 3603 Lemesos, Cyprus
Full list of author information is available at the end of the article

Background

Loquat (*Eriobotrya japonica* Lindl) is a member of the Rosaceae family that is commercially cultivated in many countries [1–3], being highly appreciated for its light, refreshing taste [2, 4]. Therefore, although initially considered as an underutilized crop, nowadays loquat can gain added value as it is available during late winter-early spring period [5]. The loquat tree has three flushes of growth per year, and the principal tree growth can be separated into 8 distinct developmental stages [2]. In particular, under Mediterranean weather conditions, the tree blooms between October and early November and its fruit develops through winter, ripening from early February until May. Fruit is usually consumed fresh, but it is also known to be used processed into jam, jelly, wine, syrup, and juice. It is also known that the leaves, flowers and fruits are traditionally used in Chinese medicine since they are linked with health-promoting properties [1, 6].

Carotenoids play an important role in loquat, as they affect organoleptic characteristics and health properties of the fruit. In particular, carotenoids are the main pigments in loquat and impact flavor acceptability, since they are precursors of important volatile flavor compounds [7]. Regarding nutraceutical properties of fruit carotenoids, a significant number of studies depicted their beneficial effect to the promotion of health, including the prevention and/or treatment of chronic and cardiovascular diseases [8]. In particular, fruits rich in carotenoids are directly connected to the prevention of inflammation and cataract [1, 9, 10] and are also known to enhance immune responses [1, 9, 11]. Carotenoid profile in loquat is influenced by maturity stage, environmental and most promptly genetic factors. Loquat cultivars have been segregated to white- and red-fleshed [10, 12]. However, segregation of loquat cultivars based on their flesh color can be confusing, since additionally the terms yellow- and orange-fleshed are being used. White-fleshed cultivars have a creamy, pale yellow color, while the terms red- and orange-fleshed can be considered as synonymous. The latter type cultivars have higher carotenoid concentrations than the lighter coloured ones [1, 5, 9, 10, 12].

Carotenoids are formed from isopentenyl diphosphate (IPP), a five-carbon compound, and dimethylallyl diphosphate (DMAPP), its allylic isomer. These compounds form geranylgeranyl diphosphate (GGPP) which in turn forms phytoene through the activity of *phytoene synthase (PSY)*. Phytoene forms lycopene via four desaturation reactions with the involvement of *phytoene desaturase (PDS)* ζ-carotene isomerase (ZISO), ζ-carotene desaturase (ZDS) and *carotene isomerase (CRTISO)* [13]. Lycopene, in turn undergoes a series of reactions to form lutein, through the ε, β-branch, the predominant carotenoid pigment in green plants [13], and violaxanthin from

zeaxanthin with the presence of *zeaxanthin epoxidase (ZEP)* though the β, β-branch (Fig. 1). This forms the xanthophyll cycle, the mechanism that enables plant adaptation to high light stress [10]. 9-cis-neoxanthin is derived from the conversion of violaxanthin by *neoxanthin synthase (NSY)*, which in turn forms the phytohormone abscisic acid through *9-cis-epoxycarotenoid dioxygenase (NCED)* activity [11, 14], which controls abiotic stress signaling pathways [14].

The carotenoid biosynthetic pathway is controlled by the presence of the key enzyme PSY (Fig. 1, [10, 11, 13, 15–17]). Regarding loquat fruit, Fu et al. [10] investigated the mechanism underlying the differentiation of carotenoids in a red-fleshed (cv. 'Luoyangqing') and a white-fleshed (cv. 'Baisha') cultivar; differences in carotenoid accumulation in the two cultivars were linked with the differential expression of *PSY1*, *CYCB*, and *BCH* genes. The aim of the current study was to monitor the carotenoid composition in peel and flesh tissue of 'Obusa' fruits, an orange-fleshed cultivar, in correlation with the progress of fruit maturity. Towards this aim, high-resolution temporal expression profiles of carotenoid biosynthetic genes in both tissues were determined by RT-qPCR and linked with individual carotenoids, quantified by LC-MS.

Methods

Fruit material and experimental design

Loquat fruits cv. 'Obusa' were harvested at ca. 10-day intervals, between March 30th and May 14th (Fig. 2), from a commercial orchard (Episkopi, Lemesos, Cyprus), owned by the first author. For each developmental stage, 30 uniform fruits were selected based on size and colour; such fruit were divided into three 10-fruit sublots, representing the biological replications. Fruit were initially used for the determination of physical dimensions and colour and subsequently for molecular and analytical analysis, as described below. The developmental stages were defined using the BBCH scale [2].

For the molecular analysis, fresh samples of both peel and flesh were flash frozen in liquid nitrogen and maintained at −80 °C until needed. For the determination of carotenoid profiles, samples were freeze-dried (Freeze Dryer-Christ Alpha 1–4 LD plus).

Quality attributes

Physical dimensions of fruit were determined with the employment of an analytical grader and an electronic calibre (IS11112, Insize). The colour parameters CIE L* (brightness or lightness; 0 = black, 100 = white), a* (−a* = greenness, +a* = redness) and b* (−b* = blueness, +b* = yellowness) were measured at the peel and at the flesh tissue per fruit with a chroma meter (CR-400, Konica Minolta).

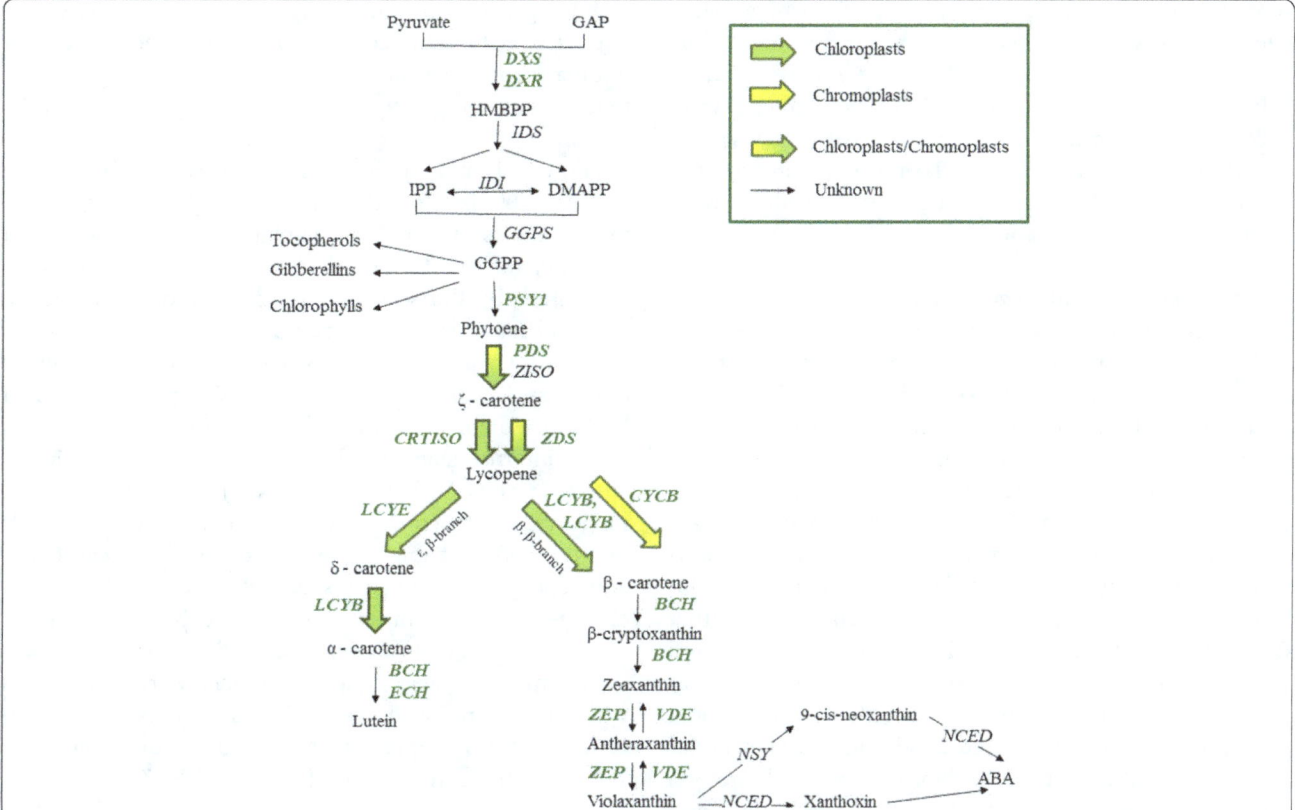

Fig. 1 Carotenoid biosynthetic pathway in loquat fruit. Genes examined are in *green italic bold letters*. The enzymes/genes are: DXS, 1-deoxy-D-xylulose 5-phosphate-synthase; DXR, DXP reductoisomerase; IDS, isopentenyl pyrophosphate synthase; IDI, isopentenyl pyrophosphate isomerase; GGPS, geranylgeranyl diphosphate synthase; PSY1, phytoene synthase; PDS, phytoene desaturase; ZISO, ζ-carotene isomerase; ZDS, ζ-carotene desaturase; CRTISO, carotene isomerase; LCYB, lycopene β-cyclase; CYCB, chromoplast-specific lycopene β-cyclase; LCYE, lycopene ε-cyclase; BCH, β-carotene hydroxylase; ECH, ε-carotene hydroxylase; ZEP, zeaxanthin epoxidase; VDE, violaxanthin de-epoxidase; NSY, neoxanthin synthase; NCED 9-cis-epoxycarotenoid dioxygenase. The metabolites are: pyruvate; GAP, D-glyceraldehyde 3-phosphate; HMBPP, (E)-4-hydroxy-3-methylbut-2-enyl diphosphate; DMAPP, dimethylallyl pyrophosphate; IPP, isopentenyl pyrophosphate; GGPP, geranylgeranyl diphosphate; Phytoene; ζ – carotene; lycopene; α – carotene; β – carotene; δ - carotene; lutein; β-cryptoxanthin; zeaxanthin; antheraxanthin; violaxanthin; 9-cis-neoxanthin; xanthoxin; ABA, abscisic acid (Figure is modified from [10, 11, 13, 16, 17])

Spectrophotometric determination of carotenoid and chlorophyll content

Twenty mL of acetone–hexane (4:6, *v*/v) were added to 100 mg of lyophilised plant material and thoroughly mixed. Upon separation of the two phases, the absorbance was determined in the supernatant at 453, 505, 645 and 663 nm. Based on Nagata and Yamashita equations, total carotenoids, chlorophyll-a and chlorophyll-b contents were determined [18].

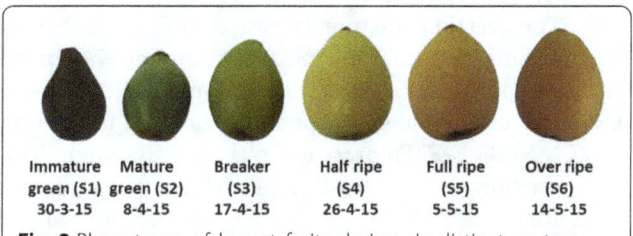

| Immature green (S1) | Mature green (S2) | Breaker (S3) | Half ripe (S4) | Full ripe (S5) | Over ripe (S6) |
| 30-3-15 | 8-4-15 | 17-4-15 | 26-4-15 | 5-5-15 | 14-5-15 |

Fig. 2 Phenotypes of loquat fruits during six distinct on-tree developmental stages (S1-S6)

Liquid chromatography mass spectrometry (LC-MS) analysis of carotenoids

Extraction and saponification of loquat carotenoids

Liquid-liquid extraction and saponification of the samples was carried out as previously described by Minguez-Mosquera & Hornero-Mendez [19]. Briefly, 0.5 g of freeze-dried tissue (peel or flesh) was homogenized in acetone-butylated hydroxytoluene - 0.1% using an UltraTurrax (Ika, Staufen, Germany) and centrifuged (2000 rpm, 10 min, 4 °C). Extraction steps were repeated until complete removal of colour in the sample. The internal standard used was β-Apo-8′-carotenal (Sigma, St Louis, MO, USA). The extracts were combined and treated with diethyl ether. A solution of NaCl (10%, *w*/v) was added to separate the phases. The lower phase was discarded and the remaining phase was washed with Na_2SO_4 (2%, *w*/v) to remove water residues. Fifty mL of a methanolic solution of KOH (20%, *w*/v) was added and left for 1 h in darkness. The organic phase was washed several times with deionized water until washings were neutral. It was then filtered through a bed of anhydrous Na_2SO_4 and evaporated until dry using a speed vacuum (Thermo Scientific Savant SPD121P). The pigments were collected with 1 mL of acetone: methanol (7:3, v/v) and stored at −20 °C until needed. To prevent isomerization and photodegradation of carotenoids, all procedures were carried out under pale light.

LC- MS analysis of loquat carotenoids

The carotenoid analysis was carried out using an Agilent 1200 HPLC equipped with a photodiode array detector and a single quadrupole mass spectrometer detector in series (6120 Quadrupole, Agilent Technologies, Santa Clara, CA, USA). Chromatographic separation was performed on a reverse-phase Poroshell 120 EC-C_{18} column (100 mm × 3 mm, 2.7 μm particle size) (Agilent Technologies) operating at 32 °C. Water with 0.05 M ammonium acetate and acetonitrile: methanol (70:30) were used as mobile phases A and B, respectively, with a flow rate of 0.7 mL min^{-1}. The linear gradient started with 60% of solvent B in A, reaching 100% solvent B at 20 min; this was maintained up to 35 min. The initial conditions were re-established at 36 min and kept under isocratic conditions up to 40 min. Injection volume was 5 μL. Detection and quantification of all carotenoids and carotenoid esters were carried out in UV-vis at 450 nm (Additional file 1: Figure S1).

The identification of carotenoids in loquat was carried out on an Agilent 1100 HPLC system equipped with a photodiode array and an ion trap mass spectrometer detector (Agilent Technologies, Waldbronn, Germany). The mass detector was a Bruker ion trap spectrometer (model HCT Ultra) equipped with an APCI (Atmospheric Pressure Chemical Ionization). The mass spectrometer parameters were as follows: positive ion mode (APCI +); source temperature, 350 °C; probe temperature, 450 °C, corona voltage, 4.0 kV; The full scan mass covered the range from m/z 100 up to m/z 1200 and the target mass was adjusted to 350. Collision-induced fragmentation experiments were performed in the ion trap using helium as the collision gas, with voltage ramping cycles from 0.3 up to 2 V. Mass spectrometry data were acquired in the positive mode, and the MS^n was carried out in the automatic mode. The identification of the peaks was performed by the extracted ion-chromatograms of the ion current at m/z values corresponding to the $[M-H]^+$ ions of the individually investigated compounds, as well as their fragmentation. Furthermore, to confirm the identification of the carotenoids and obtain a more reliable identification, samples were analyzed with the Agilent 1290 Infinity UPLC system coupled to a quadrupole (Q-TOF) mass spectrometer (6550 Accurate-Mass QTOF, Agilent Technologies) using an electrospray interface with jet stream technology. The chromatographic separation was developed under the same conditions, as described above. The optimal conditions for the electrospray interface were as follows: gas temperature: 300 °C, drying gas 11 L/min, nebulizer 65 psi, sheath gas temperature 400 °C, sheath gas flow 12 L/min. Spectra were acquired in the m/z range of 100–1100, in a positive mode and with an acquisition rate of 1.5 spectra in MS, maintaining a mass resolution over 50,000 for the mass range used. Internal mass calibration by the simultaneous acquisition of reference ions and mass drift compensation was used for obtaining low mass errors. Q-TOF MS data were processed using the Mass Hunter Qualitative Analysis software (version B.06.00). For quantification, β-apo-8′-carotenal was used as an internal standard. Lutein, β-carotene and violaxanthin (Sigma, St Louis, MO, USA) in a concentration range of 5–100 μg.mL^{-1} were used to quantify compounds in three different groups, hydroxycarotenoids, carotenes, and epoxycarotenoids, respectively. Neoxanthin, neochrome, β-cryptoxanthin epoxides and β-carotene epoxides were estimated as violaxanthin. β-Cryptoxanthin was quantitated as lutein. The *cis*-isomers were quantitated with the calibration curve of the all-*trans* isomers. Concentrations were expressed as micrograms of pigment per 100 g of sample fresh weight (μg 100 g^{-1} F.W.).

RNA extraction, cDNA synthesis and quantitative real-time RT-PCR analysis

Total RNA was extracted from three bulked biological replicates of loquat fruit material for each developmental stage (S1-S6) according to a modified cetyltrimethylammonium bromide (CTAB) protocol developed by Gambino et al. [20]. Next, RNA integrity was confirmed spectrophotometrically (Nanodrop 1000 Spectrophotometer, Thermo Scientific) followed by gel electrophoresis and samples were treated with RNase-free DNase (Cat. No. NU01a, HT Biotechnology LTD, England) to remove total gDNA, as elsewhere described [21].

Total RNA (0.5 μg) was reverse transcribed using the PrimeScript™ RT reagent Kit (Takara Bio, Japan), according to the manufacturer's protocol (Takara Bio, Japan). Subsequently, quantitative real-time RT-PCR was performed with BioRad IQ5 real-time PCR cycler (BioRad, USA). In total, three biological replicates were analyzed for each developmental stage for both loquat peel and flesh. The reaction mixture consisted of 4 μL cDNA in reaction buffer (15-fold diluted first-strand cDNA for all genes except for *VDE* and *LCYE* that were diluted 5-fold), 0.5 μL of each primer (10 pmol/μL) and 5 μL Sensi-FAST™ SYBR® & Fluorescein mix 2× (Bioline). The total reaction volume was 10 μL. The initial denaturation step was at 95 °C for 5 min, followed by 40 cycles of amplification [95 °C for 30 s, annealing temperature (Ta °C) for 30 s, and 72 °C for 30 s] and a final elongation stage at 72 °C for 5 min. Gene amplification cycle was followed by a melting curve run, carrying out 61 cycles with 0.5 °C increment from 65 to 95 °C. The annealing temperature of previously published primers for loquat carotenoid biosynthetic genes (58 to 65 °C) is shown in Additional file 1: Table S1. Loquat's *actin* gene was used as a housekeeping reference gene (*EjACT*).

Statistical analyses

Statistical analyses were carried out by comparing the averages of each developmental stage based on the analysis of variance (one way ANOVA) according to Duncan's multiple way test with a significance level of 5% ($P \leq 0,05$), using the SPSS v.17.0 statistical analysis software package.

The relative quantification and statistical analysis of gene expression levels was performed with the REST-XL software, using the pairwise fixed reallocation randomization test [22]. Gene expression levels were normalized against the *EjACT* housekeeping reference gene; the initial developmental stage (S1) both for peel and flesh tissue was used for calibration.

Results and discussion
Qualitative attributes

Fruit weight, length and width ranged between 25.3–59.1 g, 48.8–57.0 mm and 34.4–45.0 mm, respectively (Additional file 1: Table S2). Maximum fruit size and weight was recorded at stage S5, which coincides with the optimum maturity stage for harvest.

Colour parameters in the flesh ranged between 52.12–74.32 for L*, −11.51-10.67 for a* and 30.75–36.67 for b*. The corresponding values for L*, a*, b* parameters in the peel were 45.65–64.99, −17.66-12.42 and 28.20–49.19 (Additional file 1: Table S3). Previous study in loquat cultivars indicated that the a*/b* chroma ratio receive negative values in immature fruits, around zero for pale yellow-colored fruits, and positive values for orange-colored loquat fruits; thus higher ratio can be linked with higher carotenoid accumulation [1]. In our study, the a*/b* ratio in the flesh received negative values at stages S1-S3, implying the green colour, the S4 value indicates the colour break and the S5 and S6 had the higher values. Similarly, the a*/b* ratio was lower in the peel during the first three early stages (−0.60 to −0.20), close to zero in the breaker stage, while it received positive values during the last maturation stages (0.24–0.29) (Additional file 1: Table S3). Overall, the a*/b* ratio can be linked with total carotenoid accumulation and its transient increase with the progress of on-tree development, in accordance with previous studies [1, 16].

Carotenoid and chlorophyll contents

Initially, a rapid colorimetric assay was employed to screen carotenoid and chlorophyll contents. Total carotenoids varied between 0.8–6.1 and 0.1–5.9 mg 100 g^{-1} FW β-carotene equivalents in the peel and in the flesh, respectively (data not shown). The highest concentrations in Chl a and Chl b were found in the peel tissue at the immature green stage (18.0 and 12.1 mg 100 g^{-1} FW, respectively), while their contents were substantially lower in the flesh (0.6 and 0.5 mg 100 g^{-1} FW at S1 stage, respectively) and degraded thereafter with the progress of on-tree ripening. During ripening of fleshy fruits chloroplasts turn into chromoplasts; this process encompass a transient increase of carotenoids and degradation of chlorophylls [23]. Chl a is a blue-green coloured pigment and is less stable than the yellow-green Chl b.

Identification of carotenoids in loquat fruit using LC-MS

Carotenoids in the peel and in the flesh were identified and quantified (Table 1, Additional file 1: Figure S1). Thirty-two carotenoids were detected by HPLC-DAD and LC-MS techniques. Peak identification was based on their relative retention time values, their UV-Vis spectra, their mass spectra, information from the literature and comparison with authentic standards when possible. Table 1 summarizes the identification data for each carotenoid, including chromatographic and spectroscopic values.

All-*trans*-neoxanthin (peak 1) showed a characteristic UV-visible spectrum. The molecular mass of neoxanthin was confirmed by the protonated molecule at m/z 601 and by consecutive losses of three hydroxyl groups from the protonated molecule, at m/z 583, 565 and 547, verified by MS/MS. The UV-visible absorption spectrum of neochrome (peak 2) showed λmax at 397, 420 and 448 nm with high spectral fine structure (%III/II 90); these values are in agreement with previous studies in loquat [9]. All-trans-violaxanthin (peak 3) was identified by comparison with the authentic standard. The protonated molecule at m/z 601, and the fragments at m/z 583 and 565, due to the losses of hydroxyl groups and at m/z 221, all formed from 601 at both MS/MS and in-source fragmentations. Peak 8 was tentatively identified as *cis*-violaxanthin with a mass spectrum, lower λmax and spectral fine structure values similar to those of peak 3. β-diepoxy cryptoxanthin, peak 7 and 32 (only in flesh in S5 and S6)] showed the [M + H]$^+$ at m/z 585. The second order MS experiments revealed a fragment at m/z 567 due to the loss of water and the ions at m/z 221 and at m/z 205 characteristic of the epoxide group with one ion located in a ring with a hydroxyl group and another one in an unsubstituted ring respectively. All *trans*-lutein was identified by comparison with an authentic standard.

Table 1 UV/vis spectra and characteristic ions of carotenoids from six maturation stages of loquat fruits, obtained by HPLC-PDA-MS

Peak	Carotenoid	t_R (min)	λ_{max} (nm)	%III/II	%Ab/II	[M + H] + m/z	HPLC/APCI (+)-MS^n experiment m/z (% base peak)	Exact mass	Score	Error (ppm)	Molecular formula
1	All-*trans*-neoxanthin	14.1	412,436,462	70	0	601 (40), 583 (75), 565 (100)	MS^2 [601]: 583 (100), 565 (47), 547 (9), 509 (6),491 (5), 221 (41) MS^3 [601>583]: 565 (48), 547 (14)	600.4188	91.88	-2.69	$C_{40}H_{56}O_4$
2	All-*trans*-neochrome	14.6	397,420,448	90	0	601 (42), 583 (100), 565 (43)	MS^2 [601]: 583 (100), 565 (57), MS^3 [601>583]:221 (62)	600.4178	96.70	-0.89	$C_{40}H_{56}O_4$
3	All-*trans*-violaxanthin	14.8	414,438,470	75	0	601 (85), 583 (100), 565 (20)	MS^2 [601]: 583 (100), 565 (12), 509 (5), 221 (24)	600.4178			$C_{40}H_{56}O_4$
4	Not identified	16.0	466	0	0	455 (100)	MS^2 [455]: 437 (80), 399 (34)	455.3324	93.60	-2.61	$C_{33}H_{43}O$
5	Not identified	16.2	396,420,448	75	0	601 (60)	MS^2 [601]: 583 (95), 221 (100)				
6	Not identified	16.4	Not detected	–	–	601 (64)	MS^2 [601]: 583 (90), 565 (40), 491 (9), 221 (100) MS^3 [601>583]:565 (100),221 (60)				
7	β- Diepoxy-cryptoxanthin	16.7	412, 436,466	72	0	585 (100)	MS^2 [585]: 567 (45), 549 (80), 493 (37), 221 (100), 205 (10)	584.4184	95.60	-2.52	$C_{40}H_{56}O_3$
8	*Cis*-violaxanthin	17.3	324,410, 434,464	60	8	601 (81), 487 (100)	MS^2 [601]: 583 (100), 565 (42), 509 (6), 491 (23), 221 (51) MS^2 [487]: 469 (100)				
9	Not identified	17.8	378,400,424	100	0	601 (100), 351 (98)	MS^2 [601]: 583 (100), 565 (13), 509 (15), 491 (14),393 (40),221 (41) MS^2 [351]: 333 (16)				
10	All-*trans*-lutein	18.2	420,444,472	48	0	569 (5), 551 (100)	MS^2[551]: 533 (51), 495 (24), 477 (35)	568.427	99.30	-0.226	$C_{40}H_{56}O_2$
11	Not identified	18.5	444	0	0	454 (100)	MS^2[454]: 436 (7), 393 (100)				
12	Not identified	18.9	448	0	0	473 (100),539 (80)	MS^2[539]: 521 (100)				
13	Not identified	19.0	454	0	0	473 (100), 454 (14)	MS^2 [473]: 455 (43), 205 (100) MS^2 [454]: 436 (31), 393 (36)				
14	Not identified	19.3	423,438,474	60	0	551 (100)	MS^2 [551]: 533 (67), 477 (55)				
15	Not identified	19.4	448	0	0	439 (93), 403 (100)	MS^2 [439]:403 (14)				
16	Not identified	19.6	434	0	0	391 (100)	MS^2 [391]:373 (14)				
17	5,6-Epoxy-β-cryptoxanthin	19.8	416,438,466	34	0	569 (17),551 (54)	MS^2 [551]: 533 (20), 205 (21)	568.428	75.56	-5.07	$C_{40}H_{56}O_2$
18	5',6'-Epoxy-β-cryptoxanthin	19.9	419,441,470	52	0	569 (30), 551 (100)	MS^2 [551]: 533 (100), 577 (14), 459 (30), 221 (13)	568.428	89.42	-2.11	$C_{40}H_{56}O_2$
19	*Cis*-lutein	20	326,412,436,462	52	20	569 (18), 551 (100)	MS^2 [551]: 533 (100), 221 (11)	568.428	77.7	3.57	$C_{40}H_{56}O_2$
32	β- Diepoxy-cryptoxanthin	20.2	416,440, 470	87	0	585 (100)	MS^2 [583]: 567 (49), 549 (10) 221 (31), 205 (29)				
20	Not identified	20.4				446 (100) 417 (90)	MS^2 [417]:399 (100) MS^2 [446]:219 (100)				

Table 1 UV/vis spectra and characteristic ions of carotenoids from six maturation stages of loquat fruits, obtained by HPLC-PDA-MS (*Continued*)

							MS²				
21	β-Apo-8'-carotenal (IS)	21.0	450	0	0	417 (100)	MS² [417]:399 (25), 389 (32), 361 (39), 325 (100), 293 (97), 157 (88), 119 (16)	416.3079	98.23	-2.15	$C_{30}H_{40}O$
22	Citranaxanthin	21.6	470	0	0	457 (100)	MS² [457]:439 (90), 399 (49)	455.3324	97.23	-1.25	$C_{33}H_{44}O$
23	Not identified	22.4	470		0	696 (100)	MS² [696]: 534 (34), 516 (100)				
24	Not identified	22.8	420,444,470	31	0	537 (58)	MS² [537]: 467 (16), 444 (51), 365 (100)				
25	Not identified	23.3	Not detected			537 (28),430 (100)	MS² [537]:444 (100),481 (24) 413 (90)				
26	All-*trans*-β-cryptoxanthin	24.0	420, 448,472	24	0	553 (100)	MS² [553]:535 (100), 497 (41),461 (10)				
27	Not identified	24.5				664 (100)	MS² [664]: 551 (100), 496 (55)				
28	Not identified	24.9	453, 479	0	0	551 (100)	MS² [551]: 534 (15), 361 (100)				
29	Phytoene + not identified 14	28.0				545 (95), 553 (100)	MS² [545]:489 (10),395 (100) MS² [553]:535 (12)				
30	5,8-epoxy-β-carotene	28.4	405, 424, 450	26	0	553 (100)	MS² [553]:535 (55),461 (35), 221 (64), 205 (17)				
31	All-*trans*-β-carotene	36.5	424, 446,470	18	0	537	MS² [537]:444 (100)				

All *trans*-lutein (peak 10) and *cis*-lutein (peak 19) showed characteristic UV-visible spectra, with a hypsochromic shift of 8 nm for the *cis* isomer. The identification of both lutein isomers was confirmed by their mass spectra with the protonated molecule at *m/z* 569 and fragments at *m/z* 551 and *m/z* 533 due to the loss of one and two hydroxyl group respectively. The MS/MS showed, in addition, the presence of fragments at *m/z* 477 resulting from the loss of toluene ([M + H-92] $^+$) from the polyene chain. In APCI-MS, the fragment with *m/z* 551 presented a higher intensity than the protonated molecular ion (*m/z* 569). Peaks 17 and 18 were identified as mono-epoxides of β-cryptoxanthin considering their UV-vis and MS characteristics by the comparison with literature data [9, 24]. The mass spectra of both epoxides presented the protonated molecule at *m/z* 569 and fragment ion at *m/z* 551 due to the loss of a hydroxyl group. Peak 17 was designated as 5′,6′-epoxy- β-cryptoxanthin due to the presence of the mass fragment at *m/z* 205 that is consistent with the location of one epoxide group in the unsubstituted ring whilst peak 18 showed the mass fragment at *m/z* 221, indicating that the epoxide groups were in a ring with a hydroxyl group. β-Apo-8′-carotenal was identified as peak 21 (internal standard). The mass spectra presented the protonated molecule at *m/z* 417. Ions of *m/z* 399 and 389 were detected corresponding to the loss of water and carbon monoxide respectively. Elimination of toluene from the protonated molecule was observed at *m/z* 325. The use of an internal standard was recommended to estimate the losses during the extraction process. Saponification with potassium hydroxide has been an integral part of carotenoid analyses. Kimura et al. [25] showed that β-apo-8′-carotenal was completely transformed to citranaxanthin (peak 22), apparently by aldol condensation with acetone. The conversion percentage from β-apo-8′-carotenal to citranaxanthin was 98% and their sum was considered for quantification. The identification of citranaxanthin was confirmed on the basis of its protonated molecule at *m/z* 457 [26] and its characteristic UV-vis spectrum [27]. Due to the presence of the same chromophore, β-cryptoxanthin (peak 26) and β-carotene have similar UV-visible spectra. As expected, the protonated molecule was detected at *m/z* 553 and the MS/MS revealed the presence of fragment ions at *m/z* 551 and 461 corresponding to the losses of the hydroxyl group and toluene. 5,8-epoxy-β-carotene (peak 30) could not be identified by its UV-visible spectral characteristics. Mass spectra highlighted the protonated molecule at *m/z* 553 and the MS/MS showed the presence of fragment ions at *m/z* 551 and 461 corresponding to the losses of the hydroxyl group and toluene, respectively, and at *m/z* 221 that corresponds to the location of the epoxide group in the 3-hydroxy-β-ring. The mass spectra of beta-carotene, peak 31, showed the protonated molecule at *m/z* 537 and a fragment ion in the MS/MS at *m/z* 444, corresponding to the loss of the toluene from the polyene chain.

Carotenoid composition in loquat fruit

Results revealed great differences in carotenoid composition between peel and flesh. In particular, 32 carotenoids were found in loquat peel, while only eighteen carotenoids were identified in the flesh. Except for qualitative differences, the concentration of carotenoids was significantly higher in the peel than in the flesh. This was not the case when total carotenoids were determined spectrophotometrically, indicating the limitations of such colorimetric assays. Chromatograms also revealed that the major carotenoids in peel were *trans*-lutein and *trans*-β-carotene. The concentration of *trans*-lutein decreased with the progress of ripening from 1621.5 to 688.4 µg 100 g^{-1} FW. On the other hand, *trans*-β-carotene content increased drastically from 151.9 to 1096.9 µg 100 g^{-1} FW. The biosynthesis of some carotenoids such as *trans*-β-cryptoxanthin, 5,8-epoxy-β-carotene, β-diepoxy-cryptoxanthin and *cis*-violaxanthin has also been monitored (Fig. 3, Additional file 1: Table S4). Conversely, *trans*-neoxanthin and *trans*-neochrome decreased or remained stable with the progress of on-tree fruit development in the peel, while they were not detected in the flesh during the last developmental stages.

The carotenoid profiling of loquat flesh was found to be quite different from the peel. The most abundant carotenoid in mature fruits was *trans*-β-cryptoxanthin, followed by *trans*-β-carotene, compounds 18 and 31, and 5,8-epoxy-β-carotene (peak 30) (Table 1, Fig. 3, Additional file 1: Table S4). An increment in the concentration of all carotenoids during on-tree development except for *trans*-neoxanthin, *trans*-neochrome and *trans*-lutein was found (Fig. 3, Additional file 1: Table S4). Overall, a great effect of the developmental stage on the carotenoid composition was revealed.

All-*trans*-neochrome, all-*trans*-violaxanthin, β-diepoxy-cryptoxanthin, *cis*-violaxanthin, all-*trans*-lutein, 5,6-epoxy-β-cryptoxanthin, 5′,6′-epoxy-β-cryptoxanthin, all-*trans*-β-cryptoxanthin, phytoene and all-*trans*-β-carotene were previously identified in five loquat cultivars, originating from Brazil [9]. In their findings, they reported *trans*-β-carotene (19–55%), *trans*-β-cryptoxanthin (18–28%), 5,6:5,6 -diepoxy-β-cryptoxanthin (9–18%) and 5,6-epoxy-β-cryptoxanthin (7–10%) to be the main carotenoids. In the flesh, it was found that β-carotene and lutein were the major carotenoids with neoxanthin, violaxanthin, luteoxanthin, 9-*cis*-violaxanthin, phytoene, phytofluene and ζ–carotene also present.

The carotenoid quantification in our study proved that the peel had higher carotenoid concentrations than the

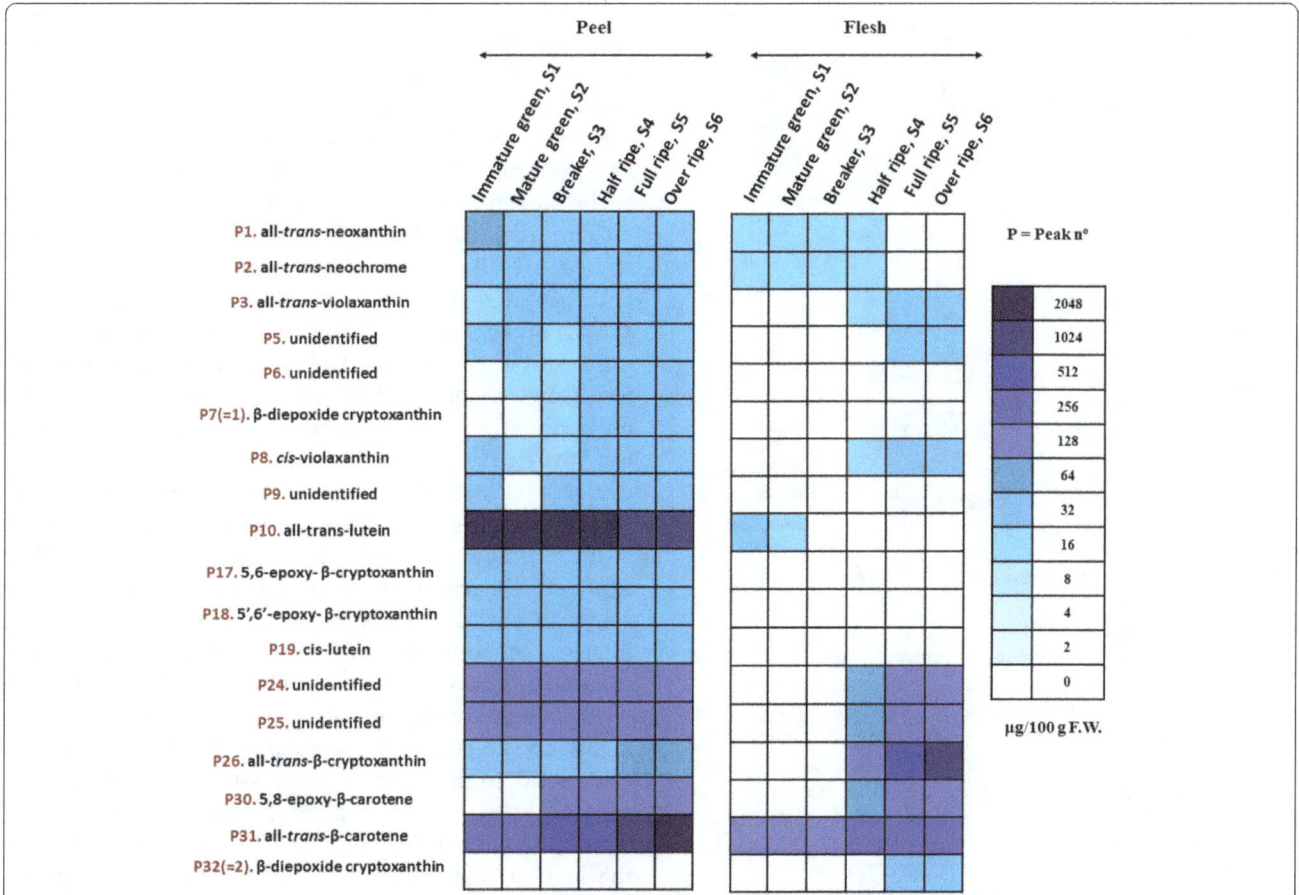

Fig. 3 Heat map of the quantification of the identified carotenoids (Table 1) in the peel and in the flesh of loquat fruit (cv. 'Obusa') during six on-tree developmental stages ($n = 3$). Results are expressed as µg 100 g^{-1} fresh weight (F.W.). A scale of colour intensity is presented as a *legend*. Actual quantification values of the identified carotenoids are shown in Additional file 1: Table S4

flesh, except for β-diepoxy-cryptoxanthin which was found from the 3rd until the 6th maturity stage in the peel, ranging from 15.0 ± 0.8 to 21.2 ± 1.7 µg 100 g^{-1} FW, as well as in flesh during the last maturity stages (S5 and S6) at 17.6 ± 1.2 and 21.4 ± 1.8 µg 100 g^{-1} FW, respectively (Fig. 3, Additional file 1: Table S4). *Trans*-neoxanthin and *trans*-neochrome appeared throughout all developmental stages in the peel as well as in the first 4 maturation stages in the flesh. On the other hand, *trans*-violaxanthin increased in the peel with the progress of on-tree ripening (15.6 ± 2.5 to 25.2 ± 4.6 µg 100 g^{-1} FW), while it was detected in the flesh during the last stages (S4-S6) with progressive increase (from 12.5 ± 0.2 to 21.5 ± 1.6 µg 100 g^{-1} FW). Similar findings and trend were observed for *cis*-violaxanthin which ranged from 16.2 ± 2.5 to 23.2 ± 3.0 µg 100 g^{-1} FW in the peel, being detectable in the flesh from S4 stage onwards (12.3 ± 0.1 to 19.1 ± 1.5 µg 100 g^{-1} FW). 5',6'-Epoxy-β-cryptoxanthin and *cis*-lutein were detected exclusively in the peel throughout all developmental stages (Fig. 3). Intriguingly, *trans*-lutein in the peel

registered the highest contents during the initial developmental stages and went descending thereafter, meanwhile it was found in detectable amounts in the flesh only at stages 1 and 2, yet substantially lower compared to the peel. Citranaxanthin and phytoene have also been identified, although they were not quantified (Table 1).

Gene expression profiles

With the aim to shed light on the carotenoid biosynthetic pathway in loquat fruit, the expression profile of thirteen known genes of the carotenoid pathway was analyzed, showing differential expression patterns in the peel and the flesh tissue (Fig. 4, Additional file 1: Tables S5–S6 and Figures S2–S3). For gene expression analyses, each tissue was examined individually, considering S1 stage as the calibrator of the tissue tested (peel or flesh).

Middle and downstream genes of the carotenoid biosynthetic pathway (*CYCB*, *LCYE*, *BCH*, *ECH* and *VDE*) showed clear differentiation in their expression levels among S2-S6 developmental stages compared with the remaining genes. In both peel and flesh, *PSY1*, *ZDS*,

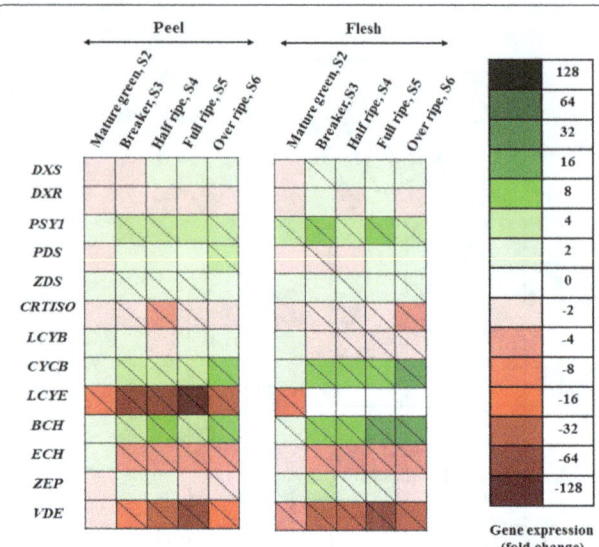

Fig. 4 Heat map of the relative expression levels of carotenoid biosynthesis genes (*DXS, DXR, PSY1, PDS, ZDS, CRTISO, LCYB, CYCB, LCYE, BCH, ECH, ZEP* and *VDE*) in loquat fruit (cv. 'Obusa'), both in the peel and in the flesh, during five on-tree developmental stages (S2-S6) (*n* = 3). Relative mRNA abundance was evaluated by real-time RT-PCR using three biological repeats. Up-regulation is indicated in *green*; down-regulation is indicated in *red*. A *diagonal line in a box* indicates a statistically significant value (*P* ≤ 0.05). A scale of colour intensity is presented as a *legend*. The first developmental stage (S1) both for the peel and the flesh was used for calibrating gene exeression values. Actual relative expression levels are shown in Additional file 1: Tables S5-S6 and Figs. S2-S3

CYCB, and *BCH* were significantly up-regulated in most of the developmental stages compared with the S1 stage respectively, whereas *CRTISO, LCYE, ECH* and *VDE* were generally down-regulated. *PDS* was significantly up-regulated at the S6 stage in the peel, while it was significantly suppressed at the S3 stage in the flesh tissue. *ZDS* presented an increase in expression levels between the S3 and S5 stage in peel, though the increase was at the S4 and S6 stage in flesh tissue. Non statistically significant changes in expression levels of *DXS, DXR* (both involved in MEP pathway) and *LCYB* were monitored in the peel in comparison with the S1 stage; in flesh *DXR* followed the same trend whereas, *DXS* presented an accumulation at the S3 and *LCYB* transcript levels decreased gradually through stages S4-S6 which could account for the low or non-detectable *trans*-lutein findings in the flesh (Figs. 1, 3 and 4). *ZEP* expression levels presented highest suppression at the S6 stage in peel, contrarily to flesh profile where it was up-regulated at the S3-S5 stages.

BCH and *CYCB* transcript levels were substantially up-regulated with the progress of on-tree fruit development both in the peel and in the flesh, registering the highest values at the last stage for both tissues (Fig. 4, Additional file 1: Tables S5, S6). Notably in flesh tissue, a

8.1- and 11.0- fold increase of *CYCB* and *BCH* transcripts during the last stage compared to the initial stage was recorded, respectively. *PSY1* presented statistically significant increases at the S3, S4 and S6 stages of the peel and throughout all stages in the flesh (S2-S6); the most prominent gene regulation in the flesh was found at the S3 and S5 stages (Additional file 1: Table S6).

Contrarily, *CRTISO, LCYE, ECH* and *VDE* demonstrated overall a down-regulation expression pattern in both peel and flesh tissue. *CRTISO* was down-regulated at stages S3-S5 in peel; a similar trend was monitored in the flesh (S3-S6 stages). *ECH* had similar expression pattern (suppression) for both peel and flesh from stage S3 onwards. *LCYE* was down-regulated in peel throughout the five developmental stages; the most abundant decrease (128 fold change) was registered at the S5 stage. Interestingly, in flesh the decrease was registered at the S2 stage and remained undetected thereafter. *VDE* expression levels were repressed throughout S3 to S6 stages in peel and throughout all developmental stages in flesh. In both peel and flesh, *VDE* presented a substantial suppression at the S5 stages (36.8 and 49.5 fold change, respectively) compared with S1 stage.

Notably, *PSY1*, known to catalyse the first step in the carotenoid formation [14], expression levels depicted a general up-regulation with the progress of on-tree fruit development both in the peel and in the flesh; however the highest transcript values in flesh tissue were monitored at S3 and S5 stages, not concomitant with total carotenoid accumulation. In another fleshy fruit (apple), Ampomah-Dwamena et al. [28] also postulated that *PSY1* expression levels had no direct correlation with carotenoid content in different genotypes. On the other hand, *PDS*, an upstream pathway gene, presented statistically significantly higher transcript levels only in the peel at the last developmental stage (S6); *PDS* expression levels have been correlated with high- and low- carotenoid content apple cultivars [28].

Trans-lutein presented appreciably high accumulation in the first four stages in the peel with a reduction at ripe and over-ripe stages, while detectable amounts in the flesh were registered only during the initial developmental stages, as chloroplasts began to develop into chromoplasts. This decrease can be attributed to the fact that (1) *LCYE* was markedly down-regulated throughout the developmental stages in both peel and flesh compared to corresponding per tissue immature stage (notably not detectable transcripts during S3-S6 of flesh was monitored), (2) *LCYB* was down-regulated over the last developmental stages in flesh (S4-S6) and (3) *ECH* mRNA expression was generally down-regulated both in the peel and in the flesh (S3-S6). Fu et al. [10] noted that lutein is showing a transient decrease in the flesh of loquat cultivars with the progress of on-tree fruit development, whereas there is no connection with *BCH* expression which appears to be up-regulated especially in the red-fleshed cultivar 'Luoyangqing'. An up-regulation of *BCH* gene expression was also monitored in our study in a similar flesh-type loquat cultivar. Ampomah-Dwamena et al. [28] showed a close correlation between *LCYE* expression and carotenoid content in apple fruit skin. This was not the case with the flesh; suggesting that down-regulation of *LCYE* is consistent with lower *trans*-lutein concentrations in the flesh.

Contrarily, the high *BCH* expression values registered in the flesh at the last two stages can be linked with the transient carotenoid accumulation of *trans*-β-cryptoxanthin in these stages, concomitant with higher mRNA expression of *CYCB* and non-detectable *LCYE* transcripts (Stages S3–S6). These findings are in accordance with Fu et al. [10], where the *BCH* values for the red-fleshed cultivar 'Luoyangqing' showed a transient increase at the breaker stage. In addition, *CYCB* expression levels were also higher after the S4 stage. Kato et al. [11] stated that the decrease of *LCYE* gene expression in Citrus fruits is related with higher contents of β-carotene as the ε, β-branch of the carotenoid pathway shifts to the β, β-branch during transition from immature to mature stage. Zhao et al. [16] found that *BCH* is responsible for high β-cryptoxanthin content in persimmon fruit, in accordance with findings in other loquat cultivars [29]. The latter study suggests a direct link between the synthesis and accumulation of β-cryptoxanthin and the abundant expression of *BCH*. The transient increase of *trans*-β-carotene towards the S6 peel stage can also be linked with the up-regulation of *CYCB* and down-regulation of *LCYE*, as elsewhere described [10]. Zhang et al. [29] also links the higher β-carotene level in loquat peel with the abundant increase of *PSY1*, as well as *CYCB* and *BCH* mRNA expression levels.

VDE expression which leads to violaxanthin biosynthesis is significantly suppressed in almost all stages, both in the flesh and in the peel compared with the calibrator (S1 stage) (thus expecting very little violaxanthin); in the case of *ZEP*, which converts violaxanthin back to precursor molecules such as zeaxanthin, the main trend is that it is induced in several stages in the flesh. This is in accordance with metabolite levels, as both *cis*- and *trans*-violaxanthin are at appreciable low concentrations and/or non-detectable during several stages in the flesh (Figs. 3, 4; Additional file 1: Table S4).

Conclusions

The carotenoid profile of 'Obusa' fruits, an orange-fleshed loquat cultivar, was elucidated during distinct on-tree developmental stages. Results indicated that carotenoid composition was greatly affected during fruit development, revealing evident differentiations between flesh and peel tissue. The major carotenoids were *trans*-lutein and *trans*-β-carotene in the peel, and *trans*-β-cryptoxanthin, *trans*-β-carotene, and 5,8-epoxy-β-carotene in the flesh. To the best of our knowledge, the presence of *cis*-lutein, citranaxanthin and 5,8-epoxy-β-carotene has not been reported previously in loquat, but only in other fruits of tropical origin [30, 31]. Furthermore, a link was attempted to be established between gene up- or down-regulation during the developmental stages of the loquat fruit, and how their expression affects carotenoid content. Elevated content of *trans*-β-carotene both in the flesh and in the peel with the progress of on-tree fruit development can be linked with the up-regulation of *CYCB*, a main carotenoid biosynthetic gene. Notably, the non-detectable amounts of *trans*-lutein in the flesh during the S3-S6 stages can be linked with the significant suppression of *LCYB* and *LCYE* expression levels during these stages. Transcripts levels of the latter gene were also significantly reduced throughout all developmental stages in the peel compared with the immature stage.

Abbreviations

ABA: Abscisic acid; *BCH*: β-carotene hydroxylase; Chl a: Chlorophyll-a; Chl b: Chlorophyll-b; *CRTISO*: Carotene isomerase; *CYCB*: Chromoplast-specific lycopene β-cyclase; DMAPP: Dimethylallyl pyrophosphate; *DXR*: DXP reductoisomerase; *DXS*: 1-deoxy-D-xylulose 5-phosphate-synthase; *ECH*: ε-carotene hydroxylase; *EjACT*: Eriobotrya japonica Actin; FW: Fresh weight; GAP: D-glyceraldehyde 3-phosphate; GGPP: Geranylgeranyl diphosphate; *GGPS*: Geranylgeranyl diphosphate synthase; HMBPP: (E)-4-hydroxy-3-methylbut-2-enyl diphosphate; *IDI*: Isopentenyl pyrophosphate isomerase; *IDS*: Isopentenyl pyrophosphate synthase; IPP: Isopentenyl pyrophosphate; *LCYB*: Lycopene β-cyclase; *LCYE*: Lycopene ε-cyclase; *NCED*: 9-cis-epoxycarotenoid dioxygenase; *NSY*: Neoxanthin synthase; *PDS*: Phytoene desaturase; *PSY1*: Phytoene synthase; *VDE*: Violaxanthin de-epoxidase; *ZDS*: ζ-carotene desaturase; *ZEP*: Zeaxanthin epoxidase; *ZISO*: ζ-carotene isomerase

Acknowledgements

The authors would like to thank Mr. George Sismanidis and Mrs. Marina Christofi for their technical assistance in the phytochemical analysis.

Funding

The authors would like to acknowledge support by Remedica through the "Cyprus University of Technology Open Access Author Fund" for covering the cost of publication fees.

Authors' contributions

GAM conceived the project and designed the experiments. VF, MH and ECG undertook the molecular experiments. FB, AM, and HMDM undertook the analytical experiments. MH, ECG, VG, VF, AM and HMDM were involved in the data analysis. MH, ECG, AM, HMDM, VG, VF, FB and GAM wrote the paper. All authors read and approved the final manuscript.

Competing interests

Loquat fruits were harvested from a commercial orchard owned by MH. Authors declare non-financial competing interests that may cause them embarrassment if they were to become public after the publication of the manuscript.

Author details
[1]Department of Agricultural Sciences, Biotechnology and Food Science, Cyprus University of Technology, 3603 Lemesos, Cyprus. [2]Quality, Safety, and Bioactivity of Plant Foods, CEBAS-CSIC, P.O. Box 164, Espinardo, Murcia, Spain.

References

1. Zhou CH, Xu CJ, Sun CD, Li X, Chen KS. Carotenoids in white-and red-fleshed loquat fruits. J Agric Food Chem. 2007;55:7822–30.

2. Martinez-Calvo J, Badenes ML, Llácer G, Bleiholder H, Hack H, Meier U. Phenological growth stages of loquat tree (Eriobotrya japonica (Thunb.) Lindl.). Ann Appl Biol. 1999;134:353–7.

3. Xu HX, Chen JW. Commercial quality, major bioactive compound content and antioxidant capacity of 12 cultivars of loquat (Eriobotrya japonica Lindl.) fruits. J Sci Food Agr. 2011;91:1057–63.

4. Pinillos V, Hueso JJ, Marcon Filho JL, Cuevas J. Changes in fruit maturity indices along the harvest season in "Algerie" loquat. Sci Hort. 2011;129:769–76.

5. Goulas V, Minas IS, Kourdoulas PM, Vicente AR, Manganaris GA. Phytochemical content, antioxidants and cell wall metabolism of two loquat (Eriobotrya japonica) cultivars under different storage regimes. Food Chem. 2014;155:227–34.

6. Ferreres F, Gomes D, Valentão P, Gonçalves R, Pio R, Chagas EA, et al. Improved loquat (Eriobotrya japonica Lindl.) cultivars: variation of phenolics and antioxidative potential. Food Chem. 2009;114:1019–27.

7. Vogel JT, Tieman DM, Sims CA, Odabasi AZ, Clark DG, Klee HJ. Carotenoid content impacts flavor acceptability in tomato (Solanum lycopersicum). J Sci Food Agric. 2010;90:2233–40.

8. Kaulmann A, Bohn T. Carotenoids, inflammation, and oxidative stress-implications of cellular signaling pathways and relation to chronic disease prevention. Nutr Res. 2014;34:907–29.

9. De Faria AF, Hasegawa PN, Chagas EA, Pio R, Purgatto E, Mercadante AZ. Cultivar influence on carotenoid composition of loquats from Brazil. J Food Comp Anal. 2009;22:196–203.

10. Fu X, Kong W, Peng G, Zhou J, Azam M, Xu C, et al. Plastid structure and carotenogenic gene expression in red- and white-fleshed loquat (Eriobotrya japonica) fruits. J Exp Bot. 2012;63:341–54.

11. Kato M. Mechanism of Carotenoid accumulation in citrus fruit. J Japan Soc Hort Sci. 2012;81:219–33.

12. Fu X, Feng C, Wang C, Yin X, Lu P, Grierson D, et al. Involvement of multiple phytoene synthase genes in tissue-and cultivar-specific accumulation of carotenoids in loquat. J Exp Bot. 2014;65:4679–89.

13. Naik PS, Chanemougasoundharam A, Paul Khurana SM, Kalloo G. Genetic manipulation of carotenoid pathway in higher plants. Curr Sci. 2003;85:1423–30.

14. Mendes AF, Soares VL, Costa MG. Carotenoid biosynthesis genomics. In: Chen C, editor. Pigments in Fruits and Vegetables. New York: Springer; 2015. p. 9–29.

15. Shumskaya M, Wurtzel ET. The carotenoid biosynthetic pathway: thinking in all dimensions. Plant Sci. 2013;208:58–63.

16. Zhao D, Zhou C, Tao J. Carotenoid accumulation and carotenogenic genes expression during two types of persimmon fruit (Diospyros kaki L.) development. Plant Mol Biol Rep. 2011;29:646–54.

17. Galpaz N, Ronen G, Khalfa Z, Zamir D, Hirschberg J. A Chromoplast-specific carotenoid biosynthesis pathway is revealed by cloning of the tomato white-flower locus. Plant Cell. 2006;18:1947–60.

18. Nagata M, Yamashita I. Simple method for simultaneous determination of chlorophyll and carotenoids in tomato fruit Masayasu N. J Jpn Soc Food Sci Technol. 1992;39:925–8.

19. Minguez-Mosquera MI, Hornero-Mendez D. Separation and quantification of the carotenoid pigments in red peppers (Capsicum annuum L.), paprika, and oleoresin by reversed-phase HPLC. J Agric Food Chem. 1993;41:1616–20.

20. Gambino G, Perrone I, Gribaudo I. A rapid and effective method for RNA extraction from different tissues of grapevine and other woody plants. Phytochem Anal. 2008;19:520–5.

21. Georgiadou EC, Goulas V, Ntourou T, Manganaris GA, Kalaitzis P, Fotopoulos V. Regulation of on-tree vitamin E biosynthesis in olive fruit during successive growing years: the impact of fruit development and environmental cues. Front Plant Sci. 2016;7:1656.

22. Pfaffl MW, Horgan GW, Dempfle L. Relative expression software tool (REST[(C)]) for group-wise comparison and statistical analysis of relative expression results in real-time PCR. Nucleic Acids Res. 2002;30:e36.

23. Lois ML, Rodriguez-Concepcion M, Gallego F, Campos N, Boronat A. Carotenoid biosynthesis during tomato fruit development:regulatory role of 1-deoxy-D-xylulose 5-phosphate synthase. Plant J. 2000;22:503–13.

24. De Rosso VV, Mercadante AZ. HPLC-PDA-MS/MS of anthocyanins and carotenoids from dovyalis and tamarillo fruits. J Agri Food Chem. 2007;55:9135–41.

25. Kimura M, Rodriguez-Amaya DB, Gody HT. Assessment of the saponification step in the quantitative determination of carotenoids and provitamins A. Food Chem. 1990;35:187–95.

26. Shlatterer J, Breithaupt DE. Xanthophylls in commercial egg yolks: quantification and identification by HPLC and LC-(APCI) MS using a C30 phase. J Agri Food Chem. 2006;54:2267–73.

27. Mitrowska K, Vicent U, Von Holst C. Separation and quantification of 15 carotenoids by reversed phase high performance liquid chromatograpy coupled to dioede array detection with isosbestic wavelength approach. J Chromatogr A. 2012;1233:44–53.

28. Ampomah-Dwamena C, Dejnoprat S, Lewis D, Sutherland P, Volz RK, Allan AC. Metabolic and gene expression analysis of apple (Malus x domestica) carotenogenesis. J Exp Bot. 2012;63:4497–511.

29. Zhang L, Zhang Z, Zheng T, Wei W, Zhu Y, Gao Y, et al. Characterization of carotenoid accumulation and carotenogenic gene expression during fruit development in yellow - and white loquat fruit. Hort Plant J. 2016;2:9–15.

30. Zanatta CF, Mercadante AZ. Carotenoid composition from the Brazilian tropical fruit camu–camu (Myrciaria dubia). Food Chem. 2007;101:1526–32.

31. De Rosso VV, Mercadante AZ. Carotenoid composition of two Brazilian genotypes of acerola (Malpighia punicifolia L.) from two harvests. Food Res Int. 2005;38:1073–7.

2

The core regulatory network of the abscisic acid pathway in banana: genome-wide identification and expression analyses during development, ripening, and abiotic stress

Wei Hu[1*†], Yan Yan[1†], Haitao Shi[3†], Juhua Liu[1†], Hongxia Miao[1], Weiwei Tie[1], Zehong Ding[1], XuPo Ding[1], Chunlai Wu[1], Yang Liu[1], Jiashui Wang[2], Biyu Xu[1*] and Zhiqiang Jin[1,2*]

Abstract

Background: Abscisic acid (ABA) signaling plays a crucial role in developmental and environmental adaptation processes of plants. However, the PYL-PP2C-SnRK2 families that function as the core components of ABA signaling are not well understood in banana.

Results: In the present study, 24 *PYL*, 87 *PP2C*, and 11 *SnRK2* genes were identified from banana, which was further supported by evolutionary relationships, conserved motif and gene structure analyses. The comprehensive transcriptomic analyses showed that banana *PYL-PP2C-SnRK2* genes are involved in tissue development, fruit development and ripening, and response to abiotic stress in two cultivated varieties. Moreover, comparative expression analyses of *PYL-PP2C-SnRK2* genes between BaXi Jiao (BX) and Fen Jiao (FJ) revealed that *PYL-PP2C-SnRK2*-mediated ABA signaling might positively regulate banana fruit ripening and tolerance to cold, salt, and osmotic stresses. Finally, interaction networks and co-expression assays demonstrated that the core components of ABA signaling were more active in FJ than in BX in response to abiotic stress, further supporting the crucial role of the genes in tolerance to abiotic stress in banana.

Conclusions: This study provides new insights into the complicated transcriptional control of *PYL-PP2C-SnRK2* genes, improves the understanding of *PYL-PP2C-SnRK2*-mediated ABA signaling in the regulation of fruit development, ripening, and response to abiotic stress, and identifies some candidate genes for genetic improvement of banana.

Keywords: Abscisic acid signaling, Abiotic stress, Banana, Fruit development and ripening, Gene expression

Background

In plants, phytohormone abscisic acid (ABA) regulates numerous developmental processes, such as seedling development, seed dormancy, and fruit ripening [1–5]. In addition, ABA plays a central role in the adaptation of plants to environmental stresses, such as drought, salinity, and cold [6, 7]. Due to the biological and agricultural importance of ABA, many studies have focused on plant responses to ABA at the level of cytology and molecular biology. Since 2009, the ABA signaling pathway began to be better understood [6]. PYR/PYL/RCARs (ABA receptors), Group A PP2Cs (negative regulators), and SnRK2s (positive regulators) were confirmed as crucial components of ABA signaling in Arabidopsis. Finally, a double negative regulatory model is constituted by these components. SnRK2s activities are repressed by direct dephosphorylation by Group A PP2Cs in the absence of ABA. When responding to developmental or environmental clues, the ABA signal induces PYR/PYL/RCAR interaction with Group A PP2Cs, including ABI1, ABI2,

---

* Correspondence: huwei2013@itbb.org.cn; biyuxu@126.com;
18689846976@163.com
†Equal contributors
[1]Key Laboratory of Biology and Genetic Resources of Tropical Crops, Institute of Tropical Bioscience and Biotechnology, Chinese Academy of Tropical Agricultural Sciences, Haikou, Hainan, China
Full list of author information is available at the end of the article

AHG3, and HAB1, leading to inhibition of Group A PP2Cs and activation of SnRK2s [6, 8–10]. This results in phosphorylation or activation of downstream targets, such as ABF/AREB/ABI5, SLAC1, and other ABA-responsive gene products [6, 11]. The ABA-mediated interaction model between PYLs and PP2Cs was validated by in vitro reconstitution in Arabidopsis protoplasts [12].

Additionally, the function of PYL-PP2C-SnRK2 genes in developmental processes and in response to ABA and abiotic stress were characterized in plants. *PYL9*, *PYL5* or *PYL8* overexpression improved drought tolerance or ABA responses in Arabidopsis [9, 13, 14]. In contrast, an ABA insensitive phenotype was observed in the quadruple mutant of pyr1 pyl1 pyl2 pyl4 [10]. Double and triple mutation of several crucial members of Group A *PP2Cs* (*ABI1*, *ABI2*, *HAB1*, *HAB2*, *AHG1*, and *PP2CA*) resulted in enhanced ABA sensitivity, indicating the negative roles of Group A *PP2Cs* in ABA signaling [15–19]. Interference of *AtPP2CA* increased tolerance to freezing stress and ABA sensitivity in Arabidopsis [20]. Mutation of abi2-1 resulted in enhanced tolerance to salt stress and ABA insensitivity in Arabidopsis [21]. Overexpression of *SnRK2.8* improved tolerance to drought stress in Arabidopsis [22]. Conversely, mutation of snrk2.2, snrk2.3, and snrk2.6 decreased drought stress tolerance and ABA responses, such as seed germination, plant growth, stomatal behavior [6]. Besides, the similar roles of *PYL* and *SnRK2* genes were also observed in rice. Overexpression of *OsPYL3* or *OsPYL9* positively regulated the ABA response during seed germination and improved drought and cold stress tolerances in rice [23]. *OsPYL/RCAR5* overexpressing rice plants showed hypersensitivity to ABA during seed germination [24]. Overexpression of *SAPK4* in rice resulted in improved germination, growth and development under salt stress both in seedlings and mature plants [25]. *OsSAPK9* was reported to improve drought tolerance and grain yield through regulating cellular osmotic potential, stomatal closure and stress-responsive gene expression in rice [26]. Interestingly, Arabidopsis plants overexpressing *OsPP108* (a Group A *PP2C* gene in rice) showed highly insensitivity to ABA and tolerance to salt and osmotic stresses during seed germination, root growth and overall seedling growth. This indicated that OsPP108 negatively regulates ABA signaling and positively regulates abiotic stress tolerance [27]. Together, this evidence suggests that Group A *PP2Cs* negatively regulate ABA signaling and negatively/positively regulate ABA-mediated biological processes; and *PYLs* and *SnRK2s* could positively regulate the response of plants to these processes.

To date, genes that encode the crucial components of ABA signaling have been identified in several species based on genome sequencing. There are 14 *PYLs* in Arabidopsis, 13 in rice, 10 in *Selaginella moellendorffi*, and 4 in *Physcomitrella patens*; 9 Group A *PP2Cs* in Arabidopsis, 10 in rice, 5 in *Selaginella moellendorffi*, and 2 in *Physcomitrella patens*; and 10 *SnRK2s* in Arabidopsis, 11 in rice, 6 in *Selaginella moellendorffi*, and 4 in *Physcomitrella patens* [6]. In spite of the economic and social importance of banana and the critical role of *PYL-PP2C-SnRK2s* in the plant development and stress responses, no information is known about the *PYL-PP2C-SnRK2* gene family in banana. Banana is the largest fruit crop and vital for food security for millions of people around the world [28, 29]. Because it is mainly cultivated as a staple food in many impoverished continents, such as Africa, banana studies have proceeded slowly [30]. Investigation of genes in the signal transduction pathways on the basis of complete genome sequences is of benefit for revealing the cellular biological processes [31]. The banana genome sequencing was finished in 2012 [32], which supplies full genome data for us to perform systematic analyses of *PYL-PP2C-SnRK2* gene families.

In this study, we identified 24 *PYLs*, 87 *PP2Cs*, and 11 *SnRK2s* from the banana genome and investigated their phylogenetic relationships, protein motifs, gene structure, and expression patterns in different tissues, in diverse stages of fruit development and ripening, and under abiotic stress. Further, we studied the interaction networks and co-expression profiles of Group A PP2Cs in response to cold, salt, and osmotic stresses. This systematic study increases the understanding of the core components of ABA signaling associated with developmental processes and abiotic stress responses and builds a solid foundation for genetic improvement of banana.

Results
Identification and phylogenetic analyses of banana PYL-PP2C-SnRK2s
To identify all PYL-PP2C-SnRK2 family members in banana, both Hidden Markov Model and BLAST searches were carried out to search the banana genome database with PYL-PP2C-SnRK2 sequences from *Arabidopsis* and rice as queries. After confirming their conserved domain using the PFAM and CDD databases, a total of 24 PYL, 87 PP2C, and 11 SnRK2 proteins were identified from the banana genome. The predicted features of the PYL, PP2C and SnRK2 proteins are summarized in Additional file 1: Table S1.

To understand the phylogenetic relationship of PYL-PP2C-SnRK2 proteins, neighbor-joining (NJ) trees were reconstructed with the complete PYL-PP2C-SnRK2 protein sequences from banana, *Arabidopsis* and rice (Figs. 1, 2, and 3). According to the phylogenetic analyses, the PYL, PP2C, and SnRK2 families were divided into 4 (group 1-4), 13 (group A-L), and 3 (group 1-3) subgroups, respectively. Some orthologous PYL-PP2C-SnRK2s between banana and

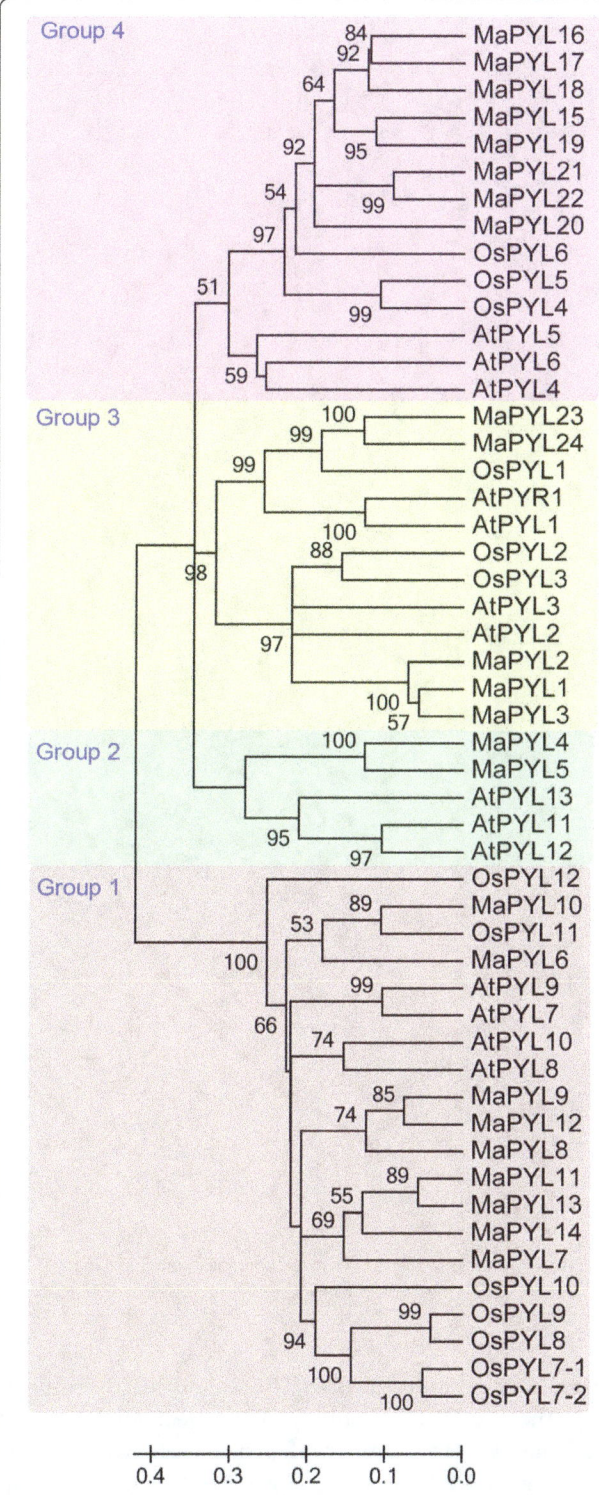

Fig. 1 Phylogenetic analysis of PYLs from banana, Arabidopsis, and rice using the complete protein sequences. The Neighbor-joining (NJ) tree was reconstructed using Clustal X 2.0 and MEGA 5.0 softwares with the pair-wise deletion option. 1000 bootstrap replicates were used to assess tree reliability

rice were identified, which implied that some ancestral PYL-PP2C-SnRK2s existed prior to the divergence of banana and rice. Generally, banana PYL-PP2C-SnRK2s showed closer relationships with PYL-PP2C-SnRK2s in rice than those in Arabidopsis, which is accordance with the current understanding of plant evolutionary history.

Conserved motifs and gene structure analyses of banana PYL-PP2C-SnRK2

To get insight into the structural features of the banana PYL-PP2C-SnRK2 proteins, conserved motifs were analyzed based on the phylogenetic relationship. Ten conserved motifs were acquired for each gene family with MEME and InterPro databases (Fig. 4). For the banana PYL family, motifs 1-3 were annotated as the START-like domain. All the identified MaPYLs contained motifs 1 and 2. The subgroup 1-3 also showed the conserved motif 3 (Fig. 4b). For the banana PP2C family, motifs 1-5 were annotated as the PPM-type phosphatase domain. Almost all of the PP2Cs contain the motifs 1, 2, 4, and 5, except for subgroup K showing motifs 1, 2, and 4. Interestingly, subgroup C specially showed motif 3, and subgroup D uniquely had motif 3, 7, 8, and 10 (Fig. 4a). For the banana SnRK2 family, motifs 1-5 were annotated as the Protein kinase domain. All the MaSnRK2s have motifs 1-5. Motif 10 was especially pronounced in subgroup 1 and motifs 8 and 9 were only found in subgroup 3 (Fig. 4c). This indicates that all the identified PYL-PP2C-SnRK2s have typical family features and the proteins classified into the same subgroup share similar amino acid sequences.

To better understand the gene structure of banana *PYL-PP2C-SnRK2s*, exon-intron organizations of these genes were tested (Fig. 5). For the banana *PYL* family, subgroups 1, 3, and 4 have 2, 0, and 1 introns, respectively; and subgroup 2 showed 0-2 introns (Fig. 5b). For the banana *PP2C* family, subgroups A, B, D, F1, G, and K contain 2-5 introns; subgroups C, E, F2, and H have 3-9 introns; and subgroup L shows 1-15 introns (Fig. 5a). For the banana *SnRK2* family, subgroups 1, 2, and 3 show 8-9, 8-13, and 8 introns, respectively (Fig. 5c). These results indicate that *PYL-PP2C-SnRK2* genes in the same subgroup show similar exon-intron organization.

Expression analyses of *PYL-PP2C-SnRK2* genes in different banana tissues

To examine the expression profiles of *PYL-PP2C-SnRK2* genes in different tissues of banana, roots, leaves, and fruits from BaXi Jiao (*Musa acuminate* L. AAA group cv. Cavendish, BX) and Fen Jiao (*Musa* ABB PisangAwak, FJ) were collected to perform trancriptomic assays (Fig. 6; Additional file 1: Tables S2; S3; S4; S5). Generally, most of the *PYL-PP2C-SnRK2* genes showed similar tissue expression

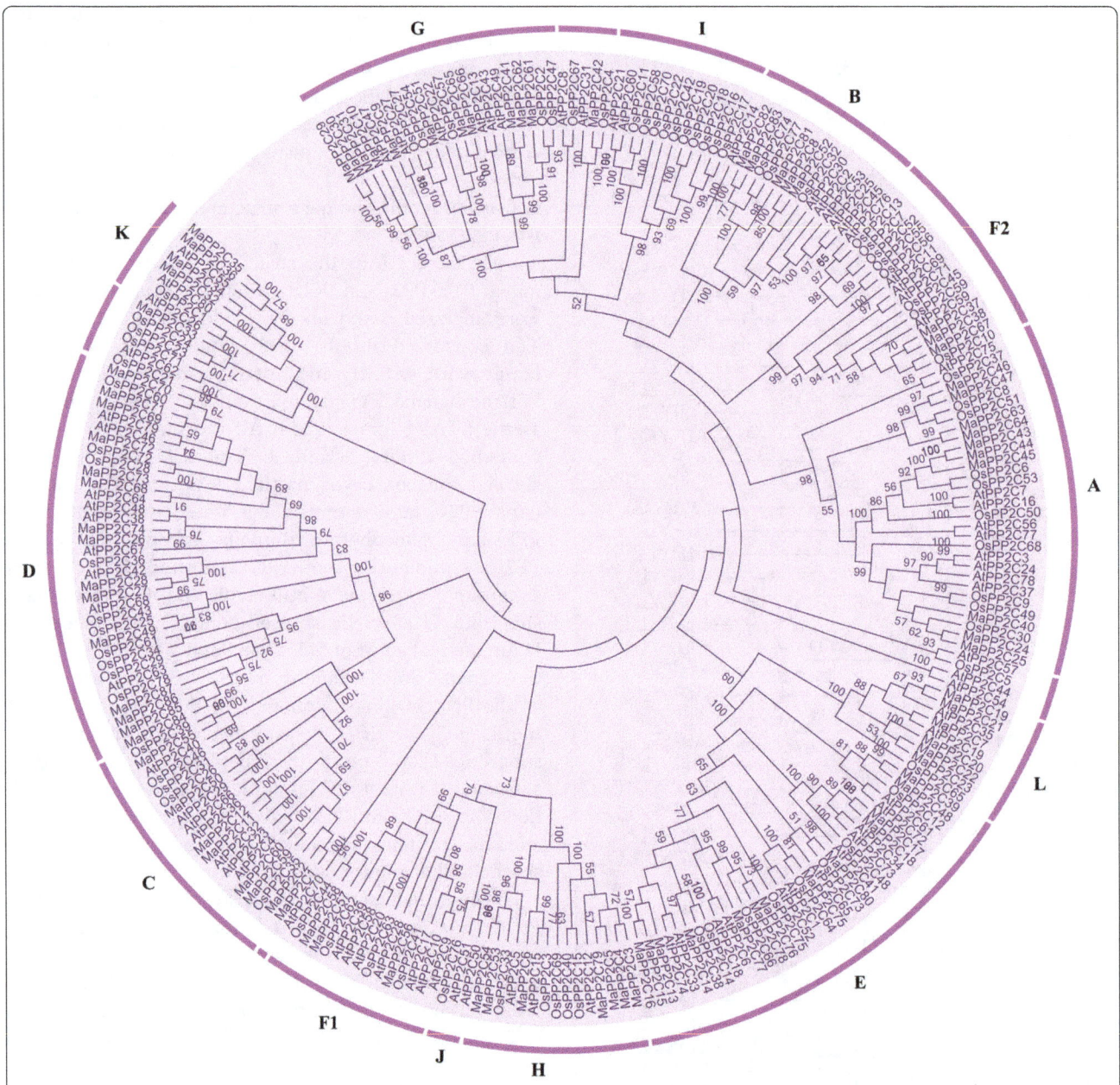

Fig. 2 Phylogenetic analysis of PP2Cs from banana, Arabidopsis, and rice using the complete protein sequences. The Neighbor-joining (NJ) tree was reconstructed using Clustal X 2.0 and MEGA 5.0 softwares with the pair-wise deletion option. 1000 bootstrap replicates were used to assess tree reliability

patterns between BX and FJ. For example, several genes (*MaPYL-14, MaPP2C-14, −34, −37, −38, −45, −47, and MaSnRK2-6*) displayed high transcript abundance (FPKM value > 20) in both BX and FJ. In contrast, some genes (*MaPYL-5, −16, −17, −18, −21,* and *MaPP2C-16, −20, −22, −23, −29, −46, −59, −63, −64, −80, −81, −84*) had low transcript abundance (FPKM value < 3) in both BX and FJ.

In addition, we also found different expression patterns of *PYL-PP2C-SnRK2* genes between BX and FJ. For the *PYL* family, the number of genes with high expression levels (FPKM value > 10) in roots and

leaves was greater in BX (10/22 and 8/21, respectively) than in FJ (7/22 and 5/20, respectively). For the *PP2C* family, the number of genes with high expression levels (FPKM value > 10) in roots and fruits was less in BX (48/86 and 19/82, respectively) than in FJ (51/86 and 33/84, respectively). This phenomenon was also observed in the tissue expression patterns of the *SnRK2* family. Taken together, the tissue expression patterns of *PYL-PP2C-SnRK2* genes in two cultivated varieties could lay a foundation for further investigation of tissue development and function.

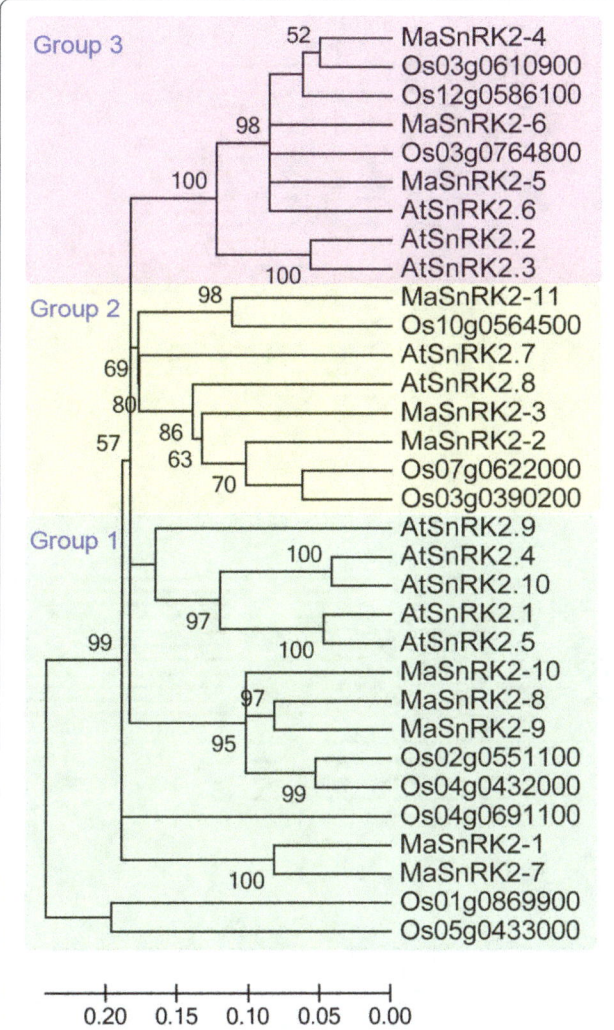

Fig. 3 Phylogenetic analysis of SnRK2s from banana, Arabidopsis, and rice using the complete protein sequences. The Neighbor-Joining (NJ) tree was reconstructed using Clustal X 2.0 and MEGA 5.0 softwares with the pair-wise deletion option. 1000 bootstrap replicates were used to assess tree reliability

0 days after flower (DAF), 20 DAF, 80 DAF, 8 days post-harvest (DPH), and 14 DPH in BX, respectively; and 7/21, 9/22, 5/17, 5/17, and 3/19 *PYL* genes showed high expression levels (FPKM value > 10) at the corresponding stages in FJ, respectively. For the PP2C family, 47/85, 45/87, 19/82, 30/83, and 27/79 *PP2C* genes showed high expression levels (FPKM value > 10) at 0 DAF, 20 DAF, 80 DAF, 8 DPH, and 14 DPH in BX, respectively; and 51/85, 52/85, 33/84, 35/84, and 28/82 *PP2C* genes showed high expression levels (FPKM value > 10) at the corresponding stages in FJ, respectively. For the SnRK2 family, 6/11, 6/11, 6/11, 7/11, and 5/10 *SnRK2* genes showed high expression levels (FPKM value > 10) at 0 DAF, 20 DAF, 80 DAF, 8 DPH, and 14 DPH in BX, respectively; and 6/11, 7/11, 7/11, 6/11, and 4/11 SnRK2 genes showed high expression levels (FPKM value > 10) at the corresponding stages in FJ, respectively. These results indicated the possible involvement of *PYL-PP2C-SnRK2* genes in banana development and ripening.

The number of *PP2C* genes in BX with high expression levels (FPKM value > 10) was more at 0 (47/85) and 20 (45/87) DAF than at subsequent stages, including 80 DAF (19/82), 8 DPH (30/83), and 14 DPH (27/79). Also, similar expression patterns for *PP2C* genes were observed in FJ. These results indicate that *PP2C* genes play an important role during early fruit development.

Notably, FJ showed more *PYL* genes with high expression levels (FPKM value > 10) than BX at 20 DAF and 3 DPH. *PP2C* genes with high expression levels (FPKM value > 10) were more in FJ than in BX during all the tested stages, except for 6 DPH. More *SnRK2* genes with high expression levels (FPKM value > 10) was also observed in FJ relative to BX at 20 and 80 DAF. These results imply that *PYL-PP2C-SnRK2* genes may be more active in FJ than in BX during fruit development and ripening stages.

A total of 17 *PYL-PP2C-SnRK2* genes, including *MaPYL-9, −10, −12, MaPP2C-7, −14, −32, −37, −45, −47, −49, −55, −67, −69, −72,* and *SnRK2-4, −5, −6,* showed high expression levels (FPKM value > 10) during all the tested stages in both BX and FJ, indicating the extensive and vital role of these genes during fruit developmental and ripening processes.

Most of the Group A *PP2Cs,* including *PP2C-24, −40, −43, −45,* and *−47,* showed high expression levels (FPKM value > 10) in the majority of the development and ripening stages of BX and FJ, whereas *PP2C-16, −20, −22, −23, −46, −59, −60, −62, −63, −82,* and *−83* had extremely low expression (FPKM value < 3) during all the stages of fruit developmental and ripening in both BX and FJ. In addition, 8, 4, 7, 7, 8 Group A *PP2C* genes showed higher expression levels (FPKM value > 10) in FJ than in BX at each stages, respectively.

Expression analyses of *PYL-PP2C-SnRK2* genes in different stages of fruit development and ripening

To get some clues on the function of the *PYL-PP2C-SnRK2* genes in fruit development and ripening of banana, total RNA was extracted during different stages of fruit development and ripening for transcriptomic analyses (Fig. 7; Additional file 1: Tables S6; S7; S8; S9).

According to the transcriptomic data, most of *PYL-PP2C-SnRK2* genes showed similar expression patterns at different stages of fruit development and ripening in both BX and FJ. Some genes showed high expression levels (FPKM value > 10) at different stages of fruit development and ripening. For the *PYL* family, 7/22, 7/22, 5/16, 4/19, and 4/17 *PYL* genes showed high expression levels (FPKM value > 10) at

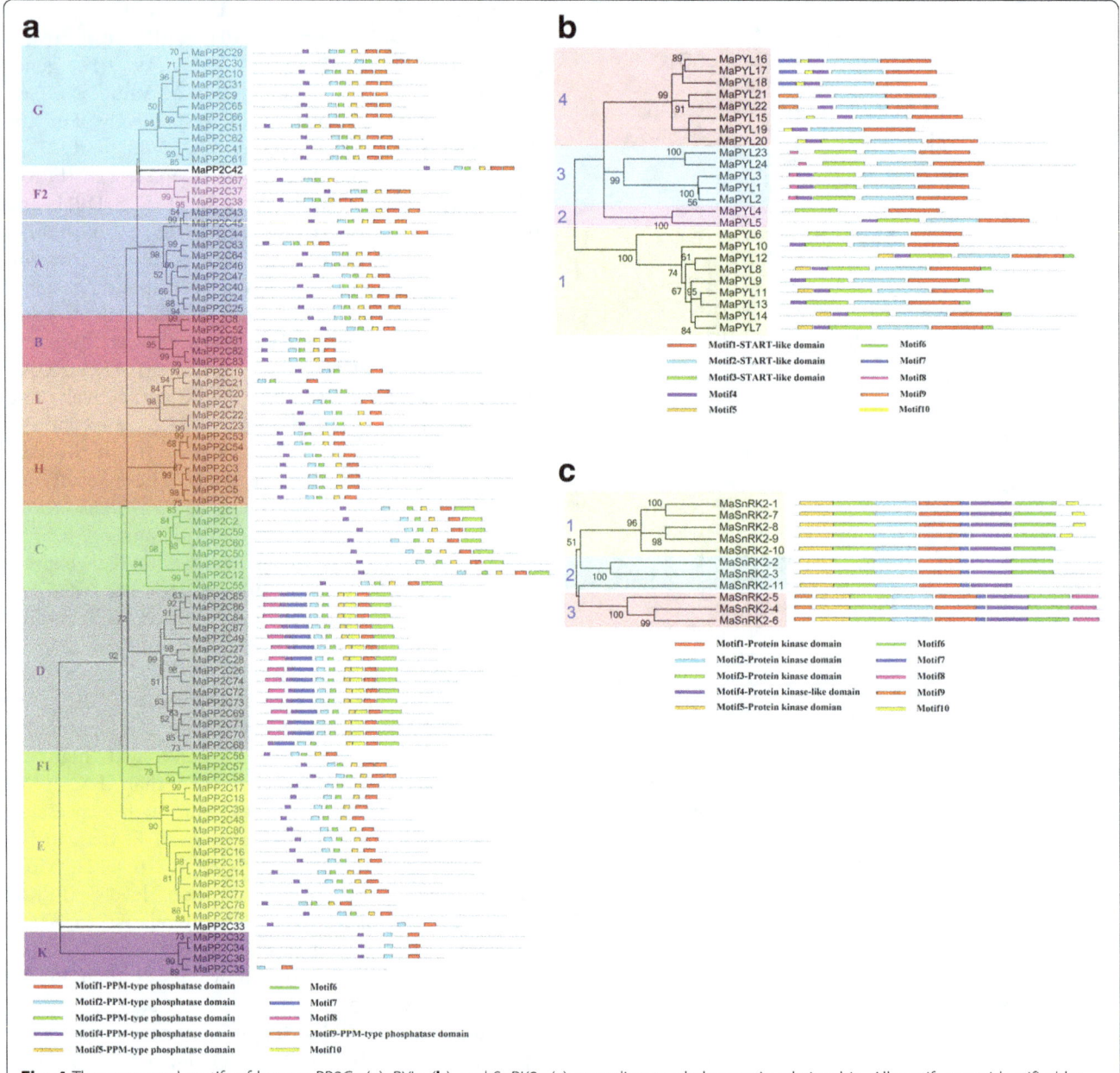

Fig. 4 The conserved motifs of banana PP2Cs (**a**), PYLs (**b**), and SnRK2s (**c**) according to phylogenetic relationship. All motifs were identified by MEME database with the complete amino acid sequences of banana PP2Cs, PYLs, and SnRK2s

Expression analyses of *PYL-PP2C-SnRK2* genes in response to cold, salt, and osmotic stresses

To gain insight into the role of *PYL-PP2C-SnRK2* genes in banana in response to abiotic stress, the leaves of banana after cold, salt, and osmotic treatments were collected for transcriptomic analyses (Fig. 8; Additional file 1: Tables S10; S11; S12; S13).

Under the cold treatment, 4/21 *PYLs*, 17/84 *PP2Cs*, and 0/11 *SnRK2s* showed significant upregulation (Log2 based fold change >1; *P*-value < 0.05) in BX, whereas 5/21 *PYLs*, 19/84 *PP2Cs*, and 2/11 *SnRK2s* were significantly upregulated in FJ. Under the salt treatment, 1/21 *PYLs*, 10/84 *PP2Cs*, and 0/11 *SnRK2s* showed significant induction in BX, while 1/21 *PYLs*, 6/84 *PP2Cs*, and 0/11 *SnRK2s* were significantly upregulated in FJ. Under the osmotic treatment, 1/21 *PYLs*, 10/84 *PP2Cs*, and 1/11 *SnRK2s* were significantly induced in BX, whereas 1/21 *PYLs*, 21/84 *PP2Cs* and 2/11 *SnRK2s* were significantly upregulated in FJ. These results suggest that the number of *PYL-PP2C-SnRK2* genes upregulated by cold and osmotic stresses was more in FJ than in BX, implying that these genes may be more active in FJ than in BX in response to cold and osmotic stresses.

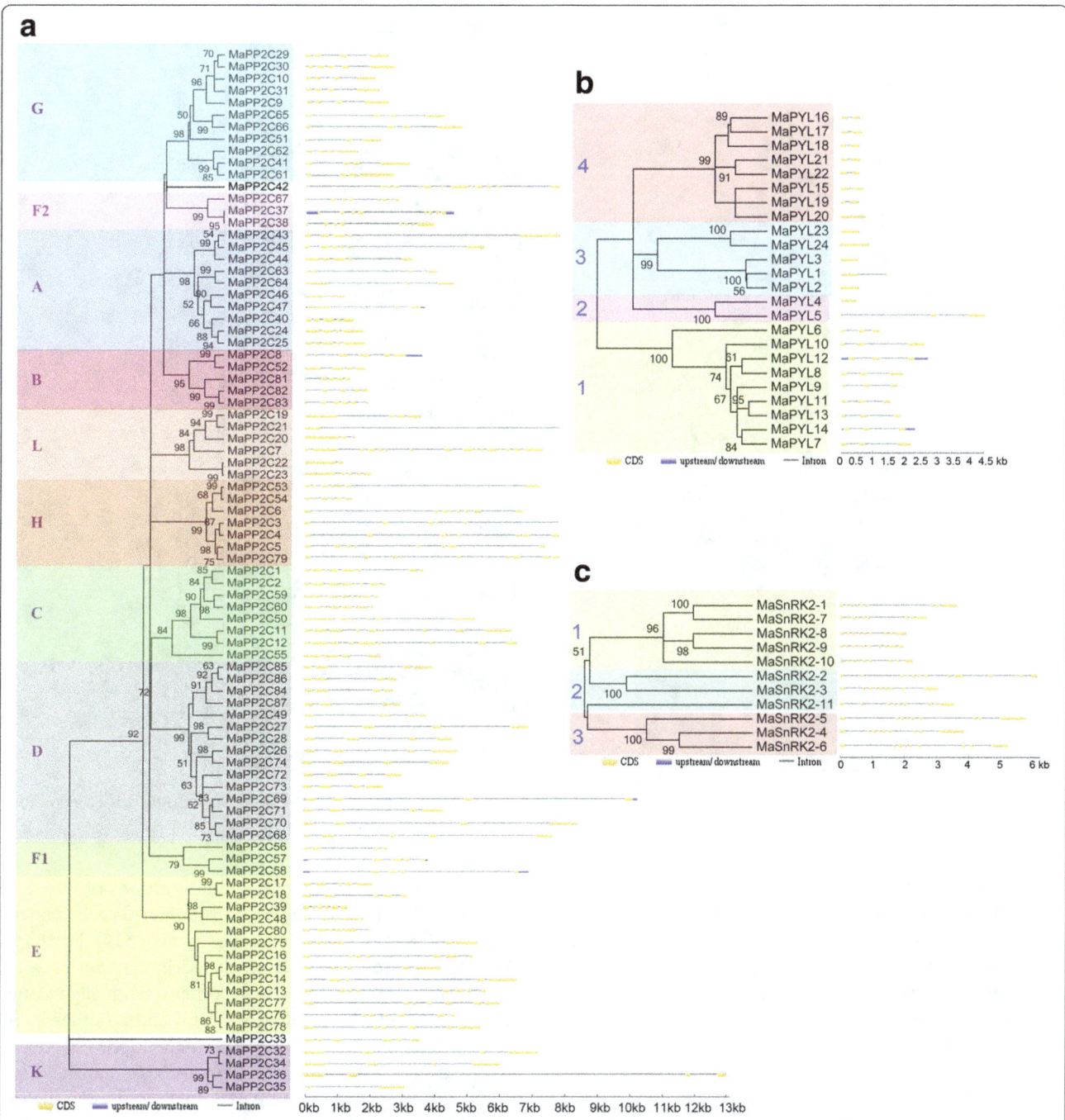

Fig. 5 Gene structure analyses of banana *PP2Cs* (**a**), *PYLs* (**b**), and *SnRK2s* (**c**) according to phylogenetic relationship. Exon-intron structure analyses were performed by GSDS database. The *blue boxes*, *yellow boxes*, and the *black lines* indicate upstream/downstream, exons, and introns, respectively

Notably, 2 *PYL* genes (*MaPYL8* and *MaPYL15*) and 12 *PP2C* genes (*MaPP2C-3, −4, −21, −52, −53, −61, −62, −74, −75, −85, −86,* and *−87*) were strongly induced (Log2 based fold change >2; *P*-value < 0.05) after cold treatment in FJ. Six *PP2C* genes (*MaPP2C-2, −8, −25, −52, −83,* and *−87*) and 1 SnRK2 genes (*MaSnRK2-11*) were strongly upregulated (Log2 based fold change >2;

P-value < 0.05) by osmotic treatments in FJ. These genes may be crucial candidates for further use to improve abiotic stress tolerance of banana.

In addition, 10 genes (*MaPYL-8, −24, MaPP2C-20, −39, −47, −52, −53, −57,* and *MaSnRK2-9, −10*), 5 genes (*MaPYL24* and *MaPP2C-67, −77, −80, −83*), and 14 genes (*MaPP2C-87, −83, −67, −52, −57, −61, −85, −25,*

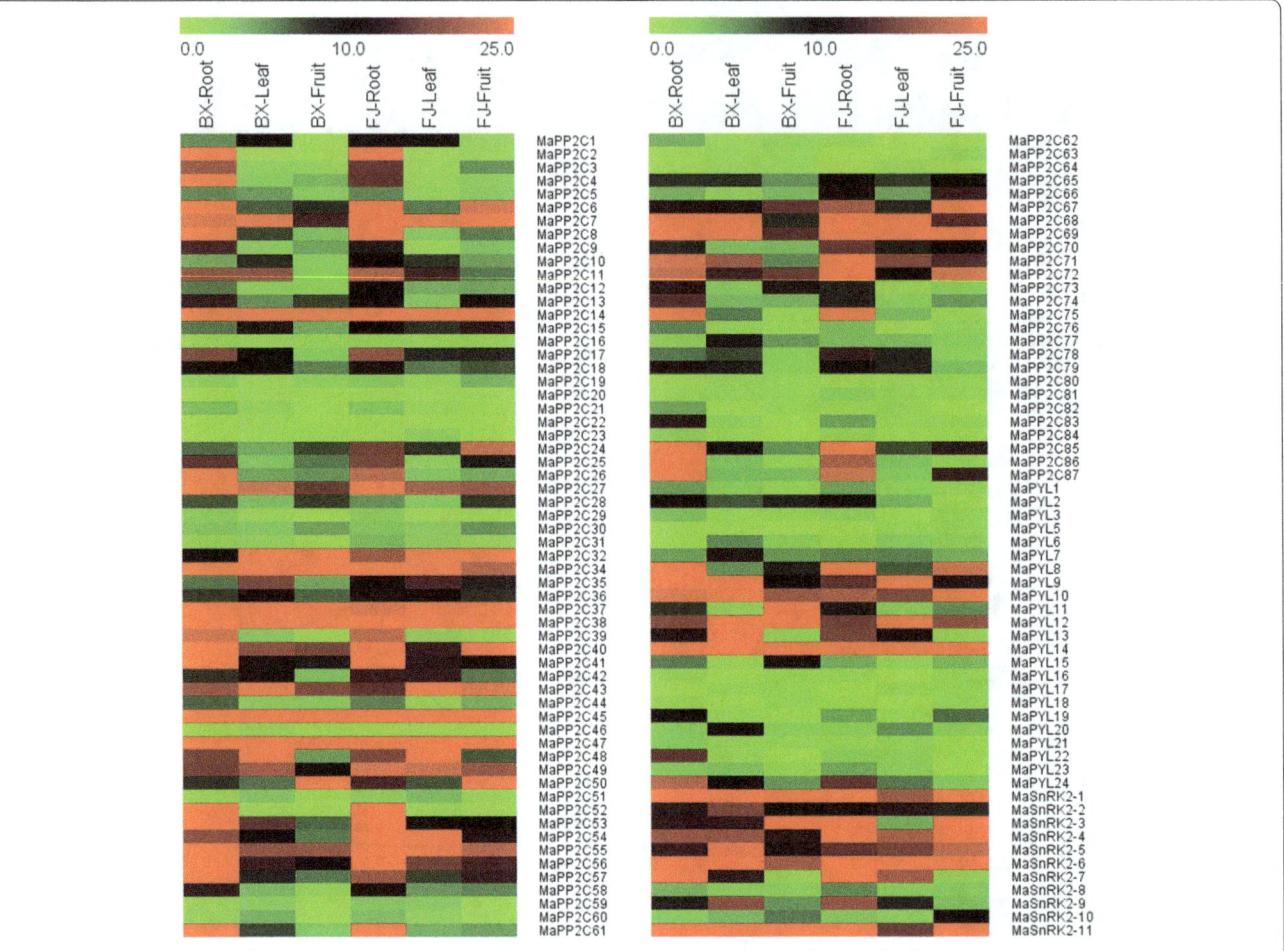

Fig. 6 Expression profiles of banana *PP2Cs*, *PYLs*, and *SnRK2s* in roots, leaves, and fruits of BX and FJ. The heat map was constructed according to the FPKM value of banana *PP2Cs*, *PYLs*, and *SnRK2s* from two independent experiments. FPKM value is shown in color as the scale

–9, –8, –51, –13, –76 and *SnRK2-9*) were significantly induced (Log2 based fold change >1; *P*-value < 0.05) by cold, salt, and osmotic treatments, respectively in FJ, but were not significantly induced in BX. These results indicate that these genes may uniquely function on the tolerance of FJ to abiotic stress.

Several Group A *PP2Cs* showed different expression patterns between BX and FJ in response to abiotic stress. *MaPP2C-25, –43, –44, –45, –46,* and *–63* were upregulated in FJ after cold treatment, whereas in BX, were downregulated or did not show any change. *MaPP2C-44* and *–63* showed upregulation in FJ after salt treatment, whereas downregulation or no change in BX. *MaPP2C43* showed induction in FJ after osmotic treatment, but showed repression in BX.

PYL-PP2C-SnRK2 interaction networks and their co-expression after abiotic stress treatment

To better understand the biological function of PYL-PP2C-SnRK2s in banana, the possible interaction networks and co-expression of Group A banana PP2Cs were investigated

based on experimentally validated interactions of Group A PP2Cs in Arabidopsis and transcriptomic data in banana (Figs. 9, 10 and 11; Additional file 1: Table S14). Firstly, an Arabidopsis Group A PP2C-mediated interaction network was created and 29 interactive proteins (with high confidence; score > 0.9), including 9 PP2Cs and 20 other interactive proteins, were identified with STRING. Secondly, homologs of these interacting proteins in banana were identified with reciprocal BLASTP analyses. Lastly, the expression profiles of the banana genes in BX and FJ under abiotic stress were extracted from RNA-seq data sets.

Under the cold and salt treatments in BX, no gene pair was found to be co-expressed (Figs. 9a and 10a). Under the osmotic treatment in BX, gene pairs HAB1:Ma6270-PYL2:Ma940/PYL6:Ma790/RCAR1:Ma460 showed uniform downregulation (Fig. 11a). Under the cold treatment in FJ, gene pairs HAB1:Ma6270-PYL4:Ma0270/PYL6:Ma790/PYL11:Ma320/PYL1:Ma780 had upregulated co-expression, whereas HAI1:Ma130-PYR1:Ma9170/RCAR1:Ma460/RCAR3:Ma490 showed co-expression of uniform downregulation (Fig. 9b). Under the salt treatment in FJ, gene pairs

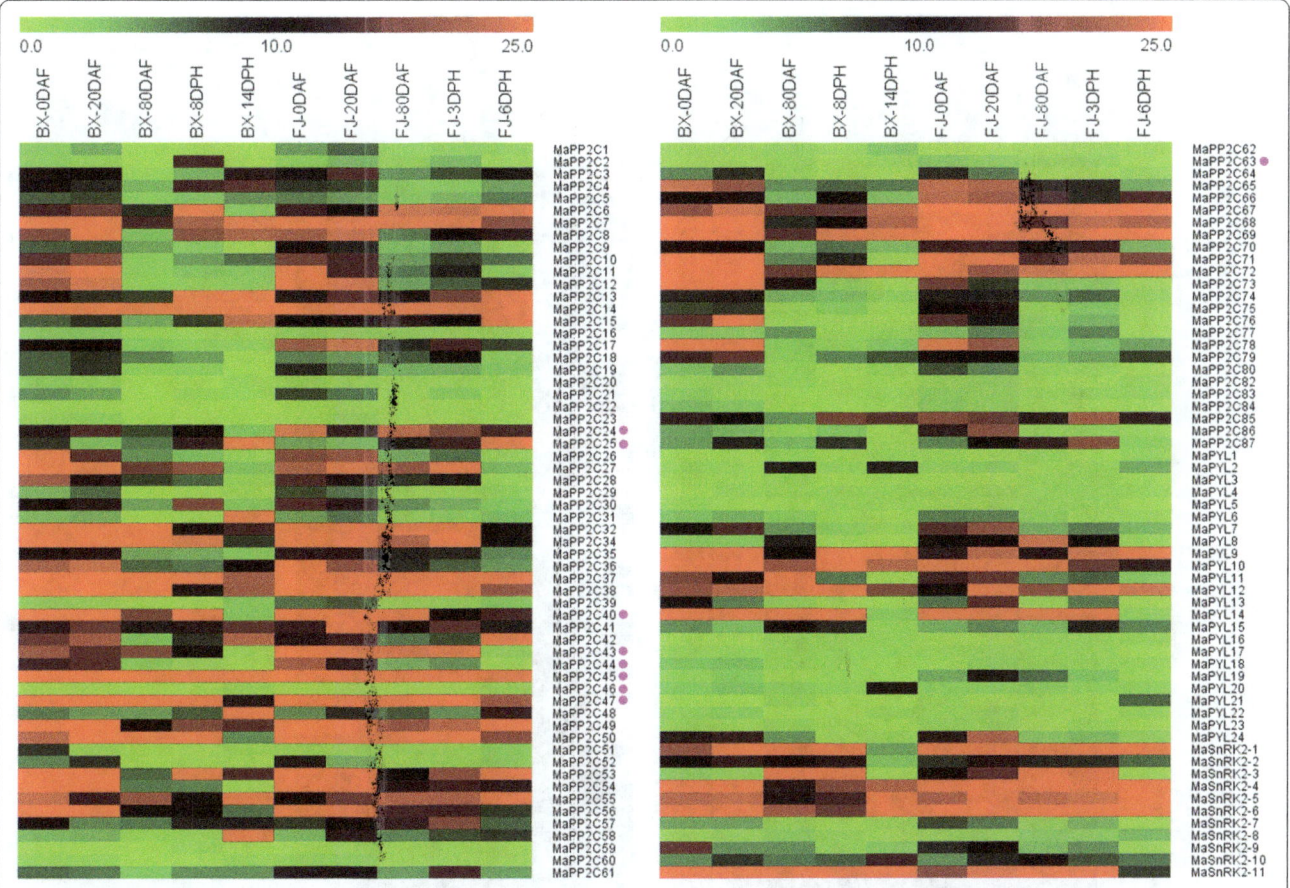

Fig. 7 Expression profiles of banana *PP2Cs*, *PYLs*, and *SnRK2s* in different stages of fruit development and ripening in BX and FJ varieties. The heat map was constructed according to the FPKM value of banana *PP2Cs*, *PYLs*, and *SnRK2s* from two independent experiments. FPKM value is shown in color as the scale. Group A *PP2Cs* are marked with purple dot

HAI2:Ma600- CIPK23:Ma540/PYL10:Ma9170, HAI3: Ma9000- PYL10:Ma9170, and PYL1:Ma780-PP2CA:Ma050 showed uniform upregulation (Fig. 10b). Under the osmotic treatment in FJ, HAI2:Ma600- CIPK23:Ma540 had upregulated co-expression (Fig. 11b). Collectively, these results suggest that more gene pairs were uniformly upregulated in FJ than in BX under cold, salt, and osmotic treatments, indicating the crucial roles of Group A PP2C-mediated network in stress signaling.

Discussion

ABA signaling plays a crucial role in regulating developmental processes and in adaptation to environmental stresses in plants [6, 7]. Investigation of the core regulatory network in the ABA pathway would advance the understanding of the roles of ABA signaling and the function of ABA-associated genes. Currently, no information is known regarding the *PYL-PP2C-SnRK2* gene family in banana. Herein, a total of 24 PYLs, 87 PP2Cs, and 11 SnRK2s were identified from the banana genome, which was classified into 4, 13, and 3 subgroups respectively according to phylogenetic relationship (Figs. 1, 2, and 3).

This classification is in accordance with previous phylogenetic analyses of PYL, PP2C, or SnRK2s in Arabidopsis, rice, *Brassica napus*, and *Brachypodium distachyon* [6, 33–35]. Moreover, the phylogenetic classification of PYL-PP2C-SnRK2 was also supported by conserved motif anslysis (Fig. 4). Conserved motif analyses showed that all the PYLs, PP2Cs, and SnRK2s had START-like, PPM-type phosphatase, and protein kinase domains, respectively, and each subfamily shared similar motifs. These typical characteristics of PYL-PP2C-SnRK2s were also observed in other plant species, such as Arabidopsis, apple, and *Brachypodium distachyon* [6, 7, 35, 36].

As one of the most popular fruits, fruit development and ripening process are crucial for banana fruit quality. ABA signaling has been demonstrated to participate in the fruit development process and ripening of many plant species, including sweet cherries, strawberry, and tomato [2–5]; however, whether *PYL-PP2C-SnRK2s* participate in fruit development and postharvest ripening of banana is unclear. In the present study, we found that more than 4/19 *MaPYLs*, 19/82 *MaPP2Cs*, and 5/10 *MaSnRK2s* showed high

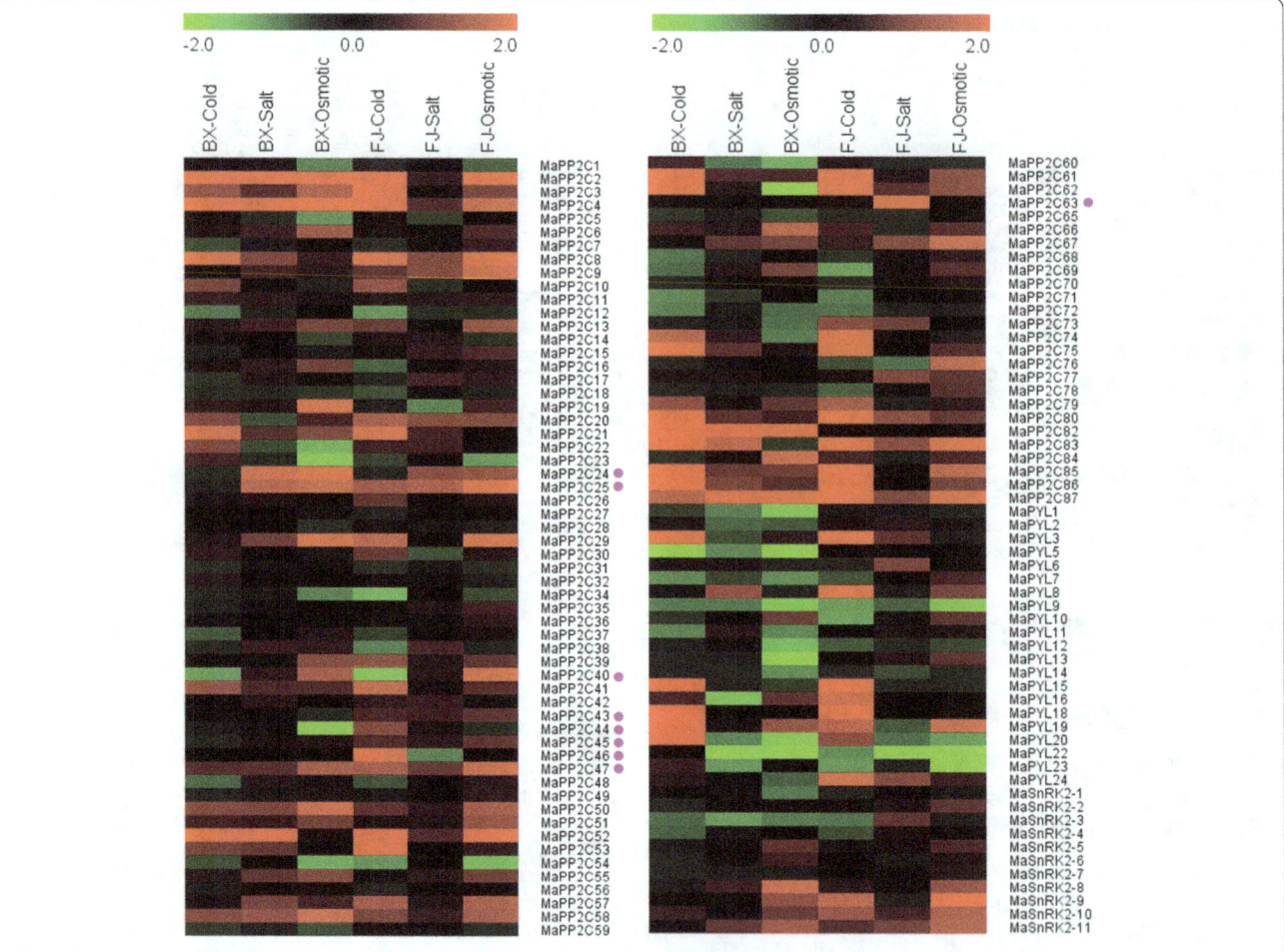

Fig. 8 Expression profiles of banana *PP2Cs*, *PYLs*, and *SnRK2s* in response to cold, salt, and osmotic treatments in BX and FJ varieties. Log2 based fold change was used to create the heat map. Fold changes in gene expression are shown in color as the scale. Group A *PP2Cs* are marked with purple dot

expression levels (FPKM value >10) in BX at any one stage of fruit development and ripening. Also, in FJ, more than 3/19 *MaPYLs*, 28/82 *PP2Cs*, and 4/11 *MaSnRK2s* showed high expression levels (FPKM value >10) at any one stage of fruit development and ripening (Fig. 7; Additional file 1: Tables S6; S7; S8; S9). Moreover, a total of 17 *PYL-PP2C-SnRK2* genes, including *MaPYL-9, −10, −12, MaPP2C-7, −14, −32, −37, −45, −47, −49, −55, −67, −69, −72*, and *SnRK2-4, −5, −6*, showed high expression levels (FPKM value > 10) during all the tested stages in both BX and FJ. Considering the negative role of PP2C in ABA signaling, we also found 11 *MaPP2C* genes (*PP2C-16, −20, −22, −23, −46, −59, −60, −62, −63, −82*, and −83) showing extremely low expression (FPKM value < 3) during all the stages of fruit developmental and ripening in both BX and FJ. These results imply that *PYL-PP2C-SnRK2* genes may be involved in the fruit development and ripening processes of banana.

The number of *PP2C* genes with high expression levels (FPKM value > 10) was more at 0 and 20 DAF than at subsequent stages in both BX and FJ, implying their regulatory role during early fruit development (Fig. 7; Additional file 1: Tables S6; S7; S8; S9). This is consistent with the expression of *CsPP2C1* that reached the first peak value at early stages during cucumber fruit development [37].

Accumulated evidences suggests that exogenous application of ABA could accelerate fruit ripening of banana [38]; however, the role of the core components of ABA signaling, *PYL-PP2C-SnRK2*, in banana development and ripening is unknown. By comparing the *PYL-PP2C-SnRK2* expression profiles at different stages of fruit development and ripening between BX and FJ, an interesting phenomenon was observed. The number of *PYL-PP2C-SnRK2* genes with high expression levels (FPKM value > 10) was more in FJ than in BX at several stages, which implied that *PYL-PP2C-SnRK2* genes may be more active in FJ than in BX during fruit development

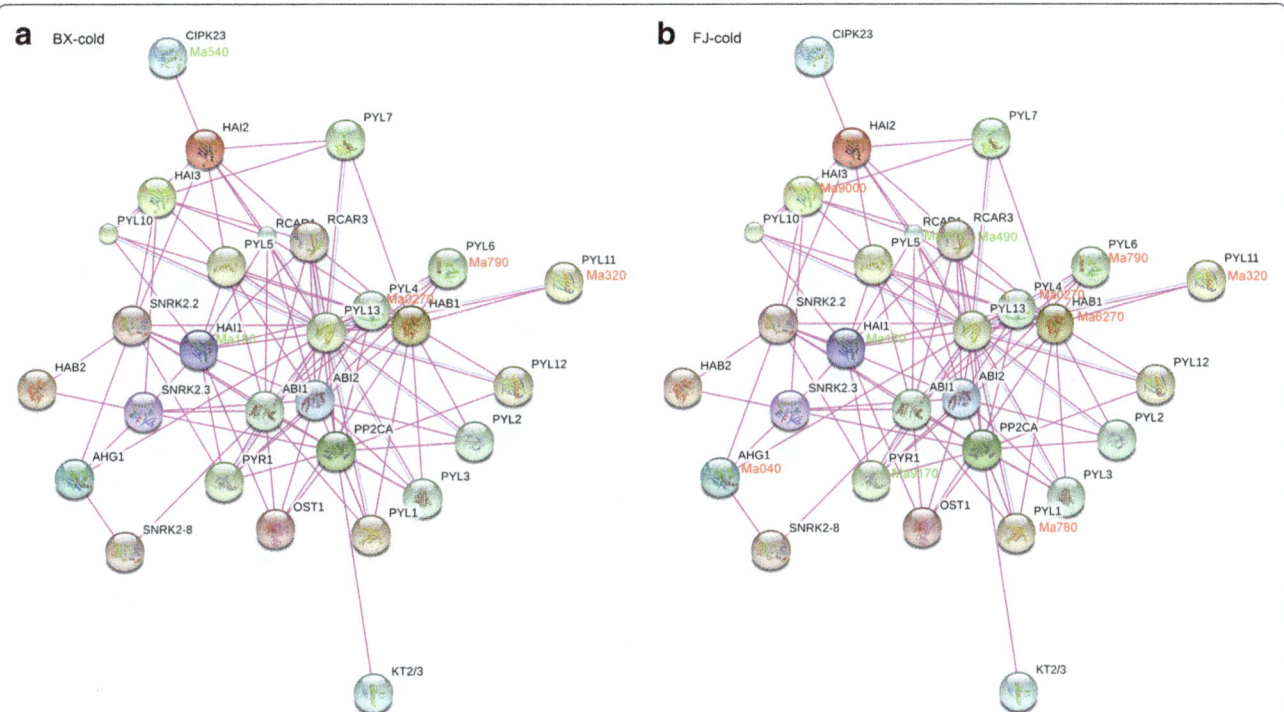

Fig. 9 Interaction network and co-expression analyses of Group A *PP2Cs* after cold treatments in BX (**a**) and FJ (**b**) and related genes in Arabidopsis. The genes marked with *red* show upregulation (Log2 based fold change >1). The genes marked with *green* show downregulation (Log2 based fold change < −1)

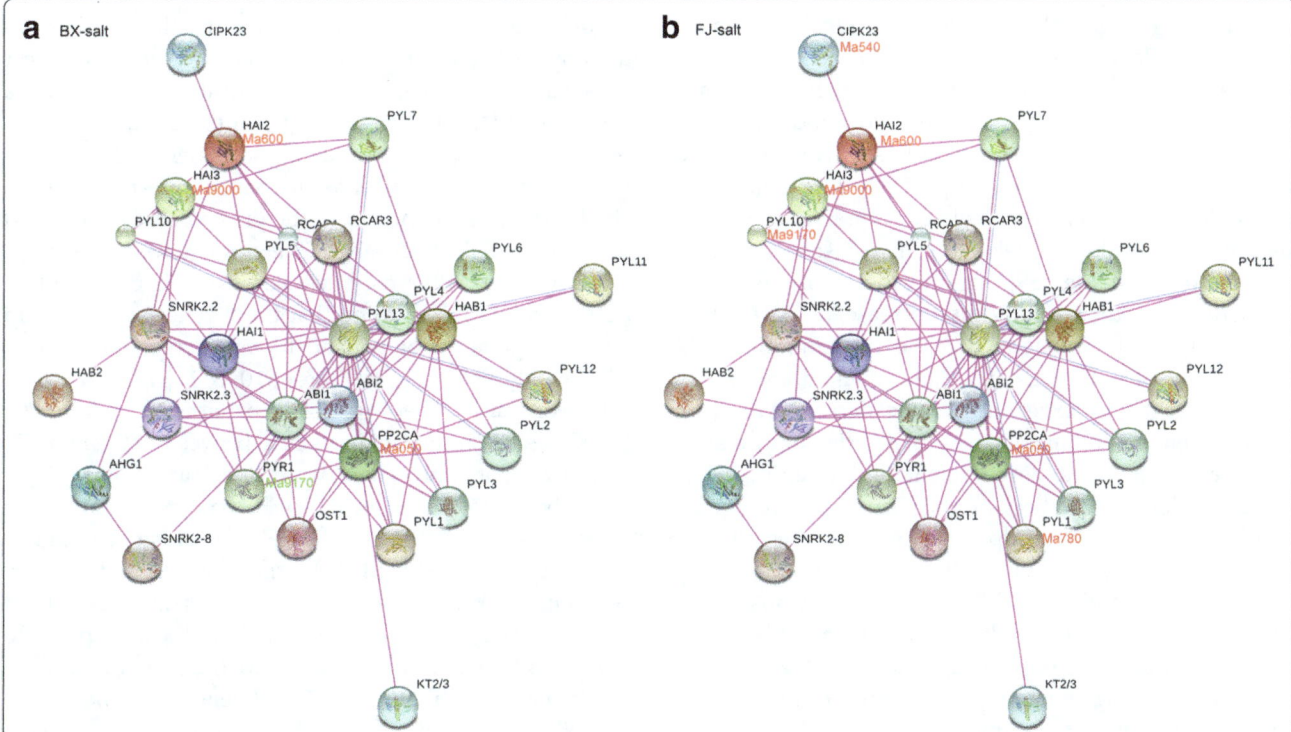

Fig. 10 Interaction network and co-expression analyses of Group A *PP2Cs* after salt treatments in BX (**a**) and FJ (**b**) and related genes in Arabidopsis. The genes marked with *red* show upregulation (Log2 based fold change >1). The genes marked with *green* show downregulation (Log2 based fold change < −1)

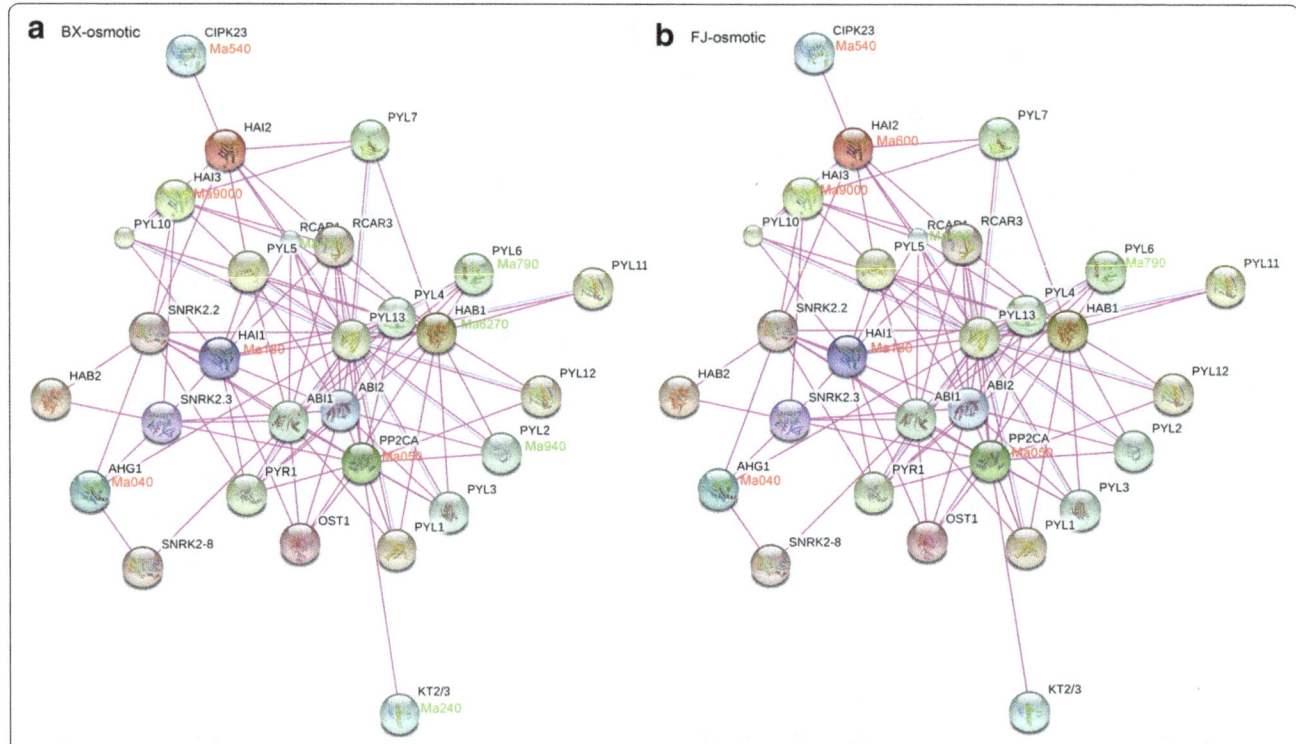

Fig. 11 Interaction network and co-expression analyses of Group A *PP2Cs* after osmotic treatments in BX (**a**) and FJ (**b**) and related genes in Arabidopsis. The genes marked with *red* show upregulation (Log2 based fold change >1). The genes marked with *green* show downregulation (Log2 based fold change < −1)

and ripening stages (Fig. 7; Additional file 1: Tables S6; S7; S8; S9). Previously, we observed that FJ ripened faster than BX during postharvest ripening. It took 8 and 14 DPH to reach more green than yellow and full yellow degrees of ripening for BX, respectively, whereas it only took 3 and 6 DPH for FJ, respectively [28, 29]. In tomato, RNA interference-mediated repression of ABA biosynthesis resulted in delay of fruit senescence and extension of shelf life [39]. In strawberry, inhibition of *FaNCED1* led to a significant decrease of ABA levels and delay of fruit ripening by gene silencing and RNA interference [40]. In grape, fruit development and quality were improved by exogenous application of ABA [41]. This evidence demonstrates that ABA signaling plays a positive role in fruit development and ripening. Additionally, down-regulation of the *FaPYR1* gene significantly delayed fruit ripening and repressed the expression of ABI1 and SnRK2 genes in strawberry, which implied that *PYL-PP2C-SnRK2* genes may positively regulate fruit development and ripening [42]. Therefore, these findings suggest that *PYL-PP2C-SnRK2*-mediated ABA signaling may contribute to fruit development and ripening in banana.

Because banana has shallow roots, permanent green canopy, and rapid growth rate, it is usually subjected to water stress caused by abiotic stress such as cold,

drought, or salt [43]. Investigation of the mechanism underlying banana response to abiotic stress is of great importance for banana breeding. Although ABA plays a predominant role in regulating plants' tolerance to abiotic stress, the role of the core components of ABA signaling, *PYL-PP2C-SnRK2*, in banana responding to abiotic stress is unknown. In the present study, we found that many *PYL-PP2C-SnRK2* genes showed transcriptional changes after cold, salt, or osmotic treatment in both BX and FJ, indicating that these genes may function on the regulation of banana tolerance to abiotic stress (Fig. 8; Additional file 1: Tables S10; S11; S12; S13).

By comparing the expression patterns of *PYL-PP2C-SnRK2* genes under abiotic stress between BX and FJ, it was clear that more genes were significantly upregulated (Log2 based fold change >1) in FJ than in BX under the cold and osmotic treatments (Fig. 8; Additional file 1: Tables S10; S11; S12; S13). Furthermore, from the interaction network and co-expression analyses, more gene pairs were uniformly upregulated in FJ than in BX in response to the osmotic, cold, and salt stresses (Figs. 9, 10, and 11; Additional file 1: Table S14). The B-genome has been considered to be related to tolerance to abiotic stresses. The banana species *M. balbisiana* with the B-genome is demonstrated to have strong resistance to drought or water stress [44, 45]. Moreover, the "ABB" banana genotypes are more tolerant to drought and other abiotic stresses than other genotypes [46]. Thus, the banana varieties based on the "ABB" genotype can be used as a crucial genetic resource for crop improvement for abiotic stress. FJ (ABB genotype), containing the B-genome, has been reported to have strong tolerance to abiotic stress [28, 29]. Much evidence confirms that *PYL-* and *SnRK2*-mediated ABA signaling play a positive role in plants response to abiotic stress [6, 9, 13, 14, 22]. Together, these findings suggest that more *PYL-PP2C-SnRK2* genes and gene pairs upregulated by abiotic stress in FJ could contribute to the tolerance of banana to abiotic stress.

Previously, Group A *PP2Cs* were demonstrated to be negative factors of ABA signaling [6, 27], whereas the function of Group A *PP2Cs* in ABA-mediated biological processes seem to be different in different species [20, 21, 27]. For example, mutation of *abi2-1* resulted in enhanced tolerance to salt stress in Arabidopsis [21], while Arabidopsis plants overexpressing *OsPP108* showed increased tolerance to salt and osmotic stresses [27]. Most of the Group A *PP2Cs* displayed high expression levels during fruit development and ripening in tomato [18]. Moreover, most of the Group A *PP2C* members were induced at transcriptional levels under osmotic, cold, salt, and drought treatments in Arabidopsis [47]. Based on our transcriptomic data, most of the Group A *PP2Cs* showed high expression levels (FPKM value > 10) in the majority of the development and ripening stages of BX and FJ, and Group A *PP2C* genes were found to be more active in FJ than in BX at transcriptional levels after cold, salt, and osmotic treatments. The function and mechanism of *PP2Cs* in ABA signaling transduction and ABA-mediated biological processes need to be further clarified in future studies.

Conclusions

In this study, we identified 24 *PYL*, 87 *PP2C*, and 11 *SnRK2* genes from banana and studied their classification and evolutionary relationships by evolutionary, conserved protein motif, and gene structure analyses. The expression analyses reveal the involvement of *PYL-PP2C-SnRK2* genes in banana fruit development, ripening, and responses to abiotic stress. Additionally, comparison of the differential expression profiles of *PYL-PP2C-SnRK2* genes between BX and FJ suggested that *PYL-PP2C-SnRK2*-mediated ABA signaling might positively regulate banana fruit ripening and responses to abiotic stress. Furthermore, interaction networks and co-expression assays demonstrated the strong transcriptional response of core components of ABA signaling in FJ responding to abiotic stress, further supporting the crucial role of the genes for banana tolerance to abiotic stress. These data will supply abundant information for functional characterization of *PYL-PP2C-SnRK2* genes,

advance the understanding of *PYL-PP2C-SnRK2*-mediated ABA signaling in the regulation of fruit development, ripening, and response to abiotic stress, and lay a solid foundation for further research on banana breeding.

Methods
Plant materials and treatments
Two banana cultivars of BaXi Jiao (*Musa acuminate* L. AAA group cv. Cavendish, BX) and Fen Jiao (*Musa* ABB PisangAwak, FJ) were used in this study. BX is widely planted in China due to its virtues of long storage and high production. FJ is widely cultivated in the Hainan province of China. FJ has stronger tolerance to abiotic stress, including drought, salt, and cold, and ripened faster than BX during postharvest ripening (unpublished data). BX and FJ seedlings at the five-leaf stage were acquired from the banana tissue culture center (Institute of Bananas and Plantains, Chinese Academy of Tropical Agricultural Sciences, Danzhou). Seedlings with consistent growth state were cultured in soil under the conditions of 70% relative humidity and 200 μmol m^{-2} s^{-1} light intensity in 16 h light/8 h dark cycle, 28 °C. Roots and leaves from the five-leaf stage plants, and fruits of 80 DAF were sampled for expression analysis in different organs. Fruits from 0 DAF (budding), 20 DAF (cutting flower) and 80 DAF (harvest stage) were collected to study the expression profiles of genes during fruit development process. Fruits from 8 DPH and 14 DPH in BX and 3 DPH and 6 DPH in FJ were sampled to investigate gene expression patterns during post-harvest ripening stages because FJ reach full yellow degree faster than BX after harvesting [28, 29]. Banana seedlings at the five-leaf stage were irrigated with 200 mM mannitol or 300 mM NaCl for 7 days to study gene expression in response to osmotic and salt stresses, respectively. Banana seedlings were incubated in 4 °C for 22 h to detect gene expression upon cold stress.

Identification and phylogenetic analyses
The whole protein sequences of banana were downloaded from the banana genome database [32]. The PYL, PP2C, and SnRK2 protein sequences from rice and *Arabidopsis* were obtained from RGAP and UniProt databases, respectively [48, 49]. The HMM profiles built from the known PYL-PP2C-SnRK2s were used as queries to search the banana dataset with HMMER software [50, 51]. BLAST was also employed to identify the predicted banana PYL-PP2C-SnRK2s with all PYL-PP2C-SnRK2s from rice and *Arabidopsis* as queries. Then, the conserved domains of predicted banana PYL-PP2C-SnRK2s were further confirmed with PFAM and CDD databases [52, 53]. The accession numbers of identified banana PYLs, PP2Cs, and SnRK2s are displayed in Table S1. The phylogenetic tree was reconstructed with the

PYL-PP2C-SnRK2 proteins from Arabidopsis, rice, and banana using MEGA 5.0 and Clustal X2.0 softwares (bootstrap values for 1000 replicates) [54, 55].

Protein properties and sequence analyses

Using the ExPASy database, the isoelectric points and molecular weights of the banana PYL-PP2C-SnRK2s were predicted [56]. MEME software was used to identify motifs of banana PYL-PP2C-SnRK2 proteins, and then the motifs were annotated with InterProScan [57, 58]. The optimum width of motifs ranged from 6 to 50, the maximum number of motifs was 10, and the other parameter settings used were default values. The *PYL-PP2C-SnRK2* gene structure was analyzed by GSDS [59]. With the help of STRING software, the Group A PP2Cs-mediated protein interactions in Arabidopsis were explored with the confidence score > 0.9 and no more than 20 interactors.

Transcriptomic analysis

Total RNA of each sample was extracted with plant RNA extraction kit (TIANGEN, China) and used for cDNA library construction. The sequencing was performed with an Illumina GAII following manufacturer's instructions. Using FASTX-toolkit, adapter sequences in the raw sequence reads were removed. After examining the sequence quality and removing low quality sequences by FastQC, clean reads were generated. Using Tophat v.2.0.10, clean reads were maped to the DH-Pahang genome (*Musa acuminate*, A-genome, 2n = 22) [32]. The transcriptome assemblies were performed by Cufflinks [60]. The RNA-seq reads status was listed in Additional file 1: Tables S2; S6; S10. Genes were scored as not expressed if the corresponding RNA-seq reads could not align to the genome. Calculation the ratio of *PYL-PP2C-SnRK2* genes with high expression levels or showing significantly changes after abiotic stress treatments was performed according to the genes that is expressed. Gene expression levels were calculated as Reads Per Kilobase of exon model per Million mapped reads (FPKM). DEGseq was used to identify differentially expressed genes (Log2 based fold change >1 or Log2 based fold change <−1; *P*-value < 0.05) in response to cold, salt, and osmotic stresses [61]. There are two biological replicates, which showed good consistency (Additional file 2: Figures S1; S2; S3; Additional file 1: Table S3; S7; S11).

Additional files

Additional file 1: Table S1. Characteristics of banana PYL, PP2C, and SnRK2 gene families. **Table S2.** Properties of transcriptome for RNA-seq analysis in different tissues. **Table S3.** FPKM value from each biological replicates of PYL-PP2C-SnRK2 genes in different tissues. **Table S4.** Expression data of the banana PYL-PP2C-SnRK2 genes in different tissues of BX and FJ varieties. **Table S5.** Statistical analyses of the expression data related to banana PYL-PP2C-SnRK2 genes in different tissues of BX and FJ

varieties. **Table S6.** Properties of transcriptome for RNA-seq analysis in different stages of fruit development and ripening. **Table S7.** FPKM value from each biological replicates of PYL-PP2C-SnRK2 genes in different stages of fruit development and ripening. **Table S8.** Expression data of the banana PYL-PP2C-SnRK2 genes in different stages of fruit development and ripening in BX and FJ varieties. **Table S9.** Statistical analyses of the expression data related to banana PYL-PP2C-SnRK2 genes in different stages of fruit development and ripenings of BX and FJ varieties. **Table S10.** Properties of transcriptome for RNA-seq analysis under cold, salt, and osmotic treatments. **Table S11.** FPKM value from each biological replicates of PYL-PP2C-SnRK2 genes in response to cold, salt, and osmotic stresses. **Table S12.** Expression data (log2-based value) of the banana PYL-PP2C-SnRK2 genes after various abiotic stress treatment in BX and FJ. **Table S13.** Statistical analyses of the expression data related to banana PYL-PP2C-SnRK2 genes in response to abiotic stress. **Table S14.** Expression data of the genes involved in Group A PP2C-mediate interaction networks under abiotic stress in BX and FJ varieties. **Table S15.** The accession numbers and gene name of PYL-PP2C-SnRK2 gene families in Arabidopsis and rice.

Additional file 2: Figure S1. Expression profiles of banana *PP2Cs, PYLs,* and *SnRK2s* in roots, leaves, and fruits of BX and FJ. The heat map was constructed according to the FPKM value of banana *PP2Cs, PYLs,* and *SnRK2s* from each replicates of two independent experiments. **Figure S2.** Expression profiles of banana *PP2Cs, PYLs,* and *SnRK2s* in different stages of fruit development and ripening in BX and FJ varieties. The heat map was constructed according to the FPKM value of banana *PP2Cs, PYLs,* and *SnRK2s* from each replicates of two independent experiments. **Figure S3.** Expression profiles of banana *PP2Cs, PYLs,* and *SnRK2s* in response to cold, salt, and osmotic treatments in BX and FJ varieties. The heat map was constructed according to the FPKM value of banana *PP2Cs, PYLs,* and *SnRK2s* from each replicates of two independent experiments.

Abbreviations
ABA: Abscisic acid; BX: BaXi Jiao; DAF: Days after flower; DPH: Days post-harvest; FJ: Fen Jiao; FPKM: Reads Per Kilobase of exon model per Million mapped reads; NJ: Neighbor-joining

Acknowledgements
Not applicable.

Funding
This work was supported by the National Nonprofit Institute Research Grant of CATAS (1630052012012), the Central Public-interest Scientific Institution Basal Research Fund for Chinese Academy of Tropical Agricultural Sciences (1,630,052,016,005, 1,630,052,016,006), the Central Public-interest Scientific Institution Basal Research Fund for Innovative Research Team Program of CATAS (17CXTD-28), and the earmarked fund for Modern Agro-industry Technology Research System (CARS-32).

Authors' contributions
BX, ZJ, and HS conceived the study. WH, YY, JL, HM, WT, ZD, XD, CW, YL, and JW performed the experiments and carried out the analysis. WH, YY, and JL designed the experiments and wrote the manuscript. All authors read and approved the final manuscript.

Competing interests
The authors declare that they have no competing interests.

Author details
[1]Key Laboratory of Biology and Genetic Resources of Tropical Crops, Institute of Tropical Bioscience and Biotechnology, Chinese Academy of Tropical Agricultural Sciences, Haikou, Hainan, China. [2]Key Laboratory of Genetic Improvement of Bananas, Hainan province, Haikou Experimental Station, China Academy of Tropical Agricultural Sciences, Haikou, Hainan, China. [3]Hainan Key Laboratory for Sustainable Utilization of Tropical Bioresources, College of Agriculture, Hainan University, Haikou, China.

References

1. Feng CZ, Chen Y, Wang C, Kong YH, Wu WH, Chen YF. Arabidopsis RAV1 transcription factor, phosphorylated by SnRK2 kinases, regulates the expression of ABI3, ABI4, and ABI5 during seed germination and early seedling development. Plant J. 2014;80(4):654–68.

2. Tijero V, Teribia N, Munoz P, Munne-Bosch S. Implication of Abscisic acid on ripening and quality in sweet cherries: differential effects during pre- and post-harvest. Front Plant Sci. 2016;7:602.

3. Li D, Mou W, Luo Z, Li L, Limwachiranon J, Mao L, Ying T. Developmental and stress regulation on expression of a novel miRNA, fan-miR73, and its target ABI5 in strawberry. Sci Rep. 2016;6:28385.

4. Mou W, Li D, Bu J, Jiang Y, Khan ZU, Luo Z, Mao L, Ying T. Comprehensive analysis of ABA effects on ethylene biosynthesis and signaling during tomato fruit ripening. PLoS One. 2016;11(4):e0154072.

5. Jia H, Jiu S, Zhang C, Wang C, Tariq P, Liu Z, Wang B, Cui L, Fang J. Abscisic acid and sucrose regulate tomato and strawberry fruit ripening through the abscisic acid-stress-ripening transcription factor. Plant Biotechnol J. 2016;14(10):2045–65.

6. Umezawa T, Nakashima K, Miyakawa T, Kuromori T, Tanokura M, Shinozaki K, Yamaguchi-Shinozaki K. Molecular basis of the Core regulatory network in ABA responses: sensing, signaling and transport. Plant Cell Physiol. 2010;51(11):1821–39.

7. Ben-Ari G. The ABA signal transduction mechanism in commercial crops: learning from Arabidopsis. Plant Cell Rep. 2012;31(8):1357–69.

8. Hu W, Yan Y, Hou XW, He YZ, Wei YX, Yang GX, He GY, Peng M. TaPP2C1, a group F2 protein Phosphatase 2C gene, confers resistance to salt stress in transgenic tobacco. PLoS One. 2015;10(6):e0129589.

9. Ma Y, Szostkiewicz I, Korte A, Moes D, Yang Y, Christmann A, Grill E. Regulators of PP2C Phosphatase activity function as Abscisic acid sensors. Science. 2009;324(5930):1064–8.

10. Park SY, Fung P, Nishimura N, Jensen DR, Fujii H, Zhao Y, Lumba S, Santiago J, Rodrigues A, Chow TFF, et al. Abscisic acid inhibits type 2C protein Phosphatases via the PYR/PYL family of START proteins. Science. 2009;324(5930):1068–71.

11. Bhaskara GB, Nguyen TT, Verslues PE. Unique drought resistance functions of the highly ABA-induced Clade a protein Phosphatase 2Cs. Plant Physiol. 2012;160(1):379–95.

12. Fujii H, Chinnusamy V, Rodrigues A, Rubio S, Antoni R, Park SY, Cutler SR, Sheen J, Rodriguez PL, Zhu JK. In vitro reconstitution of an abscisic acid signalling pathway. Nature. 2009;462(7273):660–4.

13. Santiago J, Rodrigues A, Saez A, Rubio S, Antoni R, Dupeux F, Park SY, Marquez JA, Cutler SR, Rodriguez PL. Modulation of drought resistance by the abscisic acid receptor PYL5 through inhibition of clade a PP2Cs. Plant J. 2009;60(4):575–88.

14. Saavedra X, Modrego A, Rodriguez D, Gonzalez-Garcia MP, Sanz L, Nicolas G, Lorenzo O. The nuclear Interactor PYL8/RCAR3 of Fagus Sylvatica FsPP2C1 is a positive regulator of Abscisic acid signaling in seeds and stress. Plant Physiol. 2010;152(1):133–50.

15. Allen GJ, Kuchitsu K, Chu SP, Murata Y, Schroeder JI. Arabidopsis abi1-1 and abi2-1 phosphatase mutations reduce abscisic acid-induced cytoplasmic calcium rises in guard cells. Plant Cell. 1999;11(9):1785–98.

16. Saez A, Apostolova N, Gonzalez-Guzman M, Gonzalez-Garcia MP, Nicolas C, Lorenzo O, Rodriguez PL. Gain-of-function and loss-of-function phenotypes of the protein phosphatase 2C HAB1 reveal its role as a negative regulator of abscisic acid signalling. Plant J. 2004;37(3):354–69.

17. Kuhn JM, Boisson-Dernier A, Dizon MB, Maktabi MH, Schroeder JI. The protein phosphatase AtPP2CA negatively regulates abscisic acid signal transduction in Arabidopsis, and effects of abh1 on AtPP2CA mRNA. Plant Physiol. 2006;140(1):127–39.

18. Sun HL, Wang XJ, Ding WH, Zhu SY, Zhao R, Zhang YX, Xin Q, Wang XF, Zhang DP. Identification of an important site for function of the type 2C protein phosphatase ABI2 in abscisic acid signalling in Arabidopsis. J Exp Bot. 2011;62(15):5713–25.

19. Rubio S, Rodrigues A, Saez A, Dizon MB, Galle A, Kim TH, Santiago J, Flexas J, Schroeder JI, Rodriguez PL. Triple loss of function of protein Phosphatases type 2C leads to partial constitutive response to endogenous Abscisic acid. Plant Physiol. 2009;150(3):1345–55.

20. Tahtiharju S, Palva T. Antisense inhibition of protein phosphatase 2C accelerates cold acclimation in Arabidopsis Thaliana. Plant J. 2001;26(4):461–70.

21. Ohta M, Guo Y, Halfter U, Zhu JK. A novel domain in the protein kinase SOS2 mediates interaction with the protein phosphatase 2C ABI2. Proc Natl Acad Sci U S A. 2003;100(20):11771–6.

22. Umezawa T, Yoshida R, Maruyama K, Yamaguchi-Shinozaki K, Shinozaki K. SRK2C, a SNF1-related protein kinase 2, improves drought tolerance by controlling stress-responsive gene expression in Arabidopsis Thaliana. Proc Natl Acad Sci U S A. 2004;101(49):17306–11.

23. Tian X, Wang Z, Li X, Lv T, Liu H, Wang L, Niu H, Bu Q. Characterization and functional analysis of pyrabactin resistance-like abscisic acid receptor family in rice. Rice. 2015;8(1):1–13.

24. Kim H, Hwang H, Hong JW, Lee YN, Ahn IP, Yoon IS, Yoo SD, Lee S, Lee SC, Kim BG. A rice orthologue of the ABA receptor, OsPYL/RCAR5, is a positive regulator of the ABA signal transduction pathway in seed germination and early seedling growth. J Exp Bot. 2012;63(2):1013–24.

25. Diédhiou CJ, Popova OV, Dietz KJ, Golldack D. The SNF1-type serine-threonine protein kinase SAPK4 regulates stress-responsive gene expression in rice. BMC Plant Biol. 2008;8(1):49.

26. Dey A, Samanta MK, Gayen S, Maiti MK. The sucrose non-fermenting 1-related kinase 2 gene SAPK9 improves drought tolerance and grain yield in rice by modulating cellular osmotic potential, stomatal closure and stress-responsive gene expression. BMC Plant Biol. 2016;16(1):1–20.

27. Singh A, Jha Bagri J, Pandey GK. ABA inducible rice protein phosphatase 2C confers ABA insensitivity and abiotic stress tolerance in Arabidopsis. PLoS One. 2015;10(4):156–62.

28. Hu W, Zuo J, Hou XW, Yan Y, Wei YX, Liu JH, Li MY, Xu BY, Jin ZQ. The auxin response factor gene family in banana: genome-wide identification and expression analyses during development, ripening, and abiotic stress. Front Plant Sci. 2015;6:742.

29. Hu W, Hou XW, Huang C, Yan Y, Tie WW, Ding ZH, Wei YX, Liu JH, Miao HX, Lu ZW, et al. Genome-wide identification and expression analyses of Aquaporin gene family during development and Abiotic stress in banana. Int J Mol Sci. 2015;16(8):19728–51.

30. Sreedharan S, Shekhawat UKS, Ganapathi TR. Transgenic banana plants overexpressing a native plasma membrane aquaporin MusaPIP1;2 display high tolerance levels to different abiotic stresses. Plant Biotechnol J. 2013;11(8):942–52.

31. Hirayama T, Shinozaki K. Research on plant abiotic stress responses in the post-genome era: past, present and future. Plant J. 2010;61(6):1041–52.

32. D'Hont A, Denoeud F, Aury JM, Baurens FC, Carreel F, Garsmeur O, Noel B, Bocs S, Droc G, Rouard M, et al. The banana (Musa Acuminata) genome and the evolution of monocotyledonous plants. Nature. 2012;488(7410):213–7.

33. Xue T, Wang D, Zhang S, Ehlting J, Ni F, Jakab S, Zheng C, Zhong Y. Genome-wide and expression analysis of protein phosphatase 2C in rice and Arabidopsis. BMC Genomics. 2008;9:550.

34. Yoo MJ, Ma T, Zhu N, Liu L, Harmon AC, Wang Q, Chen S. Genome-wide identification and homeolog-specific expression analysis of the SnRK2 genes in Brassica Napus guard cells. Plant Mol Biol. 2016;91(1-2):211–27.

35. Cao JM, Jiang M, Li P, Chu ZQ. Genome-wide identification and evolutionary analyses of the PP2C gene family with their expression profiling in response to multiple stresses in Brachypodium Distachyon. BMC Genomics. 2016;17:175.

36. Shao Y, Qin Y, Zou YJ, Ma FW. Genome-wide identification and expression profiling of the SnRK2 gene family in Malus Prunifolia. Gene. 2014;552(1):87–97.

37. Wang YP, Wu Y, Duan CR, Chen P, Li Q, Dai SJ, Sun L, Ji K, Sun YF, Xu W, et al. The expression profiling of the CsPYL, CsPP2C and CsSnRK2 gene families during fruit development and drought stress in cucumber. J Plant Physiol. 2012;169(18):1874–82.

38. Jiang Y, Joyce DC, Macnish AJ. Effect of Abscisic acid on banana fruit ripening in relation to the role of ethylene. J Plant Growth Regul. 2000;19(1):106–11.

39. Sun L, Sun YF, Zhang M, Wang L, Ren J, Cui MM, Wang YP, Ji K, Li P, Li Q, et al. Suppression of 9-cis-Epoxycarotenoid Dioxygenase, which encodes a key enzyme in Abscisic acid biosynthesis, alters fruit texture in transgenic tomato. Plant Physiol. 2012;158(1):283–98.

40. Jia HF, Chai YM, Li CL, Lu D, Luo JJ, Qin L, Shen YY. Abscisic acid plays an important role in the regulation of strawberry fruit ripening. Plant Physiol. 2011;157(1):188–99.

41. Cantín CM, Fidelibus MW, Crisosto CH. Application of abscisic acid (ABA) at veraison advanced red color development and maintained postharvest quality of 'crimson seedless' grapes. Postharvest Biol Technol. 2007;46(3):237–41.

42. Chai YM, Jia HF, Li CL, Dong QH, Shen YY. FaPYR1 is involved in strawberry fruit ripening. J Exp Bot. 2011;62(14):5079–89.

43. van Asten PJA, Fermont AM, Taulya G. Drought is a major yield loss factor for rainfed east African highland banana. Agric Water Manag. 2011;98(4):541–52.

44. Davey MW, Gudimella R, Harikrishna JA, Sin LW, Khalid N, Keulemans J. A draft Musa Balbisiana genome sequence for molecular genetics in polyploid, inter- and intra-specific Musa hybrids. BMC Genomics. 2013;14:683.

45. Vanhove AC, Vermaelen W, Panis B, Swennen R, Carpentier SC. Screening the banana biodiversity for drought tolerance: can an in vitro growth model and proteomics be used as a tool to discover tolerant varieties and understand homeostasis. Front Plant Sci. 2012;3:176.

46. Ravi I, Uma S, Vaganan MM, Mustaffa MM. Phenotyping bananas for drought resistance. Front Physiol. 2013;4:9.

47. Chan ZL. Expression profiling of ABA pathway transcripts indicates crosstalk between abiotic and biotic stress responses in Arabidopsis. Genomics. 2012; 100(2):110–5.

48. Bateman A, Martin MJ, O'Donovan C, Magrane M, Apweiler R, Alpi E, Antunes R, Ar-Ganiska J, Bely B, Bingley M, et al. UniProt: a hub for protein information. Nucleic Acids Res. 2015;43(D1):D204–12.

49. Kawahara Y, de la Bastide M, Hamilton JP, Kanamori H, McCombie WR, Ouyang S, Schwartz DC, Tanaka T, Wu JZ, Zhou SG, et al. Improvement of the Oryza Sativa Nipponbare reference genome using next generation sequence and optical map data. Rice. 2013;6:4.

50. Eddy SR. Accelerated profile HMM searches. PLoS Comput Biol. 2011;7(10):e1002195.

51. Finn RD, Clements J, Eddy SR. HMMER web server: interactive sequence similarity searching. Nucleic Acids Res. 2011;39:W29–37.

52. Marchler-Bauer A, Derbyshire MK, Gonzales NR, Lu SN, Chitsaz F, Geer LY, Geer RC, He J, Gwadz M, Hurwitz DI, et al. CDD: NCBI's conserved domain database. Nucleic Acids Res. 2015;43(D1):D222–6.

53. Finn RD, Coggill P, Eberhardt RY, Eddy SR, Mistry J, Mitchell AL, Potter SC, Punta M, Qureshi M, Sangrador-Vegas A, et al. The Pfam protein families database: towards a more sustainable future. Nucleic Acids Res. 2016;44(D1):D279–85.

54. Larkin MA, Blackshields G, Brown NP, Chenna R, McGettigan PA, McWilliam H, Valentin F, Wallace IM, Wilm A, Lopez R, et al. Clustal W and Clustal X version 2.0. Bioinformatics. 2007;23(21):2947–8.

55. Tamura K, Peterson D, Peterson N, Stecher G, Nei M, Kumar S. MEGA5: molecular evolutionary genetics analysis using maximum likelihood, evolutionary distance, and maximum parsimony methods. Mol Biol Evol. 2011;28(10):2731–9.

56. Gasteiger E, Gattiker A, Hoogland C, Ivanyi I, Appel RD, Bairoch A. ExPASy: the proteomics server for in-depth protein knowledge and analysis. Nucleic Acids Res. 2003;31(13):3784–8.

57. Brown P, Baxter L, Hickman R, Beynon J, Moore JD, Ott S. MEME-LaB: motif analysis in clusters. Bioinformatics. 2013;29(13):1696–7.

58. Mulder N, Apweiler R. InterPro and InterProScan: tools for protein sequence classification and comparison. Methods Mol Biol. 2007;396:59–70.

59. Hu B, Jin JP, Guo AY, Zhang H, Luo JC, Gao G. GSDS 2.0: an upgraded gene feature visualization server. Bioinformatics. 2015;31(8):1296–7.

60. Trapnell C, Roberts A, Goff L, Pertea G, Kim D, Kelley DR, Pimentel H, Salzberg SL, Rinn JL, Pachter L. Differential gene and transcript expression analysis of RNA-seq experiments with TopHat and cufflinks. Nat Protoc. 2012;7(3):562–78.

61. Wang LK, Feng ZX, Wang X, Wang XW, Zhang XG. DEGseq: an R package for identifying differentially expressed genes from RNA-seq data. Bioinformatics. 2010;26(1):136–8.

3

Ambient temperature and genotype differentially affect developmental and phenotypic plasticity in *Arabidopsis thaliana*

Carla Ibañez[1,2†], Yvonne Poeschl[3,4†], Tom Peterson[2], Julia Bellstädt[1,2], Kathrin Denk[1,2], Andreas Gogol-Döring[3,4], Marcel Quint[1,2] and Carolin Delker[1,2*] (iD)

Abstract

Background: Global increase in ambient temperatures constitute a significant challenge to wild and cultivated plant species. Forward genetic analyses of individual temperature-responsive traits have resulted in the identification of several signaling and response components. However, a comprehensive knowledge about temperature sensitivity of different developmental stages and the contribution of natural variation is still scarce and fragmented at best.

Results: Here, we systematically analyze thermomorphogenesis throughout a complete life cycle in ten natural *Arabidopsis thaliana* accessions grown under long day conditions in four different temperatures ranging from 16 to 28 °C. We used Q_{10}, GxE, phenotypic divergence and correlation analyses to assess temperature sensitivity and genotype effects of more than 30 morphometric and developmental traits representing five phenotype classes. We found that genotype and temperature differentially affected plant growth and development with variing strengths. Furthermore, overall correlations among phenotypic temperature responses was relatively low which seems to be caused by differential capacities for temperature adaptations of individual accessions.

Conclusion: Genotype-specific temperature responses may be attractive targets for future forward genetic approaches and accession-specific thermomorphogenesis maps may aid the assessment of functional relevance of known and novel regulatory components.

Keywords: Arabidopsis, Natural variation, Phenotypic plasticity, Thermomorphogenesis, Phenotyping

Background

Recurrent changes in ambient temperature provide plants with essential information about time of day and seasons. Yet, even small changes in mean ambient temperatures can profoundly affect plant growth and development resulting in thermomorphogenic changes of plant architecture [1]. In crops like rice, a season-specific increase in the mean minimum temperature of 1 °C results in a ∼ 10% reduction in grain yield [2]. Likewise, up to 10% of the yield stagnation of wheat and barley in Europe over the past two decades can be attributed to climate change [3]. Current projections indicate that mean global air temperatures will increase up to 4.8 °C by the end of the century [4, 5]. Global warming will thus have significant implications on biodiversity and future food security.

Elevated ambient temperatures affect of course also wild species in their natural habitats. Long-term phenology studies of diverse plant populations have revealed an advance in first and peak flowering and alterations in the total length of flowering times [6, 7]. Furthermore, estimates project that temperature effects alone will account for the extinction of up to one-third of all European plant species [8]. As the impact of changes in ambient temperature on crop plants and natural habitats emerge, a comprehensive understanding of temperature-mediated growth responses throughout development becomes paramount.

* Correspondence: carolin.delker@landw.uni-halle.de
†Equal contributors
[1]Institute of Agricultural and Nutritional Sciences, Martin Luther University Halle-Wittenberg, Betty-Heimann-Str. 5, 06120 Halle (Saale), Germany
[2]Department of Molecular Signal Processing, Leibniz Institute of Plant Biochemistry, Weinberg 3, 06120 Halle (Saale), Germany
Full list of author information is available at the end of the article

Our present knowledge on molecular responses to ambient temperature changes has significantly progressed by studies in *Arabidopsis thaliana*. Model thermomorphogenesis phenotypes such as hypocotyl elongation [9], hyponastic leaf movement [10], and alterations in flowering time have served in various genetic approaches to identify relevant molecular players (reviewed in [1]). In this regard, exploiting naturally occurring genetic variation in these model traits has served as a valuable tool [11–16]. Primary signaling genes/proteins seem to function in response to both temperature and light stimuli. Prominent members of this network are photoreceptors such as CRYPTOCHROME 1 (CRY1 [17]), CRY2 [18] or the recently identified thermosensor PHYTOCHROME B (phyB [19, 20]) . Further components include PHYTOCHROME INTERACTING FACTOR 4 (PIF4,] [21–23], DE-ETIOLATED 1, CONSTITUTIVELY PHOTOMORPHOGENIC 1, ELONGATED-HYPOCOTYL 5 [24–26] and EARLY FLOWERING 3 (ELF3); the latter as a component of the circadian clock [12, 13].

The investigation of signaling pathways that translate temperature stimuli into qualitative and quantitative developmental responses has so far largely been limited to either seedling development or flowering time. However, it seems likely that temperature responses in different phases of development either require variations of a canonical signaling pathway or involve at least partially specific signaling components. To enable the dissection of thermomorphogenic signaling at different developmental stages, it is vital to gather a comprehensive understanding of the diversity of temperature reactions throughout plant development.

According to basic principles of thermodynamics, temperature-induced changes in free energy will affect the rates of biochemical reactions. As these effects should occur generally, albeit to different magnitudes, non-selective phenotypic responses can be expected to occur robustly and rather independently of genetic variation. Such traits may therefore be indicative of passive, thermodynamic effects on a multitude of processes. Alternatively, robust temperature responses may be due to thermodynamic effects on highly conserved signaling elements. These may be attractive targets for classic mutagenesis screens to identify the relevant regulatory components. In contrast, natural variation in thermomorphogenesis traits is likely the consequence of variability in one or several specific signaling or response components. It may be addressed by quantitative genetic approaches to identify regulators that contribute to variable temperature responses. Such genes may represent attractive candidates for targeted breeding approaches.

In this study we aim to (i) provide a map of developmental phenotypes that are sensitive to ambient temperature effects throughout a life cycle in the model organism *A.*

thaliana, (ii) identify traits that are robustly affected by temperature with little variation among different accessions, and ask (iii) which traits are affected differentially by different genotypes and thus show natural variation in temperature responses.

To realize this, we performed a profiling of numerous developmental and morphological traits which can be sorted into five main categories: juvenile vegetative stage, adult vegetative stage, reproductive stage, morphometric parameters and yield-associated traits. Phenotypes were analyzed in a subset of ten *A. thaliana* accessions which were grown at 16, 20, 24, and 28 °C in climate-controlled environments under long day photoperiods (16 h light/ 8 h dark). Knowing that even a small randomly selected set of *A. thaliana* accessions covers a wide spectrum of genetic diversity [27], we chose to analyze commonly used lab accessions such as Col-0 and Ws-2, accessions known to react hypersensitively to elevated temperature (e.g., Rrs-7, [25, 28]), and parental lines of available mapping populations such as Bay-0, Sha, Lerand Cvi-0 which have been used in previous studies of natural variation in individual thermomorphogenesis traits [16, 18].

In addition to a meta-analysis of the phenotypic data, we provide accession-specific developmental reference maps of temperature responses that can serve as resources for future experimental approaches in the analysis of ambient temperature responses in *A. thaliana*.

Methods
Plant material and growth conditions
Phenotypic parameters (Fig. 1) were assessed in *A. thaliana* accessions that were obtained from the Nottingham Arabidopsis Stock Centre [29]. Morphological markers and time points of analyses are described in Additional file 1. Sample sizes for each accession-temperature-trait combination and detailed information on stock numbers and geographic origin of accessions are listed in Additional file 2. For seedling stage analyses, surface-sterilized seeds were stratified for 3 days in deionized water at 4 °C and subsequently placed on *A. thaliana* solution (ATS) nutrient medium [30]. Seeds were germinated and cultivated in climate-controlled growth cabinets (Percival, AR66-L2) at constant temperatures of 16, 20, 24 or 28 °C under long day photoperiods (16 h light/8 h dark) and a photosynthetically active fluence rate (PAR) of 90 $\mu mol \cdot m^{-2} \cdot sec^{-1}$ of cool white fluorescent lamps. We refrained from including a vernalization step because the primary focus of this study was to record morphology and development in response to different constant ambient temperature conditions.

Germination rates were assessed daily on seeds cultivated on horizontal plates. Hypocotyl, root length, and petiole angles were measured in 7 days old seedlings

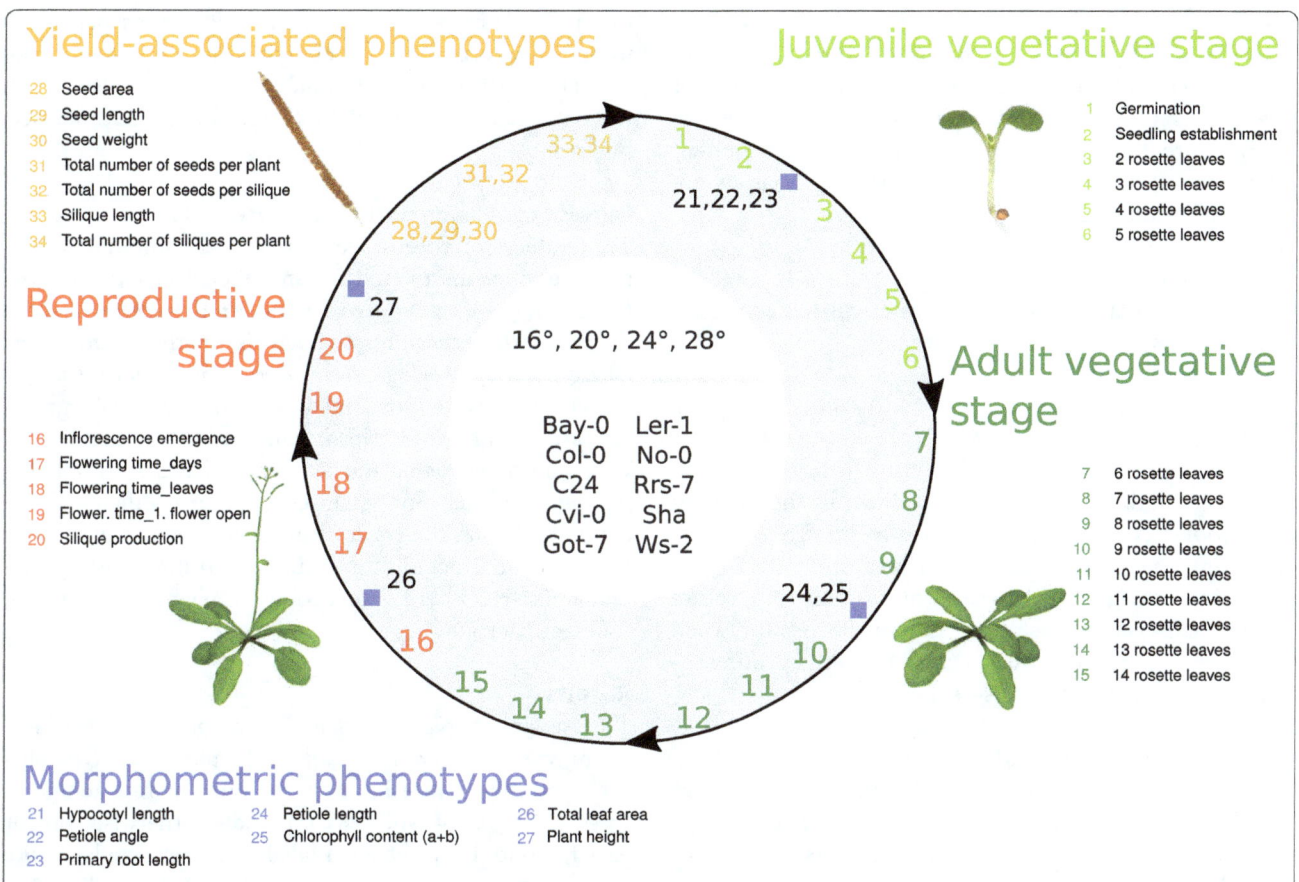

Fig. 1 Phenotypic profiling approach. Schematic representation of the accessions, cultivation temperatures (°C) and phenotype classes used in the phenotypic profiling approach. Numbers indicate individual traits listed and color-coded according to the corresponding phenotype class. *Blue squares* indicate phenotypes sorted into *'morphometric phenotypes'*. Their position is indicative for the developmental stage at time of assessment. Further information on trait values and specific time of assessments are shown in Additional file 1

grown on vertically oriented plates using ImageJ [31] and Root Detection [32].

All other analyses were performed on soil-grown plants cultivated in growth cabinets (Percival) at a PAR of 140 (+/− 20) μmol·m^{-2}·sec^{-1} and long day photoperiods (16 h light/8 h dark). After imbibition for 3 days at 4 °C, seeds were grown in individual 5 × 5 cm pots, which were randomized twice a week to minimize position effects. Randomization was performed in blocks of 60 pots alternating between 8 positions in the cultivation chamber. In addition, pots within the trays of 60 plants were randomized in groups of 10 twice a week on alternate days to the block randomization of trays. Relative humidity of growth cabinets was maintained at 70% and plants were watered by subirrigation. Plants ($n > 15$) were photographed daily for subsequent manual determination of phenotypic parameters (leaf number, total leaf area, and petiole length) using Image J [31]. "*Total leaf area*" was determined in the sense of foliar surface coverage so that overlapping leaf areas were only counted once. Determination of developmental progression largely followed the stages defined in Boyes et al.

[33]. The vegetative growth period was divided in a juvenile phase (germination to initiation of the fifth rosette leave) and an adult vegetative stage (initiation of the sixth rosette leave to floral transition). At transition to the reproductive growth phase, the number of leaves was determined by manual counting in addition to recording the number of days after germination.

Spectrophotometric determination of chlorophyll content was performed as described in [34].

Data analysis

Visualization and statistical analyses of the data were performed using the software R [35]. Box plots were generated using the *boxplot* function contained in the graphics package. Heat maps were generated using the *heatmap.2* function contained in the gplots package.

ANOVAs for a single factor (either accession or temperature) and Tukey's 'Honest Significant Difference' test as post hoc test were performed using the *anova* and *TukeyHSD* function, respectively, which are both contained in the R stats package.

Variation in phenotype expression was analyzed by 2-way ANOVA according to Nicotra [36] and Whitman and Agrawal [37] to test each phenotype for a significant effect of genotype (*G*, accession) or environment (*E*, temperature), and a significant genotype by environment interaction (GxE). Reaction norms for each analysis are shown in Additional file 3.

Q_{10} temperature coefficient

The Q_{10} temperature coefficient was calculated according to Loveys [38].

$$Q_{10} = \left(\frac{P_w}{P_c}\right)^{\frac{10}{T_w - T_c}}$$

where P_w and P_c are the trait values at the warmer and cooler temperatures, respectively. T_w and T_c represent the corresponding temperatures in °C. We computed the geometric mean of the six Q_{10} values of all pairwise temperature combinations for each phenotypic trait to avoid artifacts caused by differential reaction norms/response shapes.

Index of phenotypic divergence (P_{st})

Calculation of the index of phenotypic divergence (P_{st} [39, 40]) as a measure to quantify variation in each phenotypic trait was calculated as previously described by Storz [39] as

$$P_{st} = \frac{\sigma_b^2}{\sigma_b^2 + 2\sigma_w^2}$$

where σ_b^2 is the variance between populations, and σ_w^2 is the variance within populations. The ANOVA framework was used to partition the variances to get unbiased estimates for σ_b^2 and σ_w^2.

Using the two factorial design, two types of indices of phenotypic variation of a trait/phenotype were considered separately. The index of phenotypic divergence for genotypes (P_{st}^{gen}) at a defined temperature level can be computed to measure the effect/impact of the genotype on the variation whereas the index of phenotypic divergence for temperatures (P_{st}^{temp}) provides a measure for the effect of temperature on the observed variation for individual genotypes.

Principal component analysis (PCA)

Arithmetic means for each genotype-temperature pair were computed except for six traits (germination, 13 rosette leaves, 14 rosette leaves, silique production, chlorophyl content (a + b), and total leaf area) due to too many missing values. The remaining 28 traits contained at most eight missing values (randomly distributed).which were replaced per trait by the arithmetic

mean of the respective trait values. PCA was perfomed using the *prcomp* function contained in the R stats package. Due to the different units and scales of the traits the data was not only set to centered but also to scaled by *prcomp*.

Pairwise correlation analysis of traits

Trait values for rosette leave traits were summarized by arithmetic means to trait groups labeled *Juvenile vegetative stage (2–5 rosette leaves)* and *Adult vegetative stage (6–14 rosette leaves)*, respectively. Similarly, *Inflorescence emergence*, *Flowering time_days* and *Flower.time_1. flower open* were combined to form the trait group *Flowering time (days)*. Spearman correlation coefficients were computed using the R stats package. Additionally, *p* values for each Spearman correlation coefficient were computed using the *cor.test* function. *P* values were subsequently corrected for multiple testing using the Benjamini-Hochberg correction implemented in the multtest package.

Results

To assess phenotypic plasticity in a range of ambient temperatures, *A. thaliana* plants were cultivated in parallel throughout an entire life cycle at four different temperatures (16, 20, 24 and 28 °C) under otherwise similar growth conditions (see Methods for further details). More than 30 morphological and developmental traits were recorded representing the following five phenotype classes: juvenile vegetative, adult vegetative, and reproductive stages as well as morphometric and yield-associated phenotypes (Fig. 1 and Additional file 1).

Temperature responses in the *A. thaliana* reference accession Col-0

We first focused on Col-0 as it is a commonly used accession and serves as a point of reference for numerous genetic and physiological analyses. In Col-0, almost all phenotypes analyzed in this study were affected by the cultivation in different ambient temperatures. Only seed weight and maximum height remained constant regardless of the growth temperature (Fig. 2a, Additional file 4). Among the temperature-sensitive traits were several growth-associated phenotypes in the juvenile vegetative stage. Primary root length, hypocotyl and petiole elongation all increased with elevated temperatures which concurs with previously published data [9, 10]. As another example, yield-related traits, such as the number of siliques per plant and the number of seeds per silique decreased with an increase in ambient temperature (Fig. 2a).

As reported previously, Col-0 plants showed a decrease in developmental time until flowering with increasing ambient temperatures [11]. The transition from the vegetative to the reproductive phase at 28 °C

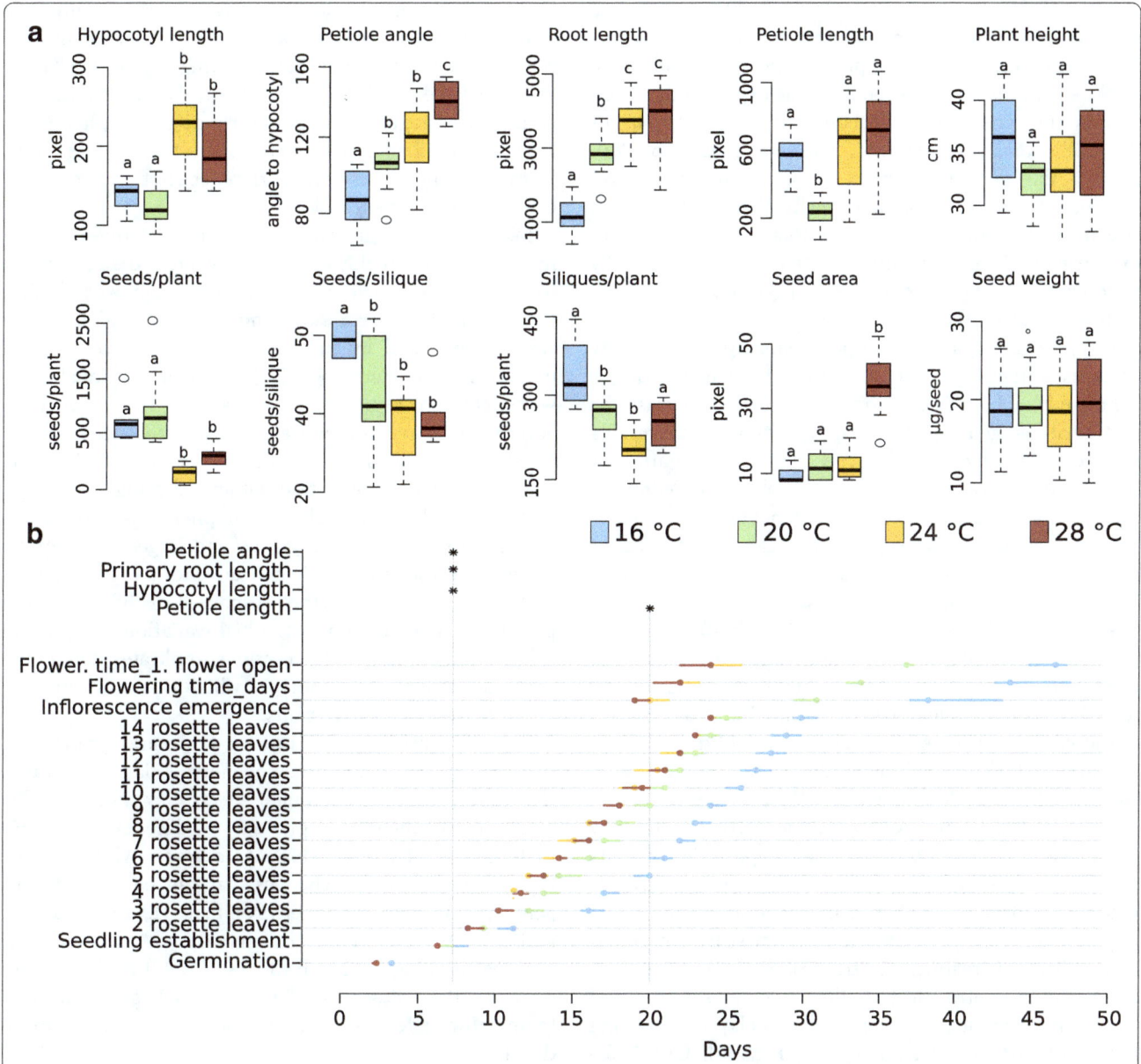

Fig. 2 Col-0 growth and development in response to different ambient temperatures. **a** Quantification of phenotypic traits recorded at different growth temperatures. *Box plots* show median and interquartile ranges (IQR), outliers (> 1.5 times IQR) are shown as *circles*. Units for each trait are specified in Additional file 16. *Different letters* denote statistical differences (*P* > 0.05) among samples as assessed by one-factorial ANOVA and Tukey HSD. **b** Summary of temperature effects on developmental timing. *Circles* denote medians, *bars* denote IQRs (*n* > 15). Times of phenotypic assessment for selected traits in (**a**) are indicated by *asterisks*

occurred about 25 days earlier than at 16 °C (Fig. 2a). Similarly, the number of rosette leaves developed at time of bolting differed by approximately 26 leaves between 28 °C and 16 °C (Additional file 4b).

The observation that only a very limited number of phenotypes were insensitive to cultivation in different temperatures clearly illustrates the fundamental impact of ambient temperature on plant growth and development.

Natural variation of temperature responses

To assess whether the observed temperature responses in Col-0 are robust among *A. thaliana* accessions or which of the responses may be affected by natural variation, phenotypic profiling was performed in nine additional *A. thaliana* accessions parallel to the analysis in Col-0 (Additional file 4, Additional file 5, Additional file 6, Additional file 7, Additional file 8, Additional file 9,

Additional file 10, Additional file 11, Additional file 12 and Additional file 13). Naturally, a panel of ten accessions does not comprehensively represent the world-wide gene pool of *A. thaliana*. However, it can be expected that even 10 randomly chosen natural accessions represent ~70% of the allelic diversity in the *A. thaliana* gene pool [27]. Hence, the general assessment of thermo-responsive development in *A. thaliana* as well as the identification and discrimination between traits that generally seem to exhibit natural variation and those that may be genetically fixed within the gene pool is a realistic aim even with a set of 10 selected accessions.

To approximate and to compare temperature sensitivity of traits among different accessions, we calculated Q_{10} values for each individual trait and phenotype class for each analyzed genotype [38]. The Q_{10} quotient represents the factor by which a trait value changes if the ambient temperature increases by 10 °C. We calculated geometric means of all possible pairwise combinations of temperatures to minimize effects potentially caused by different response curves and used the $\log_2 Q_{10}$ for visualization as to retain high resolution in the presentation of the data.

Similarly to the response observed in Col-0 (Fig. 2), all analyzed genotypes showed a temperature-induced acceleration of vegetative development as indicated by negative $\log_2 Q_{10}$ values with low variability among accessions (Fig. 3a, b, Additional file 4, Additional file 5, Additional file 6, Additional file 7, Additional file 8, Additional file 9, Additional file 10, Additional file 11, Additional file 12 and Additional file 13). Considerably higher variation was observed in $\log_2 Q_{10}$ values of traits related to reproductive stages. As all accessions investigated were principally able to flower despite the lack of an extended cold period, none of them strictly required a vernalization treatment to transition to the reproductive phase. In contrast to the other accessions, Got-7 and Rrs-7, however, showed a significant delay in flowering time with increasing temperature (Fig. 3b). Got-7, for example, did not flower within the first 85 days of cultivation when grown in 24 or 28 °C. Thus, initiated leaf senescence at bolting stage prevented accurate determination of leaf number at the onset of flowering.

A direct comparison of leaf number and time of development further corroborates a sudden increase in the overall variation among accessions at the transition to flowering (Additional file 14). However, at 16 °C and 20 °C several accessions contribute to the overall variability in the graph, whereas at 24 °C and 28 °C, C24 and Rrs-7 are the main determinants of variation due to their massive number of leaves corresponding to an extension of the vegetative growth phase (Additional file 14). Got-7 likely would increase the interspecific variation in flowering time at 24 and 28 °C, but is missing in this representation due to the lack of flowering transition within 85 days.

Here, the lack of vernalization may at least partially be a significant factor because cold treatment is explicitly recommended to induce earlier flowering for several Got-7 lines available at NASC/ABRC [41]. Natural variation in vernalization regulators may contribute to this phenotype. However, as all accessions were able to flower at temperatures of 16 and 20 °C vernalization does not seem to be an essential requirement.

Taken together, juvenile and adult vegetative development responded highly conserved, whereas the reproductive stage and yield-associated traits showed higher variation between accessions and within individual accession, as indicated by the ranges/dimensions of the box plots in Fig. 3a. Here, $\log_2(Q_{10})$ values of individual traits were summarized by phenotype classes. Thus, high variation within a phenotype class indicates that temperature effects on individual traits within that class are highly variable. The strongest variation within individual accessions was observed for morphometric phenotypes as represented by a wide whisker range and large boxes for all analyzed accessions. This is indicative for a high degree of diversity in the temperature responses of different morphometric traits. In contrast, a high variation of $\log_2(Q_{10})$ values between different accessions implicates differential responses of different genotypes which was most prominent in reproductive stage traits.

The differential variances of $\log_2 Q_{10}$ values among the two vegetative and the other phenotype classes indicated that genotype and environment effects may contribute differentially to phenotypic plasticity of different traits. We first used a 2-factorial ANOVA to assess which phenotypes show significant changes that can be attributed to genotype (G, accession), environment (E, temperature), and/or GxE interaction. Subsequently, we used the variance partitioning approach [39, 40, 42, 43] to dissect and quantify the extent of the individual genotype and temperature effects on the phenotypic variation in more detail.

Genotype, environment, and GxE interaction analysis

Each phenotypic trait was subjected to a 2-factorial ANOVA to address which of the analyzed factors (G, E, GxE) had significant effects on the trait. Reaction norm plots for each phenotype are shown in Additional file 3. Each of the analyzed traits showed significant effects of genotype, environment (temperature) and GxE interaction (Additional file 15). Surprisingly, this included all juvenile and adult vegetative stages despite their seemingly uniform impression of temperature responses given by the Q_{10} values (Fig. 3a, b).

To assess genotype and temperature contributions in a more quantitative manner, we next used a variance partitioning approach [39, 40, 42, 43]. Specifically, we calculated the index of phenotypic divergence (P_{st}, [39])

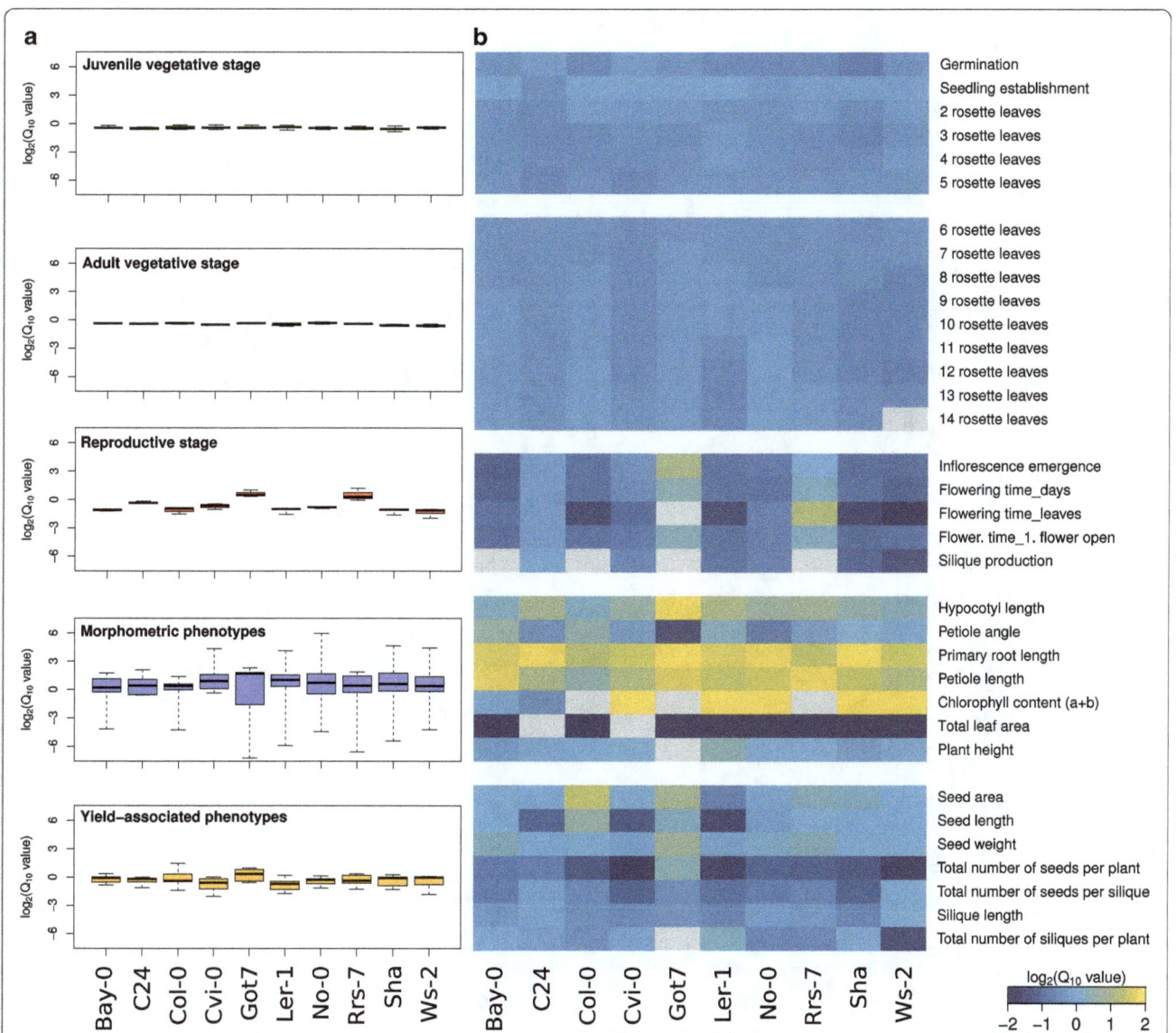

Fig. 3 Natural variation in temperature sensitivity of phenotypic traits (Q_{10}). Mean $\log_2 Q_{10}$ values for each accession (**a**) summarized in *box plots* for each phenotype class and (**b**) presented as a heatmap for all individual phenotypes. **a** *Box plots* show median and interquartile ranges (IQR), whiskers range from min. to max. Values. **b** positive (increasing) and negative (decreasing) $\log_2 Q_{10}$ values are shown in *yellow* and *blue*, respectively with a $\log_2 Q_{10}$ cut-off value of 2 for better resolution. Missing data are denoted in *light gray*

at each analyzed temperature as a measure of genotype effects P_{st}^{gen} on the trait of interest (Additional file 16a). To complement this analysis, we also estimated the variation occurring across temperatures P_{st}^{temp} for each of the analyzed accessions (Additional file 16b), which enabled us to assess the temperature effect on the trait of interest for specific genotypes.

Genotype effects

Individual P_{st}^{gen} values that provide a quantitative assessment of the genotype contribution to variation at individual temperatures showed highly variable

patterns among the different traits and phenotype classes (Additional file 16a). Regardless of the individual temperature, mean genotype effects on developmental timing throughout the vegetative phase were generally very low (Fig. 4a), supporting the results from the analysis of Q_{10} values (Fig. 3). However, genotype effects on later stages of adult vegetative development seem to increase with higher temperatures (Additional file 16a), which may be the significant effect observed in the ANOVA-based GxE interaction assessment.

Similarly, strong genotype effects at higher temperatures were also observed for reproductive traits. Here,

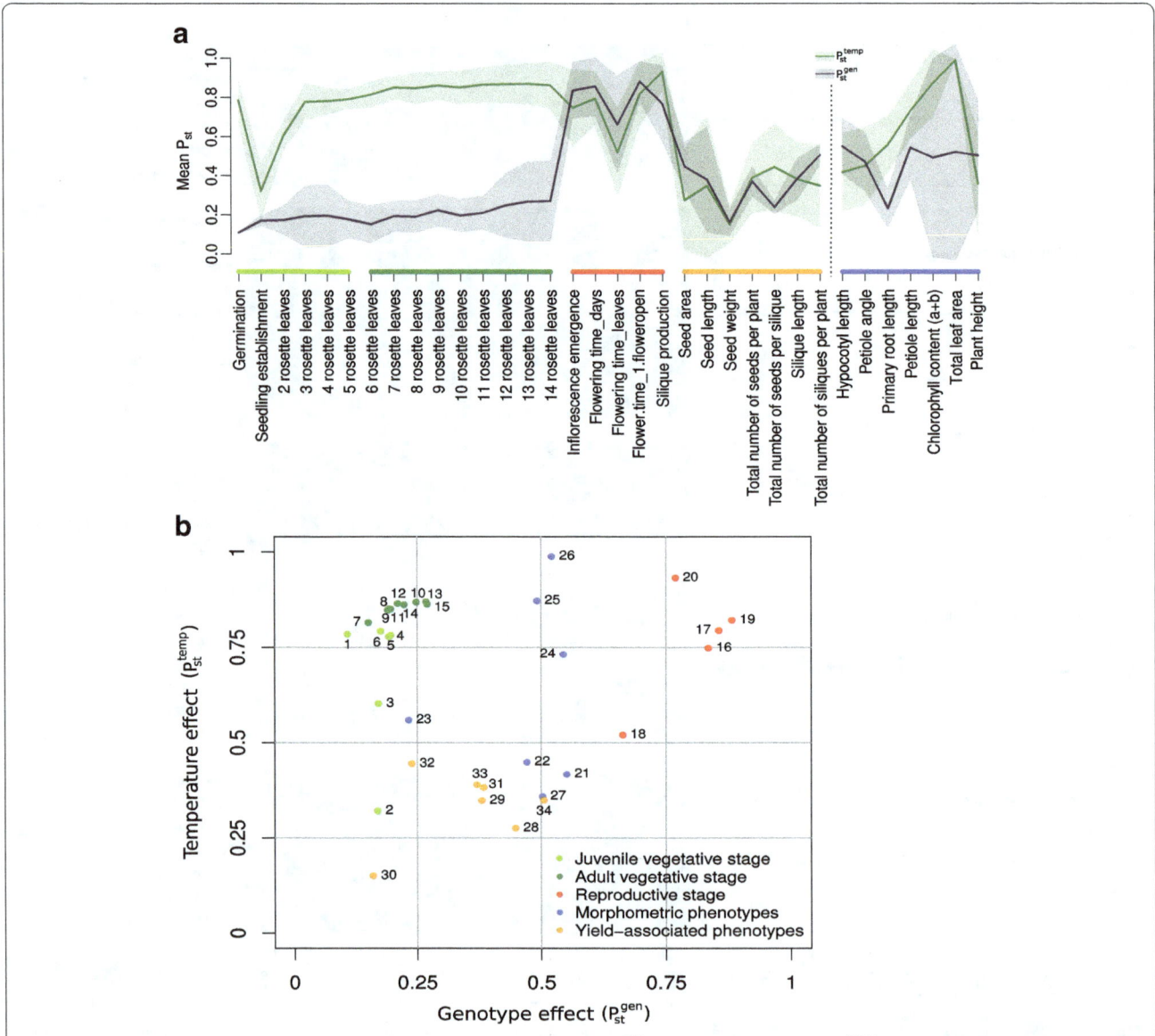

Fig. 4 Genotype and temperature effects on phenotypic variation. **a** Genotype (P_{st}^{gen}, *black*) and temperature (P_{st}^{temp}, *green*) contribution to variation. *Solid lines* show mean P_{st} values and shaded areas indicate standard deviations. **b** *Scatter plot* of mean P_{st}^{gen} and P_{st}^{temp} values over all temperatures and accessions, respectively. Phenotypes are color-coded according to the phenotype classes shown in Fig. 1 and described in Additional file 1. A heatmap of individual P_{st}^{gen} and P_{st}^{temp} values and a scatter plot including standard deviations are shown in Additional file 16

P_{st}^{gen} values at 16 °C were already considerably higher than for vegetative growth stages and increased further with elevated temperatures (Additional file 16a). A contrasting pattern of decreasing genotype effects with an increase in temperatures was observed for plant height indicating that here, natural variation in growth is higher at lower temperatures. Yield-associated phenotypes in general showed only low genotype effects on variation, indicating that under our experimental conditions variation in trait expression in this category is primarily affected by temperature (Fig. 4a).

Other phenotypes display rather differential or less gradual genotype effects among different temperatures. For example, the genotype impact on variation in hypocotyl and petiole length increases only at 28°C whereas the overall genotype contribution to variation from 16 to 24 °C remains rather low (Additional file 16a). Such patterns of genotype effects may indicate a certain buffering capacity or a threshold for natural variation effects.

In some cases, such as flowering time, a strong genotype effect seems to correlate also with a strong general temperature sensitivity as indicated by the high between-accessions variability in Q_{10} values (Figs. 4a and 3b). However, this does not seem to be a general principle. In case of root length, for example, low genotype effects were observed (Fig. 4a, b), even though the phenotype in principle was highly sensitive to a change in ambient temperature (Fig. 3b).

Temperature effects

We also used the variance partitioning approach to analyze the extent of the significant impact of temperature on phenotypic variation that was detected in the GxE interaction analysis (Additional file 15). Therefore, we calculated the index for temperature effects (P_{st}^{temp}) on the variation of phenotypic plasticity across all four temperatures within each of the ten accessions (Additional file 16b). In contrast to P_{st}^{gen}, the P_{st}^{temp} thus provides information primarily on the temperature-induced variability for each accession individually.

The heatmap representation of temperature effects (Additional file 16b) partially complements the genotype effect results. For example, variation in the timing of vegetative development was highly affected by temperature (high P_{st}^{temp}), whereas P_{st}^{gen} values were generally low (Fig. 4a, Additional file 16a, b). Interestingly, temperature effects in juvenile vegetative stages seemed to be lower (for seedling establishment and 2 rosette leave stage) than in later vegetative stages with the exception of germination which showed strong temperature effects in most accessions.

Many traits exhibit highly differential temperature effects among accessions in the sense of one accession demonstrating a particularly strong temperature effect on a specific trait, while another accession may show low to no temperature effects (e.g. chlorophyll content in Ler-1 vs. Bay-0). This is particularly obvious for yield-related traits such as total number of seeds per plant and silique as well as silique length. Here, temperature effects on phenotype variation were low for Col-0, C24 and Bay-0, whereas considerably higher P_{st}^{temp} values were determined for the other accessions (Additional file 16b). Accessions which exhibit differential temperature effects on phenotypic variation may be interesting candidates for forward genetic approaches to identify the contributing molecular regulatory components.

Comparison of temperature and genotype effects

As each phenotypic trait has been assigned a value for genotype and temperature effects, they can easily be compared to assess which of the two has a stronger influence on the phenotypic plasticity. To allow a direct comparison of effects, we compared mean values for

P_{st}^{gen} across all temperatures and P_{st}^{temp} across all accessions (Fig. 4a, b).

Temperature effects on vegetative development showed a high, largely robust impact with little variance in P_{st}^{temp} values, whereas genotype effects were generally low with diverging variances. Genotype effects peak at the transition to the reproductive phase and in some morphometric phenotypes. In general, morphometric parameters show high temperature and varying genotype effects. Phenotypes associated with late developmental stages were generally less affected by both factors indicating an overall buffering effect. Yet, variances in temperature effects tended to be high here, which may indicate genotype-specific thresholds for temperature effects (Fig. 4a, Additional file 16c). A scatter plot representation of mean P_{st}^{gen} and P_{st}^{temp} values for each trait allows further comparison of phenotypes according to the impact of both factors (Fig. 4b). While vegetative and reproductive phenotypes form tight clusters, morphometric phenotypes displayed a heterogenous pattern. In these traits, temperature responses seem to be affected by natural variation and may thus serve as candidate phenotypes for classic or quantitative forward genetic analyses.

Several yield-associated phenotypes such as total number of seeds, seed size, and seed weight showed varying degrees of temperature sensitivity, likely caused by the partially distinct temperature effects on individual accessions (Fig. 2b, Additional file 17).

The fundamental impact of temperature on the phenotypic responses is also reflected in the results of the principle component analysis (PCA). The PCA was performed on mean-centered and scaled data in order to allow integration of data with different scaling. PC1 which covered 50% of the observed variation, allowed a clear separation of samples via temperature (Fig. 5a). Here, the differentiation between 16 and 20 °C seems to be higher than the temperature changes from 20 to 24 °C and 24 to 28 °C. PC2 explained ~16% of the variation and separated samples rather by genotype. Here, Rrs-7 and Got-7 showed a clear divergence from other genotypes. Again, this separation is already clear between the temperatures 16 and 20 °C whereas a further increase in temperature contributed little more to the separation.

Correlation of phenotypic temperature responses

Finally, we analyzed putative correlations in temperature responses among different phenotypes to assess whether individual phenotype responses are indicative of temperature responses in general. As redundancies of individual phenotypes may bias the analyses several traits were combined in groups for further analyses (e.g. rosette development or flowering traits). We used the

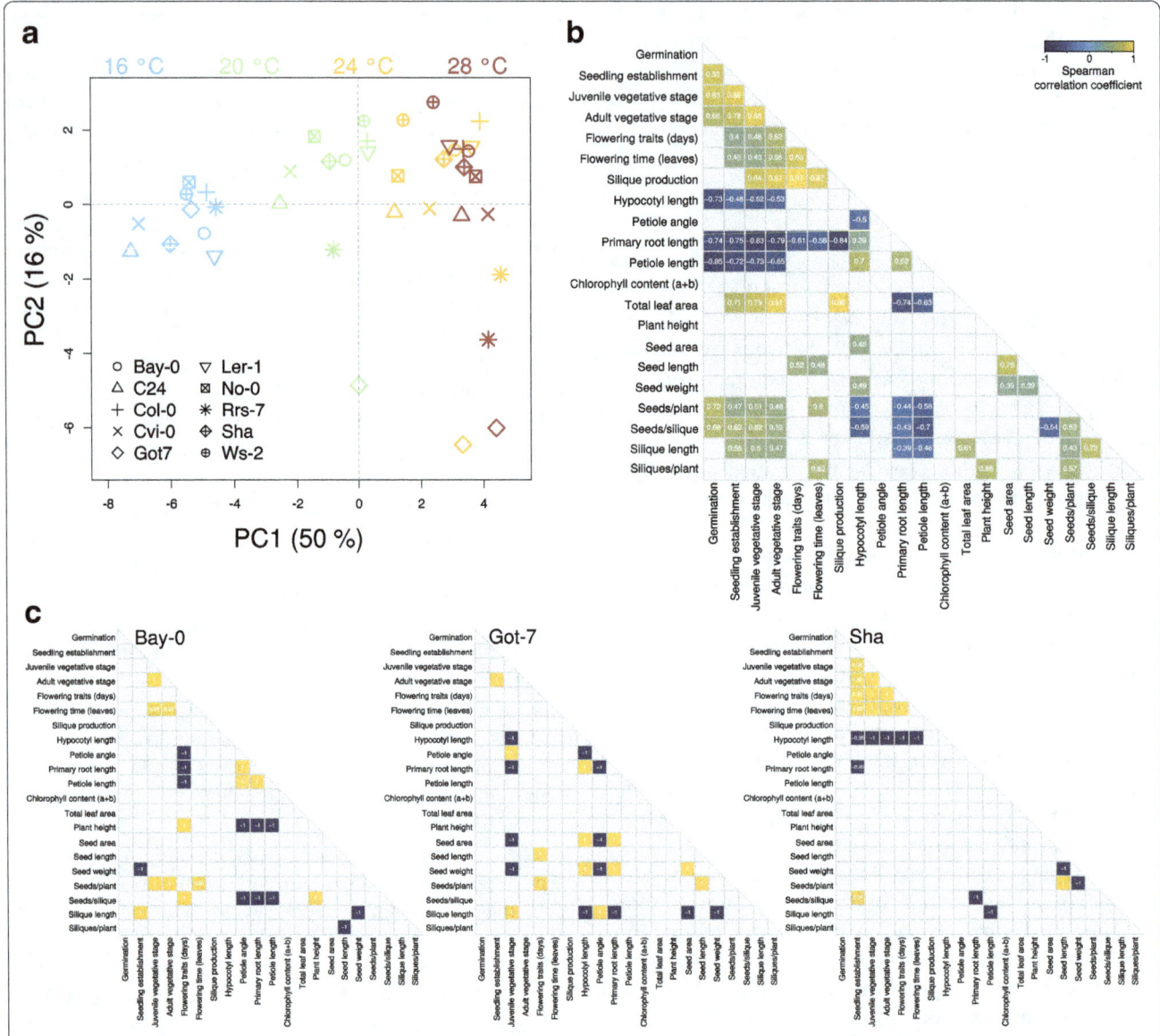

Fig. 5 Principle component and correlation analyses. **a** Phenotypic data of all temperatures and genotypes were subjected to principle component analysis (PCA). **b-c** Correlation analysis of temperature responses among individual traits or trait groups of all analyzed genotypes (**b**) or in selected individual accessions (**c**). Spearmann correlation coefficients were tested for significance and coefficients with $P < 0.05$ and $P < 0.1$ are presented in (**b**) and (**c**), respectively. Phenotype correlations for all accessions individually are shown in Additional file 18

rank-based Spearman correlation coefficients for pairwise comparisons of averaged trait (group) values among all accessions to account for potential non-linear relationships and minimize outlier effects. As to be expected from the varying degrees of genotype and temperature effects on different traits, phenotypic correlations also varied considerably. To filter for robust correlations, only significant correlations ($P < 0.05$) were retained in the analysis (Fig. 5b).

High correlations were detected among traits within the vegetative stage of development (e.g. juvenile and adult vegetative stage), and among traits within the reproductive phase (e.g. flowering traits and the onset of

silique production). In addition, temperature-induced reduction in total leaf area correlated strongly with the decrease in developmental time in vegetative and reproductive phases. Similarly, the reduction in developmental times and total leaf area were moderately correlated to the effect on several seed-associated traits (Fig. 5b).

Model temperature phenotypes such as petiole and hypocotyl length showed a positive correlation and were in turn correlated or inversely correlated with several other phenotypes or trait groups. However, temperature responses in primary root length under these experimental conditions showed an even more robust connection to many other traits. Mostly, these were inverse correlations with the exception of other seedling traits which were positively correlated with primary root lengths (Fig. 5b).

Due to the differential genotype effects on variation we also wondered whether individual genotypes may show different correlation patterns among phenotypic temperature responses. Calculation of Spearmann correlation coefficients for each individual accession is based on a maximum of four data points per phenotype or trait group which generally results in weaker interactions among samples. Thus, the *P*-value threshold was set to 0.1 in the analysis which retained only the strongest (inverse) correlations. Inspection of the correlation patterns reveals remarkable differences among accessions (Fig. 5c, Additional file 18). For instance, petiole lentgh, angle and primary root length in Bay-0 were all inversely correlated with flowering time, plant height and the number of seeds/silique, whereas in Sha, only hypocotyl lengths showed an inverse correlation with developmental timing in vegetative and reproductive stages. Got-7 even showed unique correlation patterns among early growth responses with inverse correlations among petiole angles and hypocotyl and root lengths, respectively (Fig. 5c). Accession-specific correlations among individual phenotypes should be treated cautiously as these correlations are based on far fewer data points than in the general analysis (Fig. 5b). Nevertheless, the diversity in correlation patterns may indicate differential capacities for temperature responses that result in differential activation or buffering and, thus, in different extents of physiological temperature impacts. On a molecular level this may involve different genetic factors or background-specific epistatic interactions. Elucidation of the underlying mechanisms of differential temperature responses and adaptations may provide essential tools for the modulation of crop responses to elevated ambient temperatures.

Discussion

Increased ambient temperatures have previously been shown to affect thermomorphogenesis for selected "model" phenotypes. A systematic assessment of developmental and phenotypic plasticity across a complete life cycle has, to the best of our knowledge, been lacking so far. This study aims to provide such a solid base of temperature effects on plants by consecutive profiling of plant growth and development throughout a life cycle of *A. thaliana* grown in four different ambient temperatures. Furthermore, including several distinct *A. thaliana* accessions reduced potential genotype-specific biases in the data and allowed the analysis of temperature and genotype effects on the variation observed in different phenotypic traits.

All of the 34 analyzed phenotypes were significantly affected by different growth temperatures, natural variation, and GxE interactions, illustrating the fundamental impact of ambient temperature on plant development and the high variability in responses among genotypes (Additional files 4, 5, 6, 7, 8, 9, 10, 11, 12, 13 and 15). The variance partitioning approach allowed the further dissection of phenotypes based on the extent of temperature and genotype effects. First, we identified phenotypes that were primarily affected by temperature and showed small genotype-induced variation. Second, we identified phenotypes that additionally or even predominantly showed genotype effects on the observed phenotypic variation.

Developmental timing of juvenile and adult vegetative growth was significantly affected by genotype and temperature (Additional file 15). Yet, temperature was the dominant factor in the observed variation (Figs. 4a, 5a, Additional file 16). Genotype effects on the variation observed during vegetative development, albeit significant, were limited (Fig. 4a). Accordingly, similar accelerations by increasing temperatures were observed in all analyzed genotypes (Additional files 4, 5, 6, 7, 8, 9, 10, 11, 12 and 13). This observation may be indicative for extensive thermodynamic effects on (conserved) regulatory mechanisms involved in this process. Indeed, thermomorphogenic responses are often speculated to be primarily caused by broad or general effects of free energy changes on biochemical reactions (e.g. enzyme activities). The validity of the early proposed temperature coefficient (Q_{10}) for plant development was demonstrated for germination rates and plant respiration [44, 45]. The strong temperature effect on the acceleration of developmental timing throughout the vegetative phase, which was only weakly affected by genotypes supports this theory. When adopting the terms of "passive" and "active" temperature effects as proposed by [46], timing of vegetative development would represent a passive temperature response that might be caused by thermodynamic effects on metabolic rates and enzyme activities or on highly conserved signaling/response components.

On the other hand, phenotypes that show a high degree of genotype and temperature effects might rather be influenced by one or more specific genes that contribute to trait expression in a quantitative manner. As such, these phenotypes would represent "active" temperature effects [46]. However, the involvement of specific signaling elements does not necessarily exclude influences via thermodynamics. In fact, the recently described thermosensing via phyB acts via the promotion of phyB P_{FR} to P_R conversion in a temperature-promoted manner [19, 20]. Natural variation in thermomorphogenic responses could be caused by polymorphisms in signaling or response genes ranging from alteration in gene sequence to expression level polymorphism [47, 48]. As they may provide keys to altered temperature responses that could be utilized in specific breeding approaches, identification of such genes would be of high interest.

In fact, natural allelic variation in the circadian clock components *ELF3* and in the regulation of *GIGANTEA* have recently been shown to directly affect PIF4-mediated hypocotyl elongation in response to elevated temperatures [12, 13, 49]. Therefore, PIF4 and PIF4-regulating components could be important targets of adaptation to growth in higher ambient temperatures. PIF4 and ELF3 have been shown to be involved in both, temperature-induced hypocotyl elongation and the induction of flowering [12, 13, 21, 50]. However, a lack of general correlation among seedling growth and flowering time responses may indicate that these processes are not universally regulated via the same components. Alternatively, the impact of these signaling components on diverse phenotypes may be more prominent for specific alleles which may be reflected by the diversity in correlation patterns among individual accessions (Fig. 5c, Additional file 18).

Partially, the intraspecific diversity in phenotypic changes in response to elevated ambient temperatures argue against a general explanation of morphological and developmental changes due to passive thermodynamic effects.

Exploiting natural genetic variation to identify genes that are involved in the regulation of temperature effects on specific traits can provide new leads for plant breeding. Particularly, identification of molecular determinants of temperature responses in yield-associated traits should be considered in future studies. The work presented here may inspire new approaches for temperature research in non-reference accessions as some temperature responses were much more pronounced in accessions other than Col-0 (Fig. 3b). Specific approaches will depend on the individual focus of the study which may encompass development, yield- or biomass-associated traits. Each of these aspects is fundamental from an agricultural point of view but molecular factors involved in their temperature responses may differ considerably.

Conclusion

In conclusion, our work provides a map that allows the dissection of thermomorphogenesis in phenotypic traits that are either robustly affected by temperature or traits that are differentially affected by temperature among different accessions. While robust temperature-sensitive phenotypes might indeed be caused by thermodynamic acceleration of metabolism or highly conserved signaling

events, natural genetic variation of temperature responses implicate the relevance of specific regulatory cascades that can be instrumental to future breeding approaches.

Additional files

Additional file 1: Table of recorded phenotypes and association to phenotype classes.

Additional file 2: Sample sizes, identity and geographic origin of analyzed *A. thaliana* accessions.

Additional file 3: Reaction norm plots of each phenotype for each of the analyzed genotypes.

Additional file 4: Summary of Col-0 thermomorphogenesis.

Additional file 5: Summary of Bay-0 thermomorphogenesis.

Additional file 6: Summary of C24 thermomorphogenesis.

Additional file 7: Summary of Cvi-0 thermomorphogenesis.

Additional file 8: Summary of Got-7 thermomorphogenesis.

Additional file 9: Summary of Ler-1 thermomorphogenesis.

Additional file 10: Summary of No-0 thermomorphogenesis.

Additional file 11: Summary of Rrs-7 thermomorphogenesis.

Additional file 12: Summary of Sha thermomorphogenesis.

Additional file 13: Summary of Ws-2 thermomorphogenesis.

Additional file 14: Natural variation in developmental timing. (leaves vs. days).

Additional file 15: GxE interaction analysis results.

Additional file 16: Detailed information on genotype and temperature effects on phenotypic variation.

Additional file 17: Temperature effect on yield.

Additional file 18: Correlations among temperature responses in individual accessions.

Acknowledgements
We are grateful to Ivo Grosse for critical discussion in the initial phase of this project.

Funding
This study was supported by the Leibniz Association and a grant from the Deutsche Forschungsgemeinschaft to M.Q. (Qu 141/3–1). Y.P. s supported by the German Centre for Integrative Biodiversity Research (iDiv) Halle-Jena-Leipzig, funded by the Deutsche Forschungsgemeinschaft (FZT 118).

Authors' contributions
CI, MQ, and CD designed the research and experimental setup. CI, TP, JB and KD performed the phenotypic analyses and data collection. CI, YP and CD analyzed the data. YP, AG-D, and CD designed and performed statistical analyses. CI, YP, MQ, and CD interpreted data, prepared figures and wrote the manuscript.

Competing interests
The authors declare that they have no competing interests.

Author details
[1]Institute of Agricultural and Nutritional Sciences, Martin Luther University Halle-Wittenberg, Betty-Heimann-Str. 5, 06120 Halle (Saale), Germany. [2]Department of Molecular Signal Processing, Leibniz Institute of Plant Biochemistry, Weinberg 3, 06120 Halle (Saale), Germany. [3]German Centre for Integrative Biodiversity Research (iDiv) Halle-Jena-Leipzig, Deutscher Platz 5e, 04103 Leipzig, Germany. [4]Institute of Computer Science, Martin Luther University Halle-Wittenberg, Von-Seckendorff-Platz 1, 06099 Halle (Saale), Germany.

References

1. Quint M, Delker C, Franklin KA, Wigge PA, Halliday KJ, van Zanten M. Molecular and genetic control of plant thermomorphogenesis. Nat Plants. 2016;2:15190.
2. Peng S, Huang J, Sheehy JE, Laza RC, Visperas RM, Zhong X, et al. Rice yields decline with higher night temperature from global warming. Proc Natl Acad Sci. 2004;101:9971–5.
3. Moore FC, Lobell DB. The fingerprint of climate trends on European crop yields. Proc Natl Acad Sci U S A. 2015;112:2670–5.
4. IPCC. Climate change 2013: The physical science basis. Fifth assessment report. [Internet]. UNEP/WMO; Available from: http://www.ipcc.ch/report/ar5/wg1/.
5. Lobell DB, Gourdji SM. The influence of climate change on global crop productivity. Plant Physiol. 2012;160:1686–97.
6. Fitter AH, Fitter RSR. Rapid changes in flowering time in british plants. Science. 2002;296:1689–91.
7. CaraDonna PJ, Iler AM, Inouye DW. Shifts in flowering phenology reshape a subalpine plant community. Proc Natl Acad Sci. 2014;111:4916–21.
8. Thuiller W, Lavorel S, Araújo MB, Sykes MT, Prentice IC. Climate change threats to plant diversity in Europe. Proc Natl Acad Sci. 2005;102:8245–50.
9. Gray WM, Östin A, Sandberg G, Romano CP, Estelle M. High temperature promotes auxin-mediated hypocotyl elongation in Arabidopsis. Proc Natl Acad Sci. 1998;95:7197–202.
10. Zanten M van, Voesenek LACJ, Peeters AJM, Millenaar FF. Hormone- and light-mediated regulation of heat-induced differential petiole growth in Arabidopsis. Plant Physiol 2009;151:1446–1458.
11. Balasubramanian S, Sureshkumar S, Lempe J, Weigel D. potent induction of *Arabidopsis thaliana* flowering by elevated growth temperature. PLoS Genet. 2006;2:e106.
12. Raschke A, Ibañez C, Ullrich KK, Anwer MU, Becker S, Glöckner A, et al. Natural variants of ELF3 affect thermomorphogenesis by transcriptionally modulating PIF4-dependent Auxin responses. BMC Plant Biol. 2015;15:197.
13. Box MS, Huang BE, Domijan M, Jaeger KE, Khattak AK, Yoo SJ, et al. ELF3 controls thermoresponsive growth in Arabidopsis. Curr Biol. 2015;25:194–9.
14. Zhu W, Ausin I, Seleznev A, Méndez-Vigo B, Picó FX, Sureshkumar S, et al. Natural variation identifies ICARUS1, a Universal Gene Required for Cell Proliferation and Growth at High Temperatures in *Arabidopsis thaliana*. PLoS Genet. 2015;11:e1005085.
15. Lutz U, Posé D, Pfeifer M, Gundlach H, Hagmann J, Wang C, et al. Modulation of ambient temperature-dependent flowering in *Arabidopsis thaliana* by natural variation of FLOWERING LOCUS M. PLoS Genet. 2015;11:e1005588.
16. Sanchez-Bermejo E, Balasubramanian S. Natural variation involving deletion alleles of FRIGIDA modulate temperature-sensitive flowering responses in *Arabidopsis thaliana*. Plant Cell Environ. 2016;39:1353–65.
17. Ma D, Li X, Guo Y, Chu J, Fang S, Yan C, et al. Cryptochrome 1 interacts with PIF4 to regulate high temperature-mediated hypocotyl elongation in response to blue light. Proc Natl Acad Sci. 2016;113:224–9.
18. Sanchez-Bermejo E, Zhu W, Tasset C, Eimer H, Sureshkumar S, Singh R, et al. Genetic architecture of natural variation in thermal responses of Arabidopsis. Plant Physiol. 2015;169:647–59.
19. Jung J-H, Domijan M, Klose C, Biswas S, Ezer D, Gao M, et al. Phytochromes function as thermosensors in Arabidopsis. Science. 2016;354:886–9.
20. Legris M, Klose C, Burgie ES, Costigliolo C, Neme M, Hiltbrunner A, et al. Phytochrome B integrates light and temperature signals in Arabidopsis. Science. 2016;354:897–900.
21. Koini MA, Alvey L, Allen T, Tilley CA, Harberd NP, Whitelam GC, et al. High temperature-mediated adaptations in plant architecture require the bHLH transcription factor PIF4. Curr Biol. 2009;19:408–13.
22. Franklin KA, Lee SH, Patel D, Kumar SV, Spartz AK, Gu C, et al. Phytochrome-interacting factor 4 (PIF4) regulates auxin biosynthesis at high temperature. Proc Natl Acad Sci U S A. 2011;108:20231–5.
23. Proveniers MCG, van Zanten M. High temperature acclimation through PIF4 signaling. Trends Plant Sci. 2013;18:59–64.

24. Toledo-Ortiz G, Johansson H, Lee KP, Bou-Torrent J, Stewart K, Steel G, et al. The HY5-PIF regulatory module coordinates light and temperature control of photosynthetic gene transcription. PLoS Genet. 2014;10: e1004416.
25. Delker C, Sonntag L, James GV, Janitza P, Ibañez C, Ziermann H, et al. The DET1-COP1-HY5 Pathway Constitutes a Multipurpose Signaling Module Regulating Plant Photomorphogenesis and Thermomorphogenesis. Cell Rep. 2014;9:1983–9.
26. Gangappa SN, Kumar SV. DET1 and HY5 Control PIF4-Mediated Thermosensory Elongation Growth through Distinct Mechanisms. Cell Rep. 2017;18:344–51.
27. McKhann HI, Camilleri C, Bérard A, Bataillon T, David JL, Reboud X, et al. Nested core collections maximizing genetic diversity in *Arabidopsis thaliana*. Plant J. 2004;38:193–202.
28. Delker C, Pöschl Y, Raschke A, Ullrich K, Ettingshausen S, Hauptmann V, et al. Natural variation of transcriptional auxin response networks in *Arabidopsis thaliana*. Plant Cell. 2010;22:2184–200.
29. Scholl RL, May ST, Ware DH. Seed and molecular resources for arabidopsis. Plant Physiol. 2000;124:1477–80.
30. Lincoln C, Britton J, Estelle M. Growth and development of the axr1 mutants of Arabidopsis. Plant Cell. 1990;2:1071–80.
31. ImageJ: http://imagej.nih.gov/ij/
32. RootDetection: http://www.labutils.de/rd.html
33. Boyes DC, Zayed AM, Ascenzi R, McCaskill AJ, Hoffman NE, Davis KR, et al. Growth stage-based phenotypic analysis of Arabidopsis: a model for high throughput functional genomics in plants. Plant Cell. 2001;13: 1499–510.
34. Porra RJ, Thompson WA, Kriedemann PE. Determination of accurate extinction coefficients and simultaneous equations for assaying chlorophylls a and b extracted with four different solvents: verification of the concentration of chlorophyll standards by atomic absorption spectroscopy. Biochim Biophys Acta BBA - Bioenerg. 1989;975:384–94.
35. R Core Team. R: a language and environment for statistical computing [Internet]. Vienna: R Foundation for Statistical Computing; 2015. https://www.R-project.org
36. Nicotra AB, Atkin OK, Bonser SP, Davidson AM, Finnegan EJ, Mathesius U, et al. Plant phenotypic plasticity in a changing climate. Trends Plant Sci. 2010; 15:684–92.
37. Whitman D, Agrawal A. What is phenotypic plasticity and why is it important? Phenotypic Plast. Insects. 2009: http://dx.doi.org/10.1201/b10201-2
38. Loveys BR, Atkinson LJ, Sherlock DJ, Roberts RL, Fitter AH, Atkin OK. Thermal acclimation of leaf and root respiration: an investigation comparing inherently fast- and slow-growing plant species. Glob Change Biol. 2003;9: 895–910.
39. Storz JF. Contrasting patterns of divergence in quantitative traits and neutral DNA markers: analysis of clinal variation. Mol Ecol. 2002;11:2537–51.
40. Leinonen T, Cano JM, Mäkinen H, Merilä J. Contrasting patterns of body shape and neutral genetic divergence in marine and lake populations of threespine sticklebacks. J Evol Biol. 2006;19:1803–12.
41. NASC/ABRC. https://www.arabidopsis.org/abrc/catalog/natural_accession_9.html
42. Gay L, Neubauer G, Zagalska-Neubauer M, Pons J-M, Bell DA, Crochet P-A. Speciation with gene flow in the large white-headed gulls: does selection counterbalance introgression? Heredity. 2008;102:133–46.
43. Whitlock MC. Evolutionary inference from QST. Mol Ecol. 2008;17:1885–96.
44. Hegarty TW. temperature coefficient (q10), seed germination and other biological processes. Nature. 1973;243:305–6.
45. Atkin OK, Tjoelker MG. Thermal acclimation and the dynamic response of plant respiration to temperature. Trends Plant Sci. 2003;8:343–51.
46. Penfield S, MacGregor D. Temperature sensing in plants. In: Franklin K a, Wigge P a, editors. Temperature and Plant Development. Hoboken: John Wiley & Sons, Inc; 2014. p. 1–18.
47. Delker C, Quint M. Expression level polymorphisms: heritable traits shaping natural variation. Trends Plant Sci. 2011;16:481–8.
48. Alonso-Blanco C, Aarts MGM, Bentsink L, Keurentjes JJB, Reymond M, Vreugdenhil D, et al. What has natural variation taught us about plant development, physiology, and adaptation? Plant Cell. 2009;21:1877–96.
49. de Montaigu A, Giakountis A, Rubin M, Tóth R, Cremer F, Sokolova V, et al. Natural diversity in daily rhythms of gene expression contributes to phenotypic variation. Proc Natl Acad Sci. 2015;112:905–10.
50. Kumar SV, Lucyshyn D, Jaeger KE, Alós E, Alvey E, Harberd NP, et al. Transcription factor PIF4 controls the thermosensory activation of flowering. Nature. 2012;484:242–5.

4

New insights into the roles of cucumber TIR1 homologs and miR393 in regulating fruit/seed set development and leaf morphogenesis

Jian Xu[†], Ji Li[†], Li Cui, Ting Zhang, Zhe Wu, Pin-Yu Zhu, Yong-Jiao Meng, Kai-Jing Zhang, Xia-Qing Yu, Qun-Feng Lou and Jin-Feng Chen[*]

Abstract

Background: TIR1-like proteins act as auxin receptors and play essential roles in auxin-mediated plant development processes. The number of auxin receptor family members varies among species. While the functions of auxin receptor genes have been widely studied in *Arabidopsis*, the distinct functions of cucumber (*Cucumis sativus* L.) auxin receptors remains poorly understood. To further our understanding of their potential role in cucumber development, two *TIR1-like* genes were identified and designated *CsTIR1* and *CsAFB2*. In the present study, tomato (*Sonanum lycopersicum*) was used as a model to investigate the phenotypic and molecular changes associated with the overexpression of *CsTIR1* and *CsAFB2*.

Results: Differences in the subcellular localizations of *CsTIR1* and *CsAFB2* were identified and both genes were actively expressed in leaf, female flower and young fruit tissues of cucumber. Moreover, *CsTIR1-* and *CsAFB2*-overexpressing lines exhibited pleotropic phenotypes ranging from leaf abnormalities to seed germination and parthenocarpic fruit compared with the wild-type plants. To further elucidate the regulation of *CsTIR1* and *CsAFB2*, the role of the miR393/TIR1 module in regulating cucumber fruit set were investigated. Activation of miR393-mediated mRNA cleavage of *CsTIR1* and *CsAFB2* was revealed by qPCR and semi-qPCR, which highlighted the critical role of the miR393/TIR1 module in mediating fruit set development in cucumber.

Conclusion: Our results provide new insights into the involvement of *CsTIR1* and *CsAFB2* in regulating various phenotype alterations, and suggest that post-transcriptional regulation of *CsTIR1* and *CsAFB2* mediated by miR393 is essential for cucumber fruit set initiation. Collectively, these results further clarify the roles of cucumber TIR1 homologs and *miR393* in regulating fruit/seed set development and leaf morphogenesis.

Keywords: Auxin receptor, Cucumber, *CsTIR1* and *CsAFB2*, miR393/TIR1 module, Fruit/seed set

Background

Since the identification of F-box proteins TIR1/AFB (transport inhibitor resistant1/auxin signaling F-box) as auxin receptors [1, 2], a SCF$^{TIR1/AFB}$-Aux/IAA-ARF signaling module has been well established, which sheds light on the linkage between auxin perception and gene expression [3]. In the absence of auxin, or at low concentrations, ARFs combine with Aux/IAA to form heterodimers; hence, transcription of auxin-responsive genes is not promoted until ARFs are released due to degradation of Aux/IAA by SCF$^{TIR1/AFB}$ –ubiquitin mediated degradation induced by the presence of high auxin levels [4–6]. Thus, auxins act as a "molecular glue" to stimulate the interaction between TIR1/AFB and Aux/IAA [7–9].

As essential regulators of auxin responses in plants, TIR1-like proteins have been identified in various species and are divided into four distinct phylogenetic

* Correspondence: jfchen@njau.edu.cn
[†]Equal contributors
State Key Laboratory of Crop Genetics and Germplasm Enhancement, Nanjing Agricultural University, Nanjing 210095, China

clades TIR1, AFB2 (AFB2/AFB3), AFB4 (AFB4/AFB5) and AFB6. TIR1-like proteins are involved in multiple auxin-responsive biological processes. In *Arabidopsis*, the TIR1/AFB auxin receptor family comprises six members: TIR1 and five additional AFB proteins [1, 3, 10]. The *tir1 afb* mutants of *Arabidopsis* exhibit defects in hypocotyl elongation, apical hook, and lateral root formation, leaf morphology and inflorescence architecture [11]. TIR1 and AFB2 act as positive regulators of auxin signaling by mediating auxin-dependent degradation of Aux/IAAs [12, 13]; While, the AFB4 and AFB5 are known to be the major targets of the synthetic auxin, picloram [14, 15], the in vivo roles of AFB1 and AFB3 are still unclear [12].

A similar role for TIR1 in leaf morphogenesis has also been elucidated in tomato and rice plants. Over-expression of *SlTIR1* in tomato plants resulted in altered leaf morphology [16], while suppression of *OsTIR1* increased the flag leaf inclination angle [17]. The role of TIR1 in fruit and seed development has also analyzed. Phenotypic and molecular analyses indicate that TIR1-like proteins are pivotal regulators of auxins in the fruit set process [16, 18, 19]. Further studies revealed that *SlTIR1* stimulates stenospermocarpic fruit formation in tomato plants [20]. Although the observation of diverse of degradation behaviors among TIR1/AFB-Aux/IAA complexes suggests that the existence of divergent properties among the *TIR1/AFB* genes [13], independent evidence has demonstrated functional redundancy among *TIR1/AFB* family genes. Loss-of-function analysis by generating higher order mutants in *Arabidopsis* confirmed that the TIR1/AFB proteins act redundantly to regulate diverse aspects of plant growth and development [11]. Individual knockdown of *OsTIR1* or *OsAFB2* in rice induced similar leaf morphology at the booting stage [17]. Thus, the mechanisms by which the small family of functionally redundant TIR1-like proteins mediate pleiotropic regulation processes to promote various auxin responses remain to be eucidated.

As the first genome-sequenced vegetable crop, cucumber now serves as a model organism for investigation of the *Cucurbitaceae* family. Compared with plants belonging to the *Cruciferae* and *Solanaceae* families, cucumber has distinct auxin-related developmental processes, such as determinate/indeterminate growth, tendril development and parthenocarpic fruit formation. However, few studies of the TIR1-like gene family of *Cucurbitaceae* have been reported to date. To gain insights into the roles of cucumber auxin receptors in mediating diverse developmental processes, *CsTIR1* and *CsAFB2* were cloned and functionally characterized in transgenic tomato plants. In accordance with the concepts of Dharmasiri et al. (2005)

and Bian et al. (2012) [11, 17], the functions of *CsTIR1* and *CsAFB2* were revealed in this study. Further studies indicated that miR393-mediated post-transcriptional regulation of *CsTIR1* and *CsAFB2* contributes to the fruit set and development processes in cucumber.

Results

Phylogenetic and polymorphism analysis of *CsTIR1/AFB2*

Two genes encoding proteins closely related to the *TIR1*-like gene family of auxin receptors were isolated from cucumber (GenBank ID: GX901282 and GX901283). To investigate their evolutionary relationships with well-defined auxin receptor family proteins of other plant species, a phylogenetic tree was generated using the neighbor-joining approach by MEGA 4.0. Phylogenetic analysis indicated that GX901282 is clustered to the TIR1 clade and has 76.6% similarity to *AtTIR1*, while GX901283 belongs to the AFB2 clade and has 75.8% similarity to *VvAFB2* (Fig. 1a; Additional file 1). Thus, the isolated sequences were designated *CsTIR1* and *CsAFB2* to be consistent with the nomenclature used for the homologs from other plant species. *CsTIR1*, which encodes a protein of 584 amino acid residues, contains an F-box region and six leucine-rich repeat (LRR) domains, while *CsAFB2* encodes a protein of 587 amino acid residues and comprises an F-box region and seven LRR domains (Fig. 1b).

Subcellular localizations of CsTIR1/AFB2 proteins

To determine the subcellular localization of CsTIR1 and CsAFB2 proteins, the CsTIR1-GFP and CsAFB2-GFP fusion proteins were transiently expressed in onion epidermal cells using gene gun bombardment. Laser confocal scanning of protein fluorescence revealed that the green fluorescence signal of GFP alone was detected throughout the cell (Fig. 2a), in accordance with the expected cytosolic localization of the GFP protein. Interestingly, the fluorescence of CsTIR1-GFP was not only detected in nucleus but also detected in cytolemma (Fig. 2b), that is inconsistent with Ren's result [16]. The green fluorescence signal of CsAFB2-GFP was detected in the nucleus (Fig. 2c).

Expression patterns of *CsTIR1/AFB2* in cucumber

Temporal and spatial transcriptional characteristics of *CsTIR1* and *CsAFB2* were investigated by qRT-PCR. Both *CsTIR1* and *CsAFB2* were detected in all the major organs of cucumber plants including root, stem, leaf, female/male flowers and young fruit. These two genes showed similar expression patterns, with the highest abundance in leaf and female flower tissues and relatively low mRNA levels in roots and young fruit (Fig. 3a, b). The phytohormones responses of *CsTIR1* and

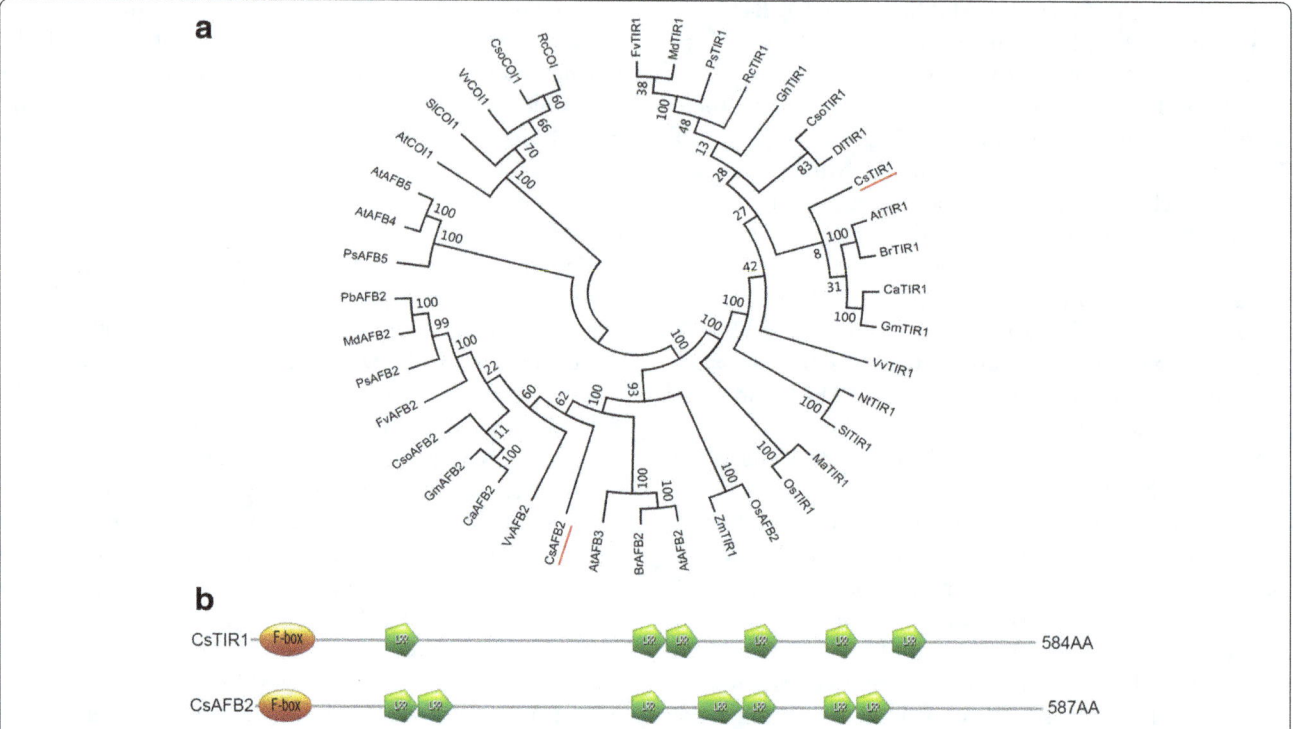

Fig. 1 Evolutionary relationships among the TIR1 protein family members and domain structure of *CsTIR1/AFB* proteins. **a** The phylogenetic tree consists of 40 protein sequences of TIR1-like auxin receptors from various land plants and was generated using the neighbour-joining method in MEGA5. Auxin receptors isolated from cucumber are labeled with a red line. **b** CsTIR1 protein contains an F-box region and six leucine-rich repeat domains, CsAFB2 protein contains an F-box region and seven leucine-rich repeat domains. The numbers on the right indicate the number of amino acid residues

CsAFB2 were investigated in exogenous hormone treatment experiments. qRT-PCR analysis showed that the transcription of *CsTIR1* was sensitive only to low (5 μM NAA) and medium (10 μM NAA) auxin concentrations (Fig. 3c). In contrast, *CsAFB2* expression was upregulated by exogenous gibberellins, cytokinin and auxin (Fig. 3d).

Treatment of cucumber ovaries (0dpa) with high concentrations of exogenous auxin (500 μM), cytokinins (400 μM), gibberellins (3000 μM) and brassinosteriod (0.2 μM) stimulated parthenocarpy. Interestingly, analysis of hormone-treated ovaries showed that expression of *CsTIR1* and *CsAFB2* was downregulated in the auxin- and cytokinins- induced parthenocarpic fruit as well as in the fruit set by pollination, while expression of the two genes was upregulated in gibberellin- and brassinosteriod- induced parthenocarpic fruit (Fig. 3e, f).

Functional analysis of *CsTIR1* and *AFB2* genes

To assess the physiological importance of the cucumber auxin receptor proteins, homozygous transgenic *Solanum lycopersicum,* cv. Micro-Tom lines overexpressing *CsTIR1* and *CsAFB2* were generated (designated *CsTIR1*-OE and *CsAFB2*-OE, respectively). qRT-PCR analysis showed that *CsTIR1* and *CsAFB2* were expressed abundantly in transgenic lines, with almost no effect on endogenous *SlTIR1* expression (Additional file 2). For each of the two genotypes, one of the most highly upregulated transgenic lines was selected for further characterization. Pleiotropic phenotypes were observed in these tomato lines (Table 1). Compact stature was the most intuitionistic alteration in the transgenic plants (Fig. 4a), which was consistent with the effects of transgenic expression of *SlTIR1* and *PslTIR1* [16, 19]. Compared with *CsAFB2*-OE lines, *CsTIR1*-OE lines exhibited reduced plant height phenotype (Fig. 4a). Distorted leaf growth resulted in severely inward curling growth status in transgenic lines (Fig. 4a, b).

Both the *CsTIR1* and the *CsAFB2* overexpression lines exhibited a convex leaf surface phenotype, compared with the smooth leaf surface of wild-type (WT) plants. Although overexpression of *SlTIR1* has been reported to increase trichrome numbers [16], no changes in the morphology and number of trichomes were observed in the leaves of *CsTIR1*-OE and *CsAFB2*-OE transgenic tomato plants. Surprisingly, *CsTIR1* overexpression reduced stomata formation, while the number was significantly reduced by *CsAFB2* overexpression (Fig. 4c). Expression analysis of related genes strongly indicated that the ARF10 protein functions as a transcriptional

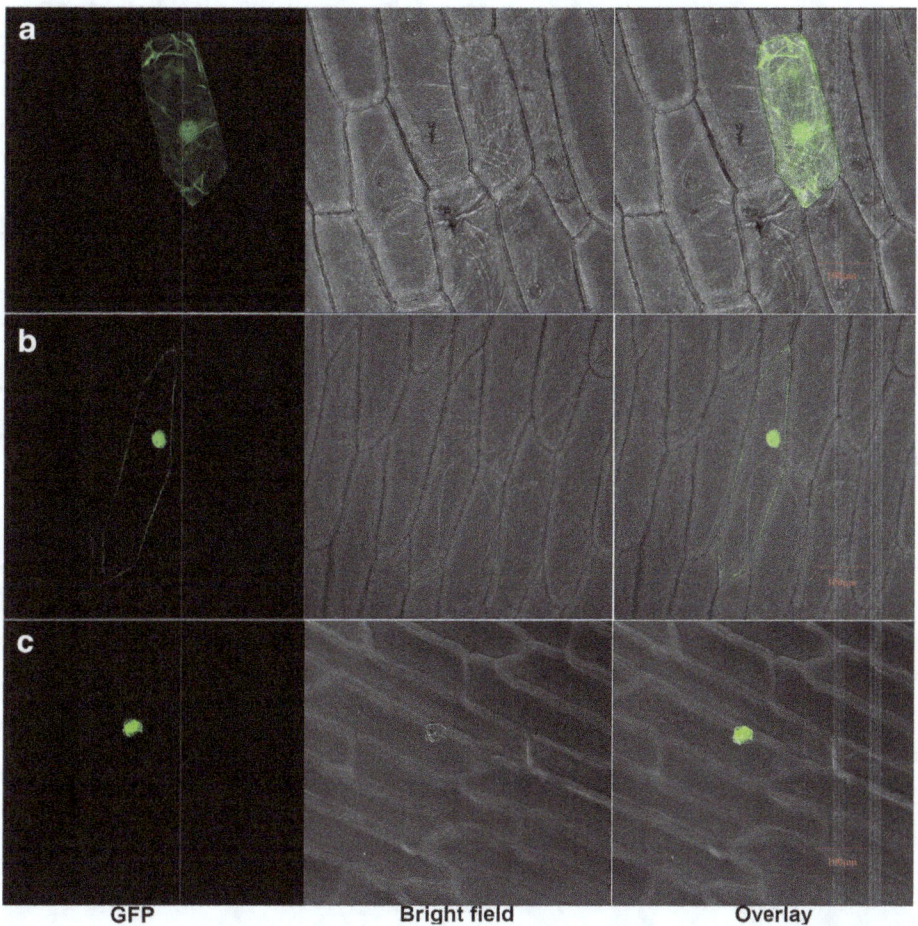

Fig. 2 Subcellular localization of CsTIR1 and CsAFB2 in onion epidermal cells. Control plasmid (*GFP*) and fusion vector constructs (*CsTIR1-GFP*, *CsAFB2-GFP*) were transformed separately into onion epidermal cells by microprojectile bombardment. **a** Subcellular localization of GFP alone. **b** Subcellular localization of the CsTIR1-GFP fusion protein. **c** Subcellular localization of the CsAFB2-GFP fusion protein. All proteins were analyzed by laser scanning confocal fluorescence microscopy (GFP). Light micrographs (Bright field) and fluorescence (GFP) images are merged (Overlay) to illustrate the different location of the three proteins. Scale bars = 100 μm

repressor of leaflet lamina outgrowth [21], while SPCH and MUTE were found to be involved in the stomata differentiation process [22, 23]. qRT-PCR analysis showed that *SlARF10* was upregulated in transgenic lines and was implicated as a positive regulator of leaf morphology. Furthermore, *SlSPCH* and *SlMUTE* were upregulated in *CsAFB2*-OE lines but down-regulated in *CsTIR1*-OE lines (Fig. 4d).

Overexpression of *CsTIR1* and *CsAFB2* resulted in reduced seed size (Fig. 5a) and fewer seeds in each fruit (Fig. 5b). Further investigations of germination properties showed that the seed germination potential was significantly reduced in transgenic lines (Fig. 5c). To assess the molecular mechanisms underlying these alterations, the expression of *AP-like* (determiner for seed size in *Arabidopsis* [24]), *STK* (a maternal role in fertilization and seed development and related to seed number [25]), *GIGANTEA* and *ELIP* (involved in the seed germination

process [26]) were investigated. Expression analysis revealed that *SlAP-like*, *SlSTK* and *SlGIGANTEA* were downregulated in the transgenic lines, while *SlELIP* was upregulated (Fig. 5d).

Although transcription analysis showed that *CsTIR1* and *CsAFB2* were downregulated during both the pollination and parthenocarpic fruit set processes, emasculation experiments suggested that overexpression of *CsTIR1* or *CsAFB2* induced facultative parthenocarpy in tomato fruit (Fig. 6a, b; Table 1). Seedless fruit phenotypes were also observed in both the *CsTIR1*-OE and *CsAFB2*-OE lines (Fig. 6c). Ovary expansion prior to anthesis may be the cause of seedlessness fruit phenotype (Fig. 6b). The *CsAFB2-OE* lines exhibited more obvious negative effects on seed set than the *CsTIR1-OE* lines because of the exposed stigma phenotype (Fig. 6a, e). Moreover, elongated fruit shape was induced by *CsAFB2* (Fig. 6d; Table 1).

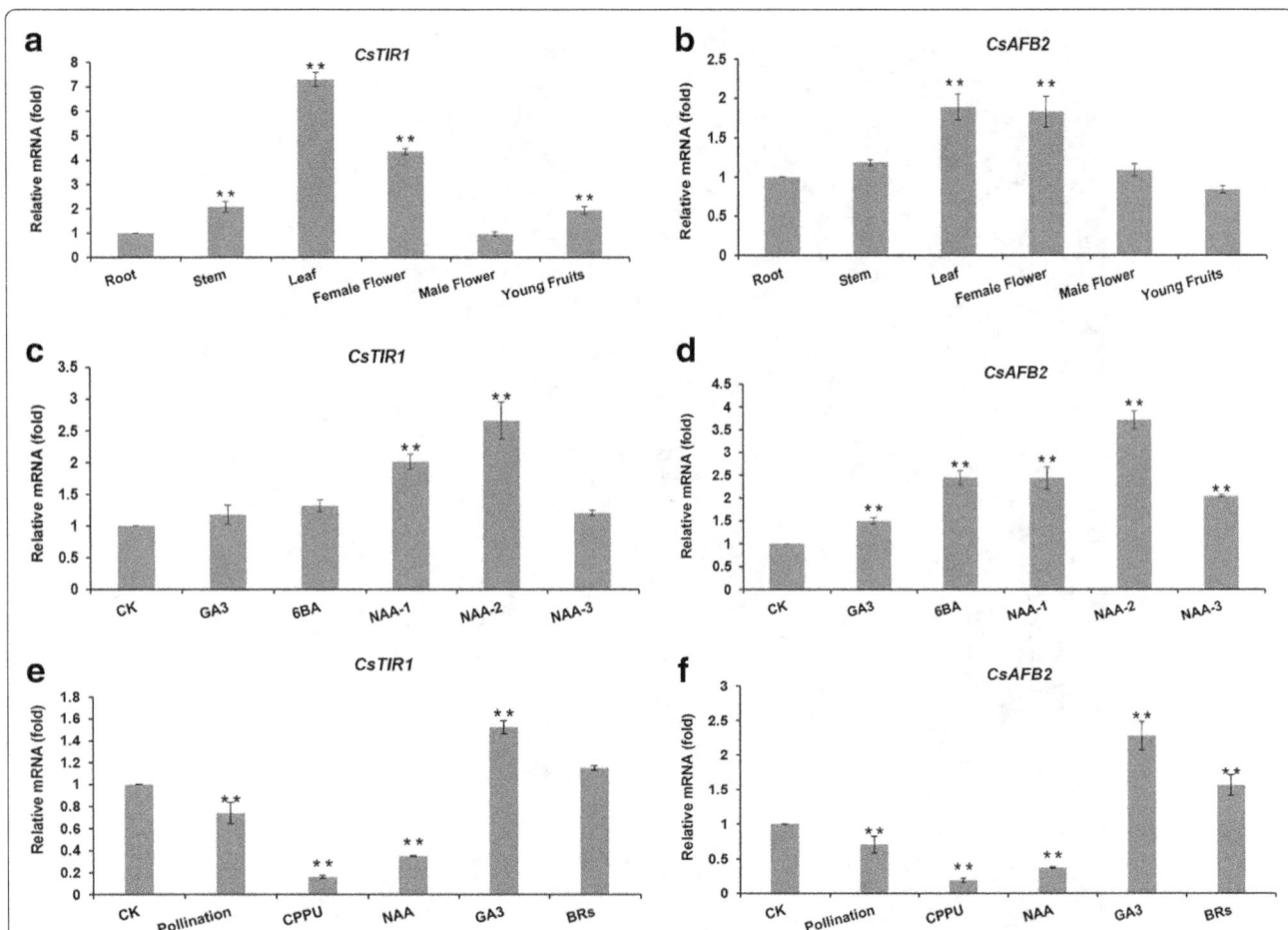

Fig. 3 Quantitative RT-PCR analysis of *CsTIR1* and *CsAFB2* expression in cucumber. **a, b** *CsTIR1* and *CsAFB2* expression in different organs of cucumber (roots, stems, leaves, female flowers, male flowers, and young fruit). **c, d** *CsTIR1* and *CsAFB2* expression in cucumber leaves treated with GA3 (10 μM), 6BA (10 μM), and different concentrations of NAA (5, 10, and 50 μM). **e, f** *CsTIR1* and *CsAFB2* expression in cucumber ovaries under pollination and treatment with CPPU (400 μM), NAA (500 μM), GA3 (3000 μM), and BRs (0.2 μM). The "young fruit" represent fruit at 0 days post-anthesis. CK, "Control". Data represent mean ± SD normalized relative to *CsActin* gene transcript levels. Root and CK expression data normalized to 1. All samples were analyzed in triplicate. Expression differences are calculated relative to expression in the root (**a, b**) and CK (**c, d, e, f**). (Student's *t* test; * $P < 0.05$; ** $P < 0.01$)

More detailed information about the phenotypes of *CsTIR1*-OE and *CsAFB2*-OE transgenic tomato plants is shown in Table 2.

Investigation of post-transcriptional regulations of *CsTIR1/AFB2* via miR393

Previous studies have demonstrated that miR393 plays a role in the regulation of TIR1/AFB expression [12, 27]. Despite the roles of miR393 and its target genes in multiple biological processes, much remains to be clarified about the function of miR393 and the miR393/TIR1

homologs module in regulating cucumber fruit set and development. MiR393 expression levels were analyzed in different early fruit developmental stages (Fig. 7). In parthenocarpic fruit, miR393 expression remained relatively stable, especially from 0 to 3 dpa (Fig. 7a), while in non-parthenocarpic fruit, miR393 expression was relatively low before anthesis but increased dramatically at the point of anthesis (0dpa). When the ovaries were pollinated manually (pollination induced fruit set), miR393 expression decreased rapidly to the low level detected in the immature ovaries (−3, −2, −1dpa). Although miR393

Table 1 Summary of phenotypes in transgenic tomato plants overexpressing *CsTIR1* and *CsAFB2*

Genotype	Line number	Phenotype
35S–*CsTIR1*	L1-L4	Severely dwarf plants, abnormal leaves, suppressed seed size and germination activity, splited staminal cone, precocious fruit set prior to anthesis, reduced seed number/per fruit, parthenocarpic fruit set
35S–*CsAFB2*	L1-L2	Compact stature of plants, abnomal leaves, suppressed seed size and germination activity, splited staminal cone, precocious fruit set prior to anthesis, altered fruit shape, reduced seed number/per fruit, parthenocarpic fruit set

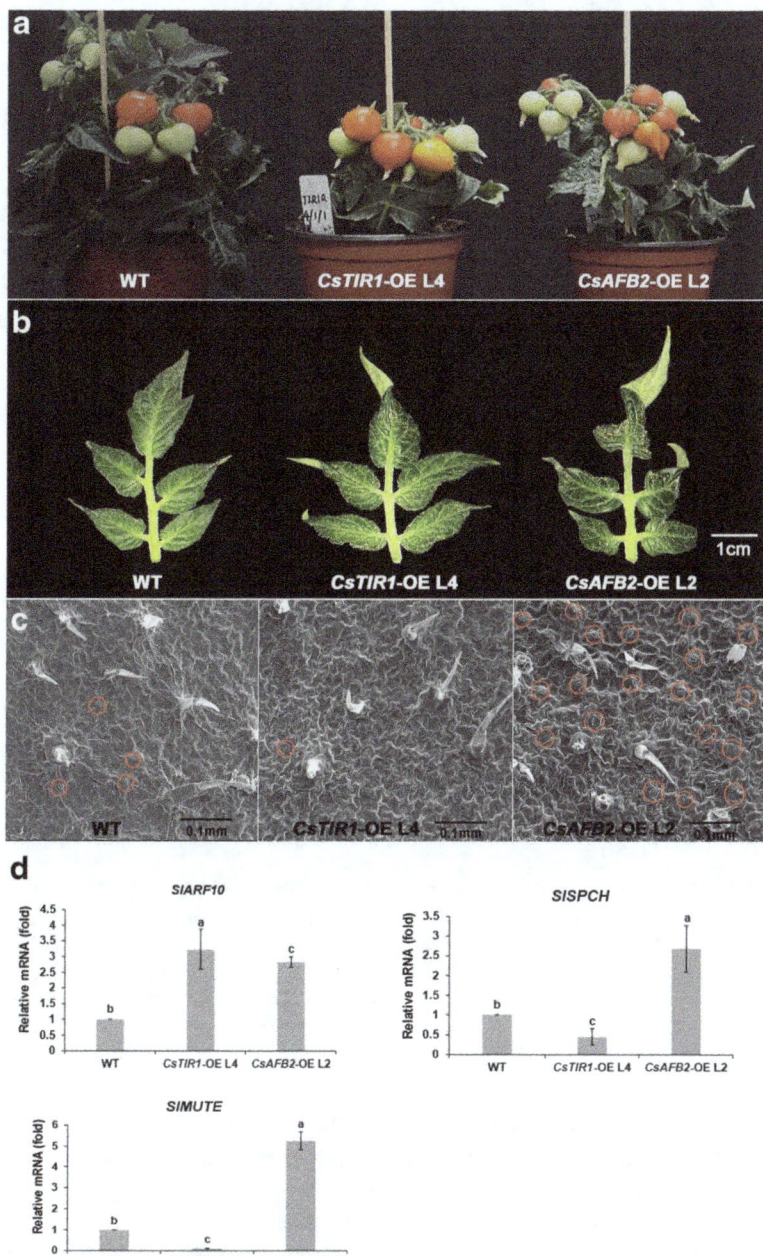

Fig. 4 Vegetative growth and leaf architecture phenotypes in wild-type and transgenic plants. **a** Both *CsTIR1*-OE L4 and *CsAFB2*-OE L2 exhibited compact stature, and *CsTIR1*-OE L4 exhibited a severely dwarfed phenotype compared with the wild-type (WT). **b** Leaves of both *CsTIR1*-OE L4 and *CsAFB2*-OE L2 exhibited distorted growth behavior. Scale bar = 1 cm. **c** Leaf surface of WT, *CsTIR1*-OE L4, and *CsAFB2*-OE L2 observed by scanning electron microscopy. Transgenic line leaf surfaces were convex compared with the WT. Stomata are labeled with red circles. Scale bar = 0.1 mm. **d** Quantitative PCR analysis of transcript accumulation of *SlARF10*, *SlSPCH*, and *SlMUTE*. Expression of *SlARF10*, *SlSPCH*, and *SlMUTE* in the WT was normalized to 1. Data represent mean ± SD of three biological replicates. Different letters above bars indicate significant differences among different genotypes (Student's *t*-test, $P < 0.05$)

expression was also downregulated in abortive fruit (bagging treatment), it was 2–4-fold higher relative to that in the pollinated fruit (Fig. 7b).

In cucumber, *CsTIR1* has three nucleotide mismatches in the miR393 recognition site, while *CsAFB2* contains two mismatches, which are thought to cause different

effects on miR393-directed mRNA cleavage (Fig. 8a). This suggests that *CsTIR1* and *CsAFB2* are regulated by miR393 in a post-transcriptional manner. To investigate miR393-mediated post-transcriptional regulations of CsTIR1/AFB2 during the fruit set process, four pairs of specific primers were designed for the detection of full-

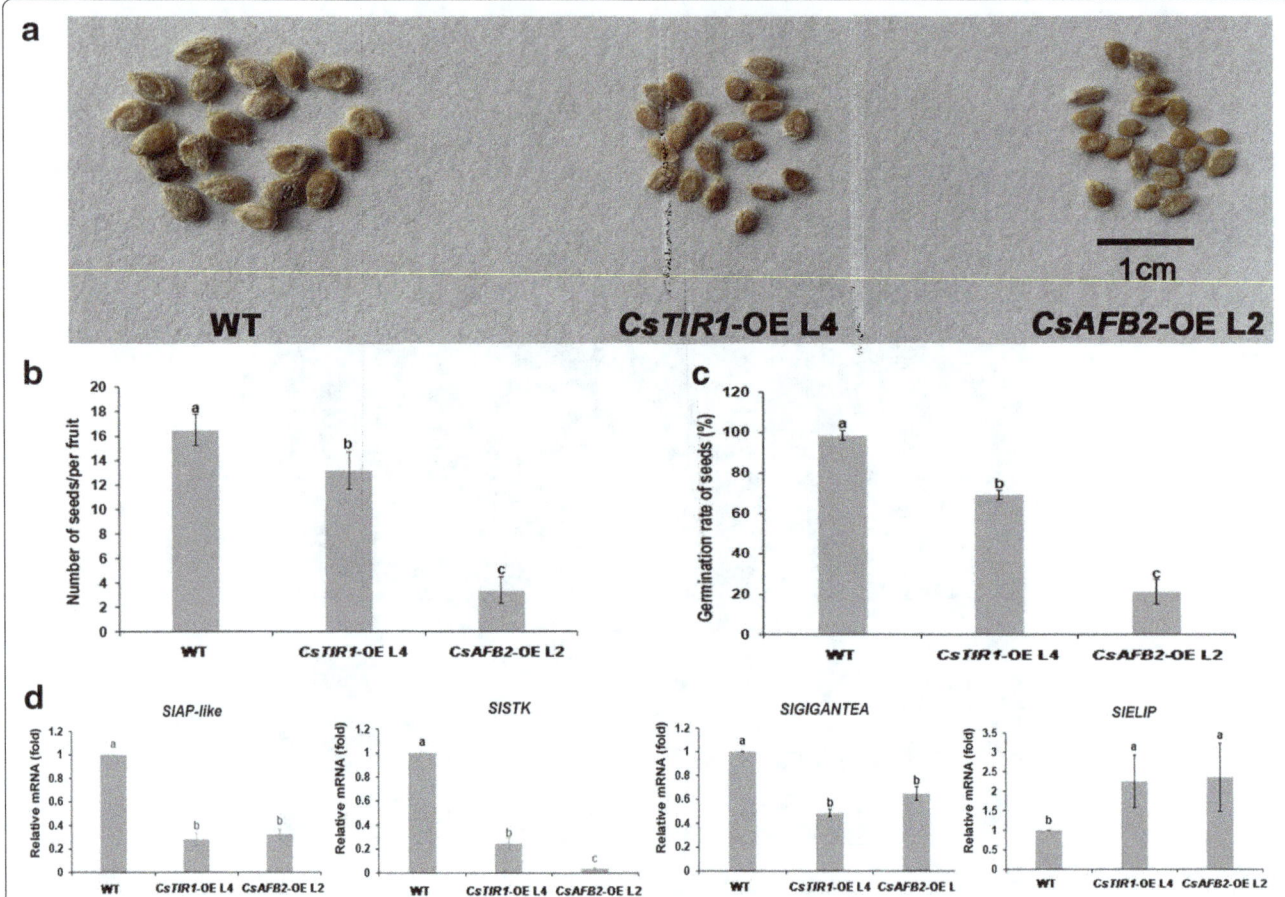

Fig. 5 Seed size, seed number and seed germination activity were changed in transgenic plants. **a** Seed size of transgenic plants was smaller than that of wild-type. Scale bar = 1 cm. **b** Seeds in each fruit of transgenic plants were fewer than the wild-type plants. **c** Seed germination potential was reduced in transgenic lines. **d** Quantitative PCR analysis of transcript accumulation of *SlAP-like*, *SlSTK*, *SlGIGANTEA* and *SlELIP*. Expression of *SlAP-like*, *SlSTK*, *SlGIGANTEA* and *SlELIP* in the WT was normalized to 1. Data represent mean ± SD of three biological replicates. Different letters above bars indicate significant differences among different genotypes (Student's *t*-test, P < 0.05)

length transcripts (complete fragments) and total mRNA (complete fragments and alternative splicing fragments) of *CsTIR1* and *CsAFB2* (Fig. 8a; Additional file 3).

Both the total and full-length mRNA transcripts of *CsAFB2* were much less abundant than those of *CsTIR1*, and the full-length mRNA of *CsAFB2* was barely detected during these fruit developmental processes. The total transcripts of *CsTIR1* and *CsAFB2* reached the highest expression levels at 0dpa, while only low levels of the full-length transcripts of these genes were detected (Fig. 8b). In accordance with this, qRT-PCR analysis showed that miR393 expression peaked at anthesis (Fig. 7b), indicating that miR393-mediated cleavage of *CsTIR1* and *CsAFB2* was active during flowering (0dpa). Despite the relatively low miR393 expression levels, some full-length mRNA was detected in both pollinated and parthenocarpic fruit set (Figs. 7 and 8b). Interestingly, although the miR393 levels in abortive fruit were relatively higher than those of set fruit (Fig.

7b), there was almost no degradation of full-length mRNA of *CsTIR1* in abortive fruit (Fig. 8b).

Discussion

As a reference species for *Cucurbitaceae* crops, cucumber exhibits various traits that may be regulated by various auxin gene networks. However, few auxin signaling transduction components of *cucumber* have been studied. To investigate the exact molecular mechanisms underlying the functions of cucumber auxin receptors in plant growth and fruit development, both *CsTIR1* and *CsAFB2* genes were isolated and the physiological and molecular consequences of their overexpression in tomato plants were evaluated.

In *Arabidopsis*, six auxin receptors have been identified, comprising *AtTIR1* and its closest paralogs *AFB1- AFB5* [10]. In tomato plants, three F-box receptor members have been identified [16], while three TIR1-like auxin receptors were found in plum [19] and there are six *TIR1/*

Fig. 6 Fruit set and parthenocarpy in transgenic plants. **a** Flowers at anthesis day in wild-type and transgenic plants. Both *CsTIR1*-OE L4 and *CsAFB2*-OE L2 exhibited ovary expansion prior to anthesis, and splited staminal cone. The exposed stigma of *CsAFB2*-OE L2 is labeled with red arrow. **b** Ovaries of transgenic lines were successfully set fruit and expansion after emasculation. **c** Wild-type seeded fruit and transgenic parthenocarpic fruit. Scale bar = 1 cm. **d** Altered fruit shape was observed in the *CsAFB2*-OE L2. Scale bar = 1 cm. **e** Comparison of parthenocarpic fruit set rate among wild-type, *CsTIR1*-OE L4, and *CsAFB2*-OE L2 plants. Data represent mean ± SD of three biological replicates with more than twenty fruits for each replicate. Different letters above bars indicate significant differences among different genotypes (Student's *t*-test, *P* < 0.05)

AFBs homologs in the rice genome [17]. In *Populus*, eight *TIR1* homologous genes (*PtrFBLs*) were identified [28]. However, only two auxin receptors belonging to the F-box *TIR1*-like gene family have been identified in cucumber [29, 30]. Cucumber has a small genome with few tandem duplications [31]. Previous studies have shown that evolutionary momentum is basically facilitated by genome duplication events, which are thought to occur frequently during the evolution of organism [32, 33]. Therefore, it is tempting to speculate that the small number of TIR1 homologs in cucumber is probably due to the absence of recent whole-genome duplications.

TIR1/AFBs show auxin-insensitive transcriptional characteristics in *Arabidopsis* and *Populus* [12, 28]. Similarly, limited responses of *CsTIR1* to NAA treatment were also detected in cucumber, although *CsAFB2*

Table 2 Phenotypes of wild-type (WT), *CsTIR1*-overxpressing (*CsTIR1*-OE) and *CsAFB2*-overexpressing (*CsAFB2*-OE) transgenic tomato plants

parameter	WT	CsTIR1-OE L1	CsTIR1-OE L2	CsTIR1-OE L3	CsTIR1-OE L4	CsAFB2-OE L1	CsAFB2-OE L2
Plant height (one month old, cm)	8.8 ± 0.2a	4.2 ± 0.3f	4.6 ± 0.2e	5.4 ± 0.1d	4.2 ± 0.2f	6.1 ± 0.3c	6.5 ± 0.1b
Plant height (three month old, cm)	20.5 ± 1.1a	9.4 ± 1.1c	9.9 ± 0.6c	8.6 ± 0.9c	8.2 ± 1.1c	13.6 ± 1.2b	12.3 ± 1.2b
Length of seed (mm)	4.4 ± 0.4a	3.0 ± 0.2b	3.1 ± 0.4b	3.2 ± 0.2b	3.2 ± 0.4b	3.3 ± 0.3b	3.3 ± 0.2b
Width of seed (mm)	2.8 ± 0.3a	1.5 ± 0.2b	1.6 ± 0.2b	1.5 ± 0.2b	1.5 ± 0.2b	1.5 ± 0.2b	1.6 ± 0.2b
Seeds per fruit (n)	16.5 ± 1.3a	13.1 ± 1.4b	13 ± 1.3b	12.9 ± 1.4b	13.2 ± 1.5b	3.5 ± 1.2c	3.4 ± 1.1c
Seeds germination rate (%)	98.7 ± 2.3%a	72 ± 6.9%b	70.6 ± 2.3%b	72 ± 6.9%b	69.3 ± 2.3%b	25.3 ± 4.6%c	21.3 ± 6.1%c
Parthenocarpic fruit set rate (%)	0c	20.8 ± 6.7%b	19.7 ± 5.9b	19.6 ± 5.5%b	20.6 ± 5.9%b	51.3 ± 11.0%a	53.3 ± 10.5%a
Horizontal diameters of fruit (cm)	2.1 ± 0.2a	2.0 ± 0.2b	1.8 ± 0.2b	1.9 ± 0.1b	1.9 ± 0.2b	1.4 ± 0.1c	1.4 ± 0.1c
Vertical diameters of fruit (cm)	3.3 ± 0.1a	2.8 ± 0.1c	2.9 ± 0.1bc	2.9 ± 0.1bc	2.9 ± 0.1bc	3.0 ± 0.1b	3.0 ± 0.1b

Values are means of 5–10 plants, ±SE. The statistical significance of mean differences was analyzed using Student's *t*-test, *P* < 0.05

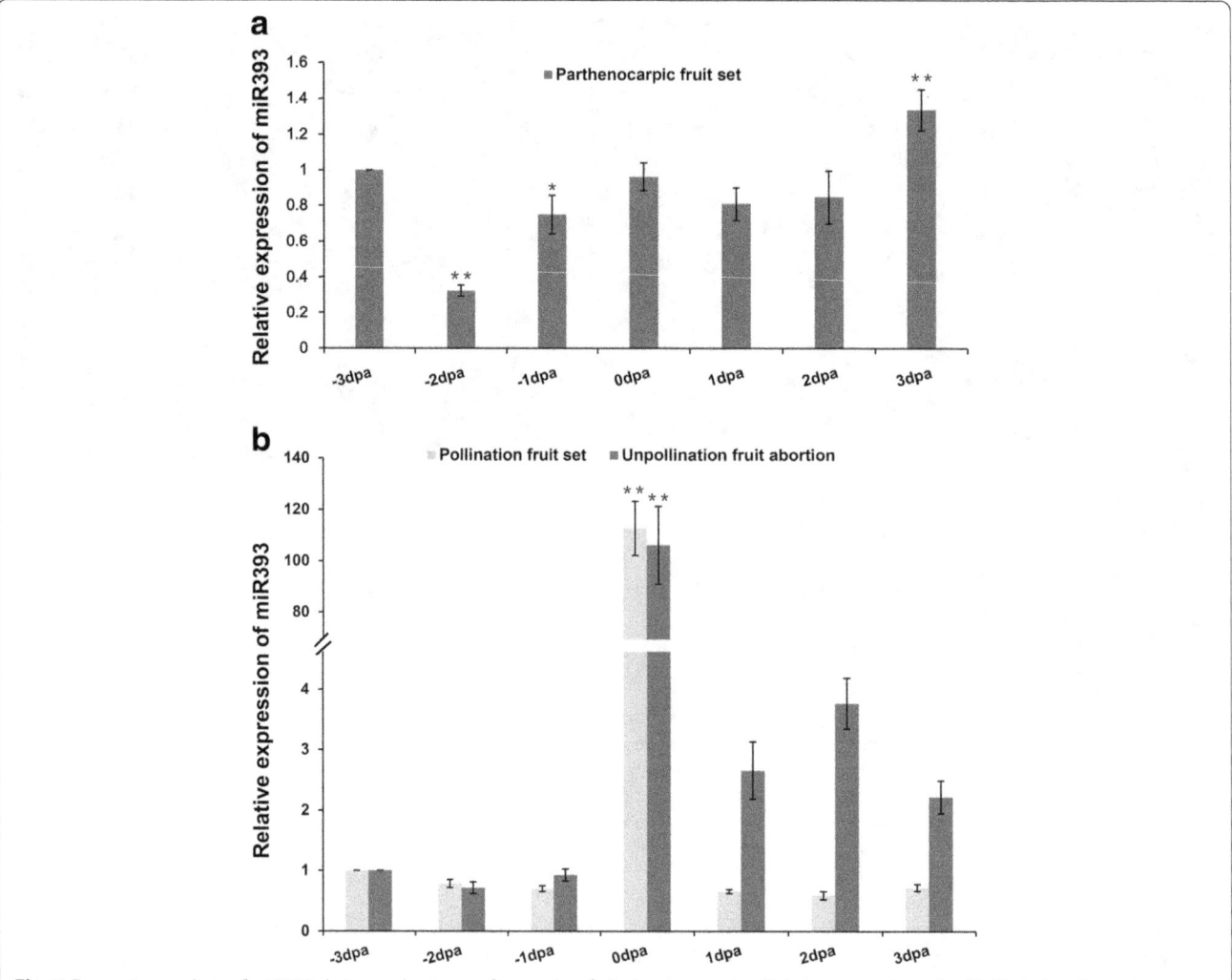

Fig. 7 Expression analysis of miR393 during early stages of cucumber fruit development. **a** Relative expression of miR393 during the early stages of parthenocarpic fruit set development (−3 dpa to 3 dpa). dpa: days post-anthesis. **b** Relative expression of miR393 during the early stages of pollinated fruit set formation and unpollinated fruit abortion (−3 dpa to 3 dpa). Expression of miR393 in −3 dpa fruit was normalized to 1. Data represent mean ± SD of three biological replicates. (Student's t-test; $*P < 0.05$; $**P < 0.01$)

showed sensitive responses to various exogenous phyto-hormones treatments (Fig. 3c, d). As intranuclear auxin receptors, TIR1/AFB proteins generally perform their functions within the nucleus [1, 16, 34]. However, subcellular localization analysis showed that cucumber CsTIR1 proteins were present not only in the nucleus but also on the cell membrane (Fig. 2b). Thus, the potential function of CsTIR1 as a membrane receptor deserves further investigation.

A series of studies suggested that the TIR1 protein may have a parallel function in leaf architecture and fruit development. Overexpression of *SlTIR1* results in pleiotropic phenotypes in tomato plants, including parthenocarpic fruit formation and leaf morphology [16]. El-Sharkawy demonstrated the critical role of *PslTIR1* as a positive regulator of auxin signaling in coordinating the development of leaves and fruit [19].

Furthermore, knockdown of either *OsTIR1* or *OsAFB2* altered the flag leaf inclination angle [17]. In this study, the highest transcription levels of *CsTIR1* and *CsAFB2* were in leaves and female flowers (Fig. 3a, b). Overexpression of the cucumber TIR1-like proteins induced curved growth of leaves and abnormal fruit expansion (Fig. 4a, b and 6a, b). Previous studies showed that trichome development is also regulated by TIR1 proteins [16]; however, no obvious alterations in trichome numbers and morphology were observed in the *CsTIR1*-OE and *CsAFB2*-OE transgenic lines. In addition, both the *CsTIR1*-OE and *CsAFB2*-OE lines exhibited decreased seed germination potential (Fig. 5c). Down-regulation of *SlGIGANTEA* in the transgenic lines is consistent with its positive role in the promotion of germination under continuous R in tomato seeds revealed in a previous study [35].

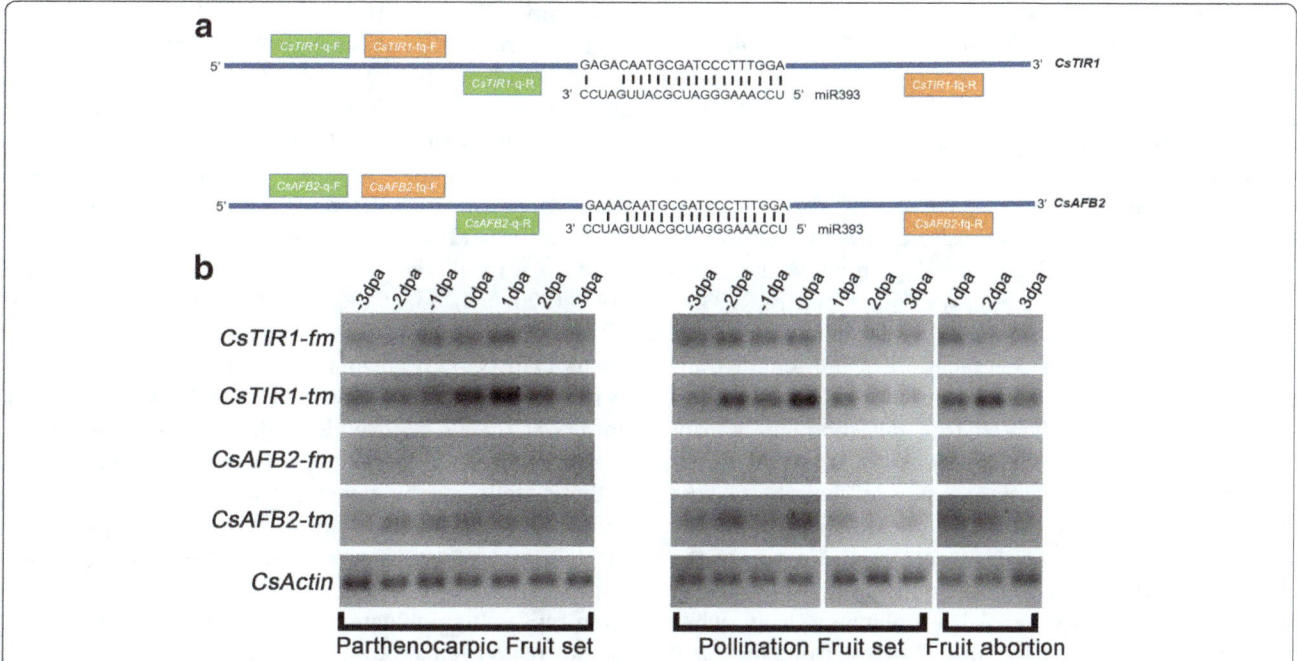

Fig. 8 Expression analysis of full-length transcripts and total transcripts of *CsTIR1* and *CsAFB2* during the cucumber parthenocarpic fruit set, pollination fruit set and fruit abortion processes. **a** Schematic diagram of specific primer location. *CsTIR1* has three nucleotide mismatches in the miR393 recognition site, while *CsAFB2* contains two mismatches in the miR393 recognition site. The orange pair of primers separated on both sides of the miR393-cleavage site were used to detect the full-length transcripts of *CsTIR1* and *CsAFB2*. The green pair of primers distributed on one side of the miR393-cleavage site were used for detection of total transcripts of *CsTIR1* and *CsAFB2*. **b** Semi-quantitative RT-PCR analysis of full-length transcript and total transcript accumulation of *CsTIR1* and *CsAFB2* during the early stages of the cucumber parthenocarpic fruit set, pollination fruit set and fruit abortion processes (−3 dpa to 3 dpa). dpa: days post-anthesis. fm: full-length mRNA (complete fragments); tm: total mRNA (complete fragments and alternative splicing fragments)

Furthermore, *SlELIP*, which has been reported to be significantly increased by inhibition of seed germination [26], was found to be upregulated in the transgenic lines. Thus, these observations indicate auxin receptor overexpression in tomato plants alters the expression of genes involved in seed germination process. Considering that silence of *AtTIR1* could also repress seed germination activity [1], therefore, it can be speculated that transcriptional balance of *TIR1* family genes is essential for seed germination.

Independent evidence suggests that TIR1/AFB proteins function as redundant auxin receptors, collectively mediating auxin-regulated responses throughout plant growth and development. Different phenotypes induced by overexpression of *CsTIR1* and *CsAFB2* were revealed in this study. *CsTIR1* and *CsAFB2* were found to have opposing regulatory functions in controlling the number of stomata, with *CsTIR1* overexpression inhibiting stomata formation, while stomata numbers were significantly increase by *CsAFB2* overexpression (Fig. 4c). Interestingly, *CsTIR1*-OE and *CsAFB2*-OE overexpression resulted in similarly opposing patterns of *SlSPCH* and *SlMUTE* expression (Fig. 4d), suggesting that overexpression of *CsTIR1* and *CsAFB2* has different effects

on expression of stomata differentiation-related genes, thereby causing the opposite phenotypes in terms of stomata number. Furthermore, *CsAFB2*-OE lines exhibited higher parthenocarpic fruit set rates and altered fruit shape compared with the *CsTIR1*-OE lines. Studies have indicated that co-receptor complexes formed by different combinations of TIR1/AFB and the Aux/IAA proteins have a wide range of auxin-binding affinities [12, 28, 36]. We speculate that this accounts for the differences in auxin-related phenotypes, including alterations in leaf architecture and fruit development that are induced by *CsTIR1* and *CsAFB2* overexpression.

The roles of the miR393/TIR1 homolog module in regulating root growth, leaf inclination, tillering, disease resistance and salt tolerance have been well characterized. In *Arabidopsis*, regulation of auxin response by miR393-targeted TIR1 is involved in normal development [37], and the nitrate-responsive miR393/AFB3 regulatory module controls the root system architecture [36]. In addition, the miRNA393/TIR1 homolog module influences flag leaf inclination, crown root initiation, seminal root development and tillering in rice [17, 38]. Moreover, miR393-guided cleavage of TIR1, AFB2, and AFB3 transcripts enhances innate immunity in response to bacterial and fungal infection in leaves [27, 39]. Downregulation of *OsTIR1* and *OsAFB2* via *OsmiR393* led to reduced tolerance to salt and drought in rice [40]. In contrast, overexpression of a miR393-resistant form of mTIR1 enhanced salt tolerance in *Arabidopsis* [41]. However, less is known about the role of the miR393/ TIR1 homolog module in the stages of fruit development. In this study, we also explored the mechanism by which the miR393/TIR1 homolog module regulates cucumber fruit development. We found that miR393regulates the degradation of *CsAFB2* in all types of fruit development, with total degradation of full-length transcripts of *CsAFB2* detected regardless of the process of parthenocarpic or pollinated fruit development or fruit abortion. Although miR393 expression levels were extremely high at anthesis, some full-length *CsTIR1* remained undegraded, probably due to the enrichment of *CsTIR1* transcripts or the incomplete cleavage of the miR393 target, *CsTIR1*. More interestingly, cleavage of *CsTIR1* was complete in 1dpa pollinated fruit, while there was no cleavage in 1dpa abortive fruit, in which even the miR393 level was much higher than that in the pollinated fruit (Fig. 8b). These findings provide an indication that the divergent functions of *CsTIR1* and *CsAFB2* are induced by the differences in cleavage regulation by miR393; however, the differences in miR393-mediated regulation of *CsTIR1* and *CsAFB2* cleavage remain to be verified through in vivo.

Conclusion

This study was focused on the identification of distinct functions of cucumber TIR1-like genes. We identified and cloned both members of this gene family in cucumber. We demonstrated the differences in subcellular localizations of *CsTIR1* and *CsAFB2* proteins and analysis of *CsTIR1* and *CsAFB2* expression indicated their involvement in leaf morphogenesis and the fruit set process in cucumber. The roles of *CsTIR1* and *CsAFB2* in these processes were further revealed by the analysis of overexpressing lines. Moreover, qPCR and semi-qPCR results support the hypothesis that the miR393/TIR1 module participates in the cucumber fruit set process. These findings further clarify the functions of cucumber auxin receptors and provide new insights into the role of cucumber TIR1 homologs and miR393 in regulating fruit/seed set development and leaf morphogenesis.

Methods
Plant materials and growth conditions

Two cucumber breeding lines, parthenocarpic line 'EC1' and non-parthenocarpic line '8419 s-1', were grown in greenhouses during the natural growing season (12 h photoperiod, 29/17 °C average day/night temperature, 85% humidity, 800 $\mu molm^{-2} s^{-1}$) at Nanjing Agricultural University (China). Both 'EC1' and '8419 s-1' were breeding materials of the Nanjing Agricultural University. Tissues (root, stem, leaf, female flower, male flower and ovary) were collected from 9-week-old '8419 s-1' plants. For hormone treatment analysis, the leaves of '8419 s-1' plants (*n* = 30 per group) were treated with GA$_3$ (10 μM), 6-BA (10 μM) and different concentrations of NAA (5, 10, and 50 μM) and collected at 6 h after treatments. Ovaries were treated by pollination and with NAA (500 μM), GA3 (3000 μM), CPPU (400 μM), or BRs (0.2 μM) at anthesis and collected at 6 h after treatments. Thirty 'EC1' plants and thirty '8419 s-1' plants were used for fruit set and development analysis. Female flowers at the 12–15th node of the main stem were isolated to prevent pollen contamination on the day before anthesis, followed by treatment as above with bagging and pollination ('EC1' bagging represents the parthenocarpic fruit set process, '8419 s-1' bagging represents unpollination fruit abortion process, '8419 s-1' pollination represents pollinated fruit set process). The ovaries of the treated female flowers were harvested at −3, − 2, −1, 0, 1, 2, 3 days post-anthesis (dpa). Non-treated leaves and fruit were used as controls. All samples were frozen in liquid nitrogen, and stored at −80 °C prior to analysis. Sampling was performed on three independent occasions.

Tomato plants (*Solanum lycopersicum*, cv MicroTom) were used in this study to generate *CsTIR1* and *CsAFB2* overexpression lines. The seeds of MicroTom were kindly provided by Professor Zhengguo Li (Genetic Engineering Research Center, Chongqing University, China). All tomato plants (10 plants per each line) were grown in an illuminated incubator under controlled conditions set as follows: 14 h day/10 h night cycle; 25/20 ° C day/night temperature; 80% relative humidity and $250 \mu molm^{-2} s^{-1}$ light intensity.

Bioinformatics analysis of *CsTIR1* and *CsAFB2*

The cucumber genome database (http://cucumber.gen-omics.org.cn/page/cucumber/index.jsp) was searched for homologs of TIR1-like auxin receptors in cucumber using the *AtTIR1* (AT3G62980.1) sequence. Blast searches revealed two cucumber genes (*Csa001802* and *Csa015043*) with the highest similarity to *AtTIR1*, and two primer pairs were designed to amplify the full-length cDNA sequences of these two genes using cDNA of cucumber fruits as a templates (Additional file 4). The PCR products were directly sequenced (Invitrogen) and sequence data were submitted to GenBank (accession number: GX901282 and GX901283). The MEGA program (version 5) was used for phylogenetic analysis with known TIR1-like auxin receptors based on homology. Structural domains were annotated using Smart (http://smart.embl-heidelberg.de/) and illustrated using prosite (http://

prosite.expasy.org/mydomains/) with default parameters. GenBank accession numbers for the sequences analysis are listed in Additional file 5.

Subcellular localization of CsTIR1-GFP and CsAFB2-GFP fusion protein

The open reading frames of *CsTIR1* and *CsAFB2* were amplified and cloned into the pGreen0029 vector without the stop codon using specific primers (Additional file 4). Recombinant plasmids (pGreen 0029-*CsTIR1-GFP* and pGreen 0029-*CsAFB2-GFP*) and a control plasmid with *GFP* alone were introduced into onion epidermal cells (obtained from Lab of Cucurbit Genetics and Germplasm Enhancement, Nanjing Agricultural University) using 1.0 μm of gold microcarriers delivered via a pneumatic particle gun (Bio-Rad, PDS-1000/He, USA) with the following bombardment conditions: vacuum, 635 mmHg; helium pressure, 1100 psi; target distance, 6 cm. After bombardment, onion epidermis cells were cultured on MS (Murashige and Skoog) medium for 14 h at 25 °C in darkness. The transformed cells were visualized using a confocal laser scanning microscope (Zeiss, LSM780, Germany).

Generation of transgenic tomato plant lines

The full-length sequences of *CsTIR1* and *CsAFB2* were amplified using gene specific primers (Additional file 4). Fragments of *CsTIR1* and *CsAFB2* were cloned into plp100 binary vector under the transcriptional control of the *35S* promoter. The constructs were then introduced into WT tomato plants (*Solanum lycopersicum* cv MicroTom) by *Agrobacterium*-mediated transformation [16]. The *Agrobacterium* (C58) was kindly provided by Professor Zhengguo Li (Genetic Engineering Research Center, Chongqing University, China). Transformed lines were selected on kanamycin (50 mg L^{-1}) and further analyzed by qRT-PCR to confirm the presence of T-DNA inserts in the transgenic lines. For each construct, more than six independent lines with consistent phenotypes were obtained. Homozygous lines from the T2 generation were used for experiments.

Electron microscopy

Segments of leaves were collected from WT and transgenic plants after 6 weeks of growth on soil. Three replicates of each sample were mounted on aluminum stubs, sputtered with gold palladium for 30s, and examined under an S-3000 N scanning electron microscope (Hitachi, Japan).

Evaluation of parthenocarpy and seed germination assays

To evaluate parthenocarpic fruit set rates of transgenic lines and WT plants, flower buds (5 plants per line) were emasculated 2 d before anthesis to prevent self-pollination. To guarantee equivalent growth conditions for parthenocarpic fruit, only seven flowers were kept per plant. The parthenocarpic fruit set rate was represented by the percentage of emasculated flowers that developed into fruit.

For seed germination assays, seeds were obtained and counted from WT and transgenic tomato lines (5 plants per line). The seeds were placed in petri dishes containing a layer of wet 3 mm filter paper (Whatman, China) and incubated in the dark at 25 °C. The number of germinated seeds was counted after 3 days. The emergence of a radicle was considered as a germination event for the calculation of the percent germination. The experiment was replicated on three independent occasions with 25 seeds per replicate. All phenotypic data was listed in Additional file 6.

Gene expression analysis

All primers used for qRT-PCR and semi-quantitative RT-PCR are listed in Additional file 3. Total RNA was isolated using TRIzol reagent (Invitrogen, USA) and treated with DNase I (Fermentas, UK) according to the to manufacturer's instructions. First-strand cDNA was synthesized using the PrimeScript™ RT-PCR Kit (TaKaRa, Japan). Real-time quantitative RT-PCR was then carried out using the SYBR® Premix Ex Taq™ Kit (TaKaRa, Japan) in a CFX96 multicolor real-time PCR detection system (Bio-Rad, USA). *Sl-actin* and *Cs-actin* were used as the internal control genes. For qRT-PCR analysis of *Cs-miR393*, U6 was used as internal control gene and the first-strand cDNA was synthesized using Mir-X miRNA First-Strand Synthesis Kit (TaKaRa, Japan). The qRT-PCR was carried out using the SYBR® Premix Ex Taq™ II Kit (TaKaRa, Japan). Quantification of mRNA and miRNA levels was based on the comparative cycle threshold (CT) method and calculated as $2^{-\Delta\Delta CT}$. Analysis was conducted on the data from three independent reactions (technical replicates) using samples from three biological replicates. All CT values were listed in Additional file 7.

Additional files

Additional file 1: Fig. S1. Comparison of the predicted amino acid sequences of cucumber CsTIR1, CsAFB2 and AtTIR1, VvAFB2.

Additional file 2: Fig. S2. Quantitative PCR assession of positive transgenic tomato lines and detection of *SlTIR1* expression in transgenic tomato lines.

Additional file 3: Table S1. The oligonucleotide primer for qRT-PCR and semi-quantitative RT-PCR.

Additional file 4: Table S2. List of primers used in construct preparation.

Additional file 5: Table S3. Sequences from TIR-like F-box proteins of Embryophyte species used to generate the phylogenetic tree.

Additional file 6: Table S4. Phenotypic data of Wild-type, CsTIR1-OE lines and CsAFB2-OE lines.

Additional file 7: Table S5. CT values of RT-PCR experiments conducted in this study.

Abbreviations

6-BA: 6-benzylamino-purine; BRs: Brassinosteroids; CPPU: N-(2-chloro-4-pyridyl)-N0-phenylurea, a diphenylurea-derived cytokinin; dpa: days post anthesis; GA3: Gibberellic acid; GFP: Green Fluorescent Protein; NAA: Naphthyl acetic acid; qRT-PCR: Quantitative real-time PCR

Acknowledgements

Not applicable.

Funding

This work was supported by the National Natural Science Foundation of China (NO: 31,430,075 and 31,672,168), the National Key Research and Development Program of China (2016YFD0101705), Special Fund for Agro-scientific Research in the Public Interest (NO: 201,403,032), National Key Research and Development Program of China (2016YFD0100204–25) and Jiangsu Agricultural Science and Technology Innovation Fund (No: CX(15)1019).

Authors' contributions

JX and JL contributed equally. JX and JL performed the experiments and drafted the manuscript. LC, TZ, ZW, PYZ, YJM and KJZ assisted with experimental procedures and data analysis. XQY and QFL supervised the experiments and assisted in writing and editing the manuscript. JFC conceived and designed the experiments, then revised the manuscript. All authors read and approved the final manuscript.

Competing interests

The authors declare that they have no competing interests.

References

1. Dharmasiri N, Dharmasiri S, Estelle M. The F-box protein TIR1 is an auxin receptor. Nature. 2005;435(7041):441–5.
2. Kepinski S, Leyser O. The Arabidopsis F-box protein TIR1 is an auxin receptor. Nature. 2005;435(7041):446–51.
3. Wang R, Estelle M. Diversity and specificity: auxin perception and signaling through the TIR1/AFB pathway. Curr Opin Plant Biol. 2014;21:51–8.
4. Korasick DA, Jez JM, Strader LC. Refining the nuclear auxin response pathway through structural biology. Curr Opin Plant Biol. 2015;27:22–8.
5. Salehin M, Bagchi R, Estelle M. SCFTIR1/AFB-based auxin perception: mechanism and role in plant growth and development. Plant Cell. 2015; 27(1):9–19.
6. Yu H, Zhang Y, Moss BL, Bargmann BOR, Wang RH, Prigge M, Nemhauser JL, Estelle M. Untethering the TIR1 auxin receptor from the SCF complex increases its stability and inhibits auxin response. Nat Plants. 2015;1(3):1–8.
7. Gray WM, Kepinski S, Rouse D, Leyser O, Estelle M. Auxin regulates SCF(TIR1)-dependent degradation of AUX/IAA proteins. Nature. 2001;414(6861):271–6.
8. Dharmasiri N, Estelle M. Auxin signaling and regulated protein degradation. Trends Plant Sci. 2004;9(6):302–8.
9. Mockaitis K, Estelle M. Auxin receptors and plant development: a new signaling paradigm. Annu Rev Cell Dev Bi. 2008;24:55–80.
10. Gagne JM, Downes BP, Shiu SH, Durski AM, Vierstra RD. The F-box subunit of the SCF E3 complex is encoded by a diverse superfamily of genes in Arabidopsis. P Natl Acad Sci USA. 2002;99(17):11519–24.
11. Dharmasiri N, Dharmasiri S, Weijers D, Lechner E, Yamada M, Hobbie L, Ehrismann JS, Jurgens G, Estelle M. Plant development is regulated by a family of auxin receptor F box proteins. Dev Cell. 2005;9(1):109–19.
12. Parry G, Calderon-Villalobos LI, Prigge M, Peret B, Dharmasiri S, Itoh H, Lechner E, Gray WM, Bennett M, Estelle M. Complex regulation of the TIR1/AFB family of auxin receptors. P Natl Acad Sci USA. 2009;106(52):22540–5.
13. Havens KA, Guseman JM, Jang SS, Pierre-Jerome E, Bolten N, Klavins E, Nemhauser JL. A synthetic approach reveals extensive Tunability of auxin signaling. Plant Physiol. 2012;160(1):135–42.
14. Walsh TA, Neal R, Merlo AO, Honma M, Hicks GR, Wolff K, Matsumura W, Davies JP. Mutations in an auxin receptor homolog AFB5 and in SGT1b confer resistance to synthetic picolinate auxins and not to 2,4-dichlorophenoxyacetic acid or indole-3-acetic acid in arabidopsis. Plant Physiol. 2006;142(2):542–52.
15. Prigge MJ, Greenham K, Zhang Y, Santner A, Castillejo C, Mutka AM, O'Malley RC, Ecker JR, Kunkel BN, Estelle M. The Arabidopsis Auxin Receptor F-Box Proteins AFB4 and AFB5 Are Required for Response to the Synthetic Auxin Picloram. G3-Genes Genom Genet. 2016;6(5):1383–90.
16. Ren ZX, Li ZG, Miao Q, Yang YW, Deng W, Hao YW. The auxin receptor homologue in Solanum Lycopersicum stimulates tomato fruit set and leaf morphogenesis. J Exp Bot. 2011;62(8):2815–26.
17. Bian HW, Xie YK, Guo F, Han N, Ma SY, Zeng ZH, Wang JH, Yang YN, Zhu MY. Distinctive expression patterns and roles of the miRNA393/TIR1 homolog module in regulating flag leaf inclination and primary and crown root growth in rice (Oryza Sativa). New Phytol. 2012;196(1):149–61.
18. El-Sharkawy I, Sherif SM, Jones B, Mila I, Kumar PP, Bouzayen M, Jayasankar S. TIR1-like auxin-receptors are involved in the regulation of plum fruit development. J Exp Bot. 2014;65(18):5205–15.
19. El-Sharkawy I, Sherif S, El Kayal W, Jones B, Li Z, Sullivan AJ, Jayasankar S. Overexpression of plum auxin receptor PsITIR1 in tomato alters plant growth, fruit development and fruit shelf-life characteristics. BMC Plant Biol. 2016;16
20. Ren ZX, Wang XM. SITIR1 is involved in crosstalk of phytohormones, regulates auxin-induced root growth and stimulates stenospermocarpic fruit formation in tomato. Plant Sci. 2016;253:13–20.
21. Hendelman A, Buxdorf K, Stav R, Kravchik M, Arazi T. Inhibition of lamina outgrowth following Solanum Lycopersicum AUXIN RESPONSE FACTOR 10 (SIARF10) derepression. Plant Mol Biol. 2012;78(6):561–76.
22. MacAlister CA, Ohashi-Ito K, Bergmann DC. Transcription factor control of asymmetric cell divisions that establish the stomatal lineage. Nature. 2007;445:537–40.
23. Pillitteri LJ, Bogenschutz NL, Torii KU. The bHLH protein, MUTE, controls differentiation of stomata and the hydathode pore in Arabidopsis. Plant Cell Physiol. 2008;49:934–43.
24. Ohto MA, Floyd SK, Fischer RL, Goldberg RB, Harada JJ. Effects of APETALA2 on embryo, endosperm, and seed coat development determine seed size in Arabidopsis. Sex Plant Reprod. 2009;22(4):277–89.
25. Mizzotti C, Mendes MA, Caporali E, Schnittger A, Kater MM, Battaglia R, Colombo L. The MADS box genes SEEDSTICK and ARABIDOPSIS Bsister play a maternal role in fertilization and seed development. Plant J. 2012;70(3):409–20.
26. Auge GA, Perelman S, Crocco CD, et al. Gene expression analysis of light-modulated germination in tomato seeds. New Phytol. 2009;183(2):301–14.
27. Navarro L, Dunoyer P, Jay F, Arnold B, Dharmasiri N, Estelle M, Voinnet O, Jones JDG. A plant miRNA contributes to antibacterial resistance by repressing auxin signaling. Science. 2006;312(5772):436–9.
28. Shu WB, Liu YL, Guo YH, Zhou HJ, Zhang J, Zhao ST, Lu MZ. A Populus TIR1 gene family survey reveals differential expression patterns and responses to 1-naphthaleneacetic acid and stress treatments. Front Plant Sci. 2015;6
29. Li J, Wu Z, Cui L, Zhang TL, Guo QW, Xu J, Jia L, Lou QF, Huang SW, Li ZG, Chen JF. Transcriptome comparison of global distinctive features between

pollination and Parthenocarpic fruit set reveals transcriptional Phytohormone cross-talk in cucumber (Cucumis Sativus L.). Plant Cell Physiol. 2014;55(7):1325–42.

30. Cui L, Zhang T, Li J, Lou Q, Chen J. Cloning and expression analysis of Cs-TIR1/AFB2: the fruit development-related genes of cucumber (*Cucumis sativus* L.). Acta Physiologiae Plantarum. 2013.

31. Huang SW, Li RQ, Zhang ZH, Li L, Gu XF, Fan W, Lucas WJ, Wang XW, Xie BY, Ni PX, Ren YY, Zhu HM, Li J, Lin K, Jin WW, Fei ZJ, Li GC, Staub J, Kilian A, der Vossen EAG v, Wu Y, Guo J, He J, Jia ZQ, Ren Y, Tian G, Lu Y, Ruan J, Qian WB, Wang MW, Huang QF, Li B, Xuan ZL, Cao JJ, Asan WZG, Zhang JB, Cai QL, Bai YQ, Zhao BW, Han YH, Li Y, Li XF, Wang SH, Shi QX, Liu SQ, Cho WK, Kim JY, Xu Y, Heller-Uszynska K, Miao H, Cheng ZC, Zhang SP, Wu J, Yang YH, Kang HX, Li M, Liang HQ, Ren XL, Shi ZB, Wen M, Jian M, Yang HL, Zhang GJ, Yang ZT, Chen R, Liu SF, Li JW, Ma LJ, Liu H, Zhou Y, Zhao J, Fang XD, Li GQ, Fang L, Li YR, Liu DY, Zheng HK, Zhang Y, Qin N, Li Z, Yang GH, Yang S, Bolund L, Kristiansen K, Zheng HC, Li SC, Zhang XQ, Yang HM, Wang J, Sun RF, Zhang BX, Jiang SZ, Wang J, Du YC LSG. The genome of the cucumber, *Cucumis sativus* L. Nat Genet. 2009;41(12):1275–U1229.

32. Bowers JE, Chapman BA, Rong JK, Paterson AH. Unravelling angiosperm genome evolution by phylogenetic analysis of chromosomal duplication events. Nature. 2003;422(6930):433–8.

33. Cui LY, Wall PK, Leebens-Mack JH, Lindsay BG, Soltis DE, Doyle JJ, Soltis PS, Carlson JE, Arumuganathan K, Barakat A, Albert VA, Ma H. dePamphilis CW. Widespread genome duplications throughout the history of flowering plants. Genome Res. 2006;16(6):738–49.

34. Tan X, Calderon-Villalobos LIA, Sharon M, Zheng CX, Robinson CV, Estelle M, Zheng N. Mechanism of auxin perception by the TIR1 ubiquitin ligase. Nature. 2007;446(7136):640–5.

35. Rutitzky M, Ghiglione H, Curá J, Casal J, Yanovsky M. Comparative genomic analysis of light-regulated transcripts in the *Solanaceae*. BMC Genomics. 2009;10:60–73.

36. Vidal EA, Araus V, Lu C, Parry G, Green PJ, Coruzzi GM, Gutierrez RA. Nitrate-responsive miR393/AFB3 regulatory module controls root system architecture in Arabidopsis Thaliana. P Natl Acad Sci USA. 2010;107(9):4477–82.

37. Chen ZH, Bao ML, Sun YZ, Yang YJ, Xu XH, Wang JH, Han N, Bian HW, Zhu MY. Regulation of auxin response by miR393-targeted transport inhibitor response protein 1 is involved in normal development in Arabidopsis. Plant Mol Biol. 2011;77(6):619–29.

38. Li X, Xia KF, Liang Z, Chen KL, Gao CX, Zhang MY. MicroRNA393 is involved in nitrogen-promoted rice tillering through regulation of auxin signal transduction in axillary buds. Sci Rep-Uk. 2016;6

39. Pinweha N, Asvarak T, Viboonjun U, Narangajavana J. Involvement of miR160/miR393 and their targets in cassava responses to anthracnose disease. J Plant Physiol. 2015;174:26–35.

40. Xia KF, Wang R, Ou XJ, Fang ZM, Tian CG, Duan J, Wang YQ, Zhang MY. OsTIR1 and OsAFB2 downregulation via OsmiR393 overexpression leads to more tillers, early flowering and less tolerance to salt and drought in Rice. PLoS One. 2012;7(1):364–73.

41. Chen ZH, Hu LZ, Han N, Hu JQ, Yang YJ, Xiang TH, Zhang XJ, Wang LL. Overexpression of a miR393-resistant form of transport inhibitor response protein 1 (mTIR1) enhances salt tolerance by increased osmoregulation and Na+ exclusion in Arabidopsis Thaliana. Plant Cell Physiol. 2015;56(1):73–83.

5

Association mapping of starch chain length distribution and amylose content in pea (*Pisum sativum* L.) using carbohydrate metabolism candidate genes

Margaret A. Carpenter[1], Martin Shaw[1], Rebecca D. Cooper[1], Tonya J. Frew[1], Ruth C. Butler[1], Sarah R. Murray[1], Leire Moya[1], Clarice J. Coyne[2] and Gail M. Timmerman-Vaughan[1*] ⓘ

Abstract

Background: Although starch consists of large macromolecules composed of glucose units linked by α-1,4-glycosidic linkages with α-1,6-glycosidic branchpoints, variation in starch structural and functional properties is found both within and between species. Interest in starch genetics is based on the importance of starch in food and industrial processes, with the potential of genetics to provide novel starches. The starch metabolic pathway is complex but has been characterized in diverse plant species, including pea.

Results: To understand how allelic variation in the pea starch metabolic pathway affects starch structure and percent amylose, partial sequences of 25 candidate genes were characterized for polymorphisms using a panel of 92 diverse pea lines. Variation in the percent amylose composition of extracted seed starch and (amylopectin) chain length distribution, one measure of starch structure, were characterized for these lines. Association mapping was undertaken to identify polymorphisms associated with the variation in starch chain length distribution and percent amylose, using a mixed linear model that incorporated population structure and kinship. Associations were found for polymorphisms in seven candidate genes plus Mendel's *r* locus (which conditions the round versus wrinkled seed phenotype). The genes with associated polymorphisms are involved in the substrate supply, chain elongation and branching stages of the pea carbohydrate and starch metabolic pathways.

Conclusions: The association of polymorphisms in carbohydrate and starch metabolic genes with variation in amylopectin chain length distribution and percent amylose may help to guide manipulation of pea seed starch structural and functional properties through plant breeding.

Keywords: *Pisum sativum*, Amylopectin, Amylose, Chain length distribution, Association mapping, Candidate genes

Background

The pulses, or grain legumes, are a subset of the legumes (Fabaceae) that accumulate starch as a storage component of the seeds. The pulses include economically important species such as pea (*Pisum sativum* L.), chickpea (*Cicer arietinum* L.), common bean (*Phaseolus vulgaris*) and lentil (*Lens culinaris* Medik.). The composition and nutritional qualities of pulses have been the subject of recent reviews [1, 2]. Pulses are relatively high in protein (16–35% dry weight) and carbohydrates (49–68%) and relatively low in oil (0.5–7%), minerals and vitamins. The carbohydrate composition of pulses includes starch (22–45%), dietary fiber (15–32%) and oligosaccharides (α-galactosides). Pulses also contain phytic acid and phenolic compounds which contribute to their nutritional qualities. Nutritionally, the pulses are characterized by the slow digestibility of their carbohydrate, which gives them relatively low glycaemic index (GI) values for carbohydrate-containing foods [2].

* Correspondence: gail.timmerman-vaughan@plantandfood.co.nz
[1]The New Zealand Institute for Plant & Food Research Limited, PO Box 4704, Christchurch, New Zealand
Full list of author information is available at the end of the article

Starch is a polymer of α-1,4-glucose moieties with occasional α-1,6-glycosidic branches. Amylose (which is primarily a linear molecule of α-1,4-glycosidic linkages with approximately 0.1% α-1,6-glycosidic branches, molecular weight 5×10^5 to 10^6) and amylopectin (which is a branched molecule with approximately 5–6% α-1,6-glycosidic branches, molecular weight 10^7 to 10^8) are the two classes of starch [3, 4]. The branching patterns of amylopectin have been described in terms of A, B and C chains. C chains contain the only reducing end in intact amylopectin and provide the central chain from which B chains branch via α-1,6-glycosidic branches. The A chains are the outer chains, and these branch off the B chains also via α-1,6-glycosidic branches. Within the starch granule, starch molecules pack together to form a semi-crystalline structure that is based on the organization of double helices from amylopectin short chains into crystalline lamellae. Two forms have been described based on x-ray diffraction, the A-form that is found in wild-type maize seeds and the B-form that is found in wild-type potato tubers. The crystalline structure of pea starch has been described as a C-form starch, consisting of both A- and B-types [5].

A major aim of the body of research on starch biochemistry and genetics has been developing the ability to manipulate starch structure [6], thereby influencing functional properties and the subsequent uses for starch in food and industrial applications [7, 8]. Making changes to amylopectin chain length distribution (CLD) is one approach. Research on starch from different botanical sources has indicated that differences in amylopectin CLD influence functional properties such as gelatinization, enthalpy change and pasting [9]. Within individual species, effects of modifying CLD on functional properties have also been shown. In both rice and maize, variation in the *ae* gene (amylose extender, starch branching enzyme IIb, SBEIIb) resulted in altered CLD and gelatinization properties [10, 11]. In another rice example using recombinant inbred lines, Luo et al. [12] compared the effects of *indica* versus *japonica* alleles at six genes on CLD and starch functional properties. They showed that the *SSIIa japonica* allele, compared to the *SSIIa indica* allele, increased short length chain proportions while decreasing intermediate length chain proportions altering viscosity properties (reduced peak viscosity and breakdown) and reducing gelatinization temperatures. The *SBEIIb japonica* allele also increased short length chain abundance, compared to *SBEIIb indica* and had an effect on pasting and thermal properties, although the effects were less marked than for the *SSIIa* alleles. In pea, different field pea cultivars have been shown to produce seed starch with different structural attributes, leading to variation in functional properties such as swelling power, gelatinization and pasting properties [13, 14].

The effects of six genes in the seed starch biosynthetic pathway on the composition, structure and physicochemical properties of pea starch have been examined based on studies of a range of mutant lines [15, 16]. Genes at the *rb* (ADP-glucose pyrophosphorylase L1 subunit, AGPL1), *rug3* (plastidial phosphoglucomutase, PGMP) and *rug4* (sucrose synthase, SuSy) loci affect substrate availability, while *rug5* (starch synthase II, StSynII), *lam* (granule bound starch synthase I, GBSSI) and *r* (starch branching enzyme I, SBEI) affect starch polymer biosynthesis and starch branching directly. Bogracheva et al. [5] found from x-ray crystallography that the A-type became more prevalent in the *rug3* mutant, and the B-type became more prevalent in the *r*, *lam* and *rug5* mutants. No A-type starch was detected for *r* mutants. The *lam*, *r* and *rug5* mutations also affected the proportion of the starch that was in amylose versus amylopectin. In comparison with wild type (30% amylose), the *lam* mutation reduced the amylose content to 4–10%, while the *r* and *rug5* mutations increased amylose content to 60–75% and 43–52%, respectively. These mutants also affected the gelatinization temperatures obtained for isolated starch with *rb*, *rug3* and *rug4*, resulting in higher peak temperatures than those of wild type. In terms of naturally occurring allelic variation, *r* locus has the greatest effect on pea phenotypes. Broadly speaking, peas are categorized as field peas (with "round" seeds) or process peas (with "wrinkled" seeds), based on whether their genotype is $R\_$ or *rr*, respectively.

In a number of species, particularly in the cereals, *Arabidopsis thaliana* and potato, roles have been established for a number of genes that encode proteins involved in starch biosynthesis and metabolism, and hence contribute to starch molecular structure and physicochemical properties [6, 17, 18]. These include enzymes involved in the pathway from sucrose to the precursor molecule ADP-glucose; in the polymerization of α-1,4-glucosyl chains (starch synthase, StSyn); in the addition of α-1,6-glucosyl linkages to produce branched starch (starch branching enzyme, SBE) leading to amylopectin; and in further debranching to rearrange the amylopectin structure (isoamylase, ISA and pullulanase, PUL). Studies of mutant lines in rice and Arabidopsis [19], as well as in vitro studies [20–22], have shown the importance of starch synthases, branching enzymes and isoamylases in determining the branch point and chain length distributions of starch. In their recent review, Sonnewald and Kossman [18] identified 46 *Arabidopsis* genes related to starch metabolism.

Association mapping studies in other plants have also demonstrated association between the properties of starch from sink tissues and allelic variation in starch pathway genes, including genes from the substrate supply, chain elongation, branching and debranching portions of the pathway. In *indica* and *japonica* rice (*Oryza sativa*), association analysis using candidate genes showed that variation in three physicochemical properties (amylose content, gel consistency, and gelatinization temperature) related to eating and cooking quality was associated with allelic variation in starch pathway genes, primarily with the *Waxy* (*Wx*, *GBSSII*) and starch synthase II-3 (*SSII-3*) genes [23]. Genes with minor effects were also identified, involved in substrate availability (two AGPase large subunit isoform genes), polymerization (*SSI*, *SSII*, *SSIII* and *SSIV-2*), branching (*SBE3*) and rearrangement and cleavage of starch branches (*PUL* and *ISA*). These authors confirmed the roles of *SBE3* and *SSII-3* through transgenic studies, showing that either repression (SBE3) or increased expression (*SSII-3*) affected all three properties. In a maize (*Zea mays*) study [24], the effect of allelic variation in six candidate genes known to control starch content identified three that affected amylose content and/or pasting properties. Two of these genes were involved in substrate supply (*sh1*, the major sucrose synthase gene; and *sh2*, the AGPase large subunit gene), with the third involved in amylopectin production (*ae1*, *SBEIIb*). In a more recent study that took a nested association mapping approach [25], associations were detected involving genes encoding sucrose synthase, β-amylase and α-amylase, but associations were not identified for candidates expected to play a major role in determining kernel composition (i.e. *Wx*, *su1* (an isoamylase-type debranching enzyme) and *sh2*). In sorghum, polymorphisms in three candidate genes involved in starch synthesis and amylopectin production (*SSIIa*, *GBSS* and *SBE*) were found to be associated with starch physicochemical properties, including gelatinization temperature and viscosity parameters [26].

Fluorophore-assisted carbohydrate electrophoresis (FACE) is an analytical method for characterising the CLD of starch (summarized in [27]), and the method for starch molecular characterization that has been used in this study. Briefly, to determine CLD using FACE, isolated starch is debranched using either ISA or PUL enzymes, the reducing ends of the debranched starch oligosaccharides are labelled with a charged fluorophore such as 8-aminonaphthalene-1,3,6-pyrenetrisulfonic acid (APTS), the labelled oligosaccharides are separated on a high resolution platform such as capillary electrophoresis with fluorescence detection, and the resulting chromatograms are analysed to provide peak areas as quantitative measures of the relative abundance of oligosaccharides with different numbers of monosaccharide units (degrees of polymerization, DP). Other methods used to characterize starch molecular structure offer greater instrumental and technical difficulties because of the requirement for the starch to be fully dissolved,

which requires use of chaotropic agents such as DMSO and LiCl. These methods include nuclear magnetic resonance spectroscopy (NMR) to estimate the ratio of α-1,4 to α-1,6-glycosidic linkages and size exclusion chromatography to estimate molecular weight, radius and size based on various detection methods [8].

Quantitative trait locus (QTL) and association mapping are two approaches for understanding the genetic basis of trait variation, which may result in the identification of molecular markers or sequence polymorphisms that are linked to a causal mutation influencing phenotype. QTL mapping is limited to family-based populations developed by crossing a limited number of parental lines, while association mapping is based on germplasm panels of diverse individual lines. Association mapping panels offer the potential for higher mapping resolution than is obtained from family-based studies resulting in reduced linkage disequilibrium (LD) between trait and marker loci [28]. Association mapping also has the potential to capture a wider range of alleles because of the greater genetic diversity in the germplasm panel. False positives can arise from association mapping in situations where there is undetected population structure or relatedness among the lines in an association mapping panel, resulting in marker-trait association that is the product of population evolutionary processes rather than linkage. Statistical approaches have been developed to identify population structure and relatedness [29, 30] based on genotypes of the contributing lines obtained using a panel of random background molecular markers, permitting the confounding effects of population structure or relatedness on association mapping to be minimized or removed through inclusion in appropriate models. Approaches that have been devised to identify the relatedness of lines in an association mapping panel [28] include structured association using the Bayesian model-based STRUCTURE program producing a Q matrix [30], Principal Components Analysis (PCA) producing a P matrix [29] and estimation of relatedness using a kinship (K) matrix [31].

In this study, a candidate gene association mapping approach has been taken to explore the genetic determination of the variation in CLD in pea. The candidate gene approach has limitations, especially the obvious one that associations will only be identified if the allelic variation contributes to trait variation that is greater than the statistical power of the experiment for detecting an effect. Associations that are detected between a trait and candidate gene polymorphism may reveal the causal variant(s) or the variant(s) that are in LD with the trait genetic determinant(s). There is substantial information available about pea starch synthesis pathways, reinforced by knowledge of starch synthesis in other plant species, facilitating the choice of candidate genes.

This study explores the range of variation that exists in a diverse collection of pea germplasm with respect to starch candidate gene allelic variation, amylose composition of extracted starch, and debranched starch CLD, one measure of the structural properties of pea starch. We describe the allelic variation in 25 pea candidate gene sequences that have been shown to be involved in carbohydrate and starch metabolism, or that are orthologues of sequences that have been shown in other species to be involved in carbohydrate and starch metabolism. Using an association mapping approach, allelic variants have been identified that show a significant association with variation in CLD or percent amylose (%amylose).

Methods
Plant material
Accessions for association mapping were chosen from among the USDA-ARS Refined Pisum Core (https://npgsweb.ars-grin.gov/gringlobal/method.aspx?id=492806) [32]. The 92 Pea Single Plant (PSP) accessions that were used for association mapping are listed in Additional file 1. The accessions were developed by selecting seeds from a single plant and were deposited with the USDA-ARS for inclusion in their PSP collection (https://npgsweb.ars-grin.-gov/gringlobal/method.aspx?id=494267). All the PSP accessions are from the USDA Western Regional Plant Introduction Station and are freely available from the USDA (https://npgsweb.ars-grin.gov/gringlobal/search.-aspx) under the Standard Material Transfer Agreement under the International Treaty on Plant Genetic Resources for Food and Agriculture, Convention on Biological Diversity. In addition, pea cultivars 'Sonata' (Dave Goulden, Plant & Food Research, Christchurch, New Zealand), 'Primo' (Cebeco, Lelystadt, The Netherlands) and breeding line SuperGreen (courtesy of Adrian Russell, Plant Research (NZ) Ltd., Lincoln, New Zealand) were grown as standards; and OSU442–15 (442–15) [33] was grown as a check line.

For starch extractions, peas were grown in two trials, a glasshouse trial held in 2010–2011 (GH2010) and a field trial held over the New Zealand summer of 2011–2012 (Field2011). No specific permissions were required for these trials which were conducted in accordance with local and national regulations. The GH2010 trial included 113 PSP lines, 'Primo', 'Sonata' and SuperGreen, grown two plants/pot in two pots; and 442–15 as the check, grown two plants/pot in eight pots. The trial consisted of two arrays of six tables with 20 pots/table, a total of 240 pots. Each array contained a complete replicate of lines, and these were positioned using a design derived from a block design, with blocks of 20 pots (one table). The trial was also designed to facilitate analysis of any trends or variation that may have resulted from laboratory processes such as starch extraction and the FACE analyses. The GH2010 trial design was generated

with CycDesigN (http://www.vsni.co.uk/software/cycdesign) using a randomized resolvable block design with blocks of six, randomized across replicates. Pots were sown in September 2010 and harvested in January 2011, over the New Zealand spring/summer period. Plants were grown in a glasshouse under natural light until mid-November 2010, when very light shade cloth (30–35% shade) was placed beneath the glasshouse roof to help to regulate glasshouse temperature. Plants were grown in a bark-sand mix containing slow release fertilizer (Osmocote® Exact Standard) and were side-dressed with Nitrophoska Blue Special (Ravensdown Fertilisers, NZ) twice. The Field2011 trial was carried out near Lincoln, Canterbury, New Zealand at approximately E172° 28′ x S43° 37′ and included 112 PSP lines, 'Primo', 'Sonata' and SuperGreen with 442–15 grown as the check. Two replicate plots (10 seeds per plot) of the 114 test lines, two replicates each of 'Primo' and 'Sonata', four replicates of SuperGreen, and 34 plots of 442–15 were laid out in two adjacent blocks of 19 by 7 plots (a total of 266 plots), separated by a tractor track. Each block contained a complete replicate of the PSP lines plus one 'Sonata', one 'Primo', two SuperGreen, and 17,442–15 lines. The positions of the lines were determined using DiGGer experimental design software (http://nswdpibiom.org/austatgen/software/), with blocks of 19 × 7, 5 × 14, 19 × 1 and 1 × 14 with no autocorrelation. Plots were single row plots of 1.2 × 1 m. Standard cultural practices for pea were practiced and irrigation was applied to avoid water deficit.

Candidate gene sequence selection
Twenty-five candidate genes (Table 1) from carbohydrate and starch metabolic pathways were selected based on a published pea starch biosynthetic pathway [15] and on the KEGG (www.genome.jp/kegg/pathway.html) Starch and Sucrose Metabolism pathway for *Arabidopsis thaliana*. For 13 of the genes, *P. sativum* cDNA sequences were directly available from GenBank. For nine of the remaining 12 candidate genes, the pea homologs of genes from other plant species were found as follows. First the relevant genes from *A. thaliana* were identified based on the KEGG Starch and Sucrose Metabolism pathway. Then the tBLASTx algorithm [34] was used to identify homologous pea sequences from among a database of 13,336 cDNA sequences obtained in our laboratory from Roche 454 sequencing of non-normalized cDNA libraries from developing pea seeds (cultivar 'Primo'; 5, 15 and 25 days after pollination) (Genbank BioProject PRJNA288408). Three additional candidate gene sequences (phosphoglucan water dikinase, PWD; chloroplastic pullulanase 1, PUL1; and invertase inhibitor, InvInh) that were not previously characterized in pea, and also were not represented in the KEGG *A.*

Table 1 Pea carbohydrate metabolism candidate genes, primer sequences for fragment amplification and fragment characterisation

Gene, EC number, (GenBank accession, locus), Genus and species	Primer(s)	Primer sequences (5' to 3' direction)	PopSet alignment[a] (location on accession)	Alignment length (bp)
Hexokinase, EC 2.7.1.1 (XM_003630659), Medicago truncatula	Ps_2048 (S), Ps_2049	F: CGGTTTACGTTCTCGTTCC R: ATCTGCCTCCAGCCAATGT	694,184,588 (1003–1079)	291
Hexokinase, EC 2.7.1.1 (XM_003630659), M. truncatula	Ps_2050 (S), Ps_2051	F: AAGCGGAGTTTTCGGAGAT R: ACCGCGATAAGCAACAATG	694,183,180 (1446–1545)	244
Phosphoglucomutase (plastidial), EC 5.4.2.2 (AJ250770, rug3), Pisum sativum	Ps_1321 (S), Ps_1322 (S)	F: GTCAACGCCAGCCGTTTC R: GGGTGTTTCCGTAAATCTTGTC	694,186,924 (566–636, 637–658)	471
Phosphoglucomutase (plastidial), EC 5.4.2.2 (AJ250770, rug3), P. sativum	Ps_1325, Ps_1328 (S)	F: AGGGTCTTGCACGATCAATG R: GGCTTCTCTCCCTGTGAA	694,182,794 (1651–1674, 1675–1752, 1753–1834)	374
Sucrose synthase, EC 2.4.1.13 (AJ012080, rug4), P. sativum	Ps_0685 (S), Ps_0689	F: TGACTGATGGTGCATTTGGT R: CGTTGGCCACAAGTAGTTCC	694,182,614 (376–401, 402–594)	378
Second sucrose synthase, EC 2.4.1.13, (AJ001071, Sus2), P. sativum	Ps_0076 (S), Ps_0079	F: ATATGTTGCTCAGGGGAAAGG R: ATTAACACGGACATACTCCAAAC	694,184,180 (297–337)	374
Sucrose phosphatase, EC 3.1.3.24 (AY651774), M. sativa	Mt_0208 (S), Mt_0211	F: GAACCAGAAATGGGACAAGG R: TGCCACTGCAGTAATTCCTCT	694,182,434 (289–373)	235
Sucrose phosphate synthase, EC 2.4.1.14 (Z56278), Vicia faba	Ps_1581, Ps_1583 (S)	F: ACAGGAAATAGAAGAACAGTGGCGCT R: AGGACGGCATTCTCCAAACGT	694,186,708 (1292–1409, 1410–1442)	569
Cell wall invertase, EC 3.2.1.26 (AF063246, bfruct1), P. sativum	Ps_0276 (S), Ps_0280	F: TGATCCTCAACTTCTGTGTAGTC R: TGCTAATGTAGGATAAACTCTGG	694,183,574 (1466–1560, 1561–1602)	303
Invertase inhibitor, putative (XM_004508064), Cicer arietinum	Ps_2042 (S), Ps_2043	F: TTAAATGAACCCCACACAGA R: TCCAGAAGCACTTTCCCATC	694,182,042 (162–556)	411
ADP glucose pyrophosphorylase L1, EC 2.7.7.27 (X96766, rb), P. sativum	Ps_0036 (S), Ps_0039	F: GGGAGCTGACTATTACCAAACTGA R: CTTGATACCTTCAGCACTCAACC	694,185,424 (1507–1567, 1568–1730)	374
ADP glucose pyrophosphorylase S2, EC 2.7.7.27 (X96765), P. sativum	Ps_0057 (S), Ps_0065 (S)	F: GGCTACTGGGAAGACATTGGTA R: GATTCTCGCGTTCTTGTCAAC	694,187,965 (1149–1269, 1270–1368)	559
UDP-glucose pyrophosphorylase, EC 2.7.7.9 (AF435969), Amorpha fruticosa	Ps_1557 (S), Ps_1562	F: AGTTGGAAATTCCTGATGGAGCCGT R: AAGAAGACAACCAGCAAGGCCTCA	694,187,311 (1529–1610)	201
UDP-glucose pyrophosphorylase, EC 2.7.7.9 (XM_003616133), M. truncatula	Ps_1556 (S), Ps_1561	F: CCGCTACCGCTACCAACCTCG R: GCAACCCATAGTTGTCCCAAGCC	694,186,492 (254–303, 304–397)	507
α-1,4-glucan phosphorylase L, EC 2.4.1.1 (Z36880), V. faba	Ps_1495 (S), Ps_1499	F: AGCTGTTGCACGATGTCCCC R: GCTCTGGGATGCACAAAGTTGGGT	694,186,062 (994–1095, 1096–1181)	491
Starch synthase II, EC 2.4.1.21 (X88790, rug5), P. sativum	Ps_1315 (S), Ps_1317 (S)	F: ACAGCATTCCTGGATTGGAA R: TTGCGAAATATTGGACTGTCA	1,206,484,033 (587–969)	521
Starch synthase II, EC 2.4.1.21 (X88790, rug5), P. sativum	Ps_1320 (S), Ps_1319 (S)	F: TTATCGCGATCATGGTTTGA R: TTGGTATTTGGCAGCAACAA	694,183,776 (1809–2299)	491
Granule bound starch synthase, EC 2.4.1.21 (X88789, lam), P. sativum	Ps_0251, Ps_0255 (S)	F: GGGTAGAAACGCCTTTTCAG R: CCTCCAGTACCTCGATTTGC	694,186,276 (1067–1191, 1192–1369)	389
Granule bound starch synthase Ib, EC 2.4.1.21 (AJ345045), P. sativum	Ps_0499 (S), Ps_0503	F: AGAAAAGTCCGCTTCTTCCA R: TTGGTCAGGGAGATTGAGAAG	694,187,140 (534–626, 627–690, 691–756)	546
Starch branching enzyme II, EC 2.4.1.18 (X80010), P. sativum	Ps_2070 (S), Ps_2071	F: AGATTTTGCTGCTCCCTACGA	694,185,214	244

Table 1 Pea carbohydrate metabolism candidate genes, primer sequences for fragment amplification and fragment characterisation (Continued)

		R: AACTTTGGCCCACATCAAAG	(709–844)	
Isoamylase, isoform 1, EC 3.2.1.68 (DQ092413), P. sativum	Ps_1512 (S), Ps_1516	F: AGGGGGAGTTTGTCAGTGCCTCA R: AGACCATGCCACTGCAGCCT	694,187,747 (1906–1959, 1960–2040, 2041–2105)	441
Isoamylase, isoform 2, EC 3.2.1.68 (DQ092414), P. sativum	Ps_0155, Ps_0152 (S)	F: GATCCTTATGTCAATAGGTCAGGTG R: CCTGAGGCTATCCAAAATCAAA	N/A [b] 1056–1135	142
Isoamylase, isoform 3, EC 3.2.1.68 (DQ092415), P. sativum	Ps_1479 (S), Ps_1483	F: TGCTTCCCACACCCCAACA R: TCGTAGGACCACTCTCAAGTAGAGCTT	694,184,796 (2220–2509)	511
Pullulanase 1, chloroplastic-like, EC 3.2.1.41 (XM_004496070), C. arietinum	Ps_1471, Ps_1472 (S)	F: TGGTGGGACACCCGTTGCTT R: TCCTGCATCTCAGCTACACCGA	694,182,256 (2170–2226)	291
Pullulanase 1, chloroplastic-like, EC 3.2.1.41 (XM_004496070), C. arietinum	Ps_1469 (S), Ps_1473	F: ACACTGGACCATCGTTGGCTTATGG R: GCACTCGCATCAGATTTTCCTTGGC	694,184,384 (2657–2717, 2718–2817, 2818–2846)	360
Beta-amylase 1-like, EC 3.2.1.2 (XM_004503530), C. arietinum	Ps_1518 (S), Ps_1521	F: CTGTGCTGCGTGGGCGTTCT R: TGGCATGTTCCAAGAGCCACCC	694,185,636 (762–810, 811–909, 910–996)	481
Beta-amylase-like, EC 3.2.1.2 (XM_003593956), M. truncatula	Ps_1523 (S), Ps_1525	F: GCTGTTCATGCTGAACCGATCAGAG R: TCTTTGTAAACACTGTCCCGACCGA	694,183,376 (436–717)	501
Beta-amylase-like, EC 3.2.1.2 (XM_003593956), M. truncatula	Ps_1475 (S), Ps_1476	F: GCGGTCCACACGATGTGCCT R: TTCATGTCTTACACTGCTTGCATGCTC	694,185,848 (1179–1699)	521
Beta-amylase like, EC 3.2.1.2 (XM_004513491), C. arietinum	Ps_1602 (S), Ps_1606	F: TGCTGCTGAACTCACTGCTGGA R: TGAATCCCAAGGGAACGGCACT	694,182,984 (1347–1596)	481
4-α glucanotransferase, EC 2.4.1.25 (XM_003602434), M. truncatula	Ps_1596 (S), Ps_1601	F: TGGGTTTGGAGGTGGTCCCG R: TTGAGCAACGGAAGCCAGCG	694,185,004 (1547–1611, 1612–1655)	311
Phosphoglucan water dikinase, chloroplastic-like, EC 2.7.9.5 (XM_004497365), C. arietinum	Ps_1575 (S), Ps_1578	F: GCTCTTCAACCCTTGCCGCTCA R: GCATGCCTATTGGGACGGTGGT	694,183,978 (912–956, 957–1040)	268
Phosphoglucan water dikinase, chloroplastic-like, EC 2.7.9.5 (XM_004497365), C. arietinum	Ps_1573 (S), Ps_1579	F: TCCAGCGCCAATGTGGAGGA R: AGGTTGGGGCCTTGTCTGAACA	694,187,529 (3119–3629)	511

[a] The GenBank PopSet alignment number

[b] Not applicable, the sequence is less than 200 bp therefore not accepted by GenBank

For each candidate gene studied, the EC number of the encoded enzyme and GenBank accession for the most similar mRNA sequence are indicated. The species related to that GenBank accession is also indicated. Genbank accessions were accessed on 9 August 2016. The primers that were used for resequencing are indicated (S). GenBank PopSet numbers for the pea candidate gene sequence fragment alignments are provided, along with alignment lengths *thaliana* Sucrose and Starch Metabolism pathway diagram, were identified from the literature on starch metabolism in other plant species and then pea homologs were found among our cDNA sequence database using tBLASTx. Searches of the in-house database were carried out using Geneious Pro version 5.5.6 created by BioMatters (www.geneious.com).

Candidate gene resequencing and sequence analysis

Primer pairs for the pea candidate gene sequences were designed using Primer3 software (http://frodo.wi.mit.edu/primer3/). The default parameters for primer design were an optimal Tm of 60 °C and an optimal length of 20 nucleotides. Two to three sets of primers were designed for each sequence. When designing primers, attempts were made to place the primer binding sites so that they flanked intron sequences to improve the chances of identifying polymorphic sites. The possible locations of introns in the pea genomic sequences for the candidate genes were estimated by aligning the pea candidate gene cDNA sequences with pea or other legume genomic sequences using the BLASTx algorithm, or if legume genomic sequences were not available, then *A. thaliana* genomic sequences were used.

Total DNA was extracted from young leaves of pea lines as described by Timmerman et al. [35]. For a minority of the PSP lines, extracted DNA did not reliably amplify during PCR. For those lines, whole genome amplification was carried out to circumvent the problem, using the illustra GenomiPhi V2 DNA amplification kit (GE Healthcare Life Sciences) following the manufacturer's instructions. The primer sequences used to identify polymorphisms are listed in Table 1. In a standard reaction, genomic DNA fragments were amplified in 15 μl containing 1× PCR buffer (various suppliers), 200 μM of each dNTP, 200 nM of each PCR primer, 0.3 U of Taq polymerase (various suppliers) and approximately 20 ng of total or genome amplified DNA. Mg^{2+} concentrations in PCR reactions were optimized where necessary.

PCR products were treated with exonuclease I and either shrimp or rAPiD alkaline phosphatase (Roche) [36] and then sequenced using BigDye ver. 3.1 (Applied Biosystems) and an ABI3130 Genetic Analyzer (Applied Biosystems). PCR primers used to prime the sequencing reactions are indicated in Table 1. Bases were called using either SeqScape ver. 2.1 (Applied Biosystems), ABI Sequence Analysis Software version 5.3, or Geneious Pro version 5.5.6 software. Alignments were constructed using ClustalX version 2.1 [37]. Polymorphisms were confirmed by visual inspection. Linkage disequilibrium among polymorphisms was calculated and plotted in R using the 'genetics' (ver 1.3.8.1) and 'Ldheatmap' packages (ver 0.99–1) [38]. For the nucleotide polymorphisms associated with CLD variation and which would generate amino acid substitutions, estimation

of the likelihood that a variant might have an effect on protein biological function was carried out using the Protein Variant Effect Analyzer (PROVEAN) ver. 1.1 [39]. Sequences are lodged as population set (PopSet) alignments (Table 1) with GenBank (sequences KM360195-KM360301, KM510517-KM513542 and KY983278-KY983354).

Starch extraction

Starch was extracted from the GH2010 and Field2011 samples based on the method described by Takeda et al. [40]. A subsample of dried pea seed (5 g) harvested from each pot or plot was soaked in 30 ml 0.2% NaOH for 2 days at 4 °C. An additional 40 ml of 0.2% NaOH was added and the soaked peas were blended for 3 × 20 s bursts using a household stick blender (200 W) then centrifuged at 3200 g for 20 min. The pellet was resuspended in 20 ml of 0.2% NaOH then sieved through three sieves with mesh sizes of 420 μm, 100 μm and 75 μm, with additional 0.2% NaOH used to ensure a good starch recovery, up to a maximum volume of 45 ml. Starch was pelleted by centrifugation of the filtrate at 3200 g for 20 min. The supernatant was discarded and any layer of non-starch material on the top of the starch pellet was removed with gentle scraping using a stainless steel spatula. The resulting starch pellet was resuspended in 40 ml water and 0.5 ml of 0.5 M MOPS to neutralize the suspension then centrifuged at 3200 g for 30 min, the supernatant discarded and any non-starch layer above the starch pellet removed by gentle scraping. The pellet was then washed three times with 40 ml water with centrifugation at 3200 g for 10 min. The final starch pellet was dried at 37 °C for 24 h, broken up with a mortar and pestle, sieved and stored at ambient temperature.

Fluorophore-assisted carbohydrate electrophoresis and data analysis

The CLD of debranched starch from each sample was estimated using FACE as described by Murray et al. [27]. For debranching and labelling of the extracted pea starch, the scaled-down protocol was followed. Debranching was carried out using 10 ± 0.5 mg starch. Labelling was carried out on a subsample of 103 μg of debranched starch per labelling reaction using 8-amino-1,3,6-pyrenetrisulfonic acid (APTS) fluorophore at 20 μg/μl. The FACE labelling reactions were diluted with water before electrophoresis to ensure that the fluorescence signal of the tallest peak fell between 7500 and 2000 relative fluorescent units (RFU) when electrophoresed on an ABI3130 Genetic Analyzer (Applied Biosystems). All analysis was done using GeneMarker software versions 1.85 and 2.2 (SoftGenetics, State College, PA, USA; www.softgenetics.com) as described by Murray et al. [27]. For each sample, degrees of polymerization (DP) between 6 and 40 were considered. Peak areas for these DPs were exported from GeneMarker, and then converted to relative abundances, expressed as molar proportions of the total peak area for that sample. For each sample, the sum of the peak area molar proportions was equal to 1.

CLD data were analysed as described by Murray et al. [27]. Briefly, a standard Poisson log-linear model for the analysis of contingency Tables [41] was used to analyse the table of samples by DP molar proportions. This approach was taken since numbers of fluorescently labelled starch chains of each DP underlie the molar proportions. To adjust for the data being proportions rather than counts, the dispersion was estimated rather than fixed at 1 (the expected value for counts). The main aim of this analysis was to explore whether there were any substantial differences in the DP distributions between lines. These effects were assessed with F-tests within the analyses of deviance that were done. Correspondence analysis [42] was used to explore patterns in the contingency tables. The results are presented as asymmetric biplots, with DP as standard coordinates where the plot is a projection of DP when treated as axes in multidimensional space (26 dimensions). Analyses were carried out in GenStat 14th edition [43].

Percent amylose estimation and data analysis

The amylose content of the pea starch samples was determined using an iodine binding assay optimized for measurement in a 96 well plate [44], with the following modifications. Pea starch samples, in 50 ml Falcon tubes, were placed in a Labconco Centrivap Concentrator (Kansas City, MO 64132, USA) and re-dried under vacuum for 2 h at ambient temperature to remove any residual moisture, before a 5 mg subsample was weighed for analysis of %amylose. The starch was dispersed in 1 ml of 90% DMSO in water by heating to 95 °C for 60 min with vortexing every 10 min, then cooled for 5 min. A 100 µl aliquot from each sample was pipetted into a 0.5 ml microfuge tube, 100 µl of I_2 solution (3.04 g I_2/L in 90% DMSO) was added and the tube vortexed for 30 s. The tubes were incubated at room temperature for 30 min before 20 µl was aliquoted in quadruplicate into a 96 well clear, flat bottomed polystyrene plate (Greiner Bio-One 655,101) and 180 µl of deionized water added to each well. The plate was shaken for 30 s before reading the absorbance. A set of ten standards containing 0, 5, 10, 15, 20, 25, 30, 50, 75 and 100% amylose was also added to each plate in triplicate. The standards were made up using amylose, Type III from potato, and amylopectin from potato (A0512 and A8515, respectively, Sigma-Aldrich, St Louis, MO, USA). Absorbance of each well was measured at 620

and 510 nm in a SpectraMax M2 platereader using SoftMax® Pro 5 software.

Calibration curves for each plate were developed by plotting the $Abs_{620nm} - Abs_{510nm}$ versus the % amylose of the calibration standards. Exploration of these curves showed clearly that the relationships were non-linear, justifying the inclusion of the quadratic term in the calibration regressions, where.

$Abs_{620nm} - Abs_{510nm} = c + b(\%amylose) + a(\%amylose^2)$.

(a = quadratic term, b = linear term, c = constant term). The resulting calibration curves all fitted very well ($R^2 > 99.65$) with the lack of fit between the means for the standards and each plate and the fitted curve being minor (<0.1% of the total variation across all plates). Parameters a, b and c were moderately variable between the plates with % CVs (standard deviation as % of mean) of 10.9, 6.5 and 9.0%, respectively. The parameters of the above equation were used to convert the $Abs_{620nm} - Abs_{510nm}$ (AbsDiff) into estimated %amylose for each well using the following equation:

$$Estimated\%Amylose = \frac{-b + \sqrt{b^2 - 4a(c - AbsDiff)}}{2a}$$

To obtain pot (GH2010) or plot (Field2011) means, data from the two trials were analysed separately. The estimated % amylose for each well was analysed using methods appropriate for percentage data. Since there were up to four wells per pot or plot (some odd data were excluded), starch extractions from replicate pots or plots for most lines, potential spatial effects from the trials, and effects relating to the plates not corrected for by the calibration regressions, an initial analysis was carried out to assess the importance of each of these. A hierarchical generalized linear modelling approach [45] was used. In this, fixed effects (lines; round vs wrinkled seeded) were fitted with a binomial distribution with a logit link and dispersion estimated. Random effects (pot (GH2010) / plots (Field2011); plates; other spatial factors) were fitted as random effects with a beta distribution and logit links. The random effects were assessed with a X^2-test of the change in likelihood on dropping a term, as implemented in GenStat's HGRTEST procedure [46]. Only important random terms were included in the final analysis. Fixed effects were assessed similarly to random effects, using GenStat's HGFTEST procedure. Mean %amylose values were obtained as predictions on the link (logit) scale, and back-transformed for presentation. All data manipulation and analyses were carried out with GenStat [47].

Association mapping analysis

Association mapping was carried out using 92 PSP lines (Additional file 1). Population structure was estimated on the basis of polymorphisms at 55 background markers, consisting of 13 SCAR markers, 12 SSR markers and 30 RAPD markers. These markers revealed 140 polymorphisms. Two approaches were used. In the first, the Bayesian, model-based approach implemented in STRUCTURE version 2.3.4 software [30] was used to determine the number of sub-populations which best represented the data, based on a no-admixture model and uncorrelated allele frequencies, and then to assign lines to subpopulations. The software was run to test from one to eight subpopulations (K) with five replicates, a burn-in period of 100,000 and then 500,000 replicates. The STRUCTURE analysis was carried out on the full set of 92 PSP lines and on a subset of 83 PSP lines that had the round seed phenotype (RR genotype at the r locus). The most likely number of subpopulations (K) was determined by plotting the Ln probability of the data (Ln P(D)) versus K. The resulting Q matrix for the most likely number of subpopulations was used for association mapping. In the second approach, a PCA of both the $n = 92$ and $n = 83$ sets of PSP lines × 55 background loci datasets was carried out using Genstat 14th ed. [43], producing a P matrix. Kinship (K) matrices were calculated for both the $n = 92$ and $n = 83$ PSP lines datasets, with the results rescaled to between 0 and 2, using the same polymorphism data, using TASSEL version 3.0.165 [31, 48].

Association mapping was carried out using the mixed linear model (MLM) function implemented in TASSEL [48]. To determine the best model to use for association mapping, the MLM + Q + K and MLM + P + K models were analysed using the 92 PSP lines dataset of 280 polymorphisms from 25 candidate genes, phenotypic variation at the r locus (round versus wrinkled seeds), and trait values that included %amylose content and CLD presented as mean peak area proportions from DP6 to DP40 for starch extracted from 92 PSP lines from the GH2010 and Field2011 trials. In the MLM analyses, the marker effects and P or Q matrices were fitted as fixed effects and the kinship matrix (K) and residual were fitted as random effects. Polymorphisms (SNPs and indels) in sequences from 25 starch and carbohydrate metabolic genes were extracted from sequence fragment alignments using TASSEL, with a required minimum minor allele frequency of 5%. For the MLM analysis, variance components were re-estimated after each marker and the compression level was optimized during the analysis. To determine which of the two models gave the best fit, quantile-quantile plots were drawn for the -$\log_{10}$ of the 281 raw p-values obtained for the traits which gave significant marker-trait associations. For the $n = 83$ round lines, association mapping was carried out using the MLM + Q + K model. Q-values [49] were calculated to provide measures of the significance of the association tests obtained from the MLM by estimating the

minimum false discovery rate (FDR) that occurs if that test is called significant. Q-values were calculated using the QVALUE software [49] implemented in R. The false discovery rate (FDR) was set to 0.05 ($n = 92$ lines) or 0.10 ($n = 83$ lines), and the bootstrap method was used to estimate π_0, the overall proportion of true null hypotheses. Characteristics of the distributions for significant associations were visualized using beanplots [50].

Results

Variation in percent amylose and starch chain length distribution

In the GH2010 and Field2011 trials, 116 and 115 pea lines respectively were grown, as well as OSU442–15 which was incorporated as a check line. Of these lines, there were 92 that yielded adequate amounts of seed from both trials for starch extraction, analysis of % amylose and FACE analysis.

From the GH2010 and Field2011 trials, the mean %amylose in starch from the wrinkled (rr) seeded lines was 63.3% (range 59.2 to 69.7) and 62.1% (range 56.7 to 69.3), respectively; while for the round (RR) seeded lines the mean %amylose was 37.2% (range 27.6 to 44.1) and 38.2% (range 34.5 to 44.6), respectively. From the GH2010 trial, most of the differences between lines were related to pea type (round versus wrinkled seeded, $p < 0.001$) with little difference between the round lines ($p = 0.821$) and relatively more between the wrinkled lines ($p = 0.038$). From the Field2011 trial, most of the major differences were again associated with the pea type ($p < 0.001$) but there were more notable differences between lines within the two types ($p = 0.014$ for round lines; $p < 0.001$ for wrinkled lines). The greater significance of the differences in terms of p-values for the %amylose values for lines from the Field2011 trial occurred because there was more consistency between replicate plots of the Field2011 trial than there had been for replicate pots of the GH2010 trial. The mean %amylose and associated 95% confidence intervals for each of the 92 lines used for associaton mapping from the GH2010 and Field2011 trials are presented in Additional file 1.

Starch was characterized for CLD by carrying out FACE on debranched starch extracted from pea seeds. Typical CLD profiles for debranched pea starch are shown in Fig. 1. Profiles are shown from both the GH2010 and Field2011 trials for six PSP lines; three each that have round (A and C) and wrinkled (B and D) seed phenotypes. The profiles have been graphed as the mean molar peak area proportion versus DP (A and B); and as difference plots, in which the value plotted is the mean molar peak area proportion for the PSP line under consideration minus the mean total molar peak area proportion for all 92 PSP lines versus DP (C and D). These plots show the general trends associated with r

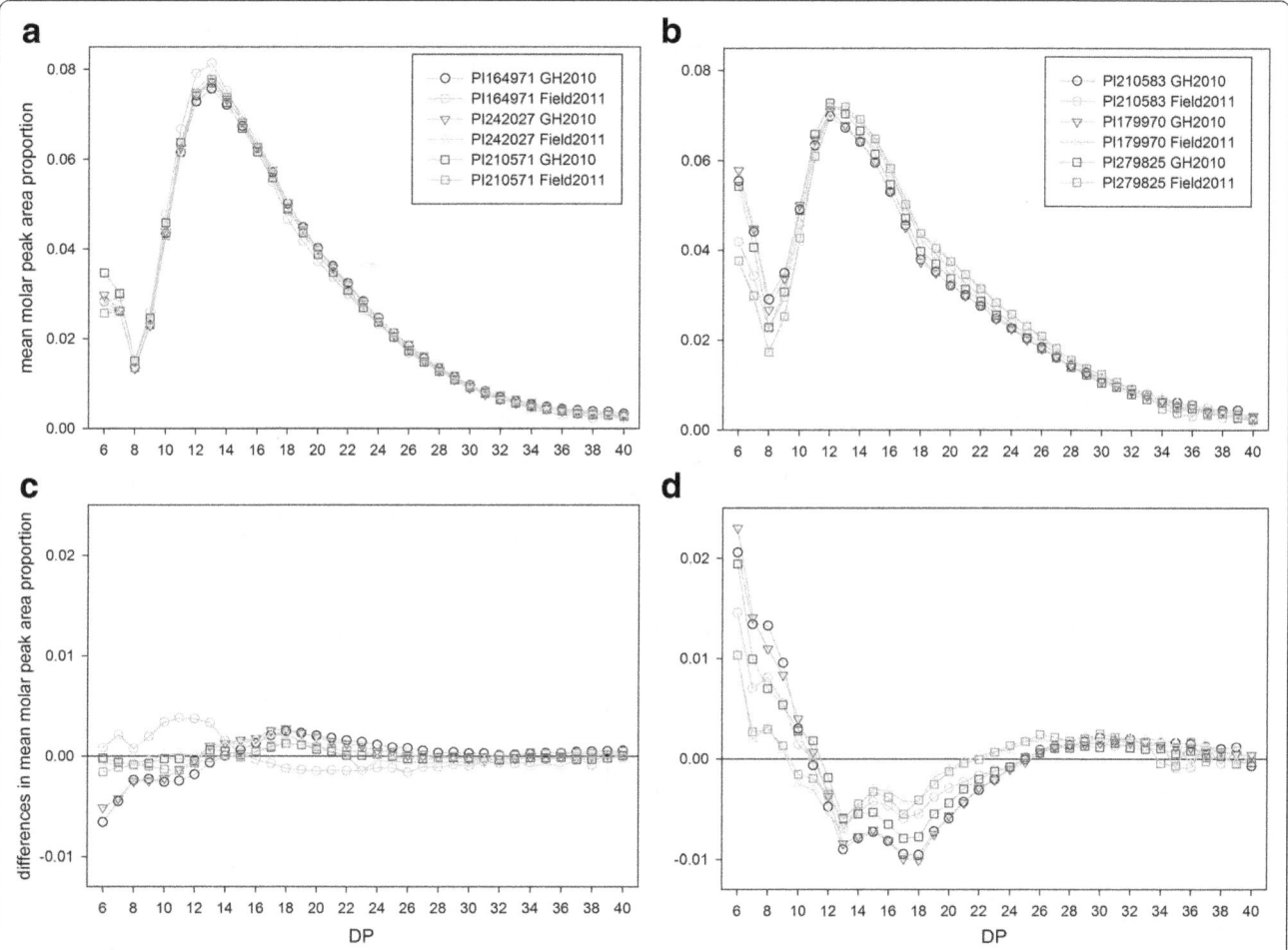

Fig. 1 Mean chain length distributions (CLDs) for debranched pea starch from three round and three wrinkled pea single plant (PSP) lines grown in the GH2010 and Field2011 trials. Profiles showing mean molar peak area proportions for starch from pea lines having: **a** the round (*RR*) and **b** the wrinkled seeded (*rr*) genotype at the *r* locus; and difference plots for **c** starch extracted from round seeded lines and **d** wrinkled seeded lines. Difference plots show each sample's mean molar peak area distribution minus the mean molar peak area distribution for the 92 PSP lines

locus phenotype. In wrinkled seeded lines the relative abundance of DP6 to DP12 chains was greater than in round lines and concomitantly the relative abundance of DP13 to DP24 chains was less in wrinkled than in round seeded lines. Analysis of the CLD resulting from the FACE using a standard Poisson log-linear model for the analysis of contingency tables found that overall the DP distributions varied significantly ($p < 0.001$) between lines.

Correspondence analysis was employed to assist with identifying differences between the CLD distributions obtained from the two trials (Fig. 2). CLD distributions from all the lines that produced sufficient seed for FACE analysis were included in these analyses of each trial (110 PSP lines plus Primo, Sonata, OSU442–15 and Supergreen for GH2010; 112 PSP lines plus Primo, Sonata, OSU442–15 and Supergreen for Field2011). For the GH2010 trial, the first two components accounted for 76% of the CLD variation, with the first component accounting for 52%. Therefore, the two-dimensional plot

shows many of the important patterns in the data. However, 24% of the variation was not associated with the first two components. Nine components were needed to explain 95% of the variation. The greatest differences between the distributions were associated with the highest and lowest DP. The first dimension separated the highest DPs (DP36 to DP39) from lower DP scores (DP10 to DP18) and also separated the wrinkled from the round seeded lines. The second dimension separated the low DP (≤9) from the rest of the distribution and showed some effect of the wrinkled versus round seeded lines. For the Field2011 trial, the first two components accounted for 90% of the variation in the data, with the first component accounting for 76%. Therefore, this two-dimensional plot captures most of the important patterns in the data. For the CLD plot distributions, the greatest differences were associated with the DP values ranging from DP6 to DP18 in dimension 1, and in dimension 2 with the low (DP6 to DP10) to the high

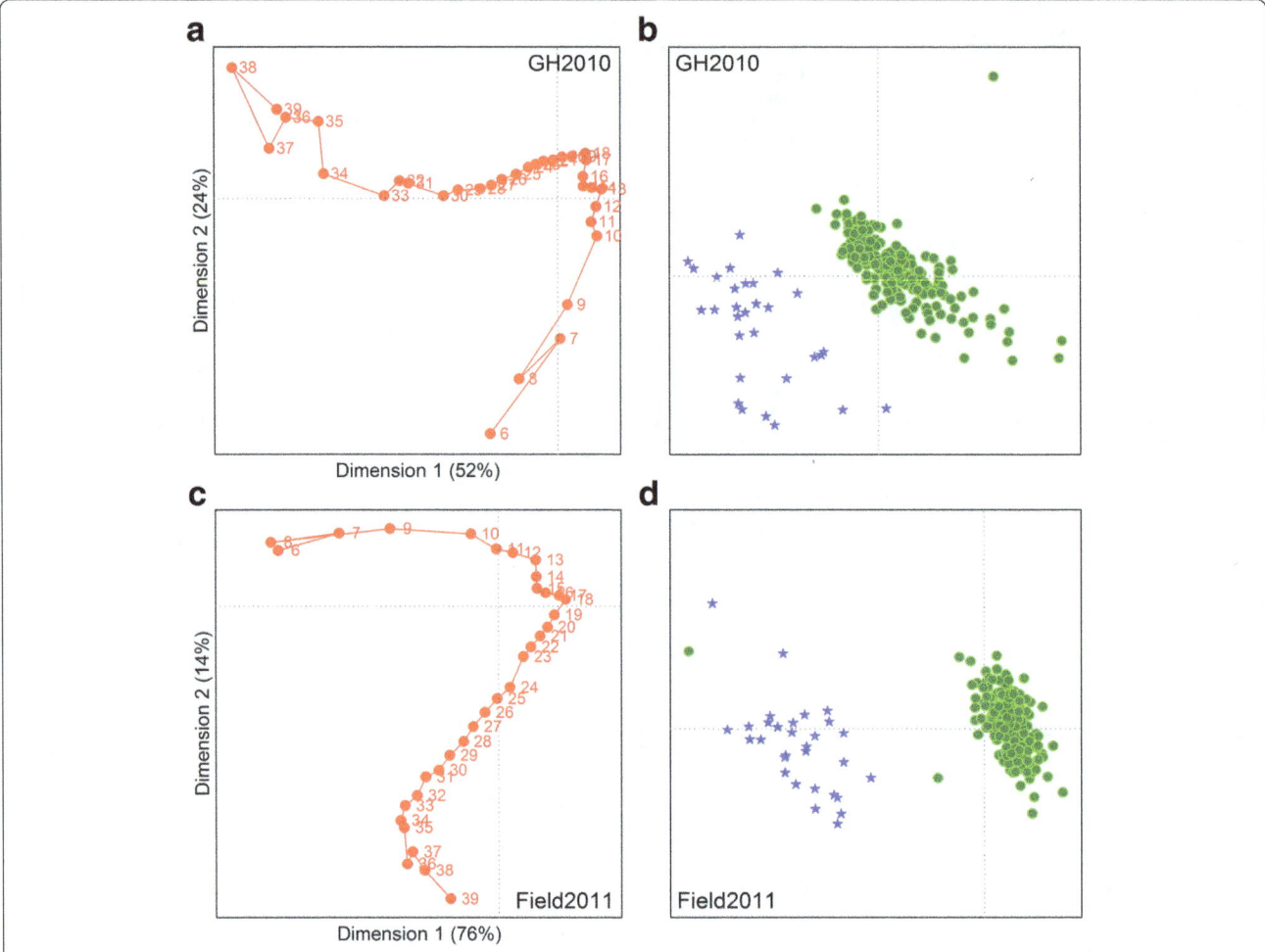

Fig. 2 Correspondence analysis bi-plots for debranched pea starch from the GH2010 (**a** and **b**) and Field2011 (**c** and **d**) trials, derived from a two-way contingency table with the variables DP (columns) and pea lines (rows), and plotted against the first and second dimensions. Degrees of polymerization (DP) are plotted in graphs **a** and **c** (*red circles*) while the pea lines in each pot (GH2010) or plot (Field2011), including controls, are plotted in graphs **b** and **d** (round seeded lines are indicated with green circles, wrinkled seeded lines with blue stars). **a** and **c** versus **b** and **d** are not plotted on the same scale since the points for the pea lines (**b** and **d**) occur near the centroid of plots A and C and occupy only a small area of the DP points

(DP39) DP values. For the analysis of pea lines, the obvious difference is the clear separation between wrinkled and round seeded lines, mostly associated with dimension 1 and seen in both trials.

Correlations between the CLD mean peak area proportions from DP6 to DP40 for the starch extracted from the PSP lines grown in the GH2010 versus Field2011 trials are presented in heat maps (Additional file 2) comparing all 92 lines (round and wrinkled, top panel) and the 83 round lines only (bottom panel). Strong correlations, both positive and negative, (range − 0.879 to 0.921) were observed when all lines were considered, no doubt because of the major effect of r locus on CLD distributions. Weaker correlations (range − 0.365 to 0.492) were seen when the mean peak area proportions at all DP values from the round lines only were compared for the GH2010 versus Field2011 trials.

Candidate gene selection, genotyping and linkage disequilibrium

Partial sequences of 25 pea candidate carbohydrate and starch metabolism genes were assessed in this study (Table 1) giving a total of 280 polymorphisms with minor allele frequencies ≥5%. The r locus phenotype (round versus wrinkled seed) was treated as the 281st polymorphism, and although the phenotypes were scored, these were treated as genotypes since the pea lines being used are inbred and segregation of the round phenotype (RR or $R\_$ genotype) was never observed in the harvested seeds. These candidate gene sequences and r locus represent 16 enzyme catalysed reaction classes in the pea carbohydrate and starch metabolic pathways (Fig. 3), including activities involved in precursor supply, chain elongation, chain branching, debranching, and phosphorylation and degradation. More than one

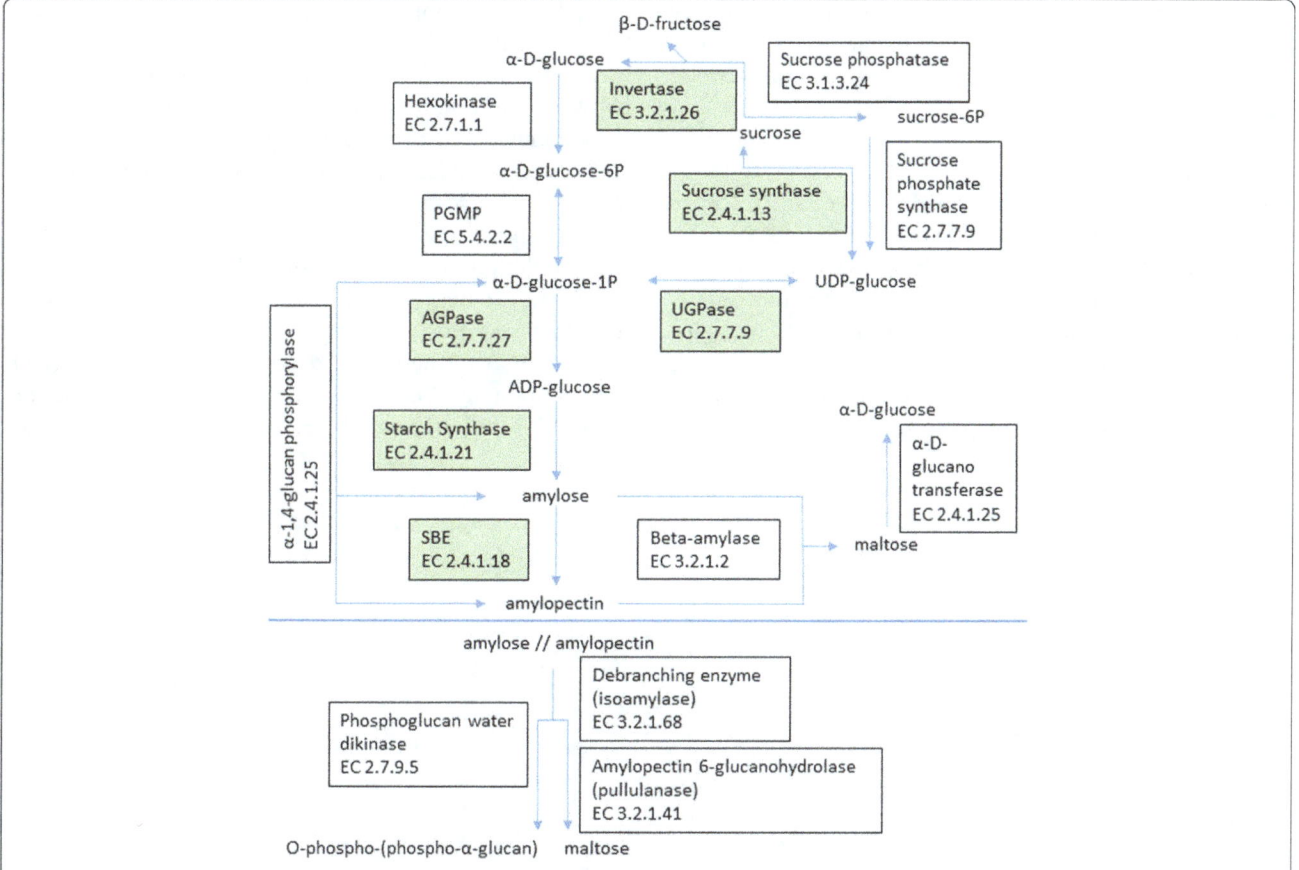

Fig. 3 A partial carbohydrate and starch metabolic pathway showing the 16 enzyme-catalyzed reaction classes for which polymorphisms in 25 candidate gene sequences and *r* locus were characterized. Enzyme classes where significant association between candidate gene polymorphism and CLD were identified are shown in *shaded boxes*. Enzyme Commission (EC) numbers are shown

candidate gene sequence was characterized for six of the enzyme classes. These were: AGPase (*AGPS2*, *AGPL1*), starch synthase (*StSynII*, *GBSSI*, *GBSSIb*), isoamylase (*ISA1*, *ISA2*, *ISA3*), sucrose synthase (*SuSy*, *SecSuSy*), invertase (*CWI*, *InvInh*) and beta-amylase-like (BAM-like, three beta-amylase-like candidate gene sequences). For some candidate genes, two genomic fragments were resequenced: *PGMP*, *StSynII*, *PUL1*, hexokinase (*Hex*), phosphoglucan water dikinase (*PWD*), UDP-glucose pyrophosphorylase (*UGPase*), and a BAM-like sequence. The candidate gene sequences for this study were either previously characterized pea genes (*n* = 13) involved in carbohydrate or starch metabolism or homologs of genes involved in carbohydrate or starch metabolism in other plant species (*n* = 12).

Linkage disequilibrium (LD) was analysed between all pairs of segregating sites with a minor allele frequency ≥ 5%, including the *r* locus phenotype, with the primary aim of understanding the chance of a correlation between polymorphisms in different candidate genes, and the secondary aim of understanding LD decay within pea genes and how this might influence the

ability to detect association between traits and polymorphisms. A heat map is presented in Additional file 3 showing LD (r^2) between the polymorphisms in the 25 candidate genes and *r*. The strong LD within sequences from a single gene is clearly observed, while in general only weak or no LD was observed between the sequences from different carbohydrate or starch metabolism genes. The strongest LD observed between different candidate gene sequences was r^2 = 0.347, observed for polymorphisms in *InvInh* and *Iso1*. The ability to understand LD decay within these pea genes using these data is limited because of the short length (≤ 569 bp) of the alignments. However, since two fragments from different regions of their respective genes were sequenced for seven of the pea candidate genes (*Hex*, *PGMP*, *UGPase*, *StSynII*, *Pul1*, *BAM-1523-1475* and *PWD*), there was the opportunity to examine the extent of LD in different parts of a single gene. The maximum LD (r^2) that was observed between polymorphisms within a fragment was 1.0, while the maximum LD observed between polymorphisms in different fragments of the same gene ranged from 0.340 (*Hex*) to 0.833 (*UGPase*). This analysis provides only limited information on LD decay in this panel of pea lines since the whole genes were not sequenced and therefore the physical distances between sites are not known.

Population structure

Two approaches, the model-based Bayesian software STRUCTURE and PCA, were taken to estimate population structure in the PSP pea lines considered in this study, based on 140 polymorphisms obtained from 55 background molecular markers. These analyses were applied to the full set of 92 PSP pea lines (containing both round and wrinkled seed types) and the subset comprising the 83 lines with the round seed phenotype.

Model-based population structure estimation using STRUCTURE with the $n = 92$ and $n = 83$ datasets distinguished three ($K = 3$) likely subpopulations. Plots for between one and eight subpopulations (K) showed that the estimated Log probability of the data (Ln P(D)), averaged over 7 replicates, peaked at $K = 3$, indicating that three subpopulations was the best estimate for both the $n = 92$ and $n = 83$ association panels (Fig. 4a and b). Subpopulation membership of individual lines is indicated in Additional file 1. The $K = 3$ Q-matrices were used in association mapping.

In the PCA of the $n = 92$ dataset, the first eight principal components accounted for >98% of the variation, with PC-1 accounting for 54.7% and PC-2 for 16.6%.

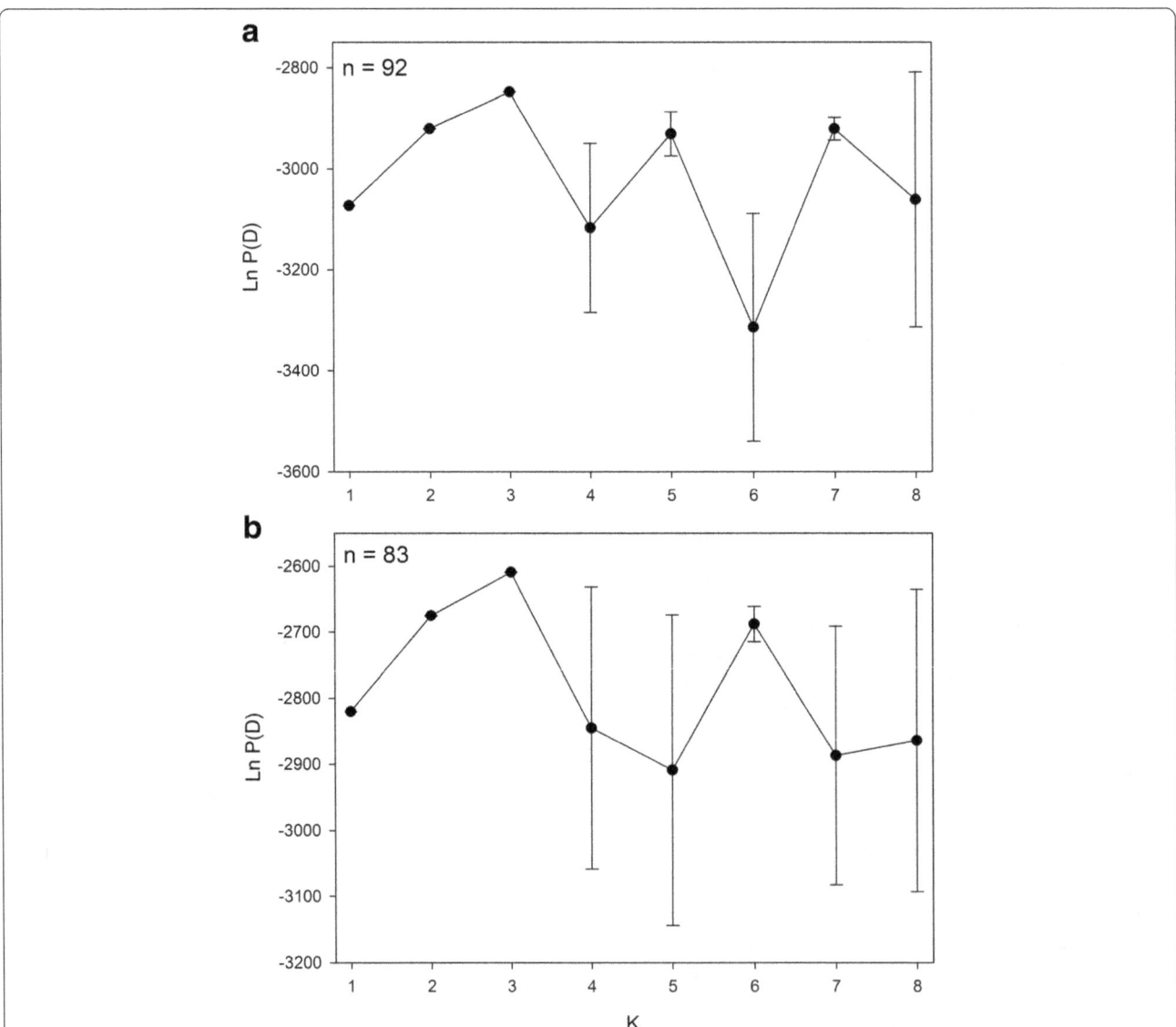

Fig. 4 Population structure estimation using STRUCTURE. Plots of the estimated Log probability of the data (Ln P(D)), averaged over 7 replicates, for between 1 and 8 subpopulations (K) obtained from analysis of 55 background markers in STRUCTURE of the $n = 92$ (**a**) and $n = 83$ (**b**) PSP lines. Error bars show standard deviations

The major inflection points in the scree plots obtained from PCA indicated three subpopulations. In addition, examination of a PCA biplot of the first two PCs shows that the PSP lines fall into three subpopulations when PC-1 is considered. Consequently, the $P = 3$ matrices were used in association mapping. PCA and scree plots for the $n = 92$ dataset are shown in Additional file 4.

Association analysis

A total of 280 polymorphisms in 25 carbohydrate and starch metabolism candidate genes, with the r locus phenotype treated as the 281st polymorphism, were tested for association with 35 CLD traits consisting of the mean molar peak area proportions at DP6 through DP40 as well as with percent amylose content of the extracted starch from the GH2010 and Field2011 trials. The association mapping analysis was carried out using the MLM + Q + K, MLM + P + K models. Quantile-quantile (Q-Q) plots of the $-\log_{10}(P)$ were drawn for the DP which gave the most significant associations (Additional file 5) to identify the best association mapping model. The Q-Q plots indicated that the MLM + Q + K model, which controls both for population structure and underlying familial relatedness, provided a similar or better overall fit than the MLM + P + K model to the expected p-values for all traits, assuming a normal distribution for p-values. Therefore, association between candidate gene polymorphisms and CLD is reported based on the MLM + Q + K model. Associations were considered as significant when they met the $\alpha < 0.05$ (for $n = 92$ lines) or $\alpha < 0.10$ (for $n = 83$ lines) criteria for minimising FDR based on Q-values. Suggestive associations that have low p-values ($p < 0.01$) or that meet the $\alpha < 0.10$ criterion are also discussed.

Since the r locus phenotype had an obvious effect on the CLD pattern of debranched isolated starch (Fig. 1), association mapping was carried out on all 92 lines that were fully phenotyped and separately on the subset of 83 round seeded lines. When all 92 lines were considered, polymorphisms in seven of the 26 genes/loci were found to be significantly associated with CLD phenotypes (Table 2). Of these seven candidate genes, three of the genes/loci were identified from both the GH2010 and Field2011 environments: r locus, *UGPase*, and *AGPS2*. The polymorphisms in the seven genes/loci were associated with CLD at DP10, DP16, DP17, DP34, and DP39 in the GH2010 environment; and with DP17, DP18 and DP23 in the Field2011 environment. The r locus had the major effect, explaining 83% of the variation at DP17 from the GH2010 trial and 89% of the variation at DP18 from the Field2011 trial. SNPs in *AGPS2* and *UGPase* were associated with CLD variation from both trials and explained from 14.3% (*AGPS2*, GH2010) to 22.2%

(*UGPase*, Field2011) of the variation at the DPs with which they were most strongly associated. For four other loci, association with CLD was detected from only a single environment each. These were: *CWI* and *SecSuSy* associated with DP39 and DP29, respectively, from the GH2010 trial; and ADP glucose pyrophosphorylase L1 subunit (*AGPL1*, *rb*) and StSynII (*rug5*) associated with DP23 and DP17, respectively, from the Field2011 trial. Suggestive associations involving an indel polymorphism in *PWD* at site 49 of the alignment were detected from both the GH2010 (DP33, $p = 6.76$ E-04, q-value = 0.081) and Field2011 (DP 9, $p = 1.23$ E-03, q-value = 0.158) trials. Some of the associations of interest involved more than one polymorphic site within a sequenced candidate gene fragment (Table 2). These sites were in strong or complete linkage disequilibrium (Additional file 4).

As expected, %amylose was associated with r locus in starch from both environments (Table 3, Figs. 5 and 6). In addition, %amylose was associated with polymorphisms in *UGPase* in the Field2011 trial.

Beanplots are presented summarising the means and distributions of the mean molar peak area proportion values for the allelic variant groups for the significant marker-trait associations for $n = 92$ PSP lines from the GH2010 and Field2011 trials (Figs. 5 and 6). For the associations detected using $n = 92$ PSP lines, the alleles for each associated SNP are presented alone, and also combined with r locus, to distinguish the effect of r and the other loci, due to the major effect which r locus has on CLD profiles. Where four genotypic classes are present, eg. *AGPS2* × r, and *AGPL1* × r, the plots show pairwise interactions illustrating additive effects between r and other loci. In the cases of three other loci (*CWI*, *UGPase*, *SecSuSy*) the pea lines with the wrinkled seeded genotype (rr) all fall within a single allelic class of the other locus, therefore only three pairwise genotypic classes occurred.

When the subset of 83 round seeded lines were considered, polymorphism in a single candidate gene, *Sec-SuSy*, was associated at FDR ≤ 0.05 with the DP29 CLD phenotype, in the GH2010 trial (Table 2). If the FDR was relaxed to $\alpha < 0.10$, then additional associations were identified involving *InvInh* and *AGPS2*, associated with DP18 and DP10, respectively, from the GH2010 trial. Beanplots are presented summarising the means and distributions of the mean peak area proportion values for the allelic variants at the three associated loci for the $n = 83$ lines (Fig. 7). For two other candidate genes, *SPS* (Field2011) and *CWI* (GH2010), polymorphisms associated with CLD were suggested because of low p-values (< 0.01) but q-values for these were >0.10. For *SPS*, the polymorphism at site 563 of the Ps_1583 alignment was associated with DP17 ($p = 7.4$ E-04, q-value = 0.171) and for *CWI*, the polymorphism at site 24

Table 2 Summary of the associations between candidate carbohydrate metabolism gene sequences and starch chain length distribution (CLD) peak areas

Environment Glasshouse 2010. All PSP lines, $n = 92$.

Gene	Sequence alignment [a]	Site(s) on alignment [b]	CLD peak with lowest p-value	p-value	R^2 (%)	Q-value
Starch branching enzyme I (r locus)	n/a [c]	n/a	DP17	7.50 E-39	83.4	1.92 E-36
Cell wall invertase	694,183,574	24 (T/G)	DP39	1.23 E-04	17.3	0.010
ADP glucose pyrophosphorylase S2 subunit	694,187,965	237 (indel)	DP10	1.83 E-04	14.3	9.13E-03
UDP glucose pyrophosphorylase	694,186,492	18 (T/G), 25 (C/G), 56 (T/G), 274 (T/G)	DP16	8.50 E-04	12.3	0.039
Second sucrose synthase	694,184,180	148 (C/T), 299 (indel)	DP34	1.52 E-03	11.8	0.015

Environment Glasshouse 2010. Round seeded PSP lines, $n = 83$.

Second sucrose synthase	694,184,180	128 (T/A), 134 (T/indel/C)	DP29	2.68E-04	16.9	0.027
Invertase inhibitor	694,182,042	357 (T/C)	DP18	3.07 E-04	15.6	0.086
ADP glucose pyrophosphorylase S2 subunit	694,187,965	517 (G/T)	DP10	1.16 E-03	12.4	0.062

Environment Field 2011. All PSP lines, $n = 92$.

Starch branching enzyme I (r locus)	n/a	n/a	DP18	8.81 E-53	88.6	1.72 E-50
UDP glucose pyrophosphorylase	694,186,492	25 (C/G), 56 (T/G), 274 (T/G)	DP17	2.62 E-06	22.2	1.02 E-04
ADP glucose pyrophosphorylase S2 subunit	694,187,965	72 (T/C)	DP23	6.58 E-04	12.1	0.027
ADP glucose pyrophosphorylase L1 subunit (rb locus)	694,185,424	9 (A/C)	DP23	8.94 E-04	12.1	0.027
Starch synthase II (rug5 locus)	1,206,484,033	307 (G/A)	DP17	1.26 E-03	15.9	0.027

[a]GenBank PopSet identification number
[b]Where more than one site is shown in this column, they all had the same p-value; (major/minor) alleles are shown
[c]Not applicable. The r locus genotype was determined by recording the round or wrinkled seed shape, so the alignment and site information are not applicable
Associations were identified using the mixed linear model approach with adjustment for population structure using Q + K matrices, implemented in the Tassel package. For all the pea lines ($n = 92$), associations that meet the $\alpha < 0.05$ criterion for minimising the false discovery rate (FDR) are shown, while for the round only pea lines ($n = 83$) associations that meet the $\alpha < 0.10$ criterion are shown

of the Ps_0276 alignment was associated with DP39 ($p = 0.7.9$ E-04, q-value = 0.221).

Associations with low p-values were detected between polymorphic sites and the CLD mean molar peak area proportions at a number of DP values. This observation is to be expected since neighbouring DPs can show strong positive correlation and also because mean molar peak area proportions must sum to 1. Hence a change in the abundance of oligosaccharide chains at one DP value or a range of DP values must result in a concomitant change elsewhere in the distribution, which can produce either positive or negative correlations (Additional file 2). To explore further the relationships between variation in candidate gene polymorphisms and the CLD curves, DP versus $-\log_{10}(p)$ plots were drawn for the $n = 92$ and $n = 83$ populations (Fig. 8). These plots reveal the regions of the CLD curves that were most strongly associated with allelic variation at the significantly associated candidate gene polymorphisms. For the $n = 92$ pea lines and r locus, similar profiles were obtained from the GH2010 and Field2011 environments, with p-value peaks obtained at DP6, DP17 and around

DP30-DP33. Likewise, similar profiles were obtained for the *UGPase* site 25 polymorphism from both environments, with p-value peaks at around DP16-DP17 and DP31-DP32. Therefore, the r locus and *UGPase* site 25 polymorphisms affected similar parts of the CLD curves, but this was not due to linkage disequilibrium between these sites ($r^2 = 0.0178$). However it does mean that the effects of r locus and UGPase on the CLD could be difficult to distinguish. There are also similarities in peak locations for the GH2010 and Field2011 curves obtained for *AGPS2* site 237 and *AGPS2* site 72, although these are less compelling than for r locus and *UGPase*.

Examination of the gene context of polymorphisms

For three of the candidate genes with significant association with CLD traits (*AGPS2*, *UGPase*, *StSynII*) polymorphisms occurred in predicted exon sequences (Table 4). The exonic SNP in *AGPS2* (site 145 in our alignment) resulted in a synonymous codon change. In *UGPase*, the SNP at base 336 of our alignment produced a synonymous codon change and occurred at the 5′ end of an exon, adjacent to an intron /exon boundary

Table 3 Summary of the most significant associations between candidate carbohydrate and starch metabolism gene sequence polymorphisms versus the percent amylose in extracted starch

Environment: Glasshouse 2010						
Gene	Sequence alignment [a]	Site on alignment (major/minor allele)	Trait	*p*-value	R[2] (%)	Q-value
Starch branching enzyme I (*r* locus)	n/a [b]	n/a	%amylose	2.26 E-33	79.0	4.29 E-31
Environment: Field trial 2011						
Starch branching enzyme I (*r* locus)	n/a	n/a	%amylose	6.85 E-47	86.2	1.90 E-44
UDP glucose pyrophosphorylase	694,186,492	18 (T/G)	%amylose	4.64 E-04	13.1	0.028

[a]GenBank PopSet number

[b]not applicable

Associations were identified using the mixed linear model approach with adjustment for population structure using Q + K matrices, implemented in the Tassel package. Only associations that meet the α < 0.05 criterion for minimising false discovery rate (FDR) are shown

predicted by homology with the *M. truncatula UGPase* mRNA (GenBank accession number XM_003616133). In *StSynII* (*rug5*) the SNP at position 307 of our alignment is predicted to produce a non-synonymous change from glycine to serine. This amino acid residue occurs at 249 in the StSynII protein sequence. Analysis of the possible functional effects of this non-synonymous SNP using PROVEAN software predicted it to be "neutral", with a score of −0.244 where a score < −2.5 predicts that a variant is likely to be "deleterious". A significantly associated SNP in *AGPL1* (at position 9) occurred in the 5′ UTR of the *AGPL1* mRNA. The remaining polymorphisms in

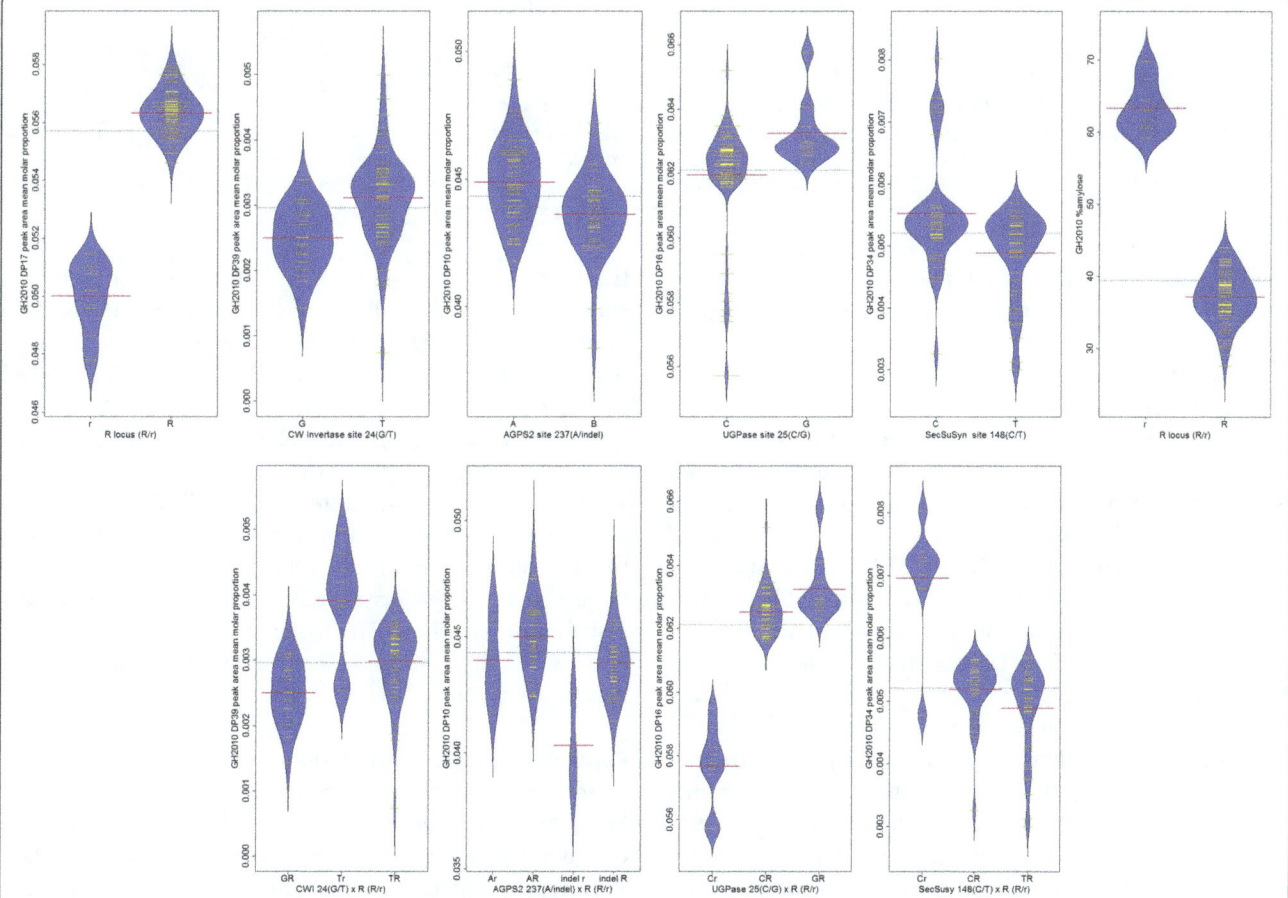

Fig. 5 Beanplots showing the distributions of trait values from the GH2010 environment and *n* = 92 PSP lines for alleles of the associated locus-trait combinations. The means and distributions of the mean molar peak area proportion values are shown, grouped by allele, or by combinations of alleles from *r* and other loci. Allelic means (*red lines*) and individual PSP lines' mean values (*yellow lines*) are shown. The second row contains plots showing pairwise interactions between alleles of *r* locus and the associated allele of the locus directly above each plot

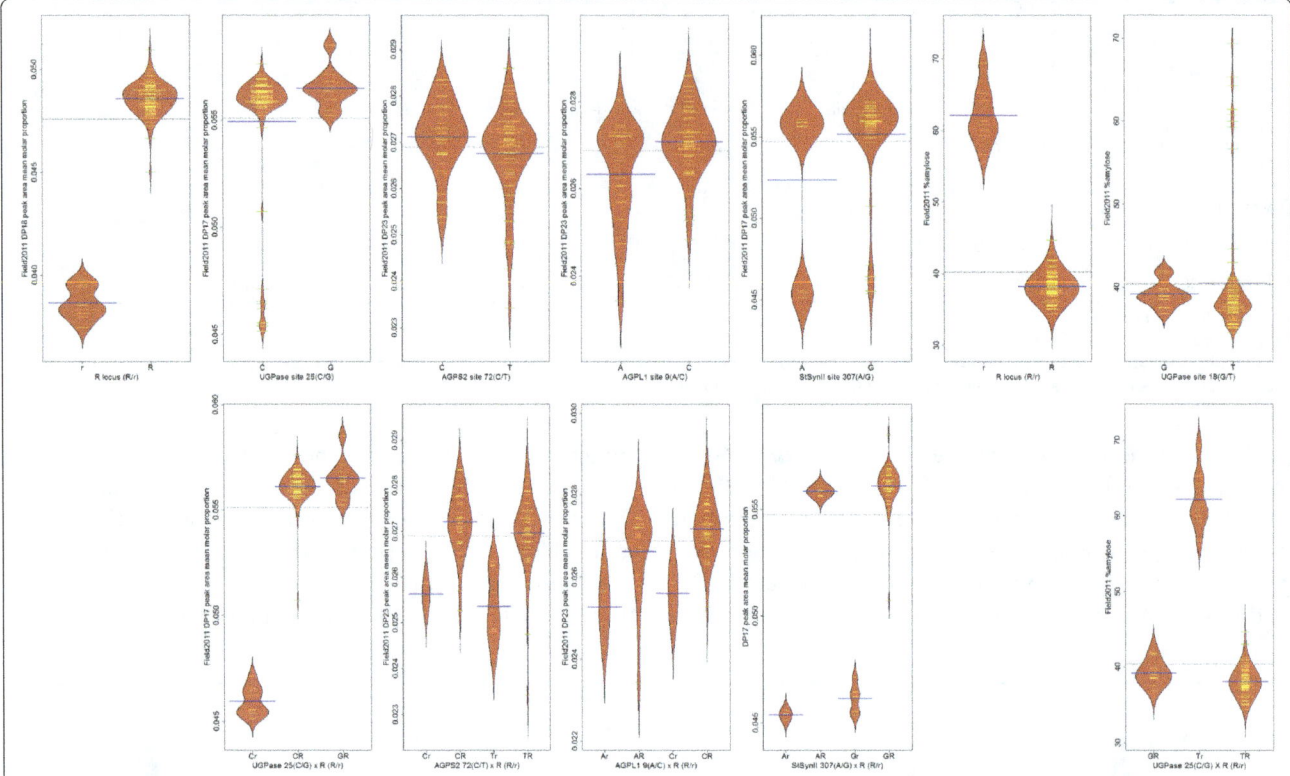

Fig. 6 Beanplots showing the distributions of trait values from the Field2011 environment and *n* = 92 PSP lines for alleles of the associated locus-trait combinations. The means and distributions of the mean molar peak area proportion values are shown, grouped by allele, or by combinations of alleles from *r* and other loci. Allelic means (*blue lines*) and individual PSP lines' mean values (*yellow lines*) are shown. The second row contains plots showing pairwise interactions between alleles of *r* locus and the associated allele of the locus directly above each plot

the candidate gene sequences that were significantly associated with starch traits occurred in intron sequences.

Discussion

Phenotypic analysis

This study has focused on variation in two pea starch characteristics, %amylose and the CLD of debranched extracted total pea starch, as determined by quantitative FACE. Although total starch was the starting material for the FACE, analysis of CLD in debranched starch in the DP range used in this study (DP6 – DP40) is generally considered to examine the amylopectin fraction, since short (mean DP ~ 15) and medium length (mean DP ~45) chains derive from amylopectin [51, 52]. In association mapping studies a relatively large number of lines must be both phenotyped and genotyped, with phenotyping based on trialling in multiple environments and using appropriate trial design and replication. Analysis of starch structural and functional properties requires a number of labor-intensive preparative and analytical steps. Starch extraction in particular is time consuming. The need to extract starch from seeds from approximately 500 samples made it necessary to start with relatively small amounts of seeds (5 g) from each

trial pot (GH2010) or plot (Field2011) and consequently starch yields were small. Hence, this research focused on understanding CLD using FACE, a method that requires only a small amount of starch and is suitable for moderate throughput.

A limited number of studies have examined CLD variation in debranched pea starch. For example, Ratnayake et al. [53] explored the differences in CLD in four field pea (round seed) lines and observed differences in the relative abundance of the DP6 peak, the DP for the largest peak of the distributions, and the shapes of the distributions in the DP16 to DP26 regions. Variation in amylose content has also been explored in small numbers of round peas [13, 14].

Population structure estimation

False associations are a potential difficulty with association mapping studies using germplasm panels because of unknown relatedness or population structure [31, 54]. As a result, apparent marker-trait association may occur when trait values and marker allele frequencies are correlated based on subpopulation or kinship, rather than being due to linkage between quantitative trait loci (QTL) and markers, leading to spurious associations.

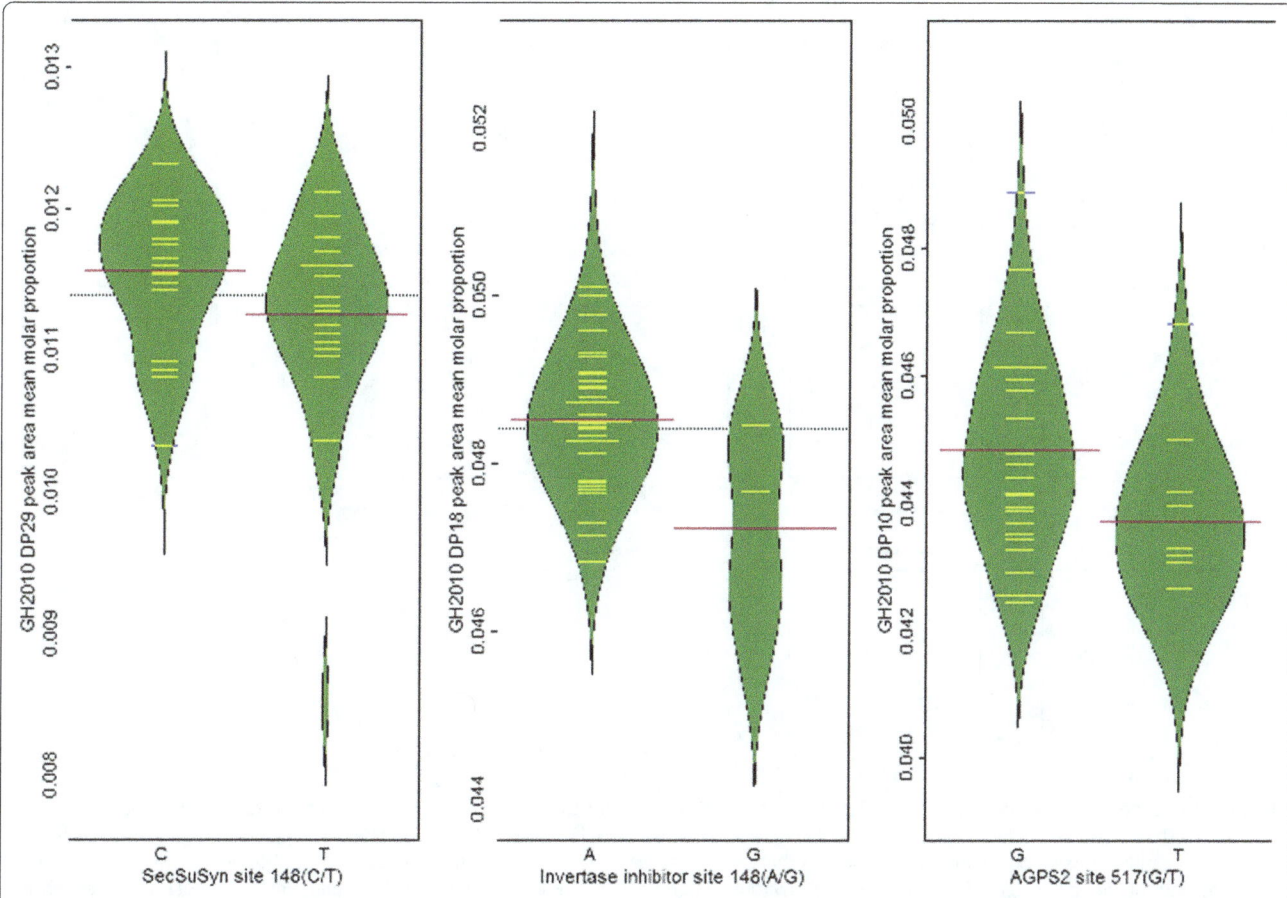

Fig. 7 Beanplots showing the distributions of trait values from the Field2011 environment and *n* = 83 round seeded PSP lines for alleles of the associated locus-trait combinations. The means and distributions of the mean molar peak area proportion values are shown, grouped by allele. Allelic means (*red lines*) and individual PSP lines' mean values (*yellow lines*) are shown

Mixed model methods are used to relate the relatedness matrix (Q, P, and/or K) to a phenotype, yielding a relatedness-based weighted average predicted phenotype.

Determination of the Q, P or K matrices relies on polymorphism information from random molecular markers that are distributed throughout the genome. In our study, we used relatively few markers (*n* = 55) and a mixture of marker types (SSRs, SCARs, and RAPD polymorphisms) that revealed 140 polymorphisms. Since our panel consisted of single-seed derived inbred lines, all markers, including RAPDs, provided homozygous genotypes. Using these markers we estimated the relatedness of pea lines to provide Q (Fig. 4), P and K matrices. Population structure estimation using STRUCTURE and PCA gave a clear answer of three subpopulations.

Prior to this study, the estimation of population structure in pea has been undertaken in collections of diverse germplasm, including studies of the John Innes *Pisum* germplasm [55], the USDA pea core collection [56], and the European germplasm collections [57]. Each of these studies estimated three populations within their respective germplasm collections, and further subdivision of populations was also indicated for the John Innes collection and European germplasm [55, 57]. The study of the USDA pea core (*n* = 285 lines) used a relatively small number of markers from a combination of SSRs (15 primer pairs), RAPDs (36 loci) and one SCAR [56], while the John Innes (*n* = 3020 lines) and European germplasm (*n* = 4538) studies relied most heavily on retrotransposon-based polymorphism markers (RBIP and SSAP), also in relatively small numbers, 45 and 27 respectively. Jing et al. [57] expressed confidence in their population structure estimation because similar results were obtained from both RBIP and SSAP marker types.

Allelic variation and association with starch physicochemical properties

In this study, we focused on association mapping of variation in debranched starch CLD and %amylose with polymorphisms in 25 candidate genes representing 16 carbohydrate and starch metabolic enzymatic reaction classes (Table 1, Fig. 3). Using the total population of *n* = 92 PSP lines, associations that met the FDR α ≤ 0.05

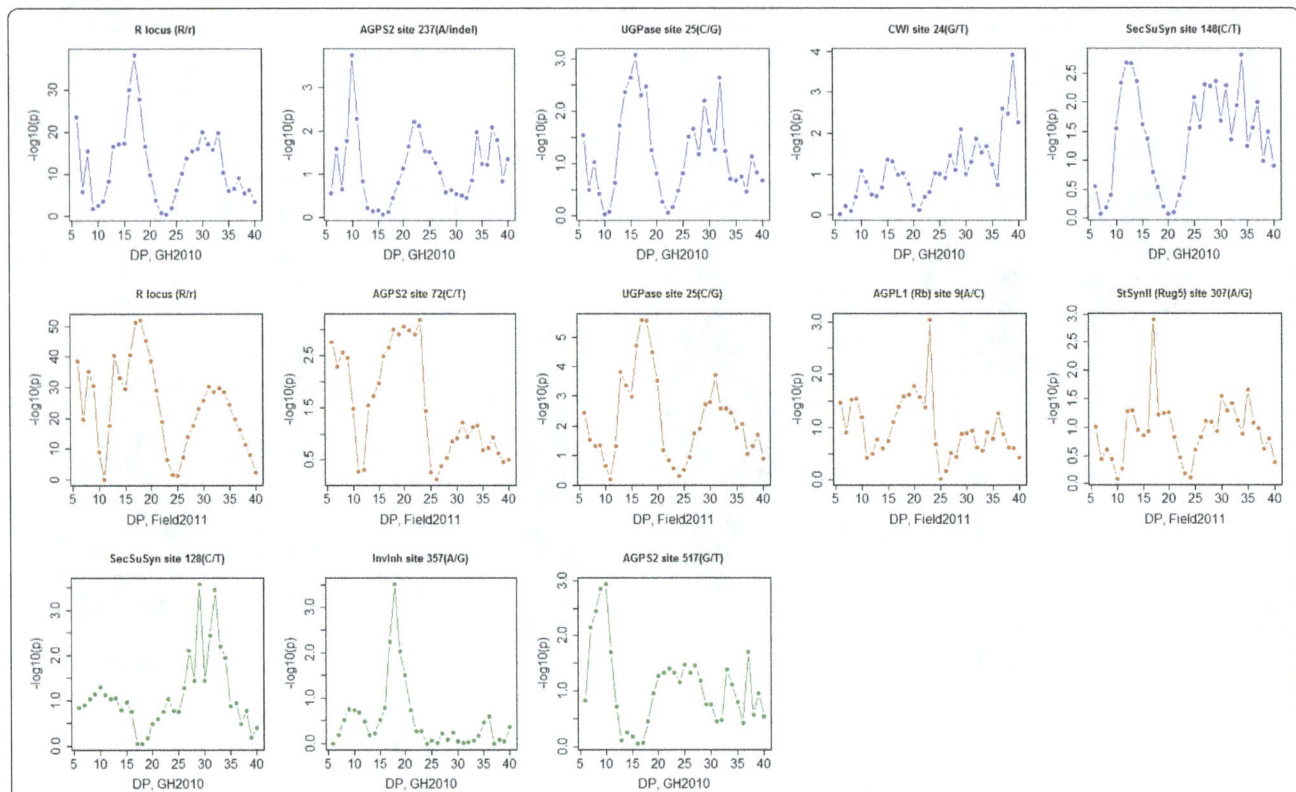

Fig. 8 Distribution of *p*-values (as $-\log_{10}(p)$) for CLD traits from DP6 to DP40, for candidate gene polymorphisms associated with variation in CLD. The graphs show the *p*-values (as $-\log_{10}(p)$) versus CLD DP values for associated polymorphisms. The top row graphs (*blue*) are plots of DP vs $-\log_{10}(p)$ for candidate gene polymorphism associations that meet or exceed the FDR < 0.05 criterion from the GH2010 environment and *n* = 92 PSP lines, and middle row (*red*) are from the Field2011 environment and *n* = 92 PSP lines, containing round and wrinkled seeded PSP lines, while bottom row (*green*) are GH2010 environment locus-trait associations (FDR < 0.10) from the *n* = 83 lines, including only round seeded PSP lines

criterion were identified for polymorphisms in seven pathway genes, representing six of the enzymatic reaction classes in Fig. 3. For the round seeded PSP lines (*n* = 83), associations that met the FDR α ≤ 0.10 criterion identified polymorphisms in three pathway genes. Taken together, eight pathway genes were associated with CLD variation and these associated candidate genes were involved in substrate availability [*CWI*, *SecSuSy*, *AGPS2*, *AGPL1* (*rb*), *UGPase* and *InvInh*], chain elongation [*StSynII* (*rug5*)] and branching [*SBEI* (*r*)].

The power of our study to identify sequence variants associated with starch CLD variation was limited. The limiting factors included: 1) the relatively small sample size (*n* = 92) we employed, 2) the limited number (25) of candidate genes in the starch biosynthetic pathway that were characterised, and 3) that the candidate genes were partially sequenced, with a bias for intron-containing regions, therefore causal variants may have occurred outside the sequenced regions and may have been in only partial LD with variants within the sequenced regions. The genome-wide association study (GWAS) offers an alternate approach to the candidate gene-based approach for identifying genes associated with complex traits such as the composition of pea seeds, although

Table 4 Exonic or 5′ UTR polymorphisms significantly associated with chain length distribution (CLD) trait variation in pea lines and their predicted effects on translation product sequence

Candidate gene	Alignment PopSet number	Site on alignment	Variant	Effects (site on reference sequence translation)
AGPS2	694,187,965	145	T/C	In exon; CTT / CTC, leucine -> leucine, synonymous
UGPase	694,186,492	336	T/C	Variant in the first base of the exon, adjacent to a predicted intron acceptor site; codon change AGT / AGC, serine -> serine, synonymous
Starch synthase II (*rug5*)	1,206,484,033	307	G/A	In exon; GGT / AGT, glycine -> serine (249), non-synonymous
AGPL1 (*rb*)	694,185,424	9	A/C	In 5′ UTR

GWAS is poor for detecting effects of minor alleles [58]. Substantially larger sample size is needed to increase the statistical power to detect effects, and methods for accurate phenotyping of pea seed compositional traits need to be appropriate for the increased throughput that is required.

CLD affects gelatinization and pasting properties of starch [8, 9]. As a generalization, an increased proportion of short chains (DP6 – DP12) results in reduced gelatinization temperature and enthalpy, and also reduces pasting temperatures and viscosities, while increased proportions of intermediate (DP13 – DP24) and long (>DP24) chains increase gelatinization temperature and enthalpy as well as pasting temperatures. The largest effect on amylopectin CLD in our study was obtained from the round (RR) versus wrinkled (rr) lines. The mean proportion of intermediate length chains was increased in round seeded lines (Fig. 8) with an associated decrease in the mean proportion of short and long chains (data not shown). Therefore, it is difficult to predict the overall effect that the r locus allelic variation would have on the thermal and pasting properties of pea amylopectin. Allelic variation in UGPase (at site 25, both trials) also affected the mean proportion of intermediate length chains, which was higher on average for the lines with the G allele at that site.

Mutational studies in pea have demonstrated the role of six pea seed starch biosynthetic genes in pea seed starch accumulation, structure and function [5, 59, 60]. In double mutant nearly isogenic lines, Lloyd et al. [60] found a modest effect on CLD of r locus (SBEI), and a smaller effect of rb locus (AGPL1) mutations. Mutations in rug5 (StSynII) were also found to affect CLD in both developing and mature pea embryos [5, 59]. In the present study, naturally occurring allelic variation in r, rb (AGPL1) and rug5 (StSynII) was associated with CLD variation (Table 2, Figs. 5 and 6). However, in our germplasm panel, we failed to detect association between CLD and polymorphisms in PGMP (rug3), GBSSI (lam) and SuSy (rug4), the other three genes identified, based on EMS mutations, in the above studies as being involved in starch metabolism.

Variation in %amylose was associated with allelic variation in r locus, the major gene responsible for the difference between round and wrinkled (high amylose) pea lines, in both trials, as expected. In starch extracted from plots from the Field2011 trial, %amylose was also associated with SNPs in UPGase. For the GH2010 trial, %amylose was only associated with r locus. This likely reflects the greater variation in %amylose observed among round seeded lines in the Field2011 trial than the GH2010 trial, in which variation between round seeded lines was not significant. The effects of mutations in pea starch biosynthetic genes on %amylose have previously

been determined, with rug3 (PGMP) and lam (GBSSI) reducing the % amylose content of starch compared with the wild type round seeded line, while rug5 (StSynII) and r (SBEI) mutants increased the percent amylose [16]. In a recent study of 50 round seeded pea lines, Jha et al. [61] have found association ($0.01 > p > 0.001$) of amylose content with SNPs in amplified fragments of the AGPL1 (rb locus), GBSSI (lam locus) and SBEII genes. However, we did not detect association in %amylose involving SNPs in these genes. This difference that may be due to the germplasm used in the different studies or to the relatively low power of both studies due to the numbers of lines used.

As CLD and amylose content affect the physicochemical properties of starches, there are parallels, as well as contradictions, between our results and those of association mapping studies which looked at physicochemical properties of starches from other species. In pea, SBEI (r locus) had the greatest effect on both amylose content and CLD, whereas in rice, polymorphisms in GBSS (Wx) had the greatest effect on amylose content, and SSII on gelatinization temperature [23], with SBE3 having only a minor effect on gelatinization temperature. In maize, polymorphisms in a sucrose synthase gene (sh1), AGPL1 (sh2) and SBEIIb (ae1) were associated with amylose content and pasting properties [24], while in pea, polymorphisms in genes from these families were associated with CLD and/or amylose content. A study in sorghum detected polymorphisms in SSII and SBE associated with physicochemical properties, similar to the results for CLD in pea, but differed from pea in that association with GBSS was also found [26]. Where the results from different species are contradictory, this reflects the fact that the associations which can be detected are limited by the variation which occurs in any germplasm collection, and by the power of the experiment.

Analysis of the gene context of the polymorphisms associated (FDR $\alpha \leq 0.05$) with CLD showed that most of these occurred in intronic sequences. For three of the genes, associated polymorphisms occurred in exonic regions (AGPS2, UGPase, StSynII), although only the mutation in StSynII resulted in a non-synonymous codon change (Table 4). Analysis of the predicted effect of the codon change in StSynII (residue 249, glycine or serine) on protein function using PROVEAN software [39] indicated that the effects were most likely to be neutral. For the UGPase gene, one of the significantly associated polymorphic SNPs occurred immediately upstream of a predicted intron acceptor cut site, hence may affect transcript splicing in lines with one or the other of the variants, a possibility that is able to be tested experimentally. Therefore, with the possible exception of the intron acceptor site variation in the UGPase gene, it is unclear whether any of the significantly associated polymorphisms identified are directly responsible for changes that would affect the CLD phenotypic variation, either through changes in coding or non-coding regions. Since only portions of the candidate genes were characterized for sequence polymorphism, it is quite possible that the significantly associated polymorphisms that were detected are in LD with the causal mutation(s) to be found in the gene regions that were not sequence characterized. It is also possible that the causal variants underlying the CLD traits do not occur in the candidate genes but in linked sequences. Our analysis of the decay of LD in this population has been limited to examining the LD within individual fragment alignments (where the maximum pairwise LD (r^2) within fragments was 1.0 or nearly), which is of limited value because of the small size of these fragments, and to examining pairwise LD in pairs of fragments amplified from seven of the pea genes characterized (Additional file 3). While the actual physical distances between the polymorphisms in the paired fragments are unknown, their relative positions on the mRNA sequences are known. A range of LD conservation was observed; seen for example by contrasting LD decay in *StSynII* (where the maximum pairwise r^2-between genomic fragments from 587 to 999 and 1809–2229 on the mRNA was 0.173) with *UGPase* (where the maximum pairwise r^2 between genomic fragments from 59 to 201 and 1529–1601 on the mRNA was 0.833).

Conclusions

We have shown that allelic variation in pea starch pathway candidate genes can have a measurable effect on amylopectin CLD. We identified polymorphisms in eight genes from the pea seed carbohydrate and starch metabolic pathway as having significant association (FDR $\leq$ 0.05 or FDR $\leq$ 0.10) with variation in debranched starch CLD, and two genes (*r* locus and *UGPase*) as having significant association with variation in %amylose. The findings were based on analysis of seed starch extracted from 92 diverse pea lines grown in two replicated trials, and on polymorphisms detected by sequencing fragments of 25 candidate genes and by scoring *r* locus. The *r* locus, which encodes SBEI, had the major effect (R^2 of 83.4–88.6%), but other genes involved in substrate availability (*CWI*, *InvInh*, *SecSuSy*, *AGPL1*, *AGPS2*, and *UGPase*) and chain elongation (*StSynII*) were also associated with CLD variation, with effects ranging from 11.8 to 22.2% of the variation in mean peak area proportion at the most strongly associated peak, as determined by quantitative FACE. Examination of the sequence context of the significantly associated polymorphisms showed that most occurred in intronic regions, but a polymorphism in *UGPase* occurred immediately upstream of a predicted intron 3′ acceptor cut site, and a polymorphism in *StSynII* produced a non-synonymous mutation that was most likely to have a neutral effect on protein function. Hence, the candidate gene polymorphisms causing CLD variation in pea seed starch are in most if not all cases likely to be in full or partial LD with the associated polymorphisms that we have detected, and these causal variants may either occur within the candidate gene or in nearby sequences. Nevertheless, this study identifies sequence polymorphisms in carbohydrate and starch pathway genes, and publicly available pea lines containing the allelic variants, that can be used for further studies of genetic determination of pea seed starch structure and function, including plant breeding.

Additional files

Additional file 1: Pea single plant (PSP) accessions used for association mapping with Names and Collection Country from the USDA-ARS GRIN database (https://npgsweb.ars-grin.gov/gringlobal/search.aspx), *r* locus phenotype (1 = round, 0 = wrinkled seeded), % amylose and CLD mean molar peak area proportions for the Field2011 and GH2010 environments.

Additional file 2: Heat maps showing correlations between chain length distribution (CLD) mean peak area proportions for debranched starch from peas grown in GH2010 versus Field2011 trials. The colour scale ranges from blue (more strongly positive correlations) to yellow (more strongly negative correlations). The top panel shows correlations for all $n = 92$ lines (round and wrinkled seed) while the bottom panel shown correlations for $n = 83$ round seed lines only.

Additional file 3: Heat map showing the extent of linkage disequilibrium (r^2) within and among polymorphisms in 32 genomic fragments representing 25 candidate genes and *r* locus. Abbreviated candidate gene names are shown along the diagonal of the Figure.

Additional file 4: Principle component analysis of population structure for $n = 92$ PI PSP lines using 55 background markers. (A) Principle components biplot for the first two PCs. Round seeded lines are indicated with clear circles and wrinkle seeded lines with red circles. (B) Scree plot of eigenvalues showing the variation in each component.

Additional file 5: Q-Q plots of the observed versus expected $-\log_{10}(P)$ for the DP-environment combinations which gave associations with the lowest *p*-values. Results for the MLM + Q + K model (blue) and MLM + P + K model (red) are shown. The DP-environment combinations are indicated on the y-axes.

Abbreviations

%amylose: Percent amylose; ae: Amylose extender; AGPL1: ADP-glucose pyrophosphorylase L1 subunit; AGPS2: ADP-glucose pyrophosphorylase S2 subunit; APTS: 8-aminonaphthalene-1, 3, 6-pyrenetrisulfonic acid; BAM: Beta-amylase; CLD: Chain length distribution; CWI: Cell wall invertase; DMSO: Dimethyl sulfoxide; DP: Degrees of polymerization; FACE: Fluorophore-assisted carbohydrate electrophoresis; Field2011: Field trial held over the New Zealand summer of 2011–2012; GBSS: Granule bound starch synthase; GH2010: Glasshouse trial held in 2010–2011; Hex: Hexokinase; InvInh: Invertase inhibitor; ISA: Isoamylase; LD: Linkage disequilibrium; MLM: Mixed linear model; PCA: Principal components analysis; PCR: Polymerase chain reaction; PGMP: Plastidial phosphoglucomutase; PSP: Pea single plant; PUL: Pullulanase; PWD: Phosphoglucan water dikinase; QTL: Quantitative trait locus/loci; RAPD: Random amplified polymorphic DNA; RFU: Relative fluorescent units; SBE: Starch branching enzyme; SCAR: Sequence characterized amplified region; SecSuSy: Second sucrose synthase; SPS: Sucrose phosphate synthase; SSR: Simple sequence repeat; StSyn: Starch synthase; SuSy: Sucrose synthase; UGPase: UDP-glucose pyrophosphorylase

Acknowledgements

We thank Ross Crowhurst, Mark Fiers and Susan Thomson for pea 454 sequence assembly, Linda Falloon, Janelle Winchester and Tina Harrison-Kirk for assistance with molecular marker genotyping, Merle Forbes for assistance with the glasshouse trial, the staff of PGG Wrightson Seeds, Kimihia station, for assistance with the field trial, and Samantha Baldwin for critical

reading of the manuscript.

Funding

The research was funded by the New Zealand Foundation for Research Science and Technology through contract C02X0805.

Authors' contributions

MAC, RDC, SRM and LM carried out FACE analyses. MS carried out the analysis of percent amylose. RDC, TJF, SRM, LM, CJC and GTV carried out sequence analysis of candidate genes and genotyping/phenotyping of background markers. MAC, SRM, TJF, LM, RDC and GTV conducted field and glasshouse work. RB carried out statistical analysis of trait data. GTV and CJC devised and selected the single plant pea core. GTV conceived the project, secured funding, carried out population structure and association mapping statistical analysis, and supervised candidate gene characterization. GTV and MAC drafted the manuscript. All authors contributed to the manuscript draft and revisions, and approved the final manuscript.

Competing interests

The authors acknowledge that they have no competing interests.

Author details

[1]The New Zealand Institute for Plant & Food Research Limited, PO Box 4704, Christchurch, New Zealand. [2]USDA-ARS Western Regional Plant Introduction Station, 59 Johnson Hall, WSU Pullman, Pullman, Washington WA 99164-6402, USA.

References

1. Chibbar RN, Ambigaipalan P, Hoover R. REVIEW: molecular diversity in pulse seed starch and complex carbohydrates and its role in human nutrition and health. Cereal Chem. 2010;87:342–52.
2. McCrory MA, Hamaker BR, Lovejoy JC, Eichelsdoerfer PE. Pulse consumption, satiety, and weight management. Adv Nutr. 2010;1:17–30.
3. Buleon A, Colonna P, Planchot V, Ball S. Starch granules: structure and biosynthesis. Int J Biol Macromol. 1998;23:85–112.
4. Kossmann J, Lloyd J. Understanding and influencing starch biochemistry. Crit Rev Plant Sci. 2000;19:171–226.
5. Bogracheva T, Cairns P, Noel T, Hulleman S, Wang T, Morris V, Ring S, Hedley C. The effect of mutant genes at the r, rb, rug3, rug4, rug5 and lam loci on the granular structure and physicochemical properties of pea seed starch. Carbohydr Polym. 1999;39:303–14.
6. Zeeman SC, Kossmann J, Smith AM. Starch: its metabolism, evolution, and biotechnological modification in plants. Annu Rev Plant Biol. 2010;61:209–34.
7. Jobling S. Improving starch for food and industrial applications. Curr Opin Plant Biol. 2004;7:210–8.
8. Wang K, Henry RJ, Gilbert RG. Causal relations among starch biosynthesis, structure, and properties. Springer Science Reviews. 2014;2:15–33.
9. Jane J, Chen Y, Lee L, McPherson A, Wong K, Radosavljevic M, Kasemsuwan T. Effects of amylopectin branch chain length and amylose content on the gelatinization and pasting properties of starch. Cereal Chem. 1999;76:629–37.
10. Liu F, Ahmed Z, Lee EA, Donner E, Liu Q, Ahmed R, Morell MK, Emes MJ, Tetlow IJ. Allelic variants of the amylose extender mutation of maize demonstrate phenotypic variation in starch structure resulting from modified protein-protein interactions. J Exp Bot. 2012;63:1167–83.
11. Nishi A, Nakamura Y, Tanaka N, Satoh H. Biochemical and genetic analysis of the effects of amylose-extender mutation in Rice endosperm. Plant Physiol. 2001;127:459–72.
12. Luo J, Jobling SA, Millar A, Morell MK, Li Z. Allelic effects on starch structure and properties of six starch biosynthetic genes in a rice recombinant inbred line population. Rice (N Y). 2015;8:15.
13. Simsek S, Tulbek MC, Yao Y, Schatz B. Starch characteristics of dry peas (Pisum Sativum L.) grown in the USA. Food Chem. 2009;115:832–8.
14. Wang S, Sharp P, Copeland L. Structural and functional properties of starches from field peas. Food Chem. 2011;126:1546–52.
15. Hedley C, Bogracheva T, Wang T. A genetic approach to studying the morphology, structure and function of starch granules using pea as a model. Starch. 2002;54:235–42.
16. Wang T, Bogracheva T, Hedley C. Starch: as simple as a, B, C? J Exp Bot. 1998;49:481–502.
17. Myers A, Morell MK, James M, Ball S. Recent progress towards understanding biosynthesis of the amylopectin crystal. J Plant Physiol. 2000;122:989–97.
18. Sonnewald U, Kossmann J. Starches–from current models to genetic engineering. Plant Biotechnol J. 2013;11:223–32.
19. Pfister B, Lu KJ, Eicke S, Feil R, Lunn JE, Streb S, Zeeman SC. Genetic evidence that chain length and branch point distributions are linked determinants of starch granule formation in Arabidopsis. Plant Physiol. 2014;165:1457–74.
20. Abe N, Asai H, Yago H, Oitome N, Itoh R, Crofts N, Nakamura Y, Fujita N. Genetic evidence that chain length and branch point distributions are linked determinants of starch granule formation in Arabidopsis. Plant Physiol. 2014;165:1457–74.
21. Brust H, Lehmann T, D'Hulst C, Fettke J. Analysis of the functional interaction of Arabidopsis starch synthase and branching enzyme isoforms reveals that the cooperative action of SSI and BEs results in glucans with polymodal chain length distribution similar to amylopectin. PLoS One. 2014;9:e102364.
22. Li C, Wu AC, Go RM, Malouf J, Turner MS, Malde AK, Mark AE, Gilbert RG. The characterization of modified starch branching enzymes: toward the control of starch chain-length distributions. PLoS One. 2015;10:e0125507.
23. Tian Z, Qian Q, Liu Q, Yan M, Liu X, Yan C, Liu G, Gao Z, Tang S, Zeng D, et al. Allelic diversities in rice starch biosynthesis lead to a diverse array of rice eating and cooking qualities. Proc Natl Acad Sci U S A. 2009;106:21760–5.
24. Wilson LM, Whitt SR, Ibanez AM, Rocheford TR, Goodman MM, Buckler ES. Dissection of maize kernel composition and starch production by candidate gene association. Plant Cell. 2004;16:2719–33.
25. Cook JP, McMullen MD, Holland JB, Tian F, Bradbury P, Ross-Ibarra J, Buckler ES, Flint-Garcia SA. Genetic architecture of maize kernel composition in the nested association mapping and inbred association panels. Plant Physiol. 2012;158:824–34.
26. Hill H, Slade Lee L, Henry RJ. Variation in sorghum starch synthesis genes associated with differences in starch phenotype. Food Chem. 2012;131:175–83.
27. Murray S, McKenzie M, Butler R, Baldwin S, Sutton K, Batey I, Timmerman-Vaughan GM. Quantitative, small-scale, fluorophore-assisted carbohydrate electrophoresis implemented on a capillary electrophoresis-based DNA sequence analyzer. Anal Biochem. 2011;413:104–13.
28. Myles S, Peiffer J, Brown PJ, Ersoz ES, Zhang Z, Costich DE, Buckler ES. Association mapping: critical considerations shift from genotyping to experimental design. Plant Cell. 2009;21:2194–202.
29. Price AL, Patterson NJ, Plenge RM, Weinblatt ME, Shadick NA, Reich D. Principal components analysis corrects for stratification in genome-wide association studies. Nat Genet. 2006;38:904–9.
30. Pritchard JK, Stephens M, Donnelly P. Inference of population structure using multilocus genotype data. Genetics. 2000;155:945–59.
31. Yu J, Pressoir G, Briggs WH, Vroh Bi I, Yamasaki M, Doebley JF, McMullen MD, Gaut BS, Nielsen DM, Holland JB, et al. A unified mixed-model method for association mapping that accounts for multiple levels of relatedness. Nat Genet. 2006;38:203–8.
32. Coyne CJ, Brown AF, Timmerman-Vaughan GM, McPhee KE. USDA-ARS refined pea core collection for 26 quantitative traits. Pisum Genetics. 2005;37:3–6.
33. Baggett J, Hampton R. Oregon B442-15 and B445-66 pea seed-borne mosaic virus-resistant breeding lines. Hort Science. 1997;12:506.
34. Altschul SF, Madden TL, Schaffer AA, Zhang J, Zhang Z, Miller W, Lipman DJ. Gapped BLAST and PSI-BLAST: a new generation of protein database search programs. Nucleic Acids Res. 1997;25:3389–402.

35. Timmerman GM, Frew TJ, Miller AL, Weeden NF, Jermyn WA. Linkage mapping of *sbm-1*, a gene conferring resistance to pea seed-borne mosaic virus, using molecular markers in *Pisum sativum*. Theor Appl Genet. 1993;85:609–15.

36. Ibrahim A, Hofman-Bang HJP, Ahring BK. Amplification and direct sequence analysis of the 23S rRNA gene from thermophilic bacteria. BioTechniques. 2001;30:414–20.

37. Larkin MA, Blackshields G, Brown NP, Chenna R, McGettigan PA, McWilliam H, Valentin F, Wallace IM, Wilm A, Lopez R, et al. Clustal W and Clustal X version 2.0. Bioinformatics. 2007;23:2947–8.

38. Shin JH, Blay S, McNeney B, Graham J. LDheatmap: An R Function for Graphical Display of Pairwise Linkage Disequilibria between Single Nucleotide Polymorphisms. J Stat Software. 2006;16:Code snippet 3.

39. Choi Y, Sims GE, Murphy S, Miller JR, Chan AP. Predicting the functional effect of amino acid substitutions and indels. PLoS One. 2012;7:e46688.

40. Takeda Y. C CT, Mizukami H, Hanashiro I. Structures of large, medium and small starch granules of barley grain. Carbohydr Polym. 1999;38:109–14.

41. McCullagh P, Nelder J. Generalized linear models. London, UK: Chapman and Hall; 1989.

42. Greenacre M. Correspondance analysis in practice. Boca Raton FL, USA: Chapman and Hall/CRC Press; 2007.

43. Committee GS. The guide to Genstat (release 14). Oxford: VSN International; 2011.

44. Kaufman RC, Wilson JD, Bean SR, Herald TJ, Shi YC. Development of a 96-well plate iodine binding assay for amylose content determination. Carbohydr Polym. 2015;115:444–7.

45. Lee Y, Nelder J, Pawitan Y. Generalized linear models with random effects: unified analysis via H-likelihood. London: Chapman and Hall/CRC Press; 2006.

46. Committee GS. Genstat reference manual (release 17). VSN International: Hemel Hempsted, UK; 2014.

47. GenStat Committee. The Guide to GenStat Command Language (Release 17). Hemel Hempsted, UK.: VSN International; 2014.

48. Bradbury PJ, Zhang Z, Kroon DE, Casstevens TM, Ramdoss Y, Buckler ES. TASSEL: software for association mapping of complex traits in diverse samples. Bioinformatics. 2007;23:2633–5.

49. Storey JD, Tibshirani R. Statistical significance for genomewide studies. Proc Natl Acad Sci U S A. 2003;100:9440–5.

50. Kampstra P. Beanplot: a boxplot alternative for visualisation of distributions. J Stat Software. 2008;28:Code Snippet 1.

51. Klucinec J, Thompson DB. Fractionation of high-amylose maize starches by differential alcohol precipitation and chromatography of the fractions. Cereal Chem. 1998;75:887–96.

52. Yao Y, Guiltinan MJ, Thompson DB. High-performance size-exclusion chromatography (HPSEC) and fluorophore-assisted carbohydrate electrophoresis (FACE) to describe the chain-length distribution of debranched starch. Carbohydr Res. 2005;340:701–10.

53. Ratnayake W, Hoover R, Shahidi F, Perera C, Jane J. Composition, molecular structure, and physicochemical properties of starches from four field pea (Pisum Sativum) cultivars. Food Chem. 2001;74:189–202.

54. Larsson SJ, Lipka AE, Buckler ES. Lessons from Dwarf8 on the strengths and weaknesses of structured association mapping. PLoS Genet. 2013;9: e1003246.

55. Jing R, Vershinin A, Grzebyta J, Shaw P, Smykal P, Marshall D, Ambrose MJ, Ellis THN, Flavell AJ. The genetic diversity and evolution of field pea (Pisum) studied by high throughput retrotransposon based insertion polymorphism (RBIP) marker analysis. BMC Evol Biol. 2010;10:44.

56. Kwon S-J, Brown AF, Hu J, McGee R, Watt C, Kisha T, Timmerman-Vaughan G, Grusak M, McPhee KE, Coyne CJ. Genetic diversity, population structure and genome-wide marker-trait association analysis emphasizing seed nutrients of the USDA pea (*Pisum sativum* L.) core collection. Genes & Genomics. 2012;34:305–320.

57. Jing R, Ambrose MA, Knox MR, Smykal P, Hybl M, Ramos A, Caminero C, Burstin J, Duc G, van Soest LJ, et al. Genetic diversity in European Pisum germplasm collections. Theor Appl Genet. 2012;125:367–80.

58. Huang X, Han B. Natural variations and genome-wide association studies in crop plants. Annu Rev Plant Biol. 2014;65:531–51.

59. Craig J, Lloyd JR, Tomlinson K, Barber L, Edwards A, Wang TL, Martin C, Hedley C, Smith AM. Mutations in the gene encoding starch synthase II profoundly alter amylopectin structure in pea embryos. Plant Cell. 1998;10:413–26.

60. Lloyd JR, Hedley C, Bull VJ, Ring SG. Determination of the effect of r and rb mutations on the structure of amylose and amylopectin in pea (Pisum Sativum L.). Carbohydr Polym. 1996;29:45–9.

61. Jha AB, Tar'an B, Diapari M, Warkentin TD. SNP variation within genes associated with amylose, total starch and crude protein concentration in field pea. Euphytica. 2015;206:459–71.

6

Transcriptomic changes reveal gene networks responding to the overexpression of a blueberry *DWARF AND DELAYED FLOWERING 1* gene in transgenic blueberry plants

Guo-qing Song[1]* and Xuan Gao[1,2]

Abstract

Background: Constitutive expression of the CBF/DREB1 for increasing freezing tolerance in woody plants is often associated with other phenotypic changes including dwarf plant and delayed flowering. These phenotypic changes have been observed when *Arabidopsis DWARF AND DELAYED FLOWERING 1* (*DDF1*) was overexpressed in *A. thaliana* plants. To date, the *DDF1* orthologues have not been studied in woody plants. The aim of this study is to investigate transcriptomic responses to the overexpression of blueberry (*Vaccinium corymbosum*) *DDF1* (herein, *VcDDF1*-OX).

Results: The *VcDDF1*-OX resulted in enhanced freezing tolerance in tetraploid blueberry plants and did not result in significant changes in plant size, chilling requirement, and flowering time. Comparative transcriptome analysis of transgenic 'Legacy-VcDDF1-OX' plants containing an overexpressed *VcDDF1* with non-transgenic highbush blueberry 'Legacy' plants revealed the *VcDDF1*-OX derived differentially expressed (DE) genes and transcripts in the pathways of cold-response, plant flowering, DELLA proteins, and plant phytohormones. The increase in freezing tolerance was associated to the expression of cold-regulated genes (CORs) and the ethylene pathway genes. The unchanged plant size, dormancy and flowering were due to the minimal effect of the *VcDDF1*-OX on the expression of DELLA proteins, flowering pathway genes, and the other phytohormone genes related to plant growth and development. The DE genes in auxin and cytokinin pathways suggest that the *VcDDF1*-OX has also altered plant tolerance to drought and high salinity.

Conclusion: A *DDF1* orthologue in blueberry functioned differently from the *DDF1* reported in *Arabidopsis*. The overexpression of *VcDDF1* or its orthologues is a new approach to increase freezing tolerance of deciduous woody plant species with no obvious effect on plant size and plant flowering time.

Keywords: Abiotic stress, Cold hardiness, C-repeat-binding factor, *DDF1*, Dehydration responsive element-binding factor, Freezing tolerance, *Vaccinium corymbosum*, Woody plant

* Correspondence: songg@msu.edu
[1]Plant Biotechnology Resource and Outreach Center, Department of Horticulture, Michigan State University, East Lansing, MI 48824, USA
Full list of author information is available at the end of the article

Background

The APETALA2/ethylene response (AP2/ERF) transcription factors play a significant role in plant responses to several abiotic stresses (e.g., cold, dehydration, and high salinity) [1]. Accordingly, there have been many recent studies on genome-wide analysis of the AP2/ERF in several plant species, including black cottonwood (*Populus trichocarpa*) [2], *Brassica oleracea* [3], carrot (*Daucus carota* L.) [4], Chinese cabbage (*Brassica rapa ssp. pekinensis*) [5], *Eucalyptus grandis* [6], *Lotus corniculatus* [7], *Medicago truncatula* [8], moso bamboo (*Phyllostachys edulis*) [9], Musa species [10], peach (*Prunus persica*) [11], physic nut (*Jatropha curcas* L.) [12], *Salix arbutifolia* [13], sweet orange (*Citrus sinensis*) [14], and tea (*Camellia sinensis*) [15]. The CBF/DREB1 (C-repeat-binding factor/dehydration responsive element-binding factor 1) genes belong to a large family of AP2/ERF transcription factors and have a conserved DNA binding domain that recognizes the dehydration-responsive element/C-repeat (DRE/CRT) *cis*-acting element in the promoters of their target genes [1, 16, 17]. Studies suggest that all plant species undergo cold acclimation through a similar process that belongs, at least partially, to the CBF/DREB1-mediated cold-response pathway [1, 18–21]. This CBF/DREB1 pathway has been documented in *Arabidopsis thaliana* [16, 22–24]. Additionally, DREB2 transcription factors function in both drought- and heat-stress responses [25–27]. As global warming poses abiotic stresses (e.g., temperature changes and drought) to numerous plant species and threatens the world's sustainable food production for a growing population, numerous studies have been done to evaluate the potential use of AP2/ERF transcription factors to enhance plant tolerance to abiotic stresses [28].

The usefulness of the CBF/DREB1 pathway genes to enhance freezing tolerance has been demonstrated in both herbaceous and woody plant species [16, 29–31]. However, modulating expression of the CBF/DREB1 pathway genes for enhancing tolerance to abiotic stress is often associated with undesirable changes in plant growth and development [32–35]. For example, the constitutive expression of a peach (*Prunus persica*) CBF1 gene (*PpCBF1*) in apple rootstock resulted in both improved freezing tolerance and altered plant growth and dormancy [36]. In another study, transgenic grape vines over-expressing a grape (*Vitis vinifera*) CBF (*VvCBF4*) showed a slight increase in freezing tolerance in non-cold-acclimated vines and dwarf phenotypes [35]. Similar response for CBF overexpression has been reported in *A. thaliana* and other species [32–35]. Ectopic expression of a CBF1 orthologue from European bilberry (*Vaccinium myrtillus*) enhanced freezing tolerance of *A. thaliana* plants and reduced rosette diameter [37]. Collectively, beyond freezing tolerance, the over-expression of CBFs is often associated with reduced plant growth (*CBF1, CBF3,* and *CBF4*) and altered developmental processes such as flowering time, leaf senescence, and plant longevity (*CBF2* and *CBF3*) [24, 32, 33, 35, 38–41]. The occurrence of these additional phenotypic changes is due to the complexity of the *CBFs*-mediated low-temperature regulatory networks and these changes are sometimes considered desirable for crop/fruit production [35, 40, 42, 43]. Of the major CBF/DREB1 transcription factors, the DRE1E_ARATH [designated as *DWARF AND DELAYED FLOWERING 1* (*DDF1*)] and DRE1F_ARATH (designated as *DDF2*) have not been studied in crops [44–48]. *Arabidopsis thaliana* plants overexpressing the *DDF1* showed dark-green leaves, dwarfism, and late flowering; concurrently, the plants displayed enhanced tolerance to cold, drought, heat and high salinity [45, 46, 48].

Overexpression of *CBF2* in *A. thaliana* is mainly associated with delayed leaf senescence and extended plant longevity; additionally, overexpression of *Muscadinia rotundifolia CBF2* gene in *Muscadinia rotundifolia* resulted in growth retardation, dwarfism, late flowering, and abiotic stress tolerance [40, 49]. In this study, we showed a blueberry-derived *CBF* (*BB-CBF*), which was initially considered to be an orthologue of *CBF2* that promoted freezing tolerance in *A. thaliana* [50], was more similar to *A. thaliana DDF1*. Overexpression of the *BB-CBF* (herein renamed as *VcDDF1*) enhanced cold tolerance in leaves and dormant buds but not in flower tissues of a southern highbush blueberry cultivar [51]. Regardless of whether *BB-CBF* is a *DDF1* or *CBF2* orthologue, further studies are needed to facilitate a better understanding of the CBF/DREB1-mediate gene networks in blueberry. Unlike *A. thaliana*, few studies have been conducted to investigate the overall impact of the overexpression of a CBF/DREB1 pathway gene on transcriptomic changes in woody plant species.

Comparative transcriptome analysis is a powerful tool used to identify differential gene expression caused by overexpression of a transgene [52]. For example, overexpression of blueberry *FLOWERING LOCUS T* (*VcFT*) in blueberry plants resulted in plant dwarfing and early flowering [53]. Transcriptome analysis of these transgenic plants revealed differentially expressed (DE) genes in flowering and phytohormone pathway genes that are involved in the phenotypic changes driven by *VcFT*-overexpression [54, 55]. The aim of this study is to elucidate transcriptomic responses to the overexpression of *VcDDF1* (herein, *VcDDF1*-OX) and predict overall performance of *VcDDF1*-OX transgenic blueberry plants. The analysis of DE genes focused on the pathways related to plant growth, flowering or freezing tolerance in blueberry such as plant flowering, CBF-mediated cold/freezing tolerance, phytohormones and DELLA proteins [51, 54, 55].

Results

VcDDF1 and VcDDF1-OX in blueberry

The *VcDDF1* was initially designated as *BB-CBF* (Gen-Bank: FJ222601.1) due to its similarity to *A. thaliana CBF2*, and this reasoning is valid when *DDF1* is not included in phylogenetic analysis [50]. However, in our recent transcriptome analysis of highbush blueberry using Trinity and Trinotate [56], the *BB-CBF* was annotated as DRE1E_ARATH (*DDF1*). Our designation of *BB-CBF* as *VcDDF1* is the result of the phylogenetic analysis of *A. thaliana* CBF/DREB1 (i.e., CBF1, CBF2, CBF3, DDF1, and DDF2) and the blueberry-derived DRE1E_ARATH, DRE1A_ARATH, DRE1B_ARATH, and DRE1F_ORYSJ, which showed that *BB-CBF* is 52.5% similar to *DDF1* compared 45.9% to *CBF2* (Fig. 1a). The *CBF2* orthologues in blueberry were then assigned to the other two gene contigs (c88132_g2 and c85919_g2 in Fig. 1a). It is interesting to note that *VcDDF1* orthologues in many other woody plants are often annotated as DREB1 due to the conserved ERF/AP2 DNA-binding domains (Fig. 1b).

To investigate the effect of *VcDDF1*-OX at transcript levels in non-acclimated floral buds, comparative transcriptome analysis was conducted in non-transgenic 'Legacy' plants and plants of a representative transgenic event 'Legacy-VcDDF1-OX' with a single copy of transgenes (named as II7 in our previous report [51]). The 'Legacy-VcDDF1-OX' showed a 145-fold increase in the expression of the *VcDDF1* in comparison to the non-transgenic 'Legacy' plants. The high *VcDDF1* expression supported our previous observation that the 'Legacy-VcDDF1-OX' transgenic event showed high freezing tolerance in electrolyte leakage assays [51].

Effect of the VcDDF1-OX on plant freezing tolerance

Constitutive expression of *VcDDF1* resulted in increased freezing tolerance in detached tissues of *A. thaliana* and blueberry plants [50, 51]. In this study, the *VcDDF1*-OX enhanced freezing tolerance in intact plants. The freezing tolerance in (45) four-year plants, one of non-transgenic 'Legacy', two of 'Legacy- pCAMBIA' events, and 41 of 'Legacy-VcDDF1' transgenic events, was investigated. The 'Legacy-VcDDF1' transgenic plants showed a significantly higher survival rate ($p = 0.000126$) than those of the non-transgenic 'Legacy' and transgenic 'Legacy-pCAMBIA' controls (Fig. 2a).

In the winter of 2015, we also investigated the freezing tolerance of 12 three-year plants for both the non-transgenic 'Legacy' and the 'Legacy-VcDDF1-OX' transgenic event. The 'Legacy-VcDDF1-OX' transgenic plants exhibited a higher plant survival rate (83.3%) than the non-transgenic plants (41.7%) (Fig. 2b). Applying a freezing shock of −12 °C for 15 min resulted in visual differences between transgenic 'Legacy-VcDDF1-OX' and non-transgenic 'Legacy' plants during the plant recovery process. For non-transgenic plant, all leaves and over 90% of the buds showed dying symptoms and died in three weeks. In contrast, for transgenic plants, about 70% of the leaves had no survival leaf tissues, in three weeks; additionally, about 25% buds died. Overall, *VcDDF1*-OX enhanced freezing tolerance of the intact 'Legacy-VcDDF1-OX' plants (named as II7 in our previous report).

Effect of the VcDDF1-OX on plant growth and flowering

The *VcDDF1*-OX did not alter the growth of transgenic blueberry plants. When four-year transgenic plants of 11 independent 'Legacy-VcDDF1' events, including the representative transgenic event 'Legacy-VcDDF1-OX', were compared with those of non-transgenic 'Legacy' and transgenic control 'Legacy-pCAMBIA' plants, all plants looked similar in plant stature and appearance (Fig. 3a) and did not show any difference ($P < 0.05$) in plant height, the number of canes, or the number of flower buds (Fig. 3b, c). These results suggest that *VcDDF1*-OX has little phenotypic effect on blueberry plant growth and floral bud formation. Therefore, *VcDDF1* do not share the designated role of *DDF1* in inducing growth retardation and dwarfism.

In relation to both non-transgenic 'Legacy' plants and transgenic control 'Legacy-pCAMBIA', delayed flowering was not found in any 'Legacy-VcDDF1' plants. For example, 'Legacy-VcDDF1-OX' and non-transgenic 'Legacy' plants did not show significant differences in the number of floral buds, the age of plant flowering, and the yearly flowering time. Moreover, *VcDDF1*-OX did not affect the chilling requirement of 'Legacy-VcDDF1-OX' plants (Fig. 3d). Taken together, *VcDDF1*-OX is not associated with significant changes in plant growth and flowering of tetraploid blueberry plants unlike overexpression of *DDF1* in *A. thaliana* [45, 46].

Profile of differentially expressed (DE) genes induced by the VcDDF1-OX

To reveal the potential roles of the *VcDDF1* at gene transcription levels, comparative transcriptome analysis was conducted between the 'Legacy-VcDDF1-OX' and non-transgenic 'Legacy' plants. The *VcDDF1*-OX in non-acclimated floral buds of the 'Legacy-VcDDF1-OX' plants resulted in 2463 DE genes and 3644 DE transcripts, of which 1668 DE genes were annotated. These DE genes were classified in 54 over-represented Gene Ontology (GO) terms ($P < 0.05$) in the analysis using the GOSlim_-Plant as the selected GO file and *A. thaliana* annotation as a reference (Fig. 4). Of the 27 over-represented GO terms in biological_process, two highly over-represented GO terms (i.e., GO:0006950-response to stress and GO:0009628-response to abiotic stimulus) revealed the potential function of the *VcDDF1*-OX in affecting plant

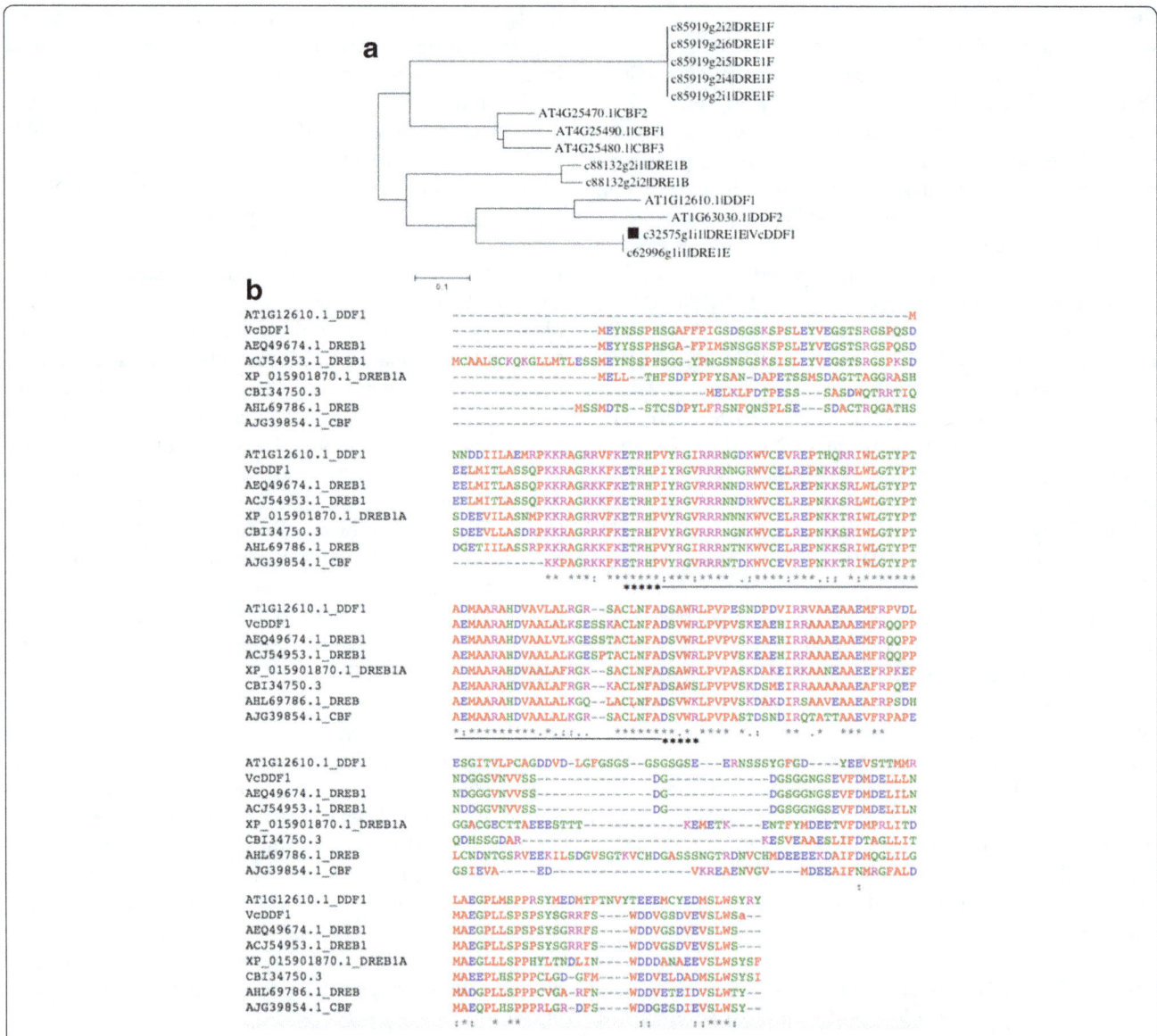

Fig. 1 Phylogenetic analysis of CBF/DREB1 proteins of blueberry and multiple protein sequence alignment of DDF1 and VcDDF1 orthologues. **a** Phylogenetic analysis of nucleotide sequences of CBF/DREB1 proteins of blueberry and *A. thaliana* using Neighbor Joining in MEGA 6.06. The bootstrap values were obtained from 500 replicates. The tree was drawn to scale, with branch length equal to substitutions per nucleotide. The black square shows the transgene *VcDDF1* and c62996-g1-i1 is another endogenous *VcDDF1*. **b** Multiple sequence alignment of DDF1, VcDDF1 and protein sequences of VcDDF1 orthologues from six plant species using Clustal Omega (https://www.ebi.ac.uk/Tools/msa/clustalo/). The ethylene-responsive element binding factor/APETELA2 (ERF/AP2) DNA-binding domain is underlined and the [ETRH and DS(A/V)WR] signatures are indicated by *. AEQ49674.1: DREB1 (*Vaccinium myrtillus*). ACJ54953.1: DREB1 (*Vaccinium vitis-idaea*). AHL69786.1: DREB (*Camellia sinensis*). AJG39854.1: CBF (*Actinidia chinensis*). XP_015901870.1: DREB1A (*Ziziphus jujube*). CBI34750.3: unnamed protein product (*Vitis vinifera*)

freezing tolerance as well as other abiotic stresses. Additionally, two other highly over-represented GO terms (i.e., GO:0007275-multicellular organismal development and GO:0009791-post-embryonic development) suggested that *VcDDF1*-OX could affect plant growth and flower development (Fig. 4a). The *VcDDF1*-OX functions at cellular levels (GO:0005263: cell) and intracellular (GO:0005622: intracellular) levels through its catalytic activity (GO:0003824), DNA binding (GO:0003677), transcription

factor activity (GO:0003700), and transcription regulator activity (GO:0030528) (Fig. 4b, c). Overall, the results suggest that *VcDDF1* is a functional DREB1 transcription factor and the *VcDDF1*-OX has an impact on gene expression of multiple pathways in blueberry.

The responses of blueberry COR genes to the *VcDDF1*-OX

We used 2445 *A. thaliana* COR genes to search for blueberry COR (VcCOR) genes in our blueberry transcriptome

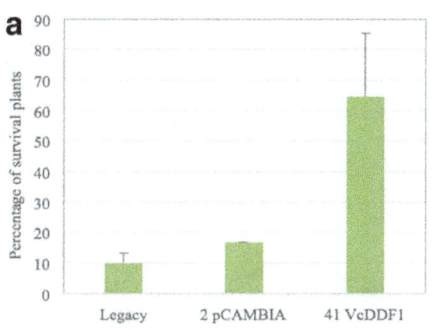

Fig. 2 Freezing tolerance in transgenic blueberry plants overexpressing a *VcDDF1*. **a** The survival rates of 3-year old blueberry plants of non-transgenic 'Legacy' (*n* = 10), transgenic 'Legacy-pCAMBIA' containing a control plasmid pCAMBIA2301 (*n* = 6), and 41 transgenic 'Legacy-VcDDF1' events (4–5 plants per event) after the exposure to unprotected environmental conditions in the winters of 2013 and 2014. The representative transgenic event 'Legacy-VcDDF1-OX' is included in the 41 events. **b** Under unprotected environmental conditions, one 4-year old 'Legacy' plant died and one 'Legacy-VcDDF1-OX' plant survived

reference Reftrinity (GenBank accession number: SRX2728597), which is developed using the RNA sequencing data of leaf, flower, and dormant bud tissues [54]. A total of 24,594 transcripts of 14,231 VcCOR genes showed similarities (e < −20) to 2181 *A. thaliana* COR genes. And 17 transcript contigs of 11 VcCOR genes are the ortholouges of *A. thaliana*

CBF1/DREB1B, *CBF2/DREB1C*, *CBF3/DREB1A*, *CBF4/DREB1D*, or *DDF1* (Fig. 1a).

In the dormant buds of 'Legacy-VcDDF1-OX' plants, 11,162 DE transcripts of 725 VcCOR genes showed high similarities (e < −20) to 1085 COR genes of *A. thaliana* (Additional file 1: Table S1). Of these DE VcCOR genes, the up-regulated *VcDDF1* and down-regulated *VcCBF2*

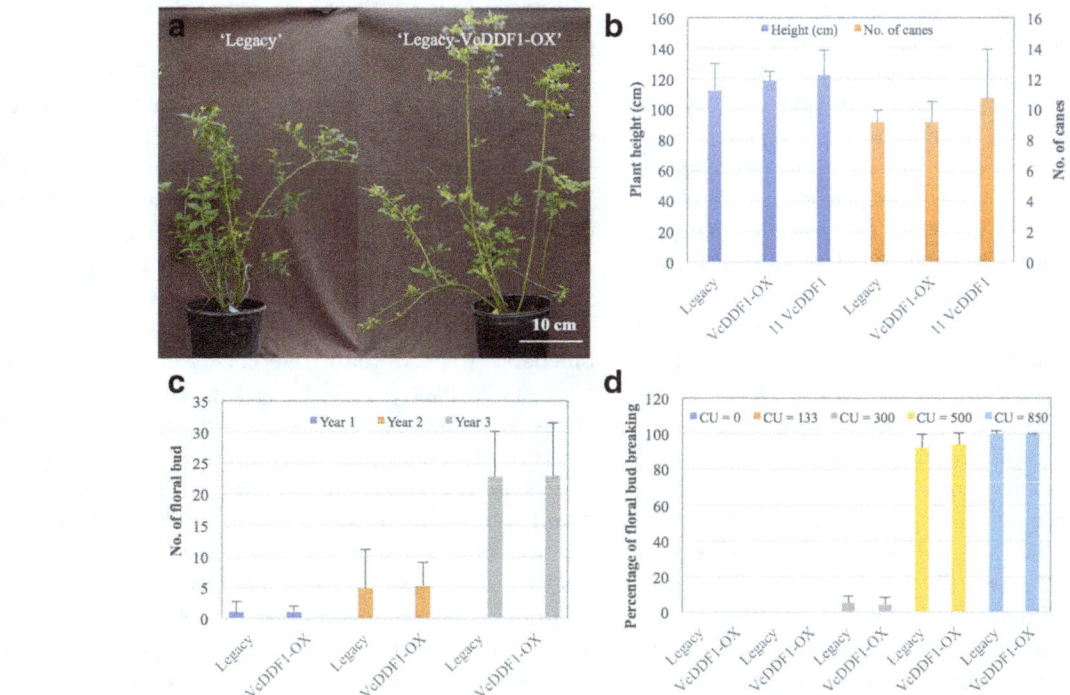

Fig. 3 Effect of overexpression of *VcDDF1* on blueberry plant growth and development. **a** Growth of 4-year old plants of 'Legacy' and 'Legacy-VcDDF1-OX' (herein VcDDF1-OX: a representative transgenic 'Legacy-VcDDF1' used for freezing tolerance assay [51] and RNA-seq analysis). **b** Average plant height and number of canes of 4-year old plants. Legacy: non-transgenic southern highbush cultivar 'Legacy', three plants. VcDDF1: transgenic 'Legacy' containing the *VcDDF1*, three plants for each of the 11 independent transgenic events. VcDDF1-OX: one representative transgenic Legacy-VcDDF1 event, three plants. **c** Average number of floral bud for 1- to 3-year old plants including 12 plants for each of 'Legacy' and 'VcDDF1-OX'. **d** Flowering of four-year old plants (five plants for each treatment) after receiving different amount of chilling. CU: chilling unit

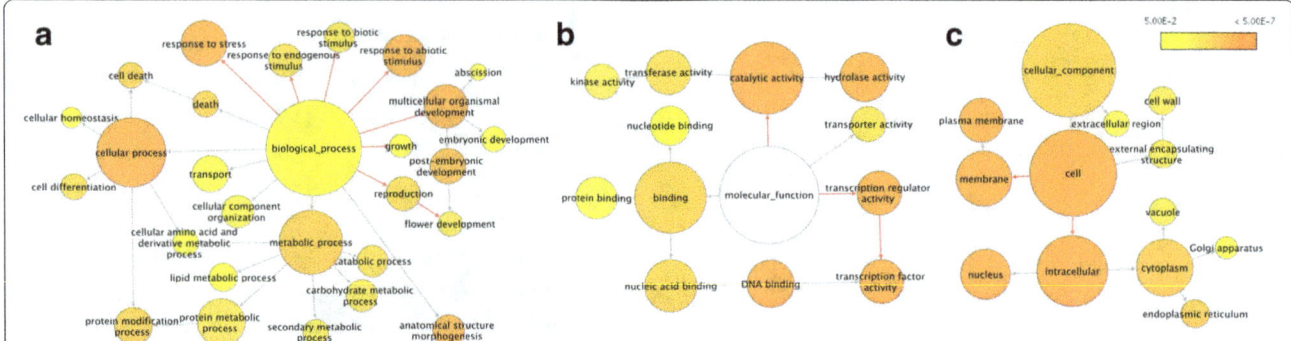

Fig. 4 Gene networks of differentially expressed genes in dormant bud tissues of 'Legacy-DDF1-OX' plants. The ontology file of GOSlim_Plants in BiNGO was used to identify overrepresented GO terms (*P* < 0.05). Bubble size and color indicates the frequency of GO term and *P*-value, respectively. Red arrows show examples of GO terms related to abiotic resistance, plant growth, and flowering (**a**), the major effect of VcDDF1-OX in molecular function (**b**), and cellular component (**c**)

are the DE CBF/DREB1 genes of blueberry. These results suggest *VcDDF1*-OX regulated some VcCOR genes, which contributed to increase freezing tolerance in intact plants of the 'Legacy-VcDDF1-OX' (Fig. 2).

The responses of blueberry floral genes to the *VcDDF1*-OX

Whereas overexpression of *DDF1* resulted in delaying *A. thaliana* plant flowering [45, 46], *VcDDF1*-OX did not result in visible changes in flowering of tetraploid blueberry plants (Fig. 3c, d). To investigate the potential impact of *VcDDF1*-OX on blueberry flowering, we searched for DE floral genes in the dormant buds of 'Legacy-VcDDF1-OX' using the floral gene list of blueberry [54]. Twenty-one floral genes derived from 44 transcripts of 32 gene contigs showed differential expression, of which seven floral genes were up-regulated and 14 were down-regulated (Table 1). This suggests *VcDDF1*-OX affects flowering pathway genes. However, none of the 21 DE floral genes showed changes above four folds. The expression of blueberry *SUPPRESSOR of OVEREXPRESSION OF CONSTANS 1* (*VcSOC1*) was reduced to 67.6% and the expression of blueberry *CONSTANS-LIKE 5* (*COL5*)-like gene was repressed to 70.5% as much as non-transgenic 'Legacy' plants. This down-regulated expression of *VcSOC1* and *VcCOL5* is theoretically associated with delayed flowering [57]. However, the blueberry *SHORT VEGETATIVE PHASE* (*SVP*)(*VcSVP*) showed a decreased expression, which in contrast theoretically promotes plant flowering. In spite of these DE floral genes, the *VcDDF1*-OX was insufficient to promote significant changes in floral bud formation, chilling requirement and flowering time of the 'Legacy-VcDDF1-OX' plants (Fig. 3c, d).

The responses of major phytohormone genes to the *VcDDF1*-OX

The *VcDDF1*-OX did not lead to dwarf 'Legacy-VcDDF1-OX' plants (Fig. 3a, b), which is inconsistent with the designated function of *DDF1* overexpression in causing dwarf *A. thaliana* plants [45, 46]. To evaluate the potential effect of the *VcDDF1*-OX on plant growth, we identified the DE pathway genes of five major phytohormones [i.e., ABA, GA, auxin (IAA), cytokinin, and ethylene] in dormant buds of 'Legacy-VcDDF1-OX' plants. Except for the ABA pathway, DE transcript contigs were found for all other pathways, including 11 for IAA, 23 for GA, five for cytokinin, and 49 for ethylene (Additional file 2: Table S2). The GA pathway has six and eight transcript contigs shared with those in the IAA and ethylene pathways, respectively, indicating the interaction of these pathway genes (Fig. 5a). In the ethylene pathway, 85 out of 86 DE transcripts showed a less than 4-fold change; and only one DE transcript contig of an orthologue of *ETHYLENE-INSENSITIVE5* (*EIN5*) was up-regulated to approximately ten fold. These DE phytohormone genes did not alter plant growth of the 'Legacy-VcDDF1-OX' plants (Fig. 3a, b).

The responses of DELLA proteins to the *VcDDF1*-OX

We found 79 transcript contigs of 47 gene contigs in the blueberry transcriptome reference Reftrinity that show high similarities (e < −20) to the five DELLA protein genes of *A. thaliana*. Of the 79 transcript contigs, two DE transcript contigs of two genes are the *RGL3* orthologues in the bud tissues of the 'Legacy-VcDDF1-OX' plants. One of them was repressed to 72.9% and another one was up-regulated to 143.8% (up-regulated by 43.8%). *VcDDF1*-OX poses little effect on the expression of DELLA protein genes in blueberry plants. This provides additional evidence to show the insignificant effect of *VcDDF1*-OX on blueberry plant growth and flowering (Fig. 3).

Confirmation of the expression of the selected DE transcripts

We designed six pairs of qRT-PCR primers, consisting of two pairs for GA and IAA pathways and one pair for

Table 1 DE floral genes in dormant bud tissues of 'Legacy-VcDDF1-OX' plants

Subject id	Floral gene	logFC	logCPM	P Value	FDR
c89508_g1_i1	ABF2	1.057	2.427	6.59E-07	0.000
c89508_g3_i4	ABF2	1.085	1.227	0.000	0.036
c89508_g1_i2	ABF3	0.891	2.990	9.35E-06	0.002
c86010_g1_i1	AGL19	−0.497	5.315	4.25E-05	0.006
c86010_g1_i2	AGL19	−0.605	4.495	0.000	0.017
c94107_g4_i5	AGL19	−0.995	1.153	0.001	0.047
c72632_g1_i1	AGL32	0.423	7.608	0.000	0.011
c97450_g4_i2	AP2	−0.593	3.993	9.23E-06	0.002
c97450_g4_i5	AP2	−0.706	2.875	0.001	0.041
c89508_g3_i3	AREB3	0.997	1.566	0.000	0.029
c99151_g2_i1	ARP6	−0.521	4.522	1.85E-05	0.003
c85121_g1_i1	ATCOL5	−0.503	5.503	6.02E-05	0.007
c91872_g2_i3	CIB1	−1.396	1.728	8.71E-08	0.000
c92899_g1_i1	CIB1	−0.592	3.203	0.000	0.016
c92899_g1_i2	CIB1	−0.659	2.907	6.03E-05	0.007
c94404_g2_i1	CIB1	−0.525	4.963	6.91E-06	0.001
c94438_g3_i2	CIB1	−0.258	8.194	4.55E-05	0.006
c80828_g1_i1	CKA3	−0.328	5.530	0.000	0.036
c84766_g4_i1	CKA3	−0.389	4.999	0.000	0.024
c88116_g1_i1	FUL	−0.402	4.725	0.000	0.036
c91613_g4_i2	GRF2	−0.321	6.129	0.001	0.037
c95520_g1_i1	OsELF3	−0.639	2.604	0.001	0.054
c95679_g4_i2	OsELF3	−0.665	3.706	2.05E-06	0.000
c96650_g1_i1[z]	OsELF3	0.610	3.533	0.000	0.027
c96822_g1_i1	OsELF3	−0.365	6.469	0.000	0.033
c96828_g2_i2	OsELF3	1.577	0.756	1.44E-05	0.002
c85043_g5_i1	OsGF14e	1.070	1.501	7.14E-05	0.008
c76027_g1_i1	PAF1	−0.496	7.167	7.35E-05	0.009
c91063_g2_i1	PRR9	1.141	1.828	6.87E-06	0.001
c86010_g1_i3	SOC1	−0.566	4.979	2.65E-06	0.001
c79187_g1_i1	SPL	−0.328	6.358	0.000	0.011
c79187_g1_i2	SPL	−0.327	6.224	0.000	0.018
c80807_g1_i1	SPL	−0.478	6.461	4.00E-08	0.000
c80807_g1_i2	SPL	−0.592	5.620	2.05E-10	0.000
c80807_g1_i3	SPL	−0.513	6.397	7.63E-11	0.000
c81320_g1_i1 [y]	SPL	0.938	1.576	0.000	0.031
c93310_g3_i1	SPL	−0.459	6.681	2.82E-10	0.000
c93310_g3_i2	SPL	−0.430	5.768	1.07E-05	0.002
c88116_g2_i1	SVP	−0.527	4.975	6.67E-06	0.001
c91377_g1_i14	SVP	−1.427	0.741	1.70E-05	0.003

Table 1 DE floral genes in dormant bud tissues of 'Legacy-VcDDF1-OX' plants *(Continued)*

c98453_g2_i3	TOE1	−0.680	3.278	3.46E-06	0.001
c98453_g2_i4	TOE1	−0.727	2.911	3.54E-05	0.005
c98453_g2_i5	TOE1	−0.536	3.998	7.14E-06	0.001
c98453_g2_i7	TOE1	−0.455	3.776	0.001	0.052
c87192_g5_i5	ZmIDS1	−0.962	1.098	0.001	0.053

[z]also annotated as PCL1_ARATH; [y] also annotated as SPL12_ARATH
LogFC: $\log_2$(fold change) = $\mathrm{Log_2}$(Legacy-VcDDF1-OX/Legacy). FDR: false discovery rate. LogCPM: $\mathrm{Log_2}$Count per million reads

each of cytokinin and ethylene pathways, to validate the DE transcripts of four phytohormone pathways. These selected DE transcripts (FDR < 0.05) often play important roles in their pathways. Of the six DE transcripts tested, qRT-PCR results and RNA-seq data of five

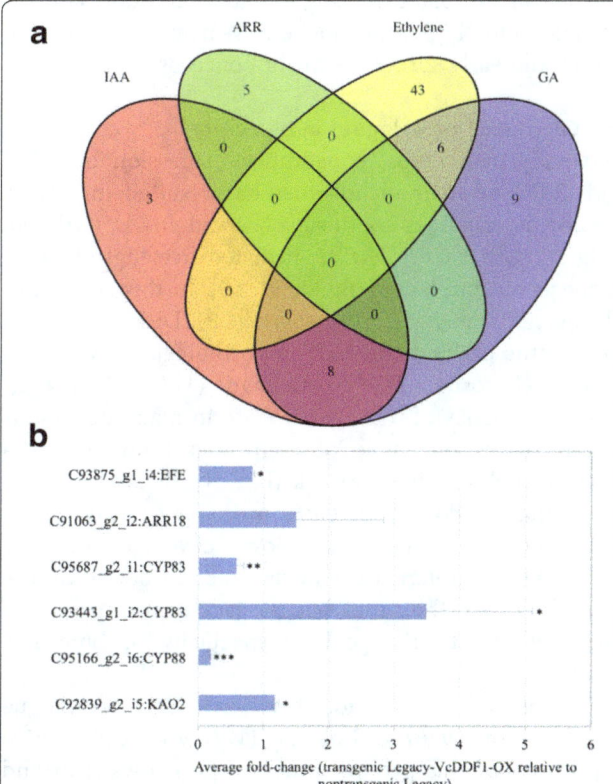

Fig. 5 Differentially expressed genes (in comparison to non-transgenic 'Legacy') in dormant bud tissues of 'Legacy-DDF1-OX' plants. **a** Major phytohormone gene contigs, i.e., gibberellin (GA), abscisic acid (ABA), ARR related genes of cytokinin, and indole acetic acid (IAA). **b** qRT-PCR analysis of representative transcripts. Eukaryotic translation initiation factor 3 subunit H is the internal control. Relative expression (fold-change) in Legacy-VcDDF1-OX was calculated by $2^{-\Delta\Delta Ct}$, $\Delta\Delta Ct = (Ct_{GOI} - Ct_{nom})_{Legacy-VcDDF1-OX} - (Ct_{GOI} - Ct_{nom})_{Legacy}$. Average fold-changes ± standard error of three biological replicates for each of 'Legacy-VcDDF1-OX' and 'Legacy' plants were plotted. Significant average fold-change determined using a Student's t-test is denoted. An asterisk (*) indicates $p < 0.05$; double asterisks (**) indicate $p < 0.01$; and triple asterisks (***) indicate $p < 0.001$

transcripts correlated very well (Fig. 5b; Additional file 3: Table S3); only one DE transcript revealed by RNA-seq did not show significant difference ($p < 0.05$) in qRT-PCR analysis (transgenic 'Legacy-VcDDF1-OX versus non-transgenic 'Legacy' samples) but it showed an increase in a regular RT-PCR analysis (Additional file 4: Fig. S1). These results suggest that our RNA-seq data analysis appears reliable for identification of DE genes.

Discussion

Plant freezing tolerance depends on many factors, such as natural environment, plant species/genotypes, plant developmental stages, acclimation state, organs, and tissues [58]. For woody fruit crops, global warming poses concerns for its impact on the phenology of plant dormancy and freezing tolerance. To address these concerns, a thorough understanding of the genetics and mechanisms of plant freezing tolerance and dormancy is needed. With highbush blueberries, freezing injuries in winter and early spring are major concerns.

The CBF/DREB1 orthologues in blueberry

In woody fruit crops, constitutive expression of *CBF1* and *CBF4* or their orthologues has resulted in similar phenotypic changes to those observed in A. thaliana, suggesting that *CBF/DREB1* mediated-freezing tolerance is conserved in plants [35, 37, 59, 60]. In this study, our phylogenetic analysis of blueberry CBF/DREB1 proteins suggest the previous *BB-CBF* (an orthologue of *CBF2*) is more likely to be a *DDF1* orthologue (*VcDDF1*) (Fig. 1). The orthologues of this *VcDDF1* are in many deciduous woody plants but none of them was annotated as a *CBF2* orthologue in GenBank (Fig. 1b). It is also interesting that we do not see *CBF1* orthologues in our transcriptome reference. The low coverage of our transcriptome reference may have contributed to the lack of *CBF1* orthologues but the lack of orthologues is probably due to the genome specificity of blueberry plants.

Regardless of distinction between *CBF2* orthologue (*BB-CBF*) or *DDF1* orthologue (*VcDDF1*), constitutive expression of this gene is anticipated for dwarfism and late flowering of transgenic plants if its designated function is conserved [45, 46, 49]. However, this is not the phenotypic change observed in transgenic *Arabidopsis* and blueberry plants, where the *VcDDF1*-OX enhanced plant freezing tolerance (Fig. 2; Fig. 3) [50, 51]. These results suggest the function of *DREB1* orthologues in different plant species may vary from their functions designated in *Arabidopsis* (Fig. 6).

Effect of the *VcDDF1*-OX on plant flowering

In A. thaliana, the delayed flowering caused by overexpression of *CBF1,2,3* was due to the increased expression of *FLOWERING LOCUS C* (*FLC*) and repressed expression of *SOC1* [61]. In this study, *VcDDF1*-OX repressed the expression of *VcSOC1*, which is similar to the previous report [61]. However, none of the orthologues of *FLC* in blueberry, including the *MADS-AFFECTING FLOWERING 2*-like gene (*VcMAF2*), *MADS-AFFECTING FLOWERING 5*-like gene (*VcMAF5*), and *VERNALIZATION1*-like gene (*VcVRN1*), showed differential expression. In addition, the expression of *VcSVP*, an orthologue of A. thaliana *SVP* that acts as a negative regulator of A. thaliana plant flowering [62], was repressed. These differences between the response of the flowering pathway genes to the *VcDDF1*-OX in blueberry and those of the overexpression of *CBF1,2,3* in A. thaliana are responsible for the unchanged flowering phenotype caused by *VcDDF1*-OX.

Effect of the *VcDDF1*-OX on DELLA protein genes

In A. thaliana the delayed flowering and growth retardation caused by the overexpression of *CBF1,2,3* are due to the changes of DELLA proteins [39, 40]. The dwarf A. thaliana plants caused by overexpression of *DDF1* was because of reducing bioactive gibberellin [46, 48]. In this study, *VcDDF1*-OX induced little change to the expression of DELLA protein genes, providing further molecular evidence to support that normal growth and flowering of 'Legacy-VcDDF1-OX' plants,

Effect of the *VcDDF1*-OX on phytohormone genes

In A. thaliana, the cold-response of *CBF1*, *CBF2*, and *CBF3* is ABA independent while the response of *CBF4* is ABA dependent [41]. Additionally, ethylene signaling can affect expression of *CBFs* [63, 64], and DELLA proteins responding to the overexpression of *CBF/DREB1* genes are GA-related [46, 48]. It seems that *CBF/DREB1* genes interact with phytohormone genes to affect plant growth and development. In this study, *VcDDF1*-OX in tetraploid blueberry plants affected gene expression of the synthesis pathways of IAA, cytokinin, GA, and ethylene but not ABA (Fig. 5a; Additional file 2: Table S2). The 49 DE transcript contigs of 36 gene contigs in the ethylene pathway could contribute to the increased freezing tolerance [63, 64]. The 23 DE transcript contigs of 16 gene in the GA pathway did not show any changes over 4-folds and may have affected the minor changes in the two DE genes of DELLA proteins. Eleven DE transcript contigs of nine genes are orthologues of two IAA pathway genes of A. thaliana, including *CYP83B1* and *AUXIN TRANSPORTER PROTEIN 1*(*AUX1*) (c90563_g2_i1); both genes are key regulators of root growth and development [65, 66]. The orthologues of two cytokinin pathway genes A. thaliana *B-TYPE RESPONSE REGULATOR18* (*ARR18*) and A. thaliana *PSEUDO-RESPONSE REGULATOR 2*

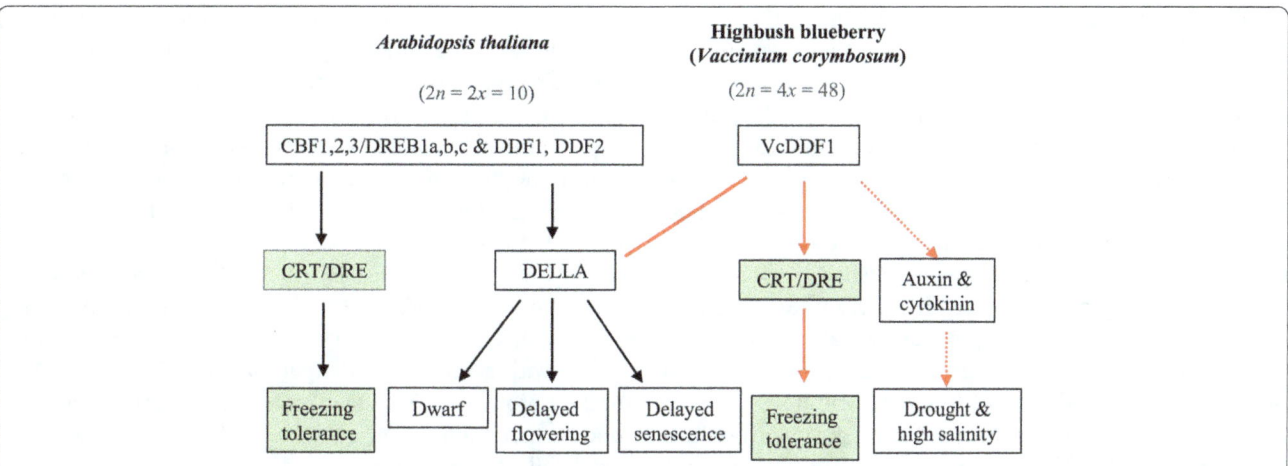

Fig. 6 Schematic diagram illustrating the effect of overexpressing a blueberry *DDF1* orthologue on plant freezing tolerance. The diagram illustrating the effect of overexpression of *CBF/DREB1* on freezing tolerance, plant growth and development in *A. thaliana* was derived from previous reports [40, 41, 46]. The overexpression of *VcDDF1* showed little effect on DELLA protein genes but could affect plant tolerance to drought and high salinity through altered gene expression in auxin and cytokinin pathways

(*APRR2*) include five DE transcript contigs, the up-regulated *ARR18* orthologue could promote root elongation [67]. The DE transcripts involved in auxin and cytokinin pathways have likely altered plant tolerance to drought or high salinity pending on further investigations [45, 46, 48]. The analysis of DE transcripts of phytohormone genes in addition to DELLA protein genes provide molecular evidence to support that *VcDDF1*-OX was not associated with dwarf and delayed flowering in tetraploid blueberry plants.

Effect of the *VcDDF1*-OX on freezing tolerance

In terms of its role in freezing tolerance, *VcDDF1* has the same function as *DDF1* and other *CBF/DREB1* genes in *A. thaliana* [48]. Based on both our previous electrolyte leakage assay of in vitro tissues [51] and freezing tolerance assay of intact plants in this study, we have demonstrated that *VcDDF1*-OX is able to enhance freezing tolerance in blueberry plants. In addition, the comparison of VcCOR genes in 'Legacy-VcDDF1-OX' plants with non-transgenic 'Legacy' has provided molecular evidence to support the role of overexpressed *VcDDF1* in enhanced freezing tolerance (Fig. 2). Of the DE orthologues of *CBF/DREB1* genes, *VcDDF1*-OX down-regulated *VcCBF2*, which did not alter plant growth and flowering.

Conclusion

In tetraploid blueberry plants, *VcDDF1*-OX resulted in enhanced freezing tolerance and normal plant growth and flowering compared to non-transgenic plants. The increased freezing tolerance is attributed to DE VcCOR genes, which are similar to the *DDF1* and the other *CBF/DREB1* genes (Fig. 6). In contrast to dwarf plant and delayed flowering associated with overexpression of *DDF1* or other *CBF/DREB1* [45, 46, 48], normal

phenotypes with regards to plant growth and flowering was due to minimal effect of overexpressed *VcDDF1* on the expression of DELLA proteins, flowering pathway genes, and other phytohormone genes related to plant growth (Fig. 6). The DE genes in phytohormone pathways of auxin and cytokinin imply that *VcDDF1*-OX might enhance plant tolerance to drought and high salinity.

This is the first known investigation of a *DDF1* orthologue in any crop. More importantly, this is the first time the overexpression of a *CBF/DREB1* orthologue was found to enhance plant freezing tolerance without altering plant growth and flowering time. This finding opens a new approach to increase freezing tolerance of deciduous woody plants by using overexpression of *VcDDF1* or its orthologues.

Method
Plant materials

A southern highbush blueberry cv. Legacy is tetraploid and needs over 800 chilling units (CU) for normal flowering. The 'Legacy' plants used in this study was original derived from the blueberry cultivar collections growing in a research field of the Horticulture Teaching and Research Center of Michigan State University. Transgenic 'Legacy' plants (herein 'Legacy-VcDDF1') contain a blueberry derived CBF gene (AVI45245.1), which was designated as *BB-CBF* [50, 51] and renamed as *VcDDF1* in this report. The 'Legacy-pCAMBIA' is a transgenic control for the *VcDDF1*. 'Legacy-VcDDF1-OX' is a representative transgenic 'Legacy-VcDDF1' that was used for RNA sequencing. The 'Legacy-VcDDF1-OX' (named as II7) contains a single copy of transgenes and showed high freezing tolerance [51]. Production of the 'Legacy-

VcDDF1' and Legacy-pCAMBIA' containing the binary vector pCAMBIA2301 was described in our previous report [51].

All non-transgenic and transgenic plants were obtained through micropropagation of in vitro cultured shoots. Plant age was determined based on the time after the shoot was rooted in soil. Rooting of in vitro cultured shoots and plant growth in the greenhouse were performed according the protocols established by Song [68]. All plants were grown normally and were fully vernalized unless otherwise mentioned. For full vernalization in winter, plants were potted and grown in a non-heated hoop house or in a secured courtyard under natural light conditions at Michigan State University, East Lansing, Michigan (latitude 42.701847, longitude –84.482170). The average low and high temperatures in January are –10.6 °C and –1.8 °C, respectively (http://www.usclimatedata.com/climate/east-lansing/michigan/united-states/usmi0248).

Plant growth and flowering

Four-year old plants were planted in 4-gal pots in 2009 and were grown in a hoop house for winters and were not pruned. These plants included 22 non-transgenic 'Legacy', 12 plants of two 'Legacy-pCAMBIA' events (6 plants per transgenic event), and 263 plants of 41 'Legacy-VcDDF1' events (5–8 plants/event). Thirty-nine selected plants, including three plants for each of the non-transgenic 'Legacy', 11 independent transgenic 'Legacy-VcDDF1' events, and one transgenic 'Legacy-pCAMBIA' event, were photographed and data was collected twice for plant height, the number of floral buds and the number of canes in October, 2012. The date of early-pink-bud of all plants, defined as the time that the first flower cluster appears, was recorded in the springs of 2009–2012. To test freezing tolerance of intact plants, 4-year old plants including, 10 non-transgenic 'Legacy' plants, six plants for each of the two 'Legacy-pCAMBIA' events, and 4–5 plants for each of the 41 'Legacy-VcDDF1' events were moved from the greenhouse to a secured courtyard under natural environmental conditions in October of 2013. The number of the survived plants was collected in May of 2015. Freezing tolerance of whole plants were tested by exposing to –12 °C for 15 min in 2012 using actively growing 4-year old plants of one 'Legacy-VcDDF1-OX' and one non-transgenic 'Legacy'. Both plants were then brought to the heated greenhouse with a temperature range of 23 °C - 30 °C under natural photoperiod for recovery. The recovery process was documented through weekly photographs for two months.

Chilling requirement of non-transgenic 'Legacy' and transgenic 'Legacy-VcDDF1-OX' plants was evaluated with five chilling treatment (i.e., 0, 133, 300, 500, and 850 CU) under controlled conditions in a hoop house in the Horticulture Teaching and Research Center at Michigan State University in the winter of 2012. For each treatment, three 'Legacy' and three 'Legacy-VcDDF1-OX' plants were used. These plants were three-year old and grown in one-gallon pots. The conversion of selected temperatures to chill units for highbush blueberry was based on the equation: total chill units = 0.5 × number of hours with temperatures below 2.4 °C and 9.2–12.4 °C + 1 × number of hours with temperatures 2.5–9.1 °C – 0.5 × number of hours with temperatures 16–18 °C -1 × number of hours with temperatures above 18 °C [69, 70]. After each chilling treatment, the plants were transferred to a heated greenhouse with a minimum temperature of 23 °C under natural photoperiod. For each plant, the number of floral buds was counted and the dates of early-pink-bud and petal fall stages were recorded. The number of the unopened floral buds was counted after eight weeks.

New non-transgenic 'Legacy' and transgenic 'Legacy-VcDDF1-OX' plants were developed through micropropagation in January of 2012 for further investigations. Twelve plants for each of the non-transgenic 'Legacy' and transgenic 'Legacy-VcDDF1-OX' were investigated from 2013 to 2016. These plants were grown in one-gallon pots in the courtyard. The number of floral buds was counted and the dates of the early-pink-bud and petal fall stages were recorded yearly. The number of fruit clusters was counted in July 2016.

RNA preparation, sequencing, and de novo transcriptome assembly

Floral buds were collected in November 2013 before the plants were exposed to a non-heated greenhouse for chilling treatments. All tissues collected were frozen immediately in liquid nitrogen and stored at –80 °C.

Total RNA isolation, RNA sequencing using the Illumina HiSeq2500 platform, de novo transcriptome assembly using the Trinity platform (trinity/20140413p1) [56] were described in our recent report [54].

Differential expression analysis and transcriptome annotation

RNA-seq reads of three biological replicates for each of 'Legacy' and 'Legacy-VcDDF1-OX' were analyzed. Two technical replicates were sequenced for each biological replicate and were combined together for analysis. The paired reads, two sets for each biological replicate, were aligned to the transcriptome reference developed for 'Legacy' [54] and the abundance of each read was estimated using the Trinity command "align_and_estimate_abundance.pl". The Trinity command "run_DE_analysis.pl –method edgeR" was used for differential expression analysis. The differentially expressed (DE) (relative to non-transgenic 'Legacy' unless other mentioned) genes or transcripts with false discovery rate (FDR) values below 0.05 were used for further analyses.

Transcriptome annotation was performed using Trinotate_v2.0 (https://trinotate.github.io).

Phylogenetic analysis of *VcDDF1*
Representative nucleotide sequences of CBF/DREB1 of five *A. thaliana CBF/DREB1* genes were retrieved using The *A. thaliana* Information Resource (TAIR) server (https://www.*arabidopsis*.org/tools/bulk/index.jsp). Orthologues of *A. thaliana CBF/DREB1* genes in blueberry were identified from our annotated transcripts. The selected transcripts were converted to amino acid sequences based on BLAST results retrieved using the NCBI server (http://blast.ncbi.nlm.nih.gov/Blast. cgi). Selected nucleotide sequences from both *A. thaliana* and blueberry were aligned using Clustal Omega multiple sequence alignment program at EBI with default parameters (http://www.ebi.ac.uk/Tools/msa/clustalo/). Phylogenetic trees were generated using MEGA 6.06 software [71].

The VcDDF1 protein sequence (AVI45245.1) was used to search for VcDDF1 orthologues using the NCBI server. The selected protein sequences were aligned using Clustal Omega multiple sequence alignment program at EBI with default parameters.

Gene network construction
Annotated transcripts were imported to Cytoscape 3.4.0 under BiNGO's default parameters with selected ontology file 'GOSlim_Plants' and selected organism *A. thaliana* [72, 73].

Identification of the selected pathway genes
Representative protein sequences of selected genes of *A. thaliana* were download from the TAIR server (https://www.arabidopsis.org/tools/bulk/sequences/index.jsp). The retrieved sequences were used to search the transcriptome reference of blueberry (herein refTrinity) using the tblastn command of BLAST+. The resultant transcripts that show e-value lower than −20 were used to screen the DE transcript list of non-acclimated floral buds.

The 2637 cold-regulated genes (CORs) identified in wild-type *A. thaliana* plants and 172 CORs differentially expressed at a warm temperature (22 °C) in transgenic *A. thaliana* plants overexpressing *CBF1*, *CBF2* or *CBF3* were obtained from Park et al. [42]. These CORs were used to identify their orthologues in blueberry (VcCORs), which was used to analyze the effect of *VcDDF1*-OX on VcCORs. The blueberry floral genes identified in our previous study [54] were used to analyze flowering pathway genes affected by *VcDDF1*-OX.

The pathway genes of major phytohormones [i.e., gibberellin (GA) [74], abscisic acid (ABA) [75], cytokinin [76], indole acetic acid (IAA) [77], and ethylene [78]] in *A. thaliana* were retrieved from TAIR_10 server based on published gene identities (Additional file 5: Table S4).

In addition, sequences of *A. thaliana* DELLA proteins were used to analyze the effect of *VcDDF1*-OX. Five *A. thaliana* DELLA proteins (Additional file 5: Table S4), including GIBBERELLIC ACID INSENSITIVE (*GAI*) (AT1G14920.1), REPRESSOR OF GA1 (*RGA1*) (AT2G01570.1), RGA-like 1 (*RGL1*) (AT1G66350.1), RGA-like 2 (*RGL2*) (AT3G03450.1), and RGA-like 3 (*RGL3*) (AT5G17490.1), were used to search for the DELLA protein genes in blueberry.

Quantitative RT-PCR (qRT-PCR) of DE transcripts
Reliability of DE genes/transcripts identified through RNA-seq was evaluated through qRT-PCR analysis of six selected transcripts (Additional file 3: Table S3). These transcripts are from the representative DE genes in auxin, ethylene, cytokinin, and GA pathways. They have high fold changes (>2) and sequence specificity (based on alignment result of different isoforms) for PCR amplification. Eukaryotic translation initiation factor 3 subunit H was the internal control (Additional file 3: Table S3).

The RNA samples used for RNA-sequencing, including samples of three biological replicates for each of 'Legacy' and 'Legacy-VcDDF1-OX', were used for cDNA preparation. Reverse transcription of RNA to cDNA was performed using SuperScript II reverse transcriptase (Invitrogen, Carlsbad, CA, USA). The resulting cDNA of one micro gram of RNA was diluted (volume 1: 4) in water and 1 μl/sample (25 ng) was used for PCR reactions.

The primers were designed using the online tool provided by Integrated DNA Technologies, Inc. (https://www.idtdna.com/Primerquest/Home/Index), where the primers were synthesized (Additional file 3: Table S3). qRT-PCR was performed in triplicate on an Agilent Technologies Stratagene Mx3005P (Agilent Technologies, Santa Clara, CA) using the SYBR Green system (Life Technologies, Carlsbad, CA). In each 25 μl reaction mixture, 25 ng cDNA, 200 nM primers and 12.5 μl of 2× SYBR Green master mix were included. The reaction conditions for all primer pairs were 95 °C for 10 min, 40 cycles of 30 s at 95 °C, 60 s at 60 °C and 60 s at 72 °C, followed by one cycle of 60 s at 95 °C, 30 s at 55 °C and 30 s at 95 °C. The specificity of the amplification reaction for each primer pair was determined by the melting curve. Transcript levels within samples were normalized to the eukaryotic translation initiation factor 3 subunit H. Fold changes were calculated using $2^{-\Delta\Delta Ct}$, where $\Delta\Delta Ct = (Ct_{GOI} - Ct_{nom})_{Legacy-VcDDF1-OX} - (Ct_{GOI} - Ct_{nom})_{Legacy}$ for each transgenic 'Legacy-VcDDF1-OX' versus a non-transgenic 'Legacy' sample ($n = 3$) [79]. In addition, regular RT-PCR was also used for selected transcripts. The reaction conditions using 50 ng cDNA per reaction for all primer pairs were 94 °C for 2 min, 35 cycles of 45 s at 94 °C, 60 s at 60 °C and 60 s at 72 °C, with a final 10 min extension at

72 °C. RT-PCR products were separated on 1.0% agarose gel containing ethidium bromide, visualized, and photographed under UV light.

Additional files

> **Additional file 1: Table S1.** Differentially expressed cold-regulated genes (CORs) of blueberry in non-acclimated floral buds. FDR (false discovery rate) < 0.05. LogFC: $\log_2$(fold change) = $\log_2$(Legacy-VcDDF1-OX/Legacy). *Some of these transcripts show similarities to multiple Arabidopsis-gene-ids that are not listed in this table
>
> **Additional file 2: Table S2.** Differentially expressed phytohormone genes (transgenic 'Legacy-VcDDF1-OX' vs. non-transgenic 'Legacy' plants) in dormant bud tissues of blueberry plants. LogFC: $\log_2$(fold change) = $\log_2$(Legacy-VcDDF1-OX/Legacy)
>
> **Additional file 3: Table S3.** Primers used for RT-PCR. FDR (false discovery rate) < 0.05. LogFC: $\log_2$(fold change) = $\log_2$(Legacy-VcDDF1-OX/Legacy)
>
> **Additional file 4: Fig. S1.** RT-PCR analysis of differentially expressed transcripts in leaf tissues of non-transgenic 'Legacy' and transgenic 'Legacy-VcDDF1-OX'. *Eukaryotic translation initiation factor 3 subunit H* is the internal control
>
> **Additional file 5: Table S4.** The pathway genes of major phytohormones [i.e., gibberellin (GA) [75], abscisic acid (ABA) [75], cytokinin [76], indole acetic acid (IAA) [77], ethylene [78], and DELLA protein genes in *A. thaliana*

Abbreviations
ABA: Abscisic acid; AP2/ERF: APETALA2/ethylene response (AP2/ERF) transcription factors; CBF/DREB1: C-repeat-binding factor/dehydration responsive element-binding factor 1; CORs: Cold-regulated genes; CPM: Count per million reads; DDF1: *DWARF AND DELAYED FLOWERING 1*; DE: Differentially expressed; FDR: False discovery rate; GA: Gibberellin; GO: Gene ontology; IAA: Indole acetic acid; OX: Overexpression; qRT-PCR: Quantitative reverse transcriptase polymerase chain reaction

Acknowledgments
The authors would thank Dr. Jeff Landgraf and Mr. Kevin Carr at Michigan State University Research Technology Support Facility for RNA sequencing, Mr. Aaron Walworth for assistance of phenotypic data collection, and Ms. QiuXia Chen for editing the manuscript. This research is partially supported by AgBioResearch Project GREEEN of Michigan State University (http://www.canr.msu.edu/research/agbioresearch/).

Funding
This research was partially supported by AgBioResearch of Michigan State University (http://www.canr.msu.edu/research/agbioresearch/).

Authors' contributions
GS conceived and supervised the study; XG and GS conducted the experiments; GS and XG analyzed the data; and GS wrote the manuscript. All authors read and approved the manuscript.

Competing interests
The authors declare that they have no competing interests.

Author details
[1]Plant Biotechnology Resource and Outreach Center, Department of Horticulture, Michigan State University, East Lansing, MI 48824, USA. [2]Key Laboratory for the Conservation and Utilization of Important Biological Resources, College of Life Sciences, Anhui Normal University, Wuhu 241000, China.

References
1. Mizoi J, Shinozaki K, Yamaguchi-Shinozaki K. AP2/ERF family transcription factors in plant abiotic stress responses. Biochim Biophys Acta. 2012;1819(2):86–96.
2. Zhuang J, Cai B, Peng RH, Zhu B, Jin XF, Xue Y, et al. Genome-wide analysis of the AP2/ERF gene family in *Populus trichocarpa*. Biochem Biophys Res Commun. 2008;371(3):468–74.
3. Thamilarasan SK, Park JI, Jung HJ, Nou IS. Genome-wide analysis of the distribution of AP2/ERF transcription factors reveals duplication and CBFs genes elucidate their potential function in *Brassica oleracea*. BMC Genomics. 2014;15:422.
4. Li MY, Xu ZS, Huang Y, Tian C, Wang F, Xiong AS. Genome-wide analysis of AP2/ERF transcription factors in carrot (*Daucus carota* L.) reveals evolution and expression profiles under abiotic stress. Mol Gen Genomics. 2015;290(6):2049–61.
5. Song X, Li Y, Hou X: Genome-wide analysis of the AP2/ERF transcription factor superfamily in Chinese cabbage (Brassica rapa ssp. pekinensis). BMC Genomics 2013, 14:573.
6. Cao PB, Azar S, SanClemente H, Mounet F, Dunand C, Marque G, et al. Genome-wide analysis of the AP2/ERF family in *Eucalyptus grandis*: an intriguing over-representation of stress-responsive DREB1/CBF genes. PLoS One. 2015;10(4):e0121041.
7. Sun ZM, Zhou ML, Xiao XG, Tang YX, Wu YM. Genome-wide analysis of AP2/ERF family genes from *Lotus corniculatus* shows LcERF054 enhances salt tolerance. Funct Integr Genomics. 2014;14(3):453–66.
8. Shu Y, Liu Y, Zhang J, Song L, Guo C. Genome-wide analysis of the AP2/ERF superfamily genes and their responses to abiotic stress in *Medicago truncatula*. Front Plant Sci. 2015;6:1247.
9. Wu H, Lv H, Li L, Liu J, Mu S, Li X, et al. Genome-wide analysis of the AP2/ERF transcription factors family and the expression patterns of DREB genes in Moso bamboo (*Phyllostachys edulis*). PLoS One. 2015;10(5):e0126657.
10. Lakhwani D, Pandey A, Dhar YV, Bag SK, Trivedi PK, Asif MH. Genome-wide analysis of the AP2/ERF family in Musa species reveals divergence and neofunctionalisation during evolution. Sci Rep. 2016;6:18878.
11. Zhang CH, Shangguan LF, Ma RJ, Sun X, Tao R, Guo L, et al. Genome-wide analysis of the AP2/ERF superfamily in peach (*Prunus persica*). Genet Mol Res. 2012;11(4):4789–809.
12. Tang Y, Qin S, Guo Y, Chen Y, Wu P, Chen Y, et al. Genome-wide analysis of the AP2/ERF Gene family in physic nut and overexpression of the JcERF011 Gene in Rice increased its sensitivity to salinity stress. PLoS One. 2016;11(3):e0150879.
13. Rao G, Sui J, Zeng Y, He C, Zhang J. Genome-wide analysis of the AP2/ERF gene family in *Salix arbutifolia*. FEBS Open Bio. 2015;5:132–7.
14. Ito TM, Polido PB, Rampim MC, Kaschuk G, Souza SG. Genome-wide identification and phylogenetic analysis of the AP2/ERF gene superfamily in sweet orange (*Citrus sinensis*). Genet Mol Res. 2014;13(3):7839–51.
15. Wu ZJ, Li XH, Liu ZW, Li H, Wang YX, Zhuang J. Transcriptome-based discovery of AP2/ERF transcription factors related to temperature stress in tea plant (*Camellia sinensis*). Funct Integr Genomic. 2015;15(6):741–52.
16. Liu Q, Kasuga M, Sakuma Y, Abe H, Miura S, Yamaguchi-Shinozaki K, et al. Two transcription factors, DREB1 and DREB2, with an EREBP/AP2 DNA binding domain separate two cellular signal transduction pathways in drought- and low-temperature-responsive gene expression, respectively, in Arabidopsis. Plant Cell. 1998;10(8):1391–406.
17. Qin F, Shinozaki K, Yamaguchi-Shinozaki K. Achievements and challenges in understanding plant abiotic stress responses and tolerance. Plant Cell Physiol. 2011;52(9):1569–82.

18. Thomashow MF, Gilmour SJ, Stockinger EJ, Jaglo-Ottosen KR, Zarka DG. Role of the Arabidopsis CBF transcriptional activators in cold acclimation. Physiol Plantarum. 2001;112(2):171–5.

19. Miller AK, Galiba G, Dubcovsky J. A cluster of 11 CBF transcription factors is located at the frost tolerance locus Fr-a(m)2 in *Triticum monococcum*. Mol Gen Genomics. 2006;275(2):193–203.

20. Navarro M, Ayax C, Martinez Y, Laur J, El Kayal W, Marque C, et al. Two EguCBF1 genes overexpressed in Eucalyptus display a different impact on stress tolerance and plant development. Plant Biotechnol J. 2011;9(1):50–63.

21. Carvallo MA, Pino MT, Jeknic Z, Zou C, Doherty CJ, Shiu SH, et al. A comparison of the low temperature transcriptomes and CBF regulons of three plant species that differ in freezing tolerance: *Solanum commersonii, Solanum tuberosum*, and *Arabidopsis thaliana*. J Exp Bot. 2011;62(11):3807–19.

22. Gilmour SJ, Zarka DG, Stockinger EJ, Salazar MP, Houghton JM, Thomashow MF. Low temperature regulation of the Arabidopsis CBF family of AP2 transcriptional activators as an early step in cold-induced COR gene expression. Plant J. 1998;16(4):433–42.

23. Stockinger EJ, Gilmour SJ, Thomashow MF. *Arabidopsis thaliana* CBF1 encodes an AP2 domain-containing transcriptional activator that binds to the C-repeat/DRE, a cis-acting DNA regulatory element that stimulates transcription in response to low temperature and water deficit. Proc Natl Acad Sci U S A. 1997;94(3):1035–40.

24. Jaglo-Ottosen KR, Gilmour SJ, Zarka DG, Schabenberger O, Thomashow MF. Arabidopsis CBF1 overexpression induces COR genes and enhances freezing tolerance. Science. 1998;280(5360):104–6.

25. Mizoi J, Shinozaki K, Yamaguchi-Shinozaki K. AP2/ERF family transcription factors in plant abiotic stress responses. Bba-Gene Regul Mech. 2012; 1819(2):86–96.

26. Sun J, Peng X, Fan W, Tang M, Liu J, Shen S. Functional analysis of BpDREB2 gene involved in salt and drought response from a woody plant *Broussonetia papyrifera*. Gene. 2014;535(2):140–9.

27. Chinnusamy V, Zhu J, Zhu JK. Cold stress regulation of gene expression in plants. Trends Plant Sci. 2007;12(10):444–51.

28. Berdeja M, Nicolas P, Kappel C, Dai ZW, Hilbert G, Peccoux A, et al. Water limitation and rootstock genotype interact to alter grape berry metabolism through transcriptome reprogramming. Hortic Res-England. 2015;2

29. Cao X, Liu Q, Rowland LJ, Hammerschlag FA. GUS expression in blueberry (Vaccinium spp.): factors influencing agrobacterium-mediated gene transfer efficiency. Plant Cell Rep. 1998;18(3–4):266–70.

30. Thomashow MF. Molecular basis of plant cold acclimation: insights gained from studying the CBF cold response pathway. Plant Physiol. 2010;154(2):571–7.

31. Artlip TS, Wisniewski ME, Arora R, Norelli JL. An apple rootstock overexpressing a peach CBF gene alters growth and flowering in the scion but does not impact cold hardiness or dormancy. Hortic Res-England. 2016;3

32. Benedict C, Skinner JS, Meng R, Chang Y, Bhalerao R, Huner NP, et al. The CBF1-dependent low temperature signalling pathway, regulon and increase in freeze tolerance are conserved in Populus spp. Plant Cell Environ. 2006; 29(7):1259–72.

33. Kasuga M, Liu Q, Miura S, Yamaguchi-Shinozaki K, Shinozaki K. Improving plant drought, salt, and freezing tolerance by gene transfer of a single stress-inducible transcription factor. Nat Biotechnol. 1999; 17(3):287–91.

34. Walworth AE, Song GQ, Warner RM. Ectopic AtCBF3 expression improves freezing tolerance and promotes compact growth habit in petunia. Mol Breed. 2014;33(3):731–41.

35. Tillett RL, Wheatley MD, Tattersall EA, Schlauch KA, Cramer GR, Cushman JC. The *Vitis vinifera* C-repeat binding protein 4 (VvCBF4) transcriptional factor enhances freezing tolerance in wine grape. Plant Biotechnol J. 2012;10(1):105–24.

36. Wisniewski M, Norelli J, Bassett C, Artlip T, Macarisin D. Ectopic expression of a novel peach (*Prunus persica*) CBF transcription factor in apple (malus x domestica) results in short-day induced dormancy and increased cold hardiness. Planta. 2011;233(5):971–83.

37. Oakenfull RJ, Baxter R, Knight MR. A C-repeat binding factor transcriptional activator (CBF/DREB1) from European bilberry (*Vaccinium myrtillus*) induces freezing tolerance when expressed in *Arabidopsis thaliana*. PLoS One. 2013;8(1):e54119.

38. Gilmour SJ, Sebolt AM, Salazar MP, Everard JD, Thomashow MF. Overexpression of the Arabidopsis CBF3 transcriptional activator mimics multiple biochemical changes associated with cold acclimation. Plant Physiol. 2000;124(4):1854–65.

39. Achard P, Gong F, Cheminant S, Alioua M, Hedden P, Genschik P. The cold-inducible CBF1 factor-dependent signaling pathway modulates the accumulation of the growth-repressing DELLA proteins via its effect on gibberellin metabolism. Plant Cell. 2008;20(8):2117–29.

40. Sharabi-Schwager M, Lers A, Samach A, Guy CL, Porat R. Overexpression of the CBF2 transcriptional activator in Arabidopsis delays leaf senescence and extends plant longevity. J Exp Bot. 2010;61(1):261–73.

41. Haake V, Cook D, Riechmann JL, Pineda O, Thomashow MF, Zhang JZ. Transcription factor CBF4 is a regulator of drought adaptation in Arabidopsis. Plant Physiol. 2002;130(2):639–48.

42. Park S, Lee CM, Doherty CJ, Gilmour SJ, Kim Y, Thomashow MF. Regulation of the Arabidopsis CBF regulon by a complex low-temperature regulatory network. Plant J. 2015;82(2):193–207.

43. Novillo F, Medina J, Rodriguez-Franco M, Neuhaus G, Salinas J. Genetic analysis reveals a complex regulatory network modulating CBF gene expression and Arabidopsis response to abiotic stress. J Exp Bot. 2011; 63(1):293–304.

44. Sakuma Y, Liu Q, Dubouzet JG, Abe H, Shinozaki K, Yamaguchi-Shinozaki K. DNA-binding specificity of the ERF/AP2 domain of Arabidopsis DREBs, transcription factors involved in dehydration- and cold-inducible gene expression. Biochem Biophys Res Commun. 2002;290(3):998–1009.

45. Magome H, Yamaguchi S, Hanada A, Kamiya Y. Oda K: dwarf and delayed-flowering 1, a novel Arabidopsis mutant deficient in gibberellin biosynthesis because of overexpression of a putative AP2 transcription factor. Plant J. 2004;37(5):720–9.

46. Magome H, Yamaguchi S, Hanada A, Kamiya Y, Oda K. The DDF1 transcriptional activator upregulates expression of a gibberellin-deactivating gene, GA2ox7, under high-salinity stress in Arabidopsis. Plant J. 2008;56(4):613–26.

47. Lehti-Shiu MD, Uygun S, Moghe GD, Panchy N, Fang L, Hufnagel DE, et al. Molecular evidence for functional divergence and decay of a transcription factor derived from whole-genome duplication in *Arabidopsis thaliana*. Plant Physiol. 2015;168(4):1717–34.

48. Kang HG, Kim J, Kim B, Jeong H, Choi SH, Kim EK, et al. Overexpression of FTL1/DDF1, an AP2 transcription factor, enhances tolerance to cold, drought, and heat stresses in *Arabidopsis thaliana*. Plant Sci. 2011;180(4): 634–41.

49. Wu J, Folta KM, Xie Y, Jiang W, Lu J, Zhang Y. Overexpression of Muscadinia rotundifolia CBF2 gene enhances biotic and abiotic stress tolerance in Arabidopsis. Protoplasma. 2016;

50. Polashock JJ, Arora R, Peng Y, Naik D, Rowland LJ. Functional identification of a C-repeat binding factor transcriptional activator from blueberry associated with cold acclimation and freezing tolerance. J Am Soc Hortic Sci. 2010;135(1):40–8.

51. Walworth AE, Rowland LJ, Polashock JJ, Hancock JF, Song GQ. Overexpression of a blueberry-derived CBF gene enhances cold tolerance in a southern highbush blueberry cultivar. Mol Breed. 2012;30(3):1313–23.

52. Du DL, Rawat N, Deng ZA, Gmitter FG. Construction of citrus gene coexpression networks from microarray data using random matrix theory. Hortic Res-England. 2015;2

53. Song GQ, Walworth A, Zhao DY, Jiang N, Hancock JF. The *Vaccinium corymbosum* FLOWERING LOCUS T-like gene (VcFT): a flowering activator reverses photoperiodic and chilling requirements in blueberry. Plant Cell Rep. 2013;32(11):1759–69.

54. Walworth AE, Chai B, Song GQ. Transcript profile of flowering regulatory genes in VcFT-overexpressing blueberry plants. PLoS One. 2016;11(6): e0156993.

55. Gao X, Walworth AE, Mackie C, Song GQ. Overexpression of blueberry FLOWERING LOCUS T is associated with changes in the expression of phytohormone-related genes in blueberry plants. Hortic Res. 2016;3:16053.

56. Haas BJ, Papanicolaou A, Yassour M, Grabherr M, Blood PD, Bowden J, et al. De novo transcript sequence reconstruction from RNA-seq using the trinity platform for reference generation and analysis. Nat Protoc. 2013;8(8):1494–512.

57. Fornara F, de Montaigu A, Coupland G: SnapShot: Control of Flowering in Arabidopsis. Cell 2010, 141(3).

58. Strimbeck GR, Schaberg PG, Fossdal CG, Schroder WP, Kjellsen TD. Extreme low temperature tolerance in woody plants. Front Plant Sci. 2015;6:884.

59. Wisniewski M, Norelli J, Artlip T. Overexpression of a peach CBF gene in apple: a model for understanding the integration of growth, dormancy, and cold hardiness in woody plants. Front Plant Sci. 2015;6:85.

60. Jin WM, Dong J, Hu YL, Lin ZP, Xu XF, Han ZH. Improved cold-resistant performance in transgenic grape (*Vitis vinifera* L.) overexpressing cold-inducible transcription factors AtDREB1b. Hortscience. 2009;44(1):35–9.

61. Seo E, Lee H, Jeon J, Park H, Kim J, Noh YS, et al. Crosstalk between cold response and flowering in Arabidopsis is mediated through the flowering-time gene SOC1 and its upstream negative regulator FLC. Plant Cell. 2009; 21(10):3185–97.

62. Lee JH, Yoo SJ, Park SH, Hwang I, Lee JS, Ahn JH. Role of SVP in the control of flowering time by ambient temperature in Arabidopsis. Genes Dev. 2007; 21(4):397–402.

63. Shi Y, Tian S, Hou L, Huang X, Zhang X, Guo H, et al. Ethylene signaling negatively regulates freezing tolerance by repressing expression of CBF and type-a ARR genes in Arabidopsis. Plant Cell. 2012;24(6):2578–95.

64. Shi Y, Ding Y, Yang S. Cold signal transduction and its interplay with phytohormones during cold acclimation. Plant Cell Physiol. 2015;56(1):7–15.

65. Bak S, Tax FE, Feldmann KA, Galbraith DW, Feyereisen R. CYP83B1, a cytochrome P450 at the metabolic branch point in auxin and indole glucosinolate biosynthesis in Arabidopsis. Plant Cell. 2001;13(1):101–11.

66. Dharmasiri S, Swarup R, Mockaitis K, Dharmasiri N, Singh SK, Kowalchyk M, et al. AXR4 is required for localization of the auxin influx facilitator AUX1. Science. 2006;312(5777):1218–20.

67. Veerabagu M, Elgass K, Kirchler T, Huppenberger P, Harter K, Chaban C, et al. The Arabidopsis B-type response regulator 18 homomerizes and positively regulates cytokinin responses. Plant J. 2012;72(5):721–31.

68. Song GQ. Blueberry (*Vaccinium corymbosum* L.). Methods Mol Biol. 2015; 1224:121–31.

69. Norvell DJ, Moore JN. An evaluation of chilling models for estimating rest requirements of highbush blueberries (Vaccinium-Corymbosum L). J Am Soc Hortic Sci. 1982;107(1):54–6.

70. Spiers JM, Marshall DA, Smith BJ, Braswell JH. Method to determine chilling requirement in blueberries. Acta Hortic. 2006;715:105–9.

71. Tamura K, Stecher G, Peterson D, Filipski A, Kumar S. MEGA6: molecular evolutionary genetics analysis version 6.0. Mol Biol Evol. 2013;30(12):2725–9.

72. Shannon P, Markiel A, Ozier O, Baliga NS, Wang JT, Ramage D, et al. Cytoscape: a software environment for integrated models of biomolecular interaction networks. Genome Res. 2003;13(11):2498–504.

73. Maere S, Heymans K, Kuiper M. BiNGO: a Cytoscape plugin to assess overrepresentation of gene ontology categories in biological networks. Bioinformatics. 2005;21(16):3448–9.

74. Regnault T, Daviere JM, Heintz D, Lange T, Achard P. The gibberellin biosynthetic genes AtKAO1 and AtKAO2 have overlapping roles throughout Arabidopsis development. Plant J. 2014;80(3):462–74.

75. Xiong L, Zhu JK. Regulation of abscisic acid biosynthesis. Plant Physiol. 2003; 133(1):29–36.

76. Hwang D, Chen HC, Sheen J. Two-component signal transduction pathways in Arabidopsis. Plant Physiol. 2002;129(2):500–15.

77. Normanly J, Bartel B. Redundancy as a way of life - IAA metabolism. Curr Opin Plant Biol. 1999;2(3):207–13.

78. Wang KL, Li H, Ecker JR. Ethylene biosynthesis and signaling networks. Plant Cell. 2002;14(Suppl):S131–51.

79. Livak KJ, Schmittgen TD. Analysis of relative gene expression data using real-time quantitative PCR and the 2(−Delta Delta C(T)) method. Methods. 2001;25(4):402–8.

7

Expression patterns of flowering genes in leaves of 'Pineapple' sweet orange [*Citrus sinensis* (L.) Osbeck] and pummelo (*Citrus grandis* Osbeck)

Melanie Pajon[1], Vicente J. Febres[1*] and Gloria A. Moore[1,2]

Abstract

Background: In citrus the transition from juvenility to mature phase is marked by the capability of a tree to flower and fruit consistently. The long period of juvenility in citrus severely impedes the use of genetic based strategies to improve fruit quality, disease resistance, and responses to abiotic environmental factors. One of the genes whose expression signals flower development in many plant species is *FLOWERING LOCUS T (FT)*.

Results: In this study, gene expression levels of flowering genes *CiFT1*, *CiFT2* and *CiFT3* were determined using reverse-transcription quantitative real-time PCR in citrus trees over a 1 year period in Florida. Distinct genotypes of citrus trees of different ages were used. In mature trees of pummelo (*Citrus grandis* Osbeck) and 'Pineapple' sweet orange (*Citrus sinensis* (L.) Osbeck) the expression of all three *CiFT* genes was coordinated and significantly higher in April, after flowering was over, regardless of whether they were in the greenhouse or in the field. Interestingly, immature 'Pineapple' seedlings showed significantly high levels of *CiFT3* expression in April and June, while *CiFT1* and *CiFT2* were highest in June, and hence their expression induction was not simultaneous as in mature plants.

Conclusions: In mature citrus trees the induction of *CiFTs* expression in leaves occurs at the end of spring and after flowering has taken place suggesting it is not associated with dormancy interruption and further flower bud development but is probably involved with shoot apex differentiation and flower bud determination. *CiFTs* were also seasonally induced in immature seedlings, indicating that additional factors must be suppressing flowering induction and their expression has other functions.

Keywords: Florigen, *FLOWERING LOCUS T*, Flowering periodicity, Gene expression, Juvenility, Time course

Background

Many *Citrus* species are characterized as having extended juvenility periods and therefore not producing flowers or fruit for many years, up to a decade or longer [1, 2]. The process that leads to flowering in citrus trees most likely involves environmental and physiological cues. In *Arabidopsis thaliana* flowering cues include phytohormones such as gibberellic acid (GA), vernalization, light, gene expression patterns and other physiological responses [3]. Floral development occurs in the shoot apical meristem; however, some of the environmental response pathways that are involved act in the leaves. *Arabidopsis* plants are known to remain in the vegetative state under short day conditions- 8 or 10 h of light. When shifted to long days- 16 h of light- genes that are involved in the flowering process are expressed in the meristem within 24 h [4]. Several studies have also demonstrated the importance of a low-temperature condition in the induction of flowering in citrus [5, 6]. This temperature condition has been shown to have an effect on the seasonal periodicity of flowering, yet there are few studies of the effect that this and other environmental cues have on the expression of genes in the flowering pathways in citrus. In *Arabidopsis* many genes involved in the flowering pathways have been

* Correspondence: vjf@ufl.edu
[1]Horticultural Sciences Department, Institute of Food and Agricultural Sciences, University of Florida, 2550 Hull Road, Gainesville, FL 32611, USA
Full list of author information is available at the end of the article

extensively studied and their roles are clearly understood [4]. The homologues of these flowering genes in perennial trees such as *Citrus* provide the foundation to further explore why the juvenility period among citrus species differs so greatly and to what extent seasonality affects genetic expression profiles.

Commercially important citrus types thrive in humid subtropical regions of the world such as Florida, Central China, Brazil and Mexico [2]. These regions have a balance of rainfall, sunlight, wind, humidity, and temperature that favors the growth and production of citrus. It is therefore important to understand how factors in these particular regions affect seasonal periodicity and consequently gene expression for flowering pathway genes. Florida in particular has high temperatures from April to October; although summer temperatures rarely exceed 40 °C. During the summer, low temperatures range from near 21 °C in northern Florida to near 27 °C in the Keys, at the southern end of the state. High temperatures during the summer average 33 to 35 °C statewide. Moderate to severe freezes do occur in Florida between November and March but the climatic conditions in this state are optimal for certain commercial citrus types [2]. Economically important citrus cultivars such as 'Pineapple' sweet orange (*Citrus sinensis* (L.) Osbeck) and pummelo (*Citrus grandis* Osbeck) are known to have long juvenility periods and seasonal flowering in Florida, underscoring that these are two separate but related processes. For these species of citrus, molecular mechanisms involved in the onset of flowering have yet to be characterized. Indeed, most such studies to date have been conducted in mandarins in Japan under climatic conditions that are quite different from those in Florida [6].

Perhaps the most widely studied flowering gene across a variety of genera is *FLOWERING LOCUS T* (*FT*). In *Arabidopsis* expression of the *FT* gene and the *SUPPRESSOR OF OVEREXPRESSION OF CONSTANS 1* (*SOC1*) gene are believed to be required for *CONSTANS* (*CO*) to induce flowering. *FT* is typically maximally expressed at the onset of floral induction [7]. Constitutive expression of the *Arabidopsis FT* gene in apple [8], soybean [9], and poplar [10] results in early flowering phenotypes and the expression levels of FT in Satsuma mandarin (*Citrus unshiu* Marc.) have been correlated with flower numbers [11]. Additionally, *FT* orthologues have been identified in a variety of genera such as rice [12], poplar [13, 14], citrus [6, 15], and many others. In citrus, the *CiFT1* homologue from Satsuma mandarin [6] was constitutively expressed in trifoliate orange and conferred a notably early-flowering phenotype [15]. Viral vectors based on *Citrus leaf blotch virus* have also been used to constitutively express *A. thaliana* and *C. sinensis FT* genes and were capable of inducing flowering in juvenile plants of a variety of citrus types [16].

Nishikawa et al. [6] conducted a quantitative real-time PCR study in Japan that included *CiFT1* and two novel homologues of *FT* in citrus (*CiFT2 and CiFT3*), although *CiFT1* and *CiFT2* are proposed to be alleles of the same locus [17]. This study provided some insight into expression patterns for several genes involved in flowering activity, yet no conclusive patterns were established for citrus trees under field conditions. Overall, some of the pathways and mechanisms of flowering in *Arabidopsis* have been shown to be conserved in other species including woody perennials, yet more research must be done to investigate specific molecular mechanisms in *Citrus*.

To summarize, for flowering to occur in citrus two events are needed: 1) plants must reach maturity and 2) they have to be exposed to the right environmental cues. Hence the main objective of the present study was to compare the seasonal periodicity of the FTs gene expression in leaves of mature and juvenile citrus. Since FT is purportedly a major determinant of flowering induction in plants and flowering seasonality is observed under Florida's conditions we set out to investigate how different its expression patterns were between mature (flowering) and juvenile (non-flowering) individuals. Furthermore, since citrus has at least two FTs, are there differences in expression levels that could indicate separate functions by age or season? Thus a 1 year study of in vivo tracking of *CiFT1*, *CiFT2* and *CiFT3* expression in various citrus trees differing in age and genotype was undertaken. Gene expression levels were compared on a month-to-month basis using the comparative C_T method of quantitative real time PCR. Leaf tissue was used for this purpose for several reasons. First, *FT* gene expression is known to occur in leaves while FT protein is translocated to the meristems [18]. Also, some of the trees analyzed were juvenile and did not produce flowers. Citrus, like many other perennial plants, has leaf and flower meristems in the same flushing stems and these different meristems cannot be easily distinguished, thus analyzing meristematic tissue, although a possibility, would have been problematic. Leaves as the examined tissue seemed appropriate since our purpose was not to compare different tissues but rather to measure gene expression as it related to the time of year and age. Thus, we looked at the production of transcripts from the selected genes throughout the year, not just at flowering.

Methods
Plant material
The citrus genotypes used in this study were 'Pineapple' sweet orange (PSO) (*Citrus sinensis* (L.) Osbeck) and pummelo (*Citrus grandis* Osbeck) (PUM). Two adult 'Pineapple' sweet orange trees of approximately 15 years of age were located in the USDA A. H. Whitmore Farm (23,402 USDA Rd., Groveland, FL 34736 (28°41′16.1″N 81°53′09.6″W; summer solstice daylight: 13:57 h, winter

solstice daylight: 10:15 h; January average high: 22 °C, average low: 9 °C; July average high: 34 °C, average low: 23 °C). The temperatures registered during the experimental period are shown in Additional file 1: Figure S1. One of the sweet orange trees was a seedless mutant and the other had seed ("seedy"). An adult pummelo of approximately 15 years of age was located in a temperature-controlled greenhouse at the University of Florida, Gainesville, FL (29°38′20.8″N 82°21′35.6″W; summer solstice daylight: 14:03 h, winter solstice daylight: 10:14 h), the temperature set to range between 18 and 35 °C with no light supplementation. Three 2-year-old PSO trees were also in the same greenhouse. Therefore a total of 6 trees were used in this study. Leaves were collected from each tree once a month for 12 months on the same day from both locations. The year collection period began on 3 July 2012 and ended on 28 June 2013. For the PSO seedless and seedy and pummelo mature trees, three different leaf samples were collected from different parts of the trees and used as biological replicates. A representative sample for each replicate included multiple leaves ranging from new growth to older growth. For the 2-year-old PSO trees, three different plants were used as biological replicates and each sample also consisted of several leaves representative of the whole tree. Collected tissue was stored in aluminum foil in a cooler with dry ice during transport to a – 80 °C freezer in the laboratory.

RNA extraction and cDNA synthesis

Total RNA was extracted using TriZol reagent (Invitrogen) according to the manufacturer's instructions followed by DNase treatment and clean up with the RNeasy Plant Mini Kit (QIAGEN). The RNA concentration and purity were determined using a NanoDrop 2000c spectrophotometer (Thermo Scientific). A criterion was employed based on OD_{260}/OD_{230} ($\geq$ 1.7), OD_{260}/OD_{280} ($\geq$ 1.7) and RNA absorbance curves to deem any sample of acceptable quality. cDNA synthesis reactions were performed as follows: a total of 16 μL consisting of 1 μg of the purified RNA, 2 μL of 50 μM random decamers (Ambion), 2 μL of dNTPs (5mM each) and RNase-free water were incubated at 80 °C for 3 min in a PTC-100 Programmable Thermal Controller (MJ Research Inc.) and then placed on ice for 3 min. Subsequently, a mixture of 1 μL (200 units) of M-MLV reverse transcriptase (Ambion), 2 μL of 10X First Strand Buffer and 1 μL (40 units) of RNase Inhibitor (Ambion) were added to the 16 μL sample for a final volume of 20 μL. The 20 μL reaction mixture underwent reverse transcription in the thermal cycler (PTC-100 Programmable Thermal Controller, MJ Research Inc.) using the following parameters: 42 °C for 1 h followed by 92 °C for 10 min. The cDNA product was stored at –20 °C. A final working 1:10 dilution was made from the 50 ng/μL cDNA stock.

Gene expression analysis

Gene expression levels were measured with reverse-transcription quantitative real-time PCR (RT-qPCR) using the StepOnePlus Real-Time PCR system (Applied Biosystems). The parameters for reactions were set as follows: comparative C_T ($\Delta\Delta C_T$) with the fast amplification of 95 °C for 20s, 40 cycles of 95 °C for 1 s, and 60 °C for 20s. A Fast 96-well Reaction Plate (0.1 mL) (MicroAmp, Applied Biosystems) was used for the reactions and each well was used to perform a 20 μL expression assay of one gene per each sample. Gene amplification was performed with 10 ng of working cDNA solution. Each reaction mixture was composed of 2 μL of cDNA, 10 μL of TaqMan Fast Universal PCR Master Mix (2X) (Applied Biosystems), 1 μL of Taq-Man probe and primer Assay Mix (20X) (Applied Biosystems) and 7 μL of RNase-free water. The 20X Assay Mix was a combination of specific TaqMan MGB probes, forward and reverse primers for each gene. For all of the assayed citrus flowering genes, the final primer concentration was 900 nM each and the final probe concentration was 250 nM. Probes were labeled with 6-carboxyfluorescein (FAM). For the endogenous control reference gene 5.8S rRNA was used, with a final primer concentration of 250 nM each and final probe concentration of 150 nM per reaction. The probe for 5.8S rRNA was labeled with 4,7,2′-trichloro-7′-phenyl-6-carboxyfluorescein (VIC). All primer and probe sequences, and sources are listed in Additional file 2: Table S1. Primers and probes for *5.8S rRNA* were designed with Primer Express Software (Applied Biosystems). The primers and probes used *for CiFT1, CiFT2* and *CiFT3* were taken from Nishikawa et al. [6]. A negative control containing all RT-qPCR reaction elements and water instead of cDNA was used for every 96-well reaction plate.

Statistical analysis

For the comparative C_T analysis, the PSO seedy sample from 3 July 2012 (PSOsd M01-1) was used as the reference sample for PSO seedy and seedless 15-year-old trees and for the PSO 2-year-old trees. The pummelo sample from 3 July 2012 (PUM M01-1) was used as the reference sample for the pummelo genotype. Quantitative real-time PCR amplification data from twelve different time points and three biological replicates for each gene was normalized with the 5.8S rRNA C_T values and the threshold was automatically set but adjusted manually when needed. The relative quantitation (RQ) values were calculated using the StepOne software version 2.3 (Applied Biosystems) and exported to Microsoft Office Excel for further analysis. Outliers were identified using Dixon's Q test at 95% confidence [19] and the quantile range outliers tool from JMP Genomics 8.2 (SAS Institute Inc. NC). The RQ data was used to calculate means and standard errors (n = 3 for PUM and 2-year-old PSO, n = 6 for 15-year-old PSO). Statistical analysis was performed using JMP Genomics model fitting of standard least square means (LS Means) and Student's t test ($P < 0.05$) (Additional file 3. Table S2).

Results and discussion

To investigate the expression patterns of genes involved in the flowering process of young and mature citrus trees in Florida, mRNA was extracted from representative samples of different species and ages and used in a quantitative real-time polymerase chain reaction (RT-qPCR) study. Expression levels were determined according to genotype, age and time points. The time points were chosen so that they would be representative of the 12 months in a year. The method used to acquire three biological replicates was based on the availability of trees of the same genotype and age. Since no young trees existed at the Groveland, Florida location we used young, greenhouse trees located at our Gainesville, Florida campus.

All three *CiFT* and internal control genes chosen for the study were detected in the leaves collected for each sample at every time point. Leaves that ranged from new growth to old growth were selected from different parts of the trees in order to get a more representative profile for the entire tree.

CiFTs expression in mature citrus

In the mature pummelo tree, the mRNA levels for *CiFT1*, *CiFT2*, and *CiFT3* were highest on 30 April 2013 compared to all other time points (Fig. 1). In general, *CiFT* levels were highest between March and June, with *CiFT2* showing the highest transcript levels, followed by *CiFT1* and then *CiFT3*. In April 2013, the 15-year-old

pummelo tree was bearing immature fruit after flowering events in late February and early March. Similarly, the mature PSO (Fig. 2) presented the highest levels of *CiFT* transcripts on 30 April 2013 with overall levels highest between April and June 2013 and July 2012. However, unlike Pummelo, *CiFT3* exhibited the highest transcription level (during April and May, Fig. 2) compared to *CiFT1/FT2* and all three *CiFT* genes reached lower relative expression levels compared to pummelo. By this time point in the field in Florida, bloom was largely over [20] and trees were bearing mature and immature fruit (the fruits were not harvested from the experimental trees). Hence, *CiFT1*, *CiFT2*, and *CiFT3* gene expression in leaves of mature plants was mostly synchronized and highest soon after flowering. This timing was surprising because in the field in Florida, flower bud induction, when signals are presumably sent to apical buds, occurs from roughly mid-October to the end of January and a release of cool temperatures is thought to be necessary for flower bud development, which in the study area occurred during the November 2012-March 2013 period, with the lowest minimum temperatures registered in March 2013 (Additional file 1: Figure S1). Furthermore, the gene expression profile for the mature pummelo grown in the greenhouse was similar to the ones seen in the field-grown PSO trees. It is worth mentioning that although the temperature in the greenhouse was controlled, it fluctuated between 18 and 35 °C throughout the year, remaining on the cooler side during the winter months (also November-March). It is possible then that, in addition to temperature, other environmental factors such as light conditions

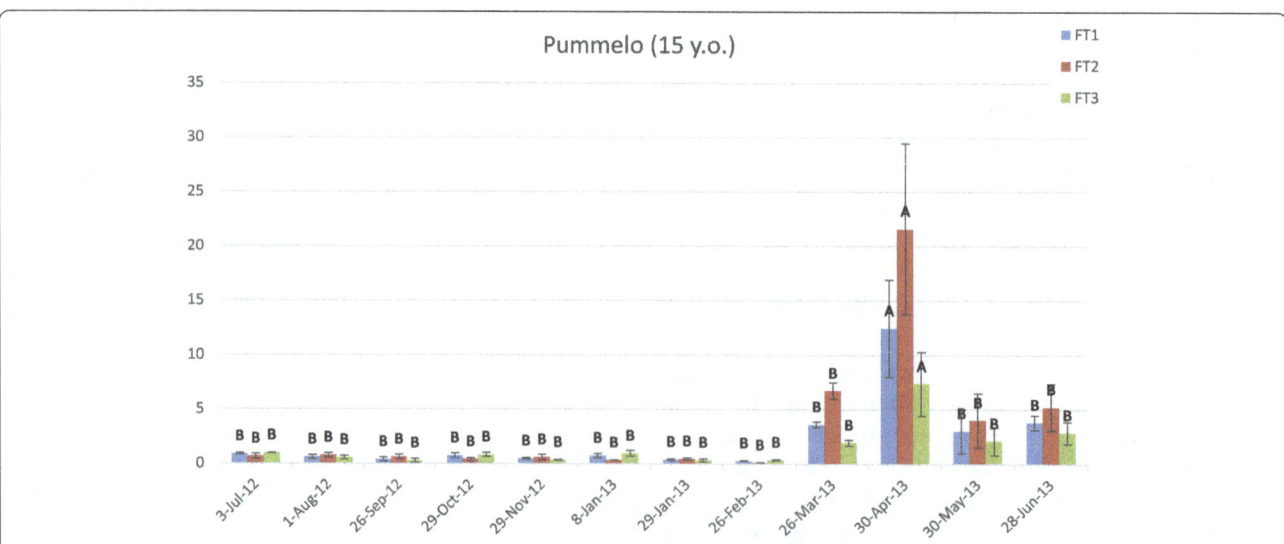

Fig. 1 Expression of *CiFT1, CiFT2, and CiFT3* in leaves of pummelo. Gene expression was quantified throughout a 12-month period by real-time PCR and evaluated using the comparative CT analysis. The vertical axis indicates the relative quantitation (RQ) of gene expression levels after each sample is compared to a reference sample from 3-Jul-12. The horizontal axis displays the collection dates. Data are means ±SE ($n = 3$). Levels A and B of the Student's t analysis are indicated. Columns with different letters are significantly different

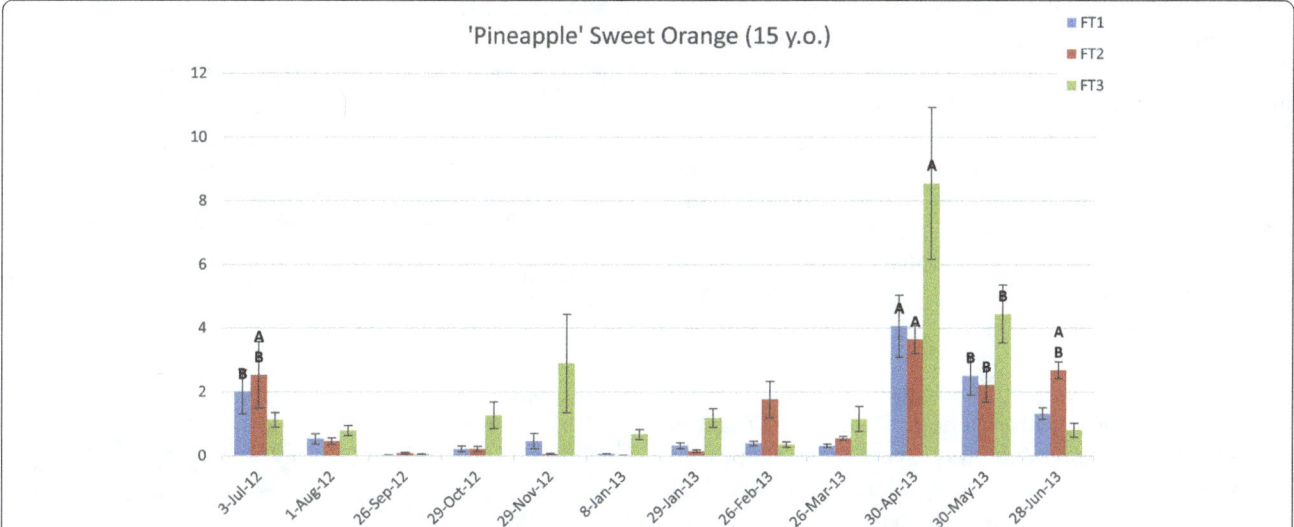

Fig. 2 Expression of *CiFT1*, *CiFT2*, and *CiFT3* in leaves of adult 'Pineapple' sweet orange. Gene expression in 15-year-old trees was quantified throughout a 12-month period by real-time PCR and evaluated using the comparative CT analysis. The vertical axis indicates the relative quantitation (RQ) of gene expression levels after each sample is compared to a reference sample from 3-Jul-12. The horizontal axis displays the collection dates. Data are means ±SE (n = 6). Levels A and B of the Student's t analysis are indicated. Columns with different letters are significantly different

also function as cues for the regulation of *FT* expression. Another possibility is that leaf expression of *FTs* in Citrus is dependent on internal signals. The results also indicate that leaf expression of *CiFTs* is not associated with dormancy interruption and further flower bud development (which occurred long before the observed expression peak in April) but may be associated with shoot apex differentiation and flower bud determination.

CiFTs expression in immature citrus

Seasonality in the expression of the *CiFT* genes was also observed in seedlings. In immature PSO, *CiFT1* transcripts were highest on 28 June 2013 (Fig. 3). *CiFT2* expression was also highest on 28 June 2013, displaying the highest transcript level of all three *CiFTs*. On the other hand, *CiFT3* expression was highest on 30 April and 28 June 2013. Overall the PSO *CiFTs* displayed higher expression levels in immature plants than in

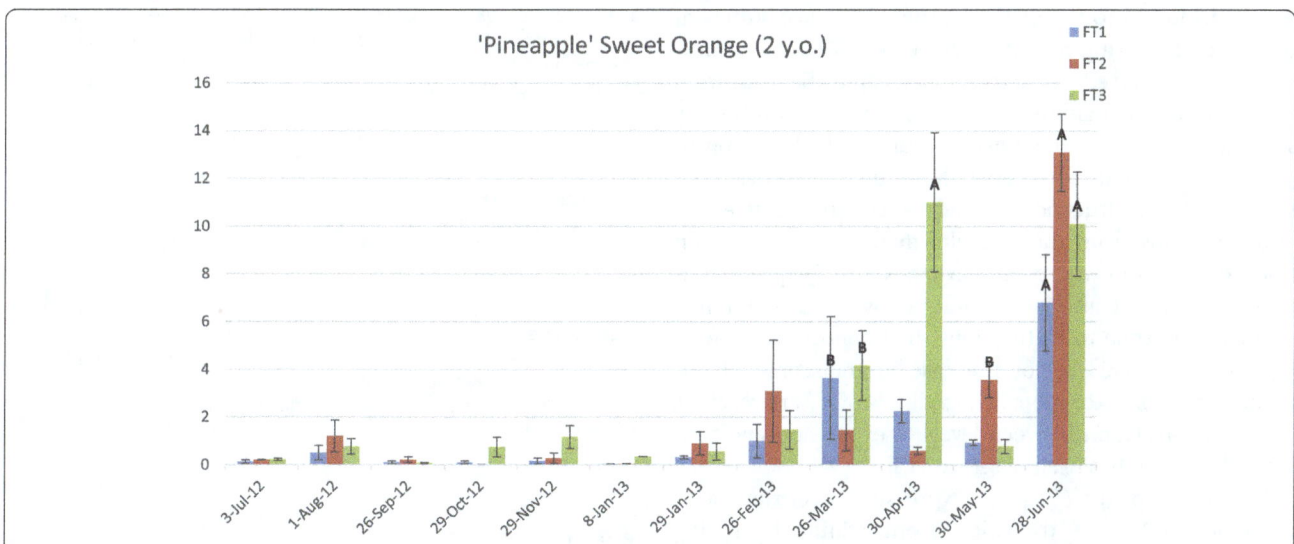

Fig. 3 Expression of *CiFT1*, *CiFT2*, and *CiFT3* in leaves of young 'Pineapple' sweet orange trees. Gene expression in 2-year-old trees was quantified throughout a 12-month period by real-time PCR and evaluated using the comparative C_T analysis. The vertical axis indicates the relative quantitation (RQ) of gene expression levels after each sample is compared to a reference sample from 3-Jul-12. The horizontal axis displays the collection dates. Data are means ±SE (n = 3). Levels A and B of the Student's t analysis are indicated. Columns with different letters are significantly different

mature trees, with fluctuations from March through June. These plants were not flowering so the function of these proteins is probably related to vegetative growth or some other function. In addition to having a role in flowering induction and breaking flowering dormancy FT seems to affect vegetative growth, leaf, flower and inflorescence architecture and thorn development [15, 21–24]. For instance overexpression of *FT* in *Poncirus trifoliata*, a citrus relative, alters tree architecture, dormancy requirements and leaf shape [15].

Conclusions

In mature citrus trees there were no differences between the leaf expression patterns of the three *CiFT* genes studied. Their expression was synchronized, peaked around April and subsided in the following months. This pattern was observed in the trees regardless of whether they were under temperature controlled conditions in the greenhouse or in the field. This was also a major difference with juvenile plants in which *CiFT3* but not *CiFT1/FT2* was induced during 30 April 2013.

FT protein, produced in the leaves is transported to the apical meristems where, through its interaction with other proteins, it triggers the formation of floral meristems [18]. Furthermore, it has been observed in various systems, including citrus, that ectopic overexpression of *FT* overcomes juvenility and induces early flowering [10, 15, 16, 23, 25, 26]. Interestingly, we observed that in leaves of immature PSO plants FT expression was highly induced, particularly during the month of June, indicating that in juvenile wild type plants high expression levels of *FT* in leaves was not associated with flowering. Assuming *CiFT* mRNA levels correlate with FT protein levels and that it gets transported to the meristem then FT is not sufficient to induce flowering and it must be preceded or accompanied by superseding factors that enable the transition from juvenile to mature. This is perhaps an evolutionary strategy to guarantee the individual has reached the appropriate size or accumulated enough resources to provide the best chances for fructification and hence reproductive survival. It seems from these results that, at the expression levels observed in wild type individuals, *CiFT* expression is necessary but not sufficient to induce flowering. The transition from juvenile to mature must first happen for flowering to occur. This is not the case in transgenic plants, perhaps because exceptionally high levels of *FT* are reached. In fact a correlation between levels of expression and timing of flowering has been observed [15].

The induction of *FT* genes expression in citrus leaves coincided with the transition from winter to spring (Additional file 1, Figure S1) and it could have been triggered by temperature (either accumulation of cold hours or the change to warmer temperatures) or increasing day length or both.

Additional files

Additional file 1: Figure S1. Temperatures registered during the experimental period in Groveland, Florida. (a) Monthly average minimum and maximum temperatures. (b) Monthly low and high temperature range. Source: Weather Warehouse (http://www.usclimatedata.com/climate/clermont/florida/united-states/usfl0086 and https://www.wunderground.com/weather/us/fl/clermont).

Additional file 2: Table S1. TaqMan MGB primers and probes.

Additional file 3: Table S2. Full Student's t test results.

Abbreviations

C_T: Cycle threshold; mRNA: Messenger ribonucleic acid; PCR: Polymerase chain reaction; PSO: 'Pineapple' sweet orange; PSOsd: 'Pineapple' sweet orange, seedy; PUM: Pummelo; RQ: Relative quantification; rRNA: Ribosomal ribonucleic acid; RT-qPCR: Reverse-transcription quantitative real-time PCR; SE: Standard error

Acknowledgements

We thank Dr. Jude Grosser and Dr. Harry Klee for their knowledge, time and effort towards improving this written work. We would also like to express our appreciation to everyone in the Moore lab. We also thank the faculty, students and staff in the Horticultural Sciences department at the University of Florida that were part of this research.

Funding

This work was funded in part by a grant to G. A. Moore by the Citrus Research and Development Foundation, Inc., grant number 5200-146.

Authors' contributions

MP designed and conducted the experiments, analyzed the data and drafted the manuscript. VJF contributed to the design and execution of the real time PCR experiments, contributed and supervised the data analysis and revised the manuscript. GAM conceived and designed the experiments and revised the manuscript. All authors read and approved the final manuscript.

Competing interests

The authors declare there is no competing interests.

Author details

[1]Horticultural Sciences Department, Institute of Food and Agricultural Sciences, University of Florida, 2550 Hull Road, Gainesville, FL 32611, USA. [2]Plant Molecular and Cellular Biology Program, University of Florida, Gainesville, FL 32611, USA.

References

1. Gmitter FG Jr, GJ W, Moore G. Citrus. In: Hammerschlag FA, Litz RE, editors. Biotechnology of perennial fruit crops. Wallingford: CAB International; 1992. p. 335–69.
2. Davies FS, Jackson LK. Floridiana collection: citrus growing in Florida. 5th ed. Gainesville: University Press of Florida; 2009.

3. Komeda Y. Genetic regulation of time to flower in *Arabidopsis thaliana*. Ann Rev Plant Biol. 2004;55:521–35.

4. Albani MC, Coupland G. Comparative analysis of flowering in annual and perennial plants. Curr Top Dev Biol. 2010;91:323–48.

5. Valiente JI, Albrigo LG. Flower bud induction of sweet orange trees [*Citrus sinensis* (L.) Osbeck]: effect of low temperatures, crop load, and bud age. J AmerSoc Hort Sci. 2004;129(2):158–64.

6. Nishikawa F, Endo T, Shimada T, Fujii H, Shimizu T, Omura M, Ikoma Y. Increased *CiFT* abundance in the stem correlates with floral induction by low temperature in Satsuma mandarin (*Citrus unshiu* Marc.). J Exp Bot. 2007;58(14):3915–27.

7. Samach A, Onouchi H, Gold SE, Ditta GS, Schwarz-Sommer Z, Yanofsky MF, Coupland G. Distinct roles of CONSTANS target genes in reproductive development of *Arabidopsis*. Science. 2000;288(5471):1613–6.

8. Yamagishi N, Sasaki S, Yamagata K, Komori S, Nagase M, Wada M, Yamamoto T, Yoshikawa N. Promotion of flowering and reduction of a generation time in apple seedlings by ectopical expression of the *Arabidopsis thaliana* FT gene using the apple latent spherical virus vector. Plant Mol Biol. 2011;75(1-2):193–204.

9. Yamagishi N, Yoshikawa N. Expression of *FLOWERING LOCUS T* from *Arabidopsis thaliana* induces precocious flowering in soybean irrespective of maturity group and stem growth habit. Planta. 2011;233(3):561–8.

10. Zhang H, Harry DE, Ma C, Yuceer C, Hsu CY, Vikram V, Shevchenko O, Etherington E, Strauss SH. Precocious flowering in trees: the *FLOWERING LOCUS T* gene as a research and breeding tool in *Populus*. J Exp Bot. 2010;61(10):2549–60.

11. Nishikawa F, Iwasaki M, Fukamachi H, Endo T. Predicting the number of flowers in Satsuma mandarin (*Citrus unshiu* Marc.) trees based on citrus *FLOWERING LOCUS T* mRNA levels. Hortic J. 2017;86(3):305–10.

12. Kojima S, Takahashi Y, Kobayashi Y, Monna L, Sasaki T, Araki T, Yano M. *Hd3a*, a rice ortholog of the *Arabidopsis* FT gene, promotes transition to flowering downstream of *Hd1* under short-day conditions. Plant Cell Physiol. 2002;43(10):1096–105.

13. Bohlenius H, Huang T, Charbonnel-Campaa L, Brunner AM, Jansson S, Strauss SH, Nilsson O. CO/FT regulatory module controls timing of flowering and seasonal growth cessation in trees. Science. 2006;312(5776):1040–3.

14. Hsu CY, Liu Y, Luthe DS, Yuceer C. Poplar *FT2* shortens the juvenile phase and promotes seasonal flowering. Plant Cell. 2006;18(8):1846–61.

15. Endo T, Shimada T, Fujii H, Kobayashi Y, Araki T, Omura M. Ectopic expression of an *FT* homolog from citrus confers an early flowering phenotype on trifoliate orange (*Poncirus trifoliata* L. Raf.). Transgenic Res. 2005;14(5):703–12.

16. Velazquez K, Aguero J, Vives MC, Aleza P, Pina JA, Moreno P, Navarro L, Guerri J. Precocious flowering of juvenile citrus induced by a viral vector based on citrus leaf blotch virus: a new tool for genetics and breeding. Plant Biotechnol J. 2016;14(10):1976–85.

17. Samach A. Congratulations, you have been carefully chosen to represent an important developmental regulator! Ann Bot. 2013;111(3):329–33.

18. Corbesier L, Vincent C, Jang S, Fornara F, Fan Q, Searle I, Giakountis A, Farrona S, Gissot L, Turnbull C, et al. FT protein movement contributes to long-distance signaling in floral induction of *Arabidopsis*. Science (New York, NY). 2007;316(5827):1030–3.

19. Rorabacher DB. Statistical treatment for rejection of deviant values - critical-values of Dixon Q parameter and related subrange ratios at the 95-percent confidence level. Anal Chem. 1991;63(2):139–46.

20. Albrigo LG. Flower bud induction advisory #6 for 2012-2013-04/09/13. In: Flower bud induction overview and advisory. Lake Alfred: Citrus Research & Education Center; 2013. http://www.crec.ifas.ufl.edu/extension/flowerbud/2013/04_09_13.shtml. Accessed July 22 2013.

21. Yarur A, Soto E, Leon G, Almeida AM. The sweet cherry (*Prunus avium*) *FLOWERING LOCUS T* gene is expressed during floral bud determination and can promote flowering in a winter-annual Arabidopsis accession. Plant Reprod. 2016;29(4):311–22.

22. Pin PA, Nilsson O. The multifaceted roles of FLOWERING LOCUS T in plant development. Plant Cell Environ. 2012;35(10):1742–55.

23. Srinivasan C, Dardick C, Callahan A, Scorza R. Plum (*Prunus domestica*) trees transformed with poplar *FT1* result in altered architecture, dormancy requirement, and continuous flowering. PLoS One. 2012;7(7):e40715.

24. Hsu CY, Adams JP, Kim HJ, No K, Ma CP, Strauss SH, Drnevich J, Vandervelde L, Ellis JD, Rice BM, et al. *FLOWERING LOCUS T* duplication coordinates reproductive and vegetative growth in perennial poplar. P Natl Acad Sci USA. 2011;108(26):10756–61.

25. Klocko AL, Ma C, Robertson S, Esfandiari E, Nilsson O, Strauss SH. FT overexpression induces precocious flowering and normal reproductive development in eucalyptus. Plant Biotechnol J. 2016;14(2):808–19.

26. Trankner C, Lehmann S, Hoenicka H, Hanke MV, Fladung M, Lenhardt D, Dunemann F, Gau A, Schlangen K, Malnoy M, et al. Over-expression of an FT-homologous gene of apple induces early flowering in annual and perennial plants. Planta. 2010;232(6):1309–24.

8

GaMYB85, an R2R3 MYB gene, in transgenic Arabidopsis plays an important role in drought tolerance

Hamama Islam Butt, Zhaoen Yang, Qian Gong, Eryong Chen, Xioaqian Wang, Ge Zhao, Xiaoyang Ge, Xueyan Zhang* and Fuguang Li*

Abstract

Background: MYB transcription factors (TFs) are one of the largest families of TFs in higher plants and are involved in diverse biological, functional, and structural processes. Previously, very few functional validation studies on R2R3 MYB have been conducted in cotton in response to abiotic stresses. In the current study, GaMYB85, a cotton R2R3 MYB TF, was ectopically expressed in Arabidopsis thaliana (Col-0) and was functionally characterized by overexpression in transgenic plants.

Results: The in-silico analysis of GaMYB85 shows the presence of a SANT domain with a conserved R2R3 MYB imperfect repeat. The GaMYB85 protein has a 257-amino acid sequence, a molecular weight of 24.91 kD, and an isoelectric point of 5.58. Arabidopsis plants overexpressing GaMYB85 exhibited a higher seed germination rate in response to mannitol and salt stress, and higher drought avoidance efficiency than wild-type plants upon water deprivation. These plants had notably higher levels of free proline and chlorophyll with subsequent lower water loss rates and higher relative water content. Germination of GaMYB85 transgenics was more sensitive to abscisic acid (ABA) and extremely liable to ABA-induced inhibition of primary root elongation. Moreover, when subjected to treatment with different concentrations of ABA, transgenic plants with ectopically expressed GaMYB85 showed reduced stomatal density, with greater stomatal size and lower stomatal opening rates than those in wild-type plants. Ectopic expression of GaMYB85 led to enhanced transcript levels of stress-related marker genes such as RD22, ADH1, RD29A, P5CS, and ABI5.

Conclusions: Our results indicate previously unknown roles of GaMYB85, showing that it confers good drought, salt, and freezing tolerance, most probably via an ABA-induced pathway. These findings can potentially be exploited to develop improved abiotic stress tolerance in cotton plants.

Keywords: MYB transcription factor, Abiotic stress tolerance, Abscisic acid

Background

Environmental factors, particularly drought stress, severely limit the production and distribution of many important agronomic crops worldwide [1]. In plants, different types of transcription factors (TFs), such as bZIP, NAC, AP2, WRKY, and MYB, control plant biological processes by modulating the initiation rate of target genes with the combined activation of a DNA binding domain, a nuclear localization signal, a transcription activation domain, and oligomerization sites in response to biotic or abiotic stresses [2]. MYB TFs were first recognized as oncogenes in animals; however, it has subsequently been found that MYB genes occur more widely in plants than in animals and fungi [3, 4]. Following the discovery of the first plant MYB gene in maize (Zea mays) [5], large numbers of MYB proteins, particularly R2R3 MYBs, have been extensively identified and characterized in different plant species, including Arabidopsis, apple, grape, maize, petunia, and snapdragon [6]. Furthermore, R2R3 MYBs are known to control abundant plant biological processes, such as hormone signal transduction,

* Correspondence: Zhangxueyan_caas@126.com; aylifug@163.com
State Key Laboratory of Cotton Biology, Institute of Cotton Research of Chinese Academy of Agricultural Science (ICR, CAAS), Anyang 455000, China

organ development, cell cycle progression, cellular morphogenesis, secondary metabolism, and stress responses [7, 8]. More recently, genome-wide studies of *Gossypium raimondii* have revealed 205 putative R2R3 MYB genes, which is a greater number than reported in any other dicot or monocot [9].

Structurally, MYB proteins have a highly conserved MYB or DBD (DNA-binding domain), located at the N terminus, which is well conserved across all eukaryotes, whereas the diverse C termini act as a *trans*-acting domain (TAD) that regulates a broad range of functions in MYB proteins [10, 11]. Moreover, each MYB repeat contains 52 amino acid residues with regularly spaced triplet tryptophan residues, which form a hydrophobic core structure. The MYB repeat structure is composed of three α-helices. Two helices form the HTH (helix-turn-helix) structure and contribute to the binding of target genes to the promoter region, whereas the third helix participates in DNA recognition [12, 13]. On the basis of their DNA-binding domain sequential repeats, MYB proteins are grouped into four types; MYB1-R, R2R3-MYB, R1R2R3-MYB (MYB3R), and 4R−MYB [14]. However, to improve drought tolerance in the relevant crops, mining of new MYB R2R3 TFs by functional genomics and comparative genomics studies is a promising strategy, as it is the largest family of MYB proteins of higher plants and well known to be involved in specific diverse biological, functional, and structural processes.

The phytohormone abscisic acid (ABA) has numerous functions, such as in seed germination inhibition and dormancy maintenance, stomatal regulation, flowering time, and adaptations to drought, cold, and salt stresses [15]. ABA is known to induce dehydration-responsive genes, and thus on encountering drought stress, elevated levels of ABA stimulate *cis*-acting and *trans*-acting factors, which in turn up-regulate the ABA-induced expression of MYB, NAC, WRKY, and bZIP TF genes [16–19], thereby contributing to the relief of stress conditions [20]. Many ABA-inducible genes contain a *cis*-element (ABRE; ACGTGG/TC), whereas dehydration and low temperature stress-inducible genes contain a different *cis*-element (DRE; TACCGACAT). Both these elements play important roles in stress management via ABA-dependent and -independent signal transduction cascades [15, 21]. Moreover, *MYC/MYB* recognition sequences are a prerequisite for transcript regulation of *RD22 and ADH1*, which is induced by high ABA levels in *Arabidopsis* under drought stress [22]. In addition, different R2R3 MYB genes (*AtMYB44, AtMYB15, AtMYB60, AtMYB96, AtMYB61,* and *GbMYB5*) have been shown to be involved in stomatal regulation via ABA in response to dehydration [23–25]. R2R3 MYB TFs are implicated in diverse plant responses to abiotic stress conditions. For example, the AtMYB102 protein is reported to be an important

component that integrates wounding, osmotic stress, and ABA transduction pathways in transgenic *Arabidopsis* [26]. Similarly, overexpression studies on R2R3 *MdSIMB1* [27], *OsMYB2* [28], *GmMYBJ1* [29], *GmMYBJ2* [30], *GbMYB5* [25], *GmMYB76* [31] and *SbMYB8* [32], have shown improved drought, salt, and cold tolerance in transgenic plants.

Cotton is an important textile fiber and oil seed crop [33]; however, the repercussions of climate change, particularly drought stress, are severely limiting its production globally [34]. Our laboratory previously released *Gossypium arboreum* sequencing data [35] and RNA-seq studies have revealed hormone crosstalk, whereby MYB TFs are implicated in modifying plant responses toward drought and NaCl stresses in different tissues [36]. Thus, in view of the current scenario, we ectopically expressed a novel *G. arboreum* R2R3 MYB gene, designated *GaMYB85*, in *Arabidopsis* to study its role under drought, salt, and freezing stress conditions. Collectively, our results suggest that the *GaMYB85* gene confers good drought tolerance in overexpressing transgenic *Arabidopsis* plants, which can potentially be employed in future cotton crop improvement.

Results
GaMYB85 characterization and phylogenetic analysis

Previously, our mRNA-seq studies of *G. arboreum* showed multiple hormone crosstalk and tissue-selective signaling; moreover, MYB transcription factors are implicated in modifying cotton plant responses toward PEG and NaCl stresses in leaf, stem, and root [36]. Therefore, characterization of a novel gene, *GaMYB85*, was performed in transgenic *Arabidopsis* plants subjected to drought and salt stresses. The full-length CDS of *GaMYB85* is 771 bp long, and codes for a protein containing 257 amino acids with an expected molecular weight of 24.91 kD and a theoretical isoelectric point of 5.58 (http://web.expasy.org). Performing SMART analysis (available online at http://smart.embl-heidelberg.de/), the deduced 257-residue polypeptide was determined to contain a SANT domain between amino acids 38 and 86. The secondary structure of GaMYB85 includes 19.53% alpha helix, 19.53% extending chain, and 60.94% random coil sequences. *GaMYB85* alignment results between cDNA and genomic sequence retrieved from the cotton genome database revealed that the gene contains no introns (Additional file 1), and is located on chromosome 13 with a cotton ID number of cotton_A_21601. The deduced GaMYB85 protein shares a high amino acid sequence homology with AtMYB85 (At4g22680). Overexpression of this protein in *Arabidopsis* resulted in the ectopic deposition of lignin in epidermal and cortical cells of the stem [37], and it was thus designated as GaMYB85 in further studies. Multiple sequence alignment revealed high similarity between GaMYB85 and retrieved R2R3 protein homologs of various dicots and monocots, including *Gossypium hirsutum* (98%), *Nicotiana tabacum* (82%), *Arabidopsis thaliana* (79%), *Theobroma cacao* (78%), *Zea mays* (84%), *Hordeum vulgare* (80%), *Oryza sativa* (70%), *Triticum aestivum* (66%), and *Vitis vinifera* (61%), which are annotated as predicted, putative, and hypothetical, and thus their functions are still unknown. However, they show the presence of conserved R2 and R3 domains in the MYB gene (http://www.ebi.ac.uk/Tools/msa/muscle/) (Fig. 1a). Moreover, in the phylogenetic tree we constructed, we also included some R2R3 MYB genes with known functions, which confer enhanced tolerance to multiple abiotic stresses. *GaMYB85* shared high similarity and clustered together with its homologs from *G. hirsutum*, *V. vinifera*, and *T. cacao*, which indicates that they might have originated from the same common ancestor. Therefore, these results suggest that *GaMYB85* is a novel gene that confers enhanced drought tolerance (Fig. 1b). *LcMYB1*, which has a single conserved SANT domain, belongs to a MYB-related protein, and was used as an outgroup in the tree.

Overexpression of *35S:GaMYB85* in plants confers ABA hypersensitivity for seed germination and root elongation

In order to gain an understanding of the possible roles of *GaMYB85* during germination and post-germination stages in response to abiotic stresses, this gene was ectopically expressed in *Arabidopsis thaliana* (Col-0) plants using the floral dip method. T_2 seeds were selected on BASTA agar plates, and subsequently the surviving to dead plant ratio was determined (Additional files 2 and 3). Ten segregating lines with a correct segregation ratio of 3:1 were selected and transcript patterns were subsequently monitored by qRT-PCR (Fig. 2b and Additional file 5). Three lines with high gene transcript levels were selected and screened on selective media until we obtained homozygous lines. The T_3 transgenic lines L3, L4, and L7 were selected for further analysis.

To examine the ABA sensitivity of transgenic plants containing ectopically expressed *35S:GaMYB85*, the seeds of L3, L4, and L7 lines and WT plants were grown directly on medium containing different concentrations of ABA (0, 0.3, 0.5, 1, 2, and 5 μM), and the germination rates of these were compared. In a primary root length elongation assay, the *35S:GaMYB85* lines were more susceptible to ABA (Fig. 3a and b) than was the WT, which indicates that ABA might be involved in the root development process, and also strengthens speculation that *35S:GaMYB85* is regulated in an ABA-dependent manner. The presence of exogenous ABA had a more pronounced effect on the germination rates of *35S:GaMYB85* lines (L3, L4, and L7) than on that of the WT (Fig. 3c). Only 25% of the seeds of the *35S:GaMYB85* lines (L3, L4, and L7) were able to germinate on 2 μM ABA MS, as against 50% germination for the WT. Furthermore, a detailed time course experiment was conducted on 1 μM ABA (Fig. 3d), in which the seeds of *35S:GaMYB85* lines showed more delayed germination rates than the WT, starting from the 4th to 8th day after the onset of germination.

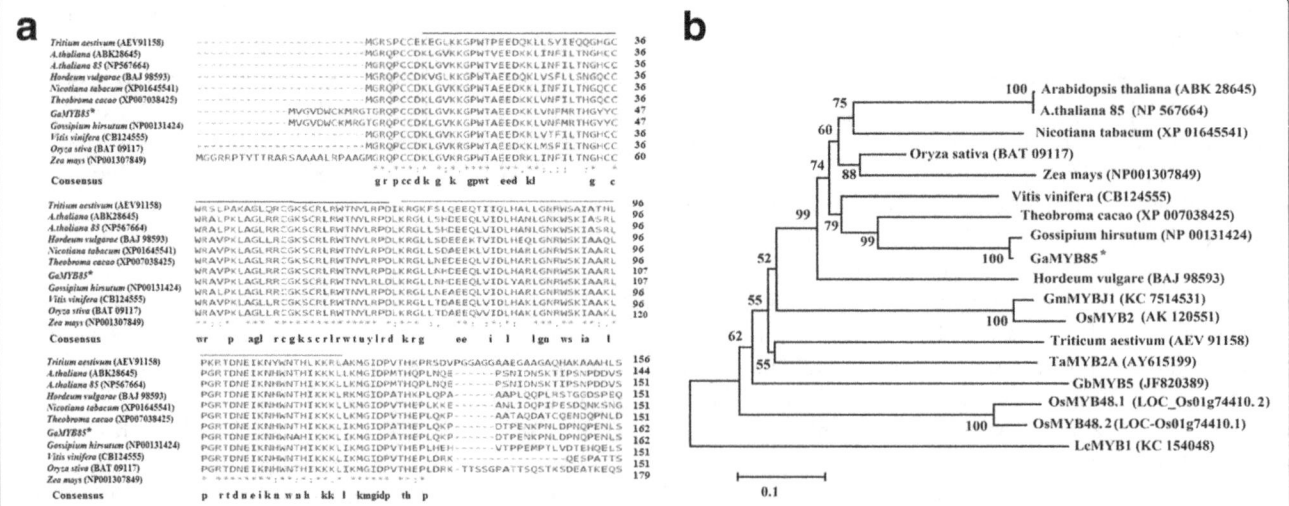

Fig. 1 The Phylogenetic analysis and predicted structure of R2R3 MYB *GaMYB85*. **a** Homologous sequences retrieved by Blast p were aligned with *GaMYB85* showing R2 and R3 repeats represented by black and red lines respectively. **b** The NJ tree analysis of *GaMYB85* proteins with homologous and known R2R3 MYB monocots and dicots sequences, along with scale bar that shows the calculated distance by multiple sequence alignment (MSA). The MSA of the respective sequences were provided in Additional file 1

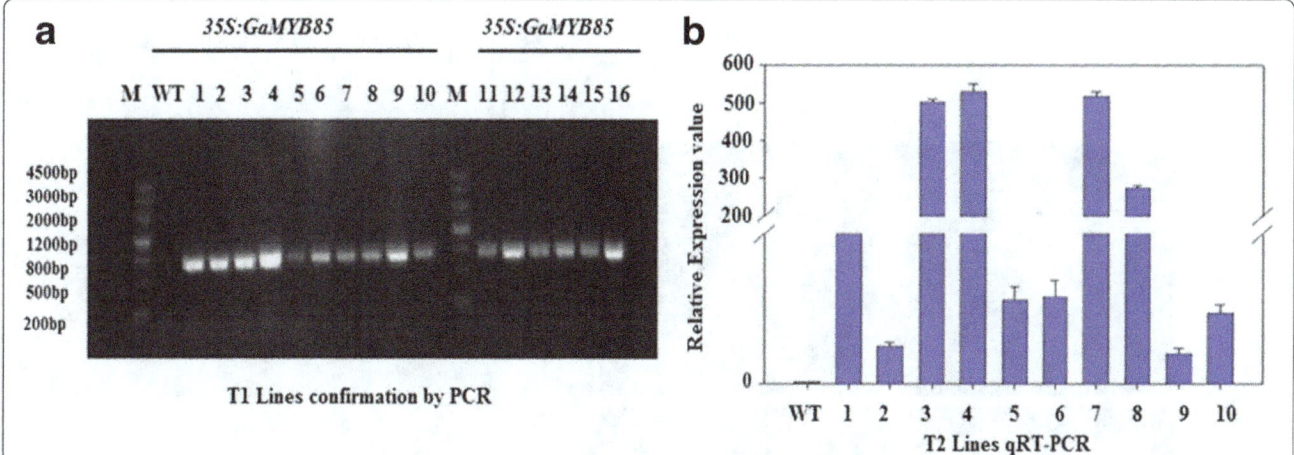

Fig. 2 Gel and qRT-PCR analysis of *GaMYB85* transgenic plants **a** The confirmation of *GaMYB85* gene CDS (771 bp) integration in T_0 generated overexpressed *Arabidopsis* plants. The DNA of WT (Col-0) was used as negative control (lane 1) and *35S:GaMYB85* DNA was used in (lanes 1–10, 11–16). M; DNA marker III, 1 kb. **b** Relative expression of *GaMYB85* gene in T_2 transgenic lines by qRT-PCR having segregating ratio of 3:1 on the selective medium, three cDNA preparations were used and error bars represented with SD value. *AtUBQ*10 gene (*Accession no:* AT4G05320) was used as an internal standard in qRT-PCR

Overexpression of *35S:GaMYB85* in plants confers enhanced salt and osmotic tolerance

For further clarification of the effects of NaCl stress on *35S:GaMYB85* plants, seed germination and post-germination growth were monitored. The results revealed 100% germination rates for control and test seeds that were grown in control MS medium. In contrast, the germination rates of *35S:GaMYB85* lines (L3, L4, and L7) were significantly higher than those of the WT on NaCl medium (50, 100, 125, and 150 mM) in a dose-dependent manner. Moreover, these plants had higher fresh weights than WT plants, which turned yellow and produced a proportion of unviable seeds (Fig. 4a, b, and c). Primary root growth is an important indicator of plant tolerance to various stress responses, and *35S:GaMYB85* transgenic seedlings showed longer roots and had greener broader rosettes than did the control when raised on 100 mM NaCl medium (Fig. 4d and e). Furthermore, compared with the WT, the germination rate and fresh weight of *35S:GaMYB85*-expressing plants were significantly higher, and growth performance was improved under 300 mM mannitol stress for 10 d (Fig. 5a, b, and c). Moreover, a root length assay on mannitol also revealed a significant difference in transgenic line root length growth on 200 and 300 mM mannitol medium from that in the WT (Fig. 5d and e). These significantly higher germination rates on simulated salt and mannitol stressed medium indicate that ectopic expression of *GaMYB85* in *Arabidopsis* confers enhanced salt and osmotic tolerance during pre- and post-seed germination stages.

Plants with *GaMYB85* overexpression show enhanced tolerance under drought and freezing stress

MYB TFs have been implicated in diverse plant responses to abiotic stress conditions [26]. For a drought assay, 7-d-old transgenic and WT seedlings were transplanted to well-hydrated soil, and then 2 weeks later, grown L3, L4, and L7 plants were subjected to drought stress. The WT plants showed lower recovery rates of 35% along with severe drought symptoms, such as leaf rolling, shrinkage, and delayed growth, when compared with the 66.66%, 77.8%, and 70% recovery rates of *GaMYB85* transgenic plants (L3, L4, and L7, respectively) following re-watering for 3 d (Fig. 6a and b). In accordance with these results, the water loss rate, which is an important indicator of drought stress evaluation, was significantly lower for transgenic lines when excised leaf weights were monitored over a 0- to 7-h time interval (Fig. 6c). The relative water content (RWC) in detached leaves of L3, L4, and L7 was approximately 71%, 81.9%, and 79.9%, respectively, whereas the RWC for WT decreased by up to 55% (Fig. 6d). Subsequently, proline and chlorophyll content were evaluated in non-stressed and drought-stressed samples. The free proline content in *GaMYB85*-expressing plants was higher under both normal and drought stress conditions. The proline content in overexpressing L3, L4, and L7 plants was significantly higher (19.8, 27.5, and 20.7 μg/g, respectively) than that in the WT (almost 9 μg/g) under drought stress (Fig. 6e and f). In addition, the total chlorophyll content under the non-stressed condition was nearly the same in WT and overexpressing L3, L4, and L7 plants. However, under drought stress, the total chlorophyll

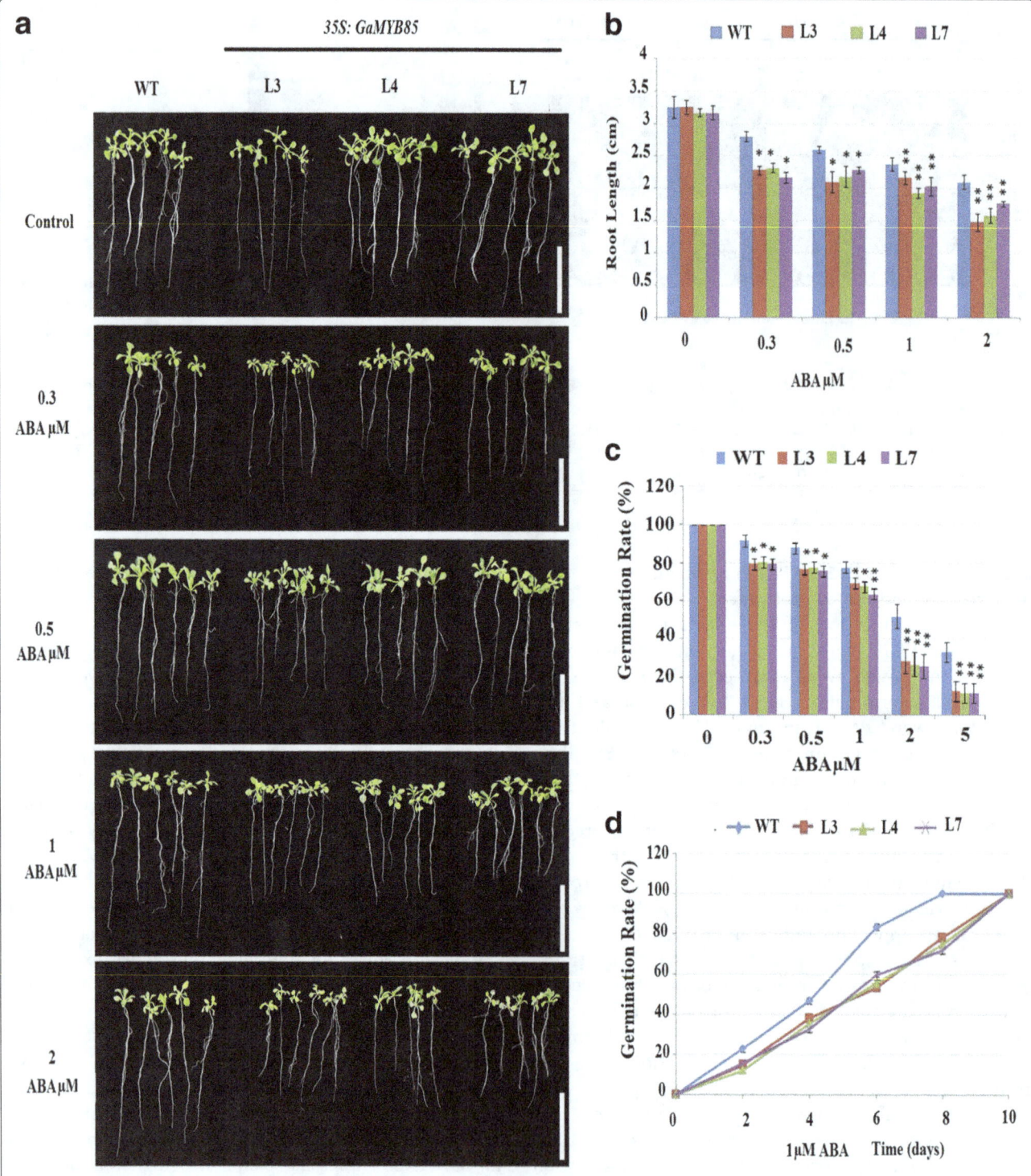

Fig. 3 Overexpressing of *GaMYB85* modulates hypersensitivity to ABA-elicited root inhibition and seed germination rate. **a** *35S:GaMYB85* plants and WT root elongation comparisons on MS with and without varied ABA conc. (µM). The seedlings were scored and photographed after 7 days. Assay was run in triplicates, Bar line = 1 cm. **b** Quantitative comparisons of root elongation assay on ABA MS with (0, 0.3, 0.5, 1 and 2 µM). The three replicates used with 30 seedlings each, * *P* < 0.05 of mean value represented by ± SE. **c** *35S:GaMYB85* plants and WT seeds germination rates with and without ABA, the results were scored at 10th day by using 50 seeds each, ± SE (*n* = 3). **d** ABA 1(µM) supplemented plates used for time course analysis, the error bars represent SE of 3 replicates

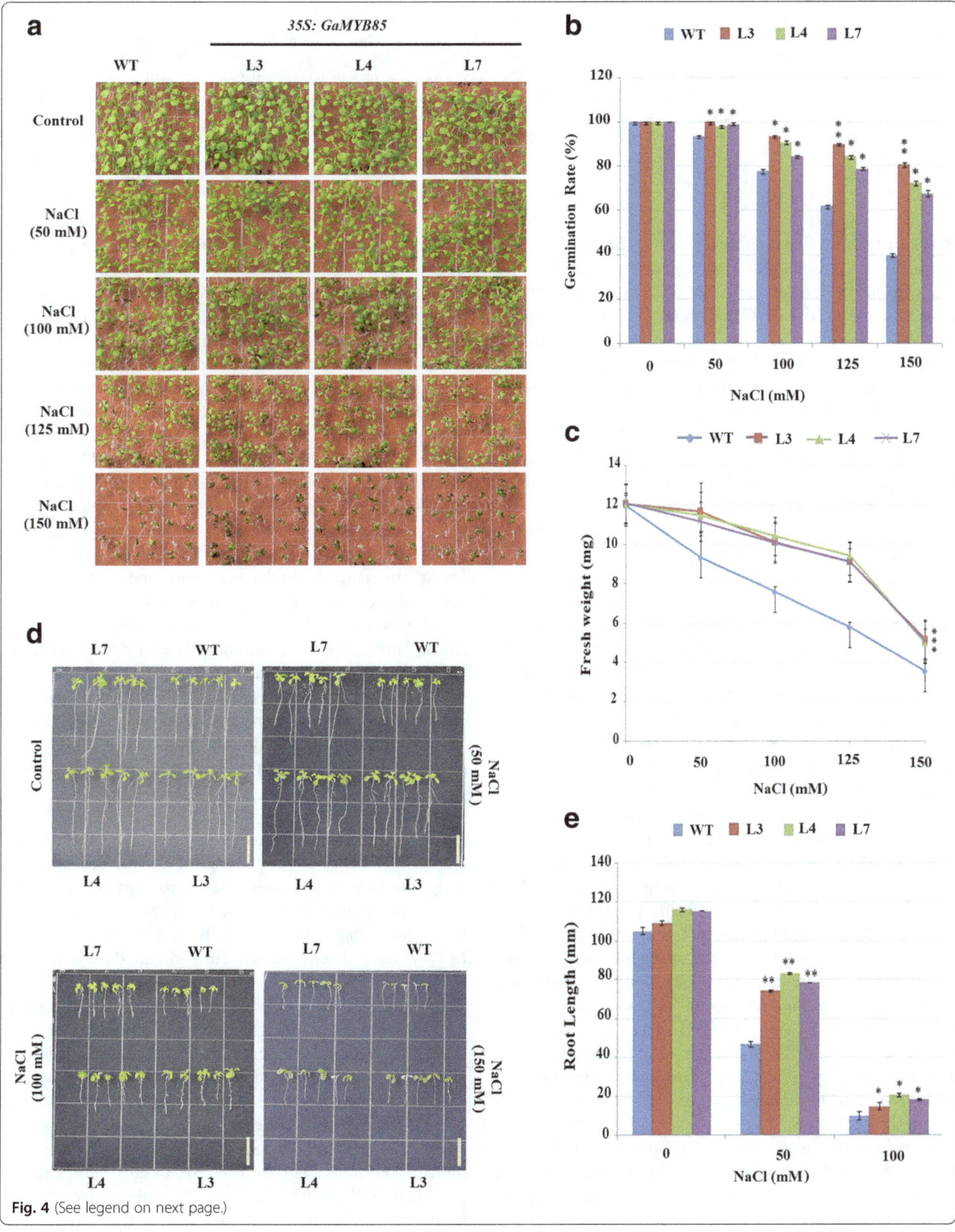

Fig. 4 (See legend on next page.)

content of *GaMYB85* plants was higher than that in WT plants. When 3-week-old *Arabidopsis* lines (L3, L4, and L7) ectopically expressing *GaMYB85* and WT plants were exposed to a − 10 °C treatment for 3 h, and subsequently revived in a 22 ± 1 °C growth room, the plants ectopically expressing *GaMYB85* showed improved recovery rates of 55.55%, 75%, and 69.44%, respectively. The adverse effect of freezing stress was more pronounced in the WT plants, which showed only 34% survival after 1 week of growth in the 22 ± 1 °C growth room (Fig. 7a and b). Overall, these results indicate that ectopically expressed *GaMYB85* in *Arabidopsis* plays an important role under drought and freezing stresses.

Reduced stomatal density, increased stomatal response, and increased stomatal size in *GaMYB85* plants

ABA production under abiotic stresses is directed to increase the sensitivity of stomatal closure, and thus to gain insight into this process, we examined the effects of ABA treatment on the stomatal opening of excised leaves of *GaMYB85* plants (Fig. 8a). When different concentrations of ABA (5 μM and 10 μM) were added to a stomata opening solution, stomatal opening was reduced, as ABA induces stomatal closure. Excised leaves of *GaMYB85* lines (L3, L4, and L7) treated with 5 μM ABA showed reduced stomatal opening rates of 40%–56%, which was further reduced to approximately 13.3%–23.3% under 10 μM ABA treatments. In contrast, WT plants had markedly higher stomatal opening rates of 73% and 50% under 5 and 10 μM ABA treatments, respectively (Fig. 8b). Moreover, WT stomatal density was also notably higher, at more than 127 per mm^2, than with 96, 111, and 112 per mm^2 for L3, L4, and L7, respectively (Fig. 8c and d). However, stomatal size (length to width dimensions) in L3, L4, and L7 was distinctly higher at 21–25 μm by 7–8.6 μm when compared with the 19 μm by 6.05 μm for the WT, as shown in Fig. 8e. Hence, *GaMYB85* reduced stomatal density and reduced stomatal opening, resulting in rapid ABA-induced stomata closure, which subsequently minimized water loss rates and led to improved drought tolerance.

Transcript analysis of abiotic stress-responsive marker genes

As *Arabidopsis thaliana* plants with constitutive overexpression of *35S:GaMYB85* have ABA-dependent enhanced drought tolerance, in addition to increased stomatal closure upon ABA treatment, it would be of interest to gain a further understanding of how *GaMYB85* affects and responds to ABA stress-responsive and signaling-related genes. To this end, eight abiotic stress-responsive genes (*RD22, RD29A, RD29B, P5CS, COR15A, CBF, ADH,* and *Rab18*) and two ABA signaling genes (*ABI5* and *ABI3*) were selected and studied. The expression levels of five stress-responsive genes (*RD29B, COR15A, CBF, Rab18,* and *ABI3*) were nearly the same in both transgenic and WT plants in ABA-treated samples (data not shown here). However, the expression levels of five genes (*RD22, RD29A, P5CS, ADH1,* and *ABI5*) were significantly and clearly higher in L3, L4, and L7 than in the WT when treated with 100 μM ABA for 6 h (Fig. 9a-e). Thus, in response to ABA treatment, the transcript levels of ABA-induced stress marker genes may positively contribute to the improved drought stress tolerance of *GaMYB85* plants.

Discussion

Drought stress is a worldwide dilemma that has severely affected the annual production of cotton crops. Transcription factors are well known to participate in the maintenance of plant homeostasis in response to various biotic and abiotic factors. MYB TFs, particularly R2R3 MYBs, have been widely studied in different plant species, but to date few of these genes have been functionally validated for cotton. On the basis of diploid cotton A sequencing data [35], a new candidate cloned R2R3 MYB gene, designated *GaMYB85*, was selected and ectopically expressed in *Arabidopsis thaliana*, and three overexpressing lines (L3, L4, and L7) were successfully generated and functionally validated for drought stress. The amino acid sequence of GaMYB85 was predicted to contain a conserved SANT-DBD and clusters phylogenetically with R2R3 MYB proteins from other monocots and dicots, indicating that the R2R3 MYB proteins that group together may retain an identical function based on high sequence similarity. Some R2R3

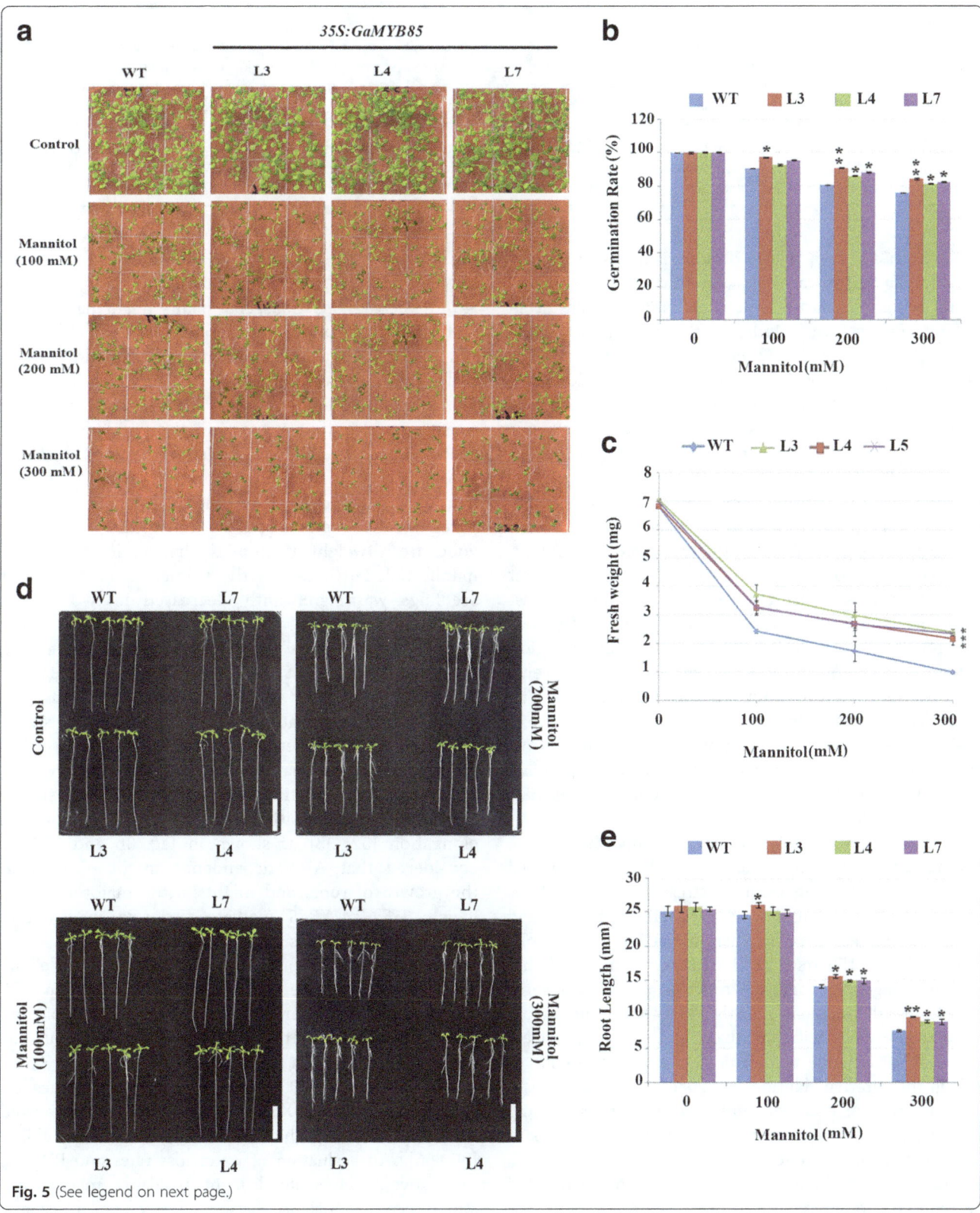

Fig. 5 (See legend on next page.)

(See figure on previous page.)
Fig. 5 Plants overexpressing *GaMYB85* perform well in response to mannitol stress **a** *35S:GaMYB85* and WT seed germination performance on MS supplemented with 0, 100, 200, and 300 mM mannitol. Germination rates were scored on the 10th day. **b** A plot of the germination rates scored on the 10th day for *35S:GaMYB85* and WT seeds germinated on the different mannitol media. Data are the means ± SE for three replicates of 50 seeds each (*$P < 0.05$). **c** Fresh weights of seedlings grown on 0–300 mM mannitol MS medium scored on the 10th day. Each treatment with 10 seedlings was performed in triplicate. Data are the means ± SE (*$P < 0.05$). **d** Comparison of root elongation of 5-day-old seedlings of *35S:GaMYB85* and WT plants transferred to 0, 100, 200, and 300 mM mannitol MS for 6 days. Values represent the data from three independent growth assays. Scale bar: 1 cm. **e** Quantitative comparison of root elongation of *35S:GaMYB85* and WT seedlings on MS supplemented with 0–300 mM mannitol (30 seedlings each). Data are the means ± SE ($n = 3$). $P < 0.05$ determine by the t-test

MYBs with a known role in drought response were also included in a tree analysis (Additional file 2). These share a high similarity with *GaMYB85* but cluster in different groups, which indicates that they might have a different number of SANT domains. Accordingly, these findings suggest that *GaMYB85* is a novel R2R3 MYB gene with a SANT-DBD domain.

Environmental stimuli like drought and salt stresses specifically initiate ABA production, which induces stomatal closure, contributing to strong and healthy roots and plants with efficient stomatal regulation and water-retaining abilities [38, 39]. Furthermore, drought-tolerant plants have higher endogenous proline content, reduced water loss rates, and improved relative water content, which enhance the response potential of transgenic plants against stress. Thus, to evaluate and dissect the possible molecular mechanisms and physiological roles underlying plant tolerance, we generated transgenic *Arabidopsis* that ectopically expressed *GaMYB85*. The important finding is that *35S:GaMYB85* lines have normal growth with respect to the WT, which is in contrast to previous studies in which transgenic overexpressing plants showed stunted growth and poor seed germination rates [40, 41].

In plants, turgor regulation plays an important role in the management of low water potential for normal seed germination rates under abiotic stress conditions. Thus, the observation that our *35S:GaMYB85* plants had higher germination rates when subjected to 300 mM mannitol and 150 mM NaCl treatments indicated that they have a general osmotic stress tolerance. The adverse effects of salt on plant growth can either result in osmotic stress or specific ion toxicity. The osmotic stress phase is attributable to dehydration or the presence of salt in the external solution, but specific ion toxicity is a consequence of salt accumulation in the transpiring leaves. Moreover, seeds germination rates are dependent on water movement into the seeds, whereas exposure to salinity causes leaf cells to lose water, which is attributed to reduced turgor pressure or decreased cell wall expansion [42–44]. These results are in accordance with the functions of *AtMYB102*, *AtMYB41*, and *GmMYBJ2*, which have been demonstrated to confer improved drought tolerance in response to ABA-dependent osmotic and

NaCl stresses [3, 26, 30]. As elevated levels of salt can interfere with the normal molecular functioning of plants [45], *35S:GaMYB85* transgenic plants might have coped with this situation via ion homeostasis mechanisms, i.e., Na^+ ion extrusion, Na^+ ion compartmentalization, and Na^+ ion reabsorption [46–48]. The roots of plants are in direct contact with the soil and more sensitive to Na^+ ion stress, and thus provide important clues regarding normal plant growth [49]. Thus, *35S:GaMYB85* plants show a dose-dependent increase in root length growth upon salt treatment when compared with WT plants, but as transgenic plants show good fresh weight it implies that ionic stress was manifested. Furthermore, the primary root growth of seedlings was significantly decreased in WT plants growing on 200 and 300 mM mannitol MS, which clearly demonstrates the high osmotic tolerance exhibited by the *35S:GaMYB85* plants [50, 51]. As ABA hypersensitivity and insensitivity of seeds can occur during germination and post-germination stages under water deficit stress, our *35S:GaMYB85* plants were confirmed to have enhanced sensitivity to ABA both at the germinating and post-germinating stages, along with hypersensitivity to ABA inhibition of root elongation [52, 53] as shown in Fig. 3b and c. It is considered that ABA dependence might up-regulate the growth of roots and might be the major route by which *35S:GaMYB85* mediates its role in drought, salt, and cold stress signaling pathways [54, 55].

Constitutive overexpression of *35S:GaMYB85* in plants also resulted in better drought tolerance than in WT plants, which is further supported by the phenotypic and physiological changes that occur in these plants, such as decreased water loss rates and higher RWC. These results are consistent with the functional studies on *GmMYBJ2* and *TaMYB19*, which have been reported to be associated with good drought tolerance ability [30, 56]. An evaluation of water loss rates and RWC is imperative to elucidate drought avoidance mechanisms and indicates a balance between water availability and transpiration rates through the stomatal apertures of transgenic plants [57, 58]. *35S:GaMYB85* plants showed slower and lower rates of water losses with higher RWC capacity than WT plants (Fig. 6c and d), which

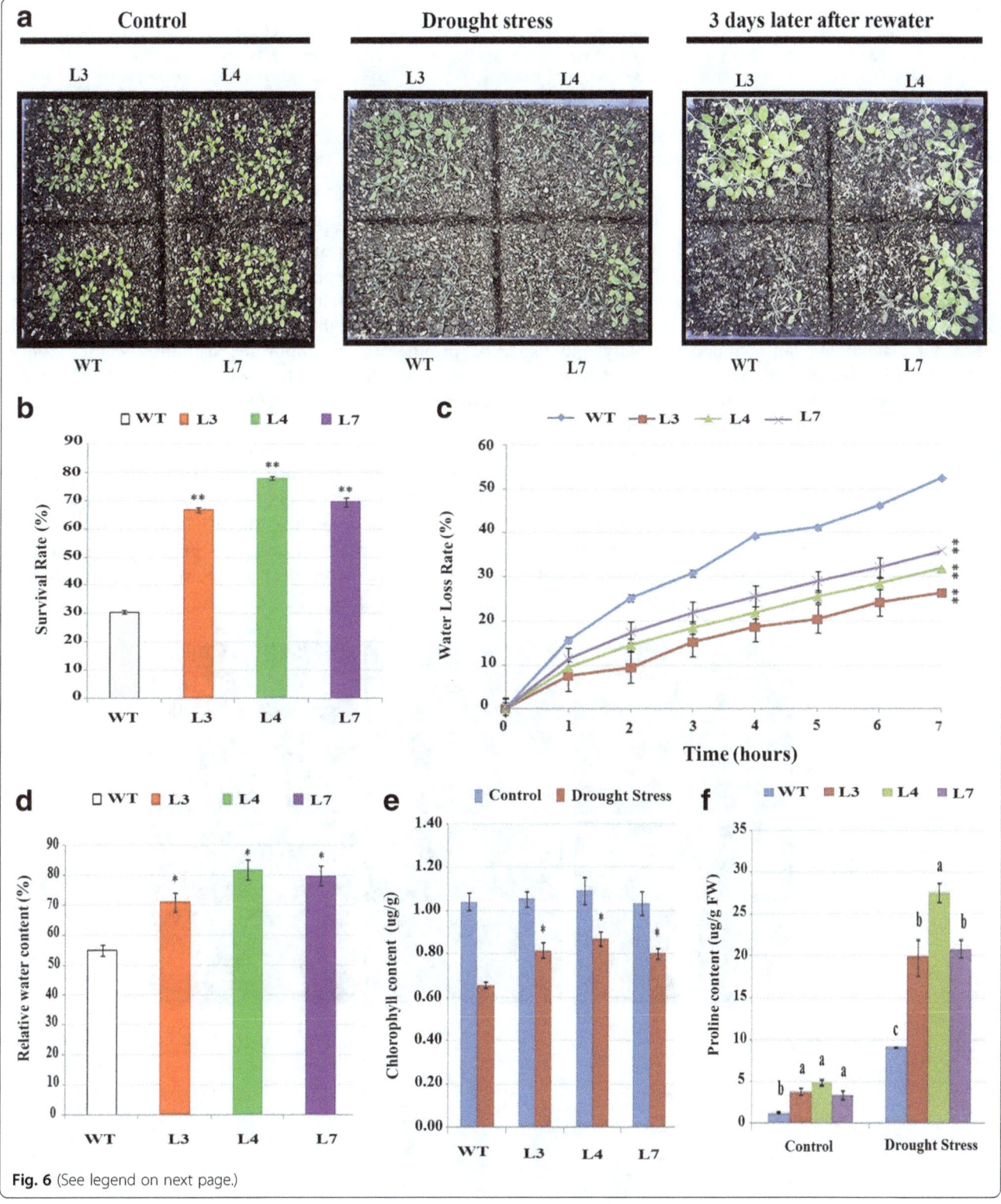

Fig. 6 (See legend on next page.)

(See figure on previous page.)

Fig. 6 The characterization of *35S:GaMYB85* under drought stress. **a** The phenotypes of *35S:GaMYB85* and WT at the initial and late stages of dehydration and 3 days after re-watering **b** Survival percentages are the mean values ± SE of three separate assays (*n = 18*). *$P < 0.05$ and **$P < 0.01$ determined by the t test. **c** Water loss rates of detached leaves of 3-week-old *35S:GaMYB85* and WT plants, expressed as percentages of initial fresh weight ± SE (*n = 10*) and (*$P < 0.05$). **d** Relative water content in detached leaves of 4-week-old *35S:GaMYB85* and WT plants. Data are the means ± SD (*n = 10*). Significant differences (*$P < 0.05$, **$P < 0.01$ and ***$P < 0.001$) were determined by Student's t-test. **e** Proline content of *35S:GaMYB85* and WT plants under normal conditions and after 14 days of water deprivation. For concentration determinations, absorbances were measured spectrophotometrically and Duncan's multiple range test was used for comparing means. Data are biological replicate means ± SE (*n = 3*)

is corroborated by recent reports on the R2R3 MYBs *GmMYBJ1* and *OsMYB48–1* [29, 59]. The total chlorophyll content of plants is linked to the physiological and transpiration efficiency of the photosynthetic machinery [60], whereas the osmoprotective molecule proline is critical for maintaining cell membrane stability, and also contributes to balancing osmotic pressure and retaining the membrane integrity of plants subjected to abiotic stresses [61]. In *35S:GaMYB85* plants, the chlorophyll and proline contents were significantly higher than those in WT plants (Fig. 6e and f), which suggests that endogenous proline levels contribute to the alleviation of water deficit stress and protect the photosynthetic apparatus from unfavorable toxic byproducts that have lethal effects on plant cells, and thereby confer

drought tolerance. Interestingly, the proline levels in *35S:GaMYB85* transgenic plants are higher, even under normal condition, which suggests that this R2R3 MYB protein might specifically protect plants and confer drought tolerance via the integrated role of proline as an important osmoprotectant. Consequently, water deficit treatment indicates that drought avoidance is not only controlled by osmotic signaling but also depends on ABA-induced regulation of stomatal movement. To confirm this supposition, we examined various stomatal parameters to determine the possible roles of *35S:GaMYB85* plants in attaining good drought tolerance. In our study, *35S:GaMYB85* plants were revealed to have greater stomatal size, lower stomatal densities, and lower rates of stomatal

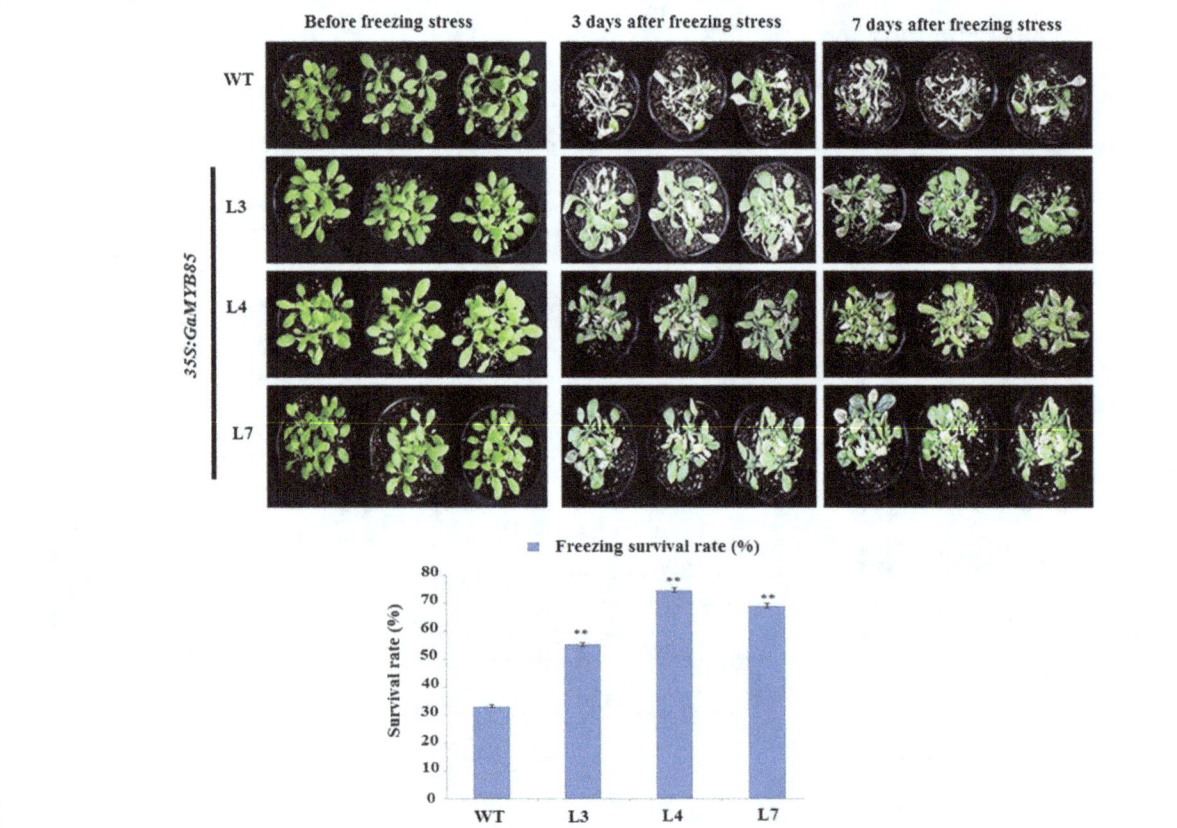

Fig. 7 Freezing stress responses of *35S:GaMYB85* and WT plants. *35S:GaMYB85* and WT plants at −10 °C treatment, the photographs were taken after 7 days of plants revival. The survival rate percentage evaluated from three separate assays, mean values ± SD (*n = 18*) and significant difference *$P < 0.05$, calculated by t-test

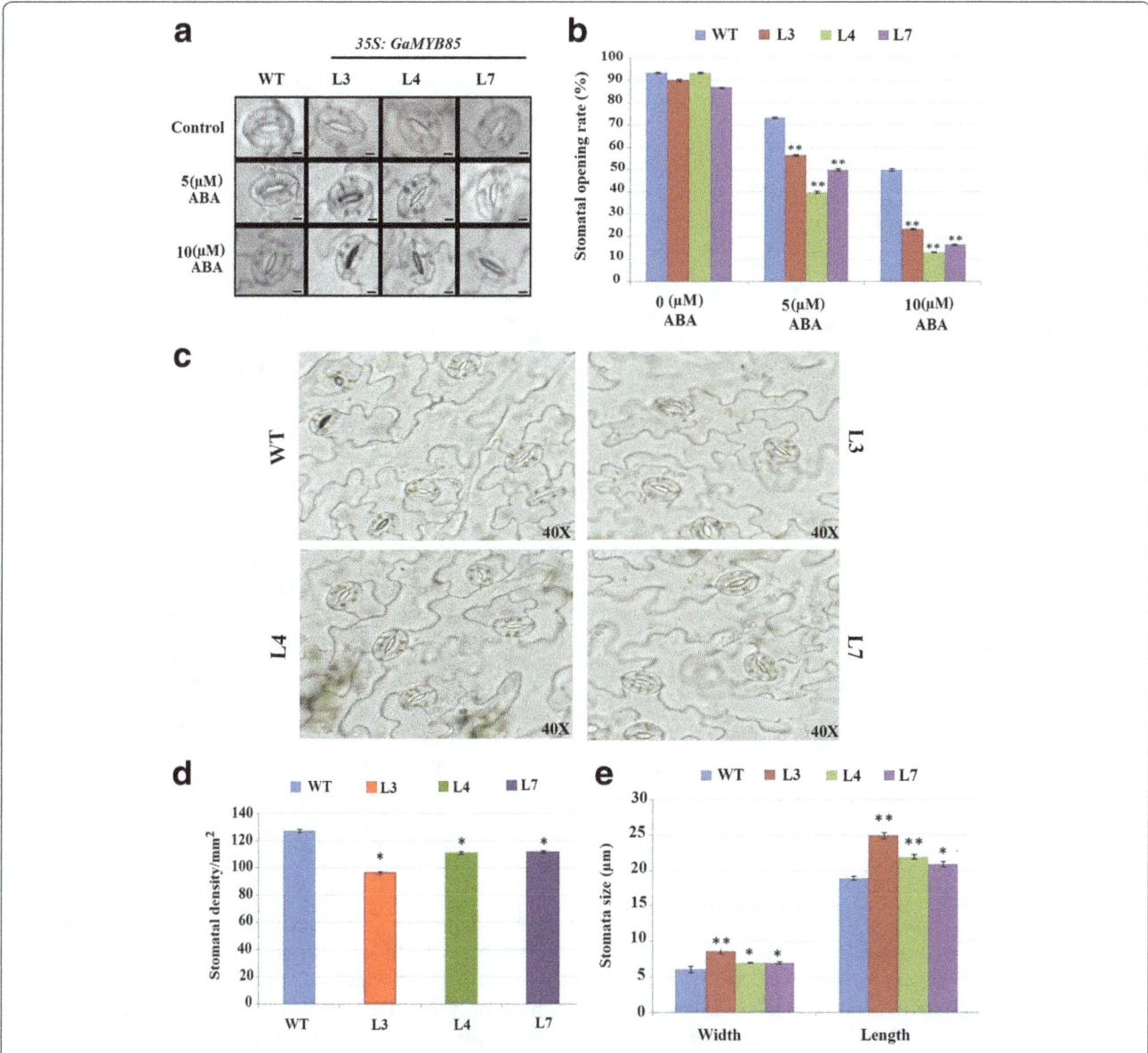

Fig. 8 ABA induces positive stomatal modulation in *35S:GaMYB85*-overexpressing lines. **a** The stomatal pores of *35S:GaMYB85* and WT were photographed following treatment with 0, 5, and 10 μM ABA. Five views from three replicate plants were observed at ×40 magnification **b** Quantitative comparisons of the stomatal pores of *35S:GaMYB85* and WT plants following treatment with 0, 5, and 10 μM ABA. The data shown are the mean values of three replicates ± SD ($n = 10$). Scale bars: approximately 1 μm **c** Leaf stomatal density of *35S:GaMYB85* and WT plants photographed under an OLYMPUS Bx51 microscope at ×40 magnification. **d** Quantitative comparisons of *35S:GaMYB85* and WT stomatal density **e** Width and length of guard cells measured using ImageJ software. Measurements were taken from the leaves of three plants (10 stomata from five microscopic views). Data are the means ± SD. *$P < 0.05$; ** $P < 0.01$

opening than those of WT plants in response to exogenous ABA (Fig. 8b–e). These results corroborate the findings of previous studies on *35S:HDG11*-transformed tobacco, which has reduced stomatal densities but increased stomatal size [62], and are in contrast to *GbMYB5* transgenic tobacco plants, which have reduced stomatal size but the same stomatal density [25]. However, both these transgenic plants have high drought tolerance as a result of superior water

retention abilities, lower transpiration rates, and reduced stomatal apertures due to elevated ABA production. Interestingly, *35S:GaMYB85* plants have reduced stomatal density but enlarged stomatal size, which can result in enhanced photosynthetic ability of plants by some unknown mechanism. This warrants further studies to gain insights into the possible function of the ABA transduction pathway in *GaMYB85*-mediated roles in reduced stomatal apertures, which

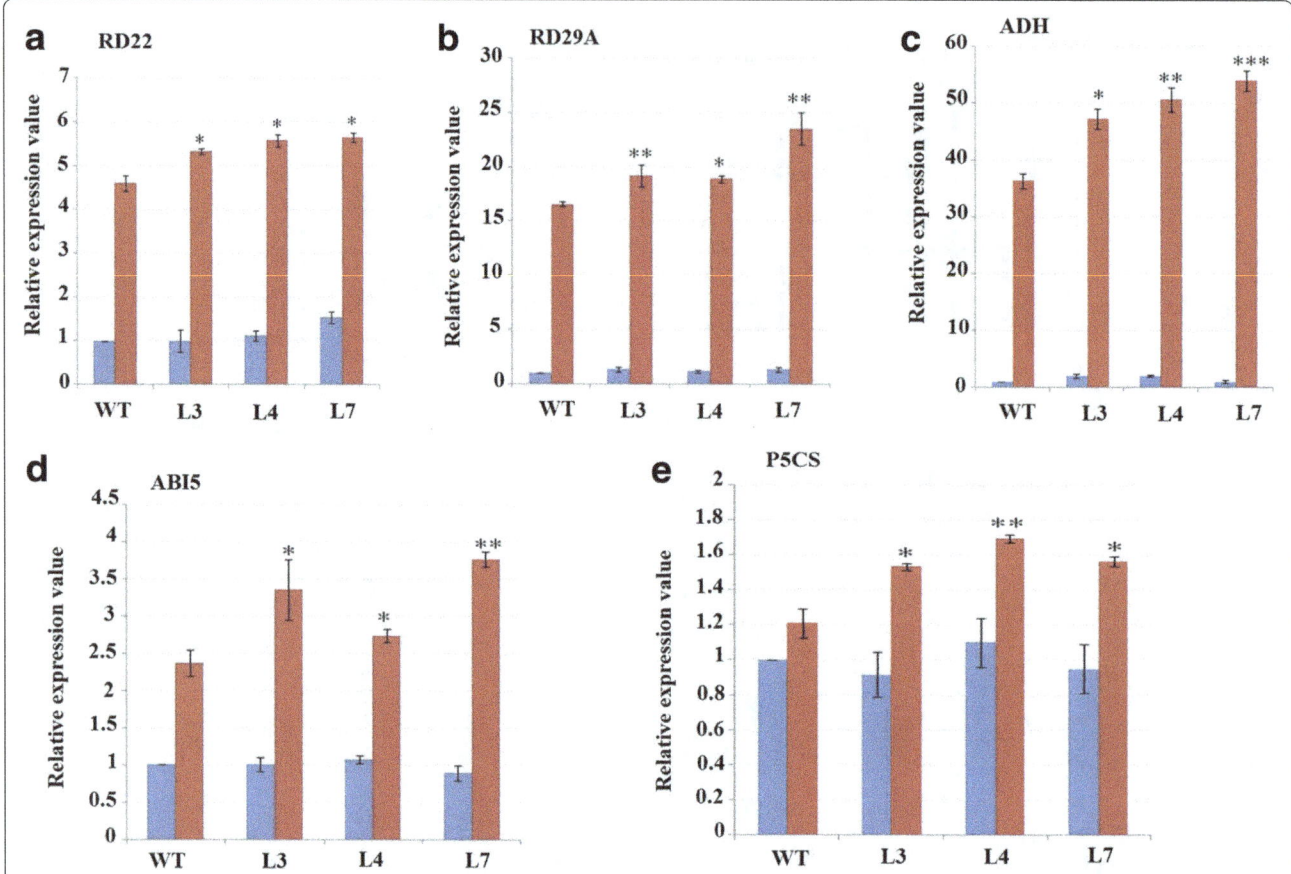

Fig. 9 Transcript levels of ABA signaling and ABA stress-responsive marker genes in *35S: GaMYB85* and WT. (**a–e**) Transcript level of ABA signaling and ABA stress-responsive marker genes in two week old seedlings on (100 μM) ABA treatment for 6 h, using qRT-PCR. *AtUQB10* was used as internal control gene, with error bars of 3 biological replicates with ± SD. *P < 0.05; ** P < 0.01 and *** P < 0.001

helps to maintain a balance between the rates of transpiration and photosynthetic capabilities of plant during drought stress responses.

Moreover, several overexpression studies have demonstrated the use of ABA-related marker genes involved in stress-related phenotypes in *Arabidopsis* [31, 56, 63]. Similarly, *35S:GaMYB85* overexpression led to the up regulation of *RD22, RD29A, ADH1, AB15,* and *P5CS*. The *AB15* gene encodes a bZIP TF, and is an ABA-responsive gene known to play a prominent role during the seed germinating period during drought stress [64, 65]. Moreover, D-pyrroline-5-carboxylase synthetase (P5CS) plays a key role in the initiation step for proline synthesis during dehydration or osmotic stress, and it has been revealed that up-regulation of the *P5CS* gene in *35S:GaMYB85* plants corroborated well with the findings of studies on *OsMYB48–1, AtLOS5,* and *TaNAC47* [52, 53, 59]. The *RD29A* gene codes for a hydrophilic protein and is known for the presence of *DRE* or related promoter motif regions. It is activated by the combined action of DRE and ABRE *cis* motif elements, which enables it to work both in ABA-dependent and

ABA-independent manners [66]. ABA-responsive expression of *RD22* is mainly due to co-expression activation of *AtMYB2/AtMYC2*. This is in contrast to *ADH1*, which only requires activated *AtMYB2* to act in a protective protein [67]. The above findings for *GaMYB85* imply that it functions in an ABA-dependent manner, and plays an important role in drought stress tolerance, which has important implications for breeding drought-tolerant cotton varieties.

Conclusions

To summarize, ectopic expression of *GaMYB85* in *Arabidopsis* led to enhanced drought avoidance, elevated survival rates, higher accumulation of the compatible osmolyte proline, good germination rates under osmotic and salt stress, reduced water loss rates with improved RWC, increased ABA sensitivity at the germination and post-germination stages, a reduction in the number of ABA-induced stomatal apertures, and up-regulated transcription of ABA-mediated stress-responsive marker genes. In the light of the results of this study, we speculate that *35S: GaMYB85* could confer good drought, salt, and freezing tolerance in cotton crops. Our observations on the activity of the R2R3 MYB transcription factor *GaMYB85* in transgenic *Arabidopsis* plants greatly enhances our perception about its possible function in response to environmental cues, and it can be considered as a novel candidate gene for future developments in cotton breeding.

Methods

Transgenic plant construction and screening

The presence of the *CaMV-35S* promoter driven pCAMBIA3300-*GaMYB85* construct in *A. tumefaciens* strain GV3101 was confirmed using a *GaMYB85* sense and antisense primer pair (Additional file 3), and was ectopically expressed in *A. thaliana* (Col-0) using the floral dip method with minor modifications [68]. The infiltration media used for transformation comprised 2.215 g/L MS containing 50 g/L of sucrose, 0.5 g/L of MES, (330 μL/L) of Silwet-77 and 0.01 mg/L of 6-BA at pH 5.7. The infiltrated *Arabidopsis* was grown under the following conditions: 16 h light/8 h dark, 60% RH, and 22°C. Transgenic T_0 seedlings were screened using 1 mg/L BASTA, which was sprayed twice on 7-d-old seedlings grown in soil pots (Additional file 4). The surviving seedlings were set to obtain T_1 seeds, which were then confirmed for T-DNA integration in transgenic plants using a PCR Mighty Amp Genotyping Kit (Takara, China) as shown in (Fig. 2a), using the sense *35S* primer and the antisense gene primer. Wild-type (WT) plants were used as a control. Ten T_2 lines with the correct segregation ratio (3:1) were selected and confirmed by qRT-PCR (Fig. 2b). Three T_3 homozygous lines were selected and used for functional validation investigations (Additional file 5).

Germination and root elongation assays

To evaluate germination in response to simulated osmotic, salt, and ABA stresses, seeds from *35S:GaMYB85* transgenic lines (L3, L4, and L7) and WT were sterilized, stratified, and sown in triplicate on half-strength MS supplemented with different mannitol concentrations (0,100, 200, and 300 mM), exogenous ABA (0, 0.3, 0.5, 1, 2, and 5 μM) and NaCl (0, 50, 100, 125, and 150 mM). The plants were scored based on non-green phenotype or dead cotyledons on the 10th day postgermination [69]. Furthermore, for seedling root length elongation assay on mannitol-, NaCl-, and ABA-supplemented MS, the methods described by [63] were followed. Three independent experiments were conducted and significant differences were determined using Student's t-test.

Drought and freezing tolerance assay

T_3 homozygous transgenic line (L3, L4, and L7) and WT seeds were sterilized with 10% bleach for 10 min, rinsed with water, and stratified at 4 °C for 3 days before the seeds were transferred to MS. One-week-old seedlings were transferred from MS to pots containing a well-watered vermiculite and humus mix, and grown for 2 weeks. Thereafter, water was withheld for 2 weeks to simulate drought stress treatment. Subsequently, pots were re-watered and plant revival rates were scored after 3 d. Plants showing green healthy leaves after water replenishment were scored as surviving. The ratios of surviving to dead plants were calculated and the experiment was repeated three times [62]. For the freezing tolerance experiment, 3-week-old overexpressing transgenic lines and control WT plants were subjected to a temperature of −10°C for 3 h. Subsequently, survival rates were calculated as previously described by [52]. All experiments were repeated three times using seeds of WT plants and the three transgenic lines independently.

Measurements of transpirational WLR and RWC

The water loss rate was determined as described by [70]. Detached leaves from 4-week-old test and control plants were immediately weighed to determine fresh weights (FW). Thereafter, at various time intervals, weights were recorded while samples were retained on a bench at 22 °C RT with a humidity level of 45% ± 5%. Fresh weights were calculated relative to plant initial weights. The relative water content of leaves was calculated using the method described by [71]. RWC percentage = (FW − DW)/ (TW − DW) × 100. After determining fresh weights, the leaves were submerged in water for 4 h and turgid weights (TW) were recorded. Dry weights (DW) of leaves were recorded by drying samples at 80 °C for approximately 72 h.

Proline and chlorophyll contents

Endogenous proline content in drought-stressed transgenic and WT plants was measured using a colorimetric PRO Kit (Jiancheng Bioengineering Institute, Nanjing, China). Three biological repeats were used, and proline absorbances at a wavelength of 520 nm were measured for determination of proline concentration.

Chlorophyll content was calculated using the formula described by [72] as follows: ((O.D 665 nm × 13.95 - O.D 649 nm × 6.88) + (O.D 649 nm × 24.96 - O.D 665 nm × 7.32))/(sample weight). The sample extraction was performed using 0.1 g of rosette leaves in 1.5 mL 95% ethanol at RT in a dark room. The absorbances of extracted chlorophyll were measured at 649 nm and 665 nm. Test and control samples were measured three times and the results were averaged. The means were compared using Duncan's multiple range test.

Stomatal measurements and determination of opening rate in response to abscisic acid treatment

To investigate if stomatal pore closure is sensitive to ABA, we examined the effect of ectopic expression of *GaMYB85* on stomatal opening rate of leaves under a microscope. The stomatal pores were initially opened by placing detached leaves from 3-week-old plants in a stomata opening solution (10 mM $CaCl_2$, 50 mM KCl, and 5 mM MES, pH 6.15) [73] for 2 h in an illuminated growth chamber at 95% RH. Stomatal apertures were determined after 2.5 h. Two concentrations of ABA (5 and 10 µM) were applied, whereas control experiments were performed without using ABA. Samples of young leaf epidermis of transgenic and WT plants were peeled and observed under an OLYMPUS Bx51 microscope for stomata density evaluation and measurement of guard cell width to length ratio. IMAGEJ 1.51d software [74] was used to measure the size of guard cells.

Expression analysis of marker genes by qRT-PCR

For transcript analysis of stress-responsive marker genes, 2-week-old transgenic and WT seedlings treated with 100 µM ABA were used. RNA extraction was performed using an RNA-prep Pure Plant kit (Tiangen, China), followed by first-strand cDNA synthesis using Prime-Script RT Master Mix (Takara, Clontech, China). Quantitative RT-PCR was carried out using a 7900HT detection system (Applied Biosystems), using SYBR Premix ExTaq™ (Takara, Clontech, China) according to manufacturer's protocol. *AtUBQ10* (Accession no: AT4G05320) was used as a control gene. Three biological replicates were run using independent cDNA preparations and three technical replicates, and relative transcripts were computed using the $2^{-\Delta\Delta Ct}$ method based on CT values [75].

Sequence and phylogenetic analysis

GaMYB85 was retrieved from the cotton genome project database (http: //cgp.genomics.org.cn/page/species/blast.jsp.) and designated *GaMYB85* (cotton_A_21601). Motif analysis was performed using SMART (http://smart.embl-heidelberg.de/), whereas intron–exon analysis was carried out using the GENE Structure Display Server available online (http://gsds.cbi.pku.edu.cn). GaMYB85 protein physio-chemical properties, such as MW and pI, were evaluated using the ExPASy proteomic portal (http://www.expasy.org/proteomics). Monocot and dicot R2R3 MYB homologs were retrieved using the Blastp tool (https://www.ncbi.nlm.nih.gov/) and known R2R3 MYB were used in the construction of a cladogram based on the neighbor-joining method using MEGA 6.0. The multiple sequence alignment of *GaMYB85* R2R3 homologs was performed using the EMBL-EBI Muscle online tool (http://www.ebi.ac.uk/Tools/msa/muscle/).

Statistical analysis of data

Statistical analysis was performed on data derived from three independent replicate experiments, using analysis of variance and Student's t-test. Duncan's multiple range test was applied to determine variation among treatment means of test and control lines.

Additional files

Additional file 1: Bioinformatics sequence analysis of *GaMYB85* protein.

Additional file 2: *Alignment of GaMYB85 with the deduced homologous amino acid sequences of R2R3 MYB retrieved from Blastp NCBI and known R2R3 MYB.*

Additional file 3: The primers sequences used for *GaMYB85* study.

Additional file 4: *35S:GaMYB85* over-expressed positive transgenic plants screening in BASTA screening at T_0 and T_3 stages.

Additional file 5: Survival percentage of *35S:GaMYB85* transgenic plants in 6% BASTA selection medium.

Abbreviations
ABA: Abscisic acid; DBD: DNA binding domain;; DW: Dry weight; FW: Fresh weight; *GaMYB85*: Gene in *Gossipium arboreum*;; kD: kilo Dalton; mM: millimolar; MW: Molecular weight; OD: Optical density; pI: Isoelectric point; RT: Room temperature; TF: Transcription factor; TW: Turgid weight; WT: Wild-type; µM: micromolar

Acknowledgements
We thank Prof. Jia Li (School of Life Sciences, Lanzhou University, Lanzhou) for providing us *A. tumefaciens* strain *GV3101* and seeds of *A. thaliana* respectively. This work was supported by the Major Program of Joint Funds (Sinkiang) of the National Natural Science Foundation of China (grant U1303282).

Funding
This work was supported by the Major Program of Joint Funds (Sinkiang) of the National Natural Science Foundation of China (grant U1303282).

Authors' contributions
FL, ZX and ZY conceived the work and revised the manuscript. HIB designed the research, performed the experiments, and drafted the manuscript. EC, QG, GXY, GZ and XW helped design the work and analysis, revise the manuscript and contributed reagents and analysis tools. Z Y and GXY helped revised the manuscript with constructive discussions. All authors read and approved the final manuscript.

Competing interests
The authors declare that they have no competing interests.

References

1. Bohnert HJ, Gong QQ, Li PH, Ma SS. Unraveling abiotic stress tolerance mechanisms - getting genomics going. Curr Opin Plant Biol. 2006;9(2):180–8.

2. Du H, Zhang L, Liu L, Tang XF, Yang WJ, Wu YM, Huang YB, Tang YX. Biochemical and molecular characterization of plant MYB transcription factor family. Biochemistry-Moscow+. 2009;74(1):1–11.

3. Lippold F, Sanchez DH, Musialak M, Schlereth A, Scheible WR, Hincha DK, Udvardi MK. AtMyb41 regulates transcriptional and metabolic responses to osmotic stress in Arabidopsis. Plant Physiol. 2009;149(4):1761–72.

4. Klempnauer KH, Gonda TJ, Bishop JM. Nucleotide sequence of the retroviral leukemia gene v-myb and its cellular progenitor c-myb: the architecture of a transduced oncogene. Cell. 1982;31(2 Pt 1):453–63.

5. Paz-Ares J, Ghosal D, Wienand U, Peterson PA, Saedler H. The regulatory c1 locus of Zea Mays encodes a protein with homology to myb proto-oncogene products and with structural similarities to transcriptional activators. EMBO J. 1987;6(12):3553–8.

6. Hichri I, Barrieu F, Bogs J, Kappel C, Delrot S, Lauvergeat V. Recent advances in the transcriptional regulation of the flavonoid biosynthetic pathway. J Exp Bot. 2011;62(8):2465–83.

7. Baldoni E, Genga A, Cominelli E. Plant MYB transcription factors: their role in drought response mechanisms. Int J Mol Sci. 2015;16(7):15811–51.

8. Dubos C, Stracke R, Grotewold E, Weisshaar B, Martin C, Lepiniec L. MYB transcription factors in Arabidopsis. Trends Plant Sci. 2010;15(10):573–81.

9. He QL, Jones DC, Li W, Xie FL, Ma J, Sun RR, Wang QL, Zhu SJ, Zhang BH. Genome-wide identification of R2R3-MYB genes and expression analyses during abiotic stress in Gossypium Raimondii. Sci Rep-Uk. 2016;6

10. Williams CE, Grotewold E. Differences between plant and animal myb domains are fundamental for DNA binding activity, and chimeric Myb domains have novel DNA binding specificities. J Biol Chem. 1997;272(1):563–71.

11. Jia L, Clegg MT, Jiang T. Evolutionary dynamics of the DNA-binding domains in putative R2R3-MYB genes identified from rice subspecies indica and japonica genomes. Plant Physiol. 2004;134(2):575–85.

12. Kanei-Ishii C, Sarai A, Sawazaki T, Nakagoshi H, He DN, Ogata K, Nishimura Y, Ishii S. The tryptophan cluster: a hypothetical structure of the DNA-binding domain of the myb protooncogene product. J Biol Chem. 1990;265(32):19990–5.

13. Lipsick JS. One billion years of Myb. Oncogene. 1996;13(2):223–35.

14. Stracke R, Werber M, Weisshaar B. The R2R3-MYB gene family in Arabidopsis Thaliana. Curr Opin Plant Biol. 2001;4(5):447–56.

15. Finkelstein RR, Gampala SS, Rock CD. Abscisic acid signaling in seeds and seedlings. Plant Cell. 2002;14(Suppl):S15–45.

16. Shinozaki K, Yamaguchi-Shinozaki K. Molecular responses to dehydration and low temperature: differences and cross-talk between two stress signaling pathways. Curr Opin Plant Biol. 2000;3(3):217–23.

17. Lu PL, Chen NZ, An R, Su Z, Qi BS, Ren F, Chen J, Wang XC. A novel drought-inducible gene, ATAF1, encodes a NAC family protein that negatively regulates the expression of stress-responsive genes in Arabidopsis. Plant Mol Biol. 2007;63(2):289–305.

18. Zhao BY, Hu YF, Li JJ, Yao X, Liu KD. BnaABF2, a bZIP transcription factor from rapeseed (Brassica napus L.), enhances drought and salt tolerance in transgenic Arabidopsis. Bot Stud. 2016;57.

19. Ren XZ, Chen ZZ, Liu Y, Zhang HR, Zhang M, Liu QA, Hong XH, Zhu JK, Gong ZZ. ABO3, a WRKY transcription factor, mediates plant responses to abscisic acid and drought tolerance in Arabidopsis. Plant J. 2010;63(3):417–29.

20. Qin F, Shinozaki K, Yamaguchi-Shinozaki K. Achievements and challenges in understanding plant abiotic stress responses and tolerance. Plant & cell physiology. 2011;52(9):1569–82.

21. Chen M, Wang QY, Cheng XG, Xu ZS, Li LC, Ye XG, Xia LQ, Ma YZ. GmDREB2, a soybean DRE-binding transcription factor, conferred drought and high-salt tolerance in transgenic plants. Biochem Biophys Res Commun. 2007;353(2):299–305.

22. Abe H, Urao T, Ito T, Seki M, Shinozaki K, Yamaguchi-Shinozaki K. Arabidopsis AtMYC2 (bHLH) and AtMYB2 (MYB) function as transcriptional activators in abscisic acid signaling. Plant Cell. 2003;15(1):63–78.

23. Jung C, Seo JS, Han SW, Koo YJ, Kim CH, Song SI, Nahm BH, Choi YD, Cheong JJ. Overexpression of AtMYB44 enhances stomatal closure to confer abiotic stress tolerance in transgenic Arabidopsis. Plant Physiol. 2008;146(2):623–35.

24. Chen YH, Zhang XB, Wu W, Chen ZL, Gu HY, Qu LJ. Overexpression of the wounding-responsive gene AtMYB15 activates the shikimate pathway in Arabidopsis. J Integr Plant Biol. 2006;48(9):1084–95.

25. Chen TZ, Li WJ, Hu XH, Guo JR, Liu AM, Zhang BL. A cotton MYB transcription factor, GbMYB5, is positively involved in plant adaptive response to drought stress. Plant Cell Physiol. 2015;56(5):917–29.

26. Denekamp M, Smeekens SC. Integration of wounding and osmotic stress signals determines the expression of the AtMYB102 transcription factor gene. Plant Physiol. 2003;132(3):1415–23.

27. Wang RK, Cao ZH, Hao YJ. Overexpression of a R2R3 MYB gene MdSIMYB1 increases tolerance to multiple stresses in transgenic tobacco and apples. Physiol Plant. 2014;150(1):76–87.

28. Yang A, Dai XY, Zhang WH. A R2R3-type MYB gene, OsMYB2, is involved in salt, cold, and dehydration tolerance in rice. J Exp Bot. 2012;63(7):2541–56.

29. Su LT, Li JW, Liu DQ, Zhai Y, Zhang HJ, Li XW, Zhang QL, Wang Y, Wang QY. A novel MYB transcription factor, GmMYBJ1, from soybean confers drought and cold tolerance in Arabidopsis Thaliana. Gene. 2014;538(1):46–55.

30. Su LT, Wang Y, Liu DQ, Li XW, Zhai Y, Sun X, Li XY, Liu YJ, Li JW, Wang QY. The soybean gene, GmMYBJ2, encodes a R2R3-type transcription factor involved in drought stress tolerance in Arabidopsis thaliana. Acta Physiol Plant. 2015;37(7).

31. Liao Y, Zou HF, Wang HW, Zhang WK, Ma B, Zhang JS, Chen SY. Soybean GmMYB76, GmMYB92, and GmMYB177 genes confer stress tolerance in transgenic Arabidopsis plants. Cell Res. 2008;18(10):1047–60.

32. Yuan Y, Qi LJ, Yang J, Wu C, Liu YJ, Huang LQ: A Scutellaria Baicalensis R2R3-MYB gene, SbMYB8, regulates flavonoid biosynthesis and improves drought stress tolerance in transgenic tobacco (vol 120, pg 961, 2015). Plant Cell Tiss Org 2015, 120(3):973–973.

33. Brown DC. Cotton: origin, history, technology, and production. Agr Hist. 2000;74(4):823–4.

34. Saleem MF, Raza MAS, Ahmad S, Khan IH, Shahid AM. Understanding and mitigating the impacts of drought stress in cotton- a review. Pak J Agr Sci. 2016;53(3):609–23.

35. Li FG, Fan GY, Wang KB, Sun FM, Yuan YL, Song GL, Li Q, Ma ZY, Lu CR, Zou CS, et al. Genome sequence of the cultivated cotton Gossypium Arboreum. Nat Genet. 2014;46(6):567–72.

36. Zhang XY, Yao DX, Wang QH, Xu WY, Wei Q, Wang CC, Liu CL, Zhang CJ, Yan H, Ling Y, et al. mRNA-seq Analysis of the Gossypium arboreum transcriptome Reveals Tissue Selective Signaling in Response to Water Stress during Seedling Stage. PLoS One. 2013;8(1).

37. Zhong RQ, Lee CH, Zhou JL, McCarthy RL, Ye ZH. A battery of transcription factors involved in the regulation of secondary Cell Wall biosynthesis in Arabidopsis. Plant Cell. 2008;20(10):2763–82.

38. Hetherington AM, Woodward FI. The role of stomata in sensing and driving environmental change. Nature. 2003;424(6951):901–8.

39. Malamy JE. Intrinsic and environmental response pathways that regulate root system architecture. Plant Cell and Environment. 2005;28(1):67–77.

40. Cheng LQ, Li XX, Huang X, Ma T, Liang Y, Ma XY, Peng XJ, Jia JT, Chen SY, Chen Y, et al. Overexpression of sheepgrass R1-MYB transcription factor LcMYB1 confers salt tolerance in transgenic Arabidopsis. Plant Physiol Bioch. 2013;70:252–60.

41. Ma T, Li ML, Zhao AG, Xu X, Liu GS, Cheng LQ. LcWRKY5: an unknown function gene from sheepgrass improves drought tolerance in transgenic Arabidopsis. Plant Cell Rep. 2014;33(9):1507–18.

42. Sabbagh E, Lakzayi M, Keshtehgar A, Rigi K: The effect of salt stress on respiration, PSII function, chlorophyll, carbohydrate and nitrogen content in crop plants. 2014.

43. Muhammad J, DEOG BL, KWANG YJ, Muhammad A, SHEONG CL, EUI SR. EFFECT OF SALT (NACL) STRESS ON GERMINATION AND EARLY SEEDLING GROWTH OF FOUR VEGETABLES SPECIES. J Cent Eur Agric. 2006;6(2):273–82.

44. Zhu JK. Regulation of ion homeostasis under salt stress. Curr Opin Plant Biol. 2003;6(5):441–5.

45. Chinnusamy V, Jagendorf A, Zhu JK. Understanding and improving salt tolerance in plants. Crop Sci. 2005;45(2):437–48.

46. de Oliveira AB, Mendes Alencar NL, Gomes-Filho E: Comparison between the water and salt stress effects on plant growth and development. 2013.

47. Munns R, Tester M. Mechanisms of salinity tolerance. Annu Rev Plant Biol. 2008;59:651–81.

48. Jakab G, Ton J, Flors V, Zimmerli L, Metraux JP, Mauch-Mani B. Enhancing Arabidopsis salt and drought stress tolerance by chemical priming for its abscisic acid responses. Plant Physiol. 2005;139(1):267–74.

49. JAMIL M, DEOG BAE L, KWANG YONG J, ASHRAF M, SHEONG CHUN L, EUI SHIK R. Effect of salt (NaCl) stress on germination and early seedling growth of four vegetables species. J Cent Eur Agric. 2006;7(2):273–82.

50. Sun X, Li Y, Cai H, Bai X, Ji W, Ding X, Zhu Y. The Arabidopsis AtbZIP1 transcription factor is a positive regulator of plant tolerance to salt, osmotic and drought stresses. J Plant Res. 2012;125(3):429–38.

51. Min H, Zheng J, Wang J. Maize ZmRAV1 contributes to salt and osmotic stress tolerance in transgenic arabidopsis. J Plant Biol. 2014;57(1):28–42.

52. Zhang LN, Zhang LC, Xia C, Zhao GY, Jia JZ, Kong XY. The novel wheat transcription factor TaNAC47 enhances multiple abiotic stress tolerances in transgenic plants. Front Plant Sci. 2016;6

53. Yue YS, Zhang MC, Zhang JC, Tian XL, Duan LS, Li ZH. Overexpression of the AtLOS5 gene increased abscisic acid level and drought tolerance in transgenic cotton. J Exp Bot. 2012;63(10):3741–8.

54. Hu HH, Dai MQ, Yao JL, Xiao BZ, Li XH, Zhang QF, Xiong LZ. Overexpressing a NAM, ATAF, and CUC (NAC) transcription factor enhances drought resistance and salt tolerance in rice. P Natl Acad Sci USA. 2006;103(35):12987–92.

55. Ko JH, Yang SH, Han KH. Upregulation of an Arabidopsis RING-H2 gene, XERICO, confers drought tolerance through increased abscisic acid biosynthesis. Plant J. 2006;47(3):343–55.

56. Zhang LC, Liu GX, Zhao GY, Xia C, Jia JZ, Liu X, Kong XY. Characterization of a wheat R2R3-MYB transcription factor gene, TaMYB19, involved in enhanced abiotic stresses in Arabidopsis. Plant Cell Physiol. 2014;55(10):1802–12.

57. Cominelli E, Galbiati M, Vavasseur A, Conti L, Sala T, Vuylsteke M, Leonhardt N, Dellaporta SL, Tonelli C. A guard-cell-specific MYB transcription factor regulates stomatal movements and plant drought tolerance. Curr Biol. 2005;15(13):1196–200.

58. Lu PT, Kang M, Jiang XQ, Dai FW, Gao JP, Zhang CQ. RhEXPA4, a rose expansin gene, modulates leaf growth and confers drought and salt tolerance to Arabidopsis. Planta. 2013;237(6):1547–59.

59. Xiong HY, Li JJ, Liu PL, Duan JZ, Zhao Y, Guo X, Li Y, Zhang HL, Ali J, Li ZC. Overexpression of OsMYB48-1, a Novel MYB-Related Transcription Factor, Enhances Drought and Salinity Tolerance in Rice. PLoS One. 2014;9(3).

60. Li GJ, Nasar V, Yang YX, Li W, Liu B, Sun LJ, Li DY, Song FM. Arabidopsis poly(ADP-ribose) glycohydrolase 1 is required for drought, osmotic and oxidative stress responses. Plant Sci. 2011;180(2):283–91.

61. Ashraf M, Foolad MR. Roles of glycine betaine and proline in improving plant abiotic stress resistance. Environ Exp Bot. 2007;59(2):206–16.

62. Yu H, Chen X, Hong YY, Wang Y, Xu P, Ke SD, Liu HY, Zhu JK, Oliver DJ, Xiang CB. Activated expression of an Arabidopsis HD-START protein confers drought tolerance with improved root system and reduced stomatal density. Plant Cell. 2008;20(4):1134–51.

63. Ding ZH, Li SM, An XL, Liu XJ, Qin HM, Wang D. Transgenic expression of MYB15 confers enhanced sensitivity to abscisic acid and improved drought tolerance in Arabidopsis Thaliana. J Genet Genomics. 2009;36(1):17–29.

64. Lopez-Molina L, Mongrand S, Chua NH. A postgermination developmental arrest checkpoint is mediated by abscisic acid and requires the AB15 transcription factor in Arabidopsis. P Natl Acad Sci USA. 2001;98(8):4782–7.

65. Mittal A, Gampala SSL, Ritchie GL, Payton P, Burke JJ, Rock CD. Related to ABA-Insensitive3(ABI3)/Viviparous1 and AtABI5 transcription factor coexpression in cotton enhances drought stress adaptation. Plant Biotechnol J. 2014;12(5):578–89.

66. Narusaka Y, Nakashima K, Shinwari ZK, Sakuma Y, Furihata T, Abe H, Narusaka M, Shinozaki K, Yamaguchi-Shinozaki K. Interaction between two cis-acting elements, ABRE and DRE, in ABA-dependent expression of Arabidopsis rd29A gene in response to dehydration and high-salinity stresses. Plant J. 2003;34(2):137–48.

67. Seo PJ, Xiang FN, Qiao M, Park JY, Lee YN, Kim SG, Lee YH, Park WJ, Park CM. The MYB96 transcription factor mediates abscisic acid signaling during drought stress response in Arabidopsis. Plant Physiol. 2009;151(1):275–89.

68. Clough SJ, Bent AF. Floral dip: a simplified method for agrobacterium-mediated transformation of Arabidopsis Thaliana. Plant J. 1998;16(6):735–43.

69. Qin YX, Wang MC, Tian YC, He WX, Han L, Xia GM. Over-expression of TaMYB33 encoding a novel wheat MYB transcription factor increases salt and drought tolerance in Arabidopsis. Mol Biol Rep. 2012;39(6):7183–92.

70. Dhanda SS, Sethi GS. Inheritance of excised-leaf water loss and relative water content in bread wheat (Triticum Aestivum). Euphytica. 1998;104(1):39–47.

71. Parida AK, Dagaonkar VS, Phalak MS, Umalkar GV, Aurangabadkar LP. Alterations in photosynthetic pigments, protein and osmotic components in cotton genotypes subjected to short-term drought stress followed by recovery. Plant Biotechnol Rep. 2007;1(1):37–48.

72. Zhao FL, Ma JH, Li LB, Fan SL, Guo YN, Song MZ, Wei HL, Pang CY, Yu SX. GhNAC12, a neutral candidate gene, leads to early aging in cotton (Gossypium Hirsutum L). Gene. 2016;576(1):268–74.

73. Pei ZM, Kuchitsu K, Ward JM, Schwarz M, Schroeder JI. Differential abscisic acid regulation of guard cell slow anion channels in Arabidopsis wild-type and abi1 and abi2 mutants. Plant Cell. 1997;9(3):409–23.

74. Schneider CA, Rasband WS, Eliceiri KW. NIH image to ImageJ: 25 years of image analysis. Nat Methods. 2012;9(7):671–5.

75. Livak KJ, Schmittgen TD. Analysis of relative gene expression data using real-time quantitative PCR and the 2(T)(−Delta Delta C) method. Methods. 2001;25(4):402–8.

9

Differentially expressed genes during the imbibition of dormant and after-ripened seeds – a reverse genetics approach

Farzaneh Yazdanpanah[1], Johannes Hanson[2,3], Henk W.M. Hilhorst[1] and Leónie Bentsink[1*] (iD)

Abstract

Background: Seed dormancy, defined as the incapability of a viable seed to germinate under favourable conditions, is an important trait in nature and agriculture. Despite extensive research on dormancy and germination, many questions about the molecular mechanisms controlling these traits remain unanswered, likely due to its genetic complexity and the large environmental effects which are characteristic of these quantitative traits. To boost research towards revealing mechanisms in the control of seed dormancy and germination we depend on the identification of genes controlling those traits.

Methods: We used transcriptome analysis combined with a reverse genetics approach to identify genes that are prominent for dormancy maintenance and germination in imbibed seeds of *Arabidopsis thaliana*. Comparative transcriptomics analysis was employed on freshly harvested (dormant) and after-ripened (AR; non-dormant) 24-h imbibed seeds of four different *DELAY OF GERMINATION* near isogenic lines (*DOGNILs*) and the Landsberg *erecta* (L*er*) wild type with varying levels of primary dormancy. T-DNA knock-out lines of the identified genes were phenotypically investigated for their effect on dormancy and AR.

Results: We identified conserved sets of 46 and 25 genes which displayed higher expression in seeds of all dormant and all after-ripened *DOGNILs* and L*er*, respectively. Knock-out mutants in these genes showed dormancy and germination related phenotypes.

Conclusions: Most of the identified genes had not been implicated in seed dormancy or germination. This research will be useful to further decipher the molecular mechanisms by which these important ecological and commercial traits are regulated.

Keywords: *Arabidopsis thaliana*, *Delay of germination*, Knockout lines, Seed performance, Transcriptromics

Background

Freshly matured seeds usually exhibit primary dormancy, a trait defined as the failure of viable seeds to germinate under favourable conditions [8]. Seed dormancy plays a crucial role in the survival of plant species, but is also important for agricultural practice to prevent pre-harvest sprouting under cool, high humidity conditions [24]. Primary dormancy can be released by either cold stratification, which is a low-temperature treatment of imbibed seeds, or by an extended period of dry seed storage (after-ripening; AR) [9].

The transition from dormancy to germination is a critical step in the life cycle of plants [25]. The plant hormone abscisic acid (ABA) has long been known to play a major role in the establishment and maintenance of seed dormancy and the inhibition of seed germination, whereas gibberellins (GAs) and several other hormones, including brassinosteroids, ethylene, and cytokinins, have been shown to promote seed germination [38]. However, it is especially the balance between ABA and GA that controls the decision to germinate or not [19]. Mutations in genes regulating ABA levels or -sensitivity result in a reduced degree of seed dormancy [34, 35]. Whereas GA biosynthesis or sensing mutants result in a block of germination [23, 31, 33]. This hormonal control is also integrated with

* Correspondence: leonie.bentsink@wur.nl
[1]Wageningen Seed Laboratory, Laboratory of Plant Physiology, Wageningen University, Droevendaalsesteeg 1, 6708, PB, Wageningen, The Netherlands
Full list of author information is available at the end of the article

the seed's responses to environmental conditions, such as light [45], temperature [55, 60] and nutrients [40].

Recent advances in gene expression analysis using microarrays allow genome-wide expression studies to characterize seed dormancy and germination [10, 20, 21, 26, 27, 36, 41, 52]. Carrera et al. [12] used a targeted transcriptomics approach in imbibed non-dormant mutants (*aba1* and *abi1*) compared to wild-type seeds that were or were not after-ripened. They concluded that, in Arabidopsis, after-ripening and dormancy are controlled by genetically separate pathways, and that ABA only affects the induction and maintenance of dormancy in imbibed seeds, but not after-ripening. The work also showed that application of exogenous ABA to after-ripened seeds does not mimic dormant seed states with respect to gene expression profiles. Recently it was shown that seed dormancy maintenance in the imbibed state was mainly controlled at the transcriptional level [3] and that transcriptional differences between dormant and non-dormant seed become visible already at early imbibition [17, 48].

Despite extensive research on dormancy and germination, many questions about the molecular mechanisms controlling these traits remain unanswered, likely due to its genetic complexity and the large environmental effects which are characteristic of these quantitative traits. Employing whole-genome scans for quantitative trait loci (QTL) is a common approach to identify genes involved in complex phenotypes. Particular attention in this method is given to the role of natural variation in the regulation of traits related to plant adaptation. Natural variation has been used to identify loci that control seed dormancy in nature. QTL analyses on six Recombinant Inbred Line (RIL) populations have identified eleven *DELAY OF GERMINATION* (*DOG*) QTL of which nine have been confirmed by near isogenic lines (NILs). The different *DOG* loci affect dormancy mainly by distinct genetic pathways as was concluded from the absence of strong epistatic interactions in the QTL analysis. This finding was confirmed by transcriptome analyses in freshly harvested dry seeds of the main *DOG*NILs, these lines showed distinct expression patterns compared to their genetic background Landsberg *erecta* (L*er*). The genes identified in the different *DOG*NILs represent largely different gene ontology profiles [6].

Here we aim at identifying genes that are required for dormancy maintenance and germination of imbibed seeds. Moreover, we focus on what is in common between the different pathways. For this the transcriptome of freshly harvested (dormant) and after-ripened (AR; non-dormant) 24-h imbibed seeds of the same set of *DOG*NILs and L*er* was investigated. We have identified sets of 46 and 25 genes that were up-regulated in seeds of all dormant (D-up) and all after-ripened (AR-up) *DOG*NILs and L*er*, respectively. We have investigated their role in seed

performance by analysing knock-out (KO) mutants in these genes. With seed performance we refer to the capacity of seeds to germinate under various environmental conditions. Traits that contribute to seed performance are seed dormancy, seed longevity (as estimated in an accelerated aging test) and germination under stress conditions, such as high salt, osmotic stress and ABA treatment [32]. In this study we have characterised several genes affecting seed performance.

Methods

Plant material

The near isogenic lines of four *DELAY OF GERMINATION* (*DOG*) loci; *NILDOG1*-Cvi, *NILDOG2*-Cvi, *NILDOG3*-Cvi and *NILDOG6*-Kas-2 and Landsberg *erecta* (L*er*) were earlier described by Bentsink et al. [6]. Although for some of the *DOG* loci several NILs, containing introgression fragments from different accessions, were available we have chosen the ones with the strongest phenotypic effects. T-DNA insertional mutant lines and Columbia-0 (Col-0; N60000) were ordered from the European Arabidopsis Stock Center (NASC, www.arabidopsis.info). Details (SALK/SAIL entry, AGI code, knock out number and encoded protein) of T-DNA lines are provided in Additional file 1: Table S1.

Growth conditions

NILs: Seeds were sown in Petri dishes on water-saturated filter paper, followed by a 4-day cold treatment at 4 °C, and transferred to an acclimated room at 25 °C with 16 h light/8 h dark for 2 days before planting in 7-cm pots with standard soil. Plants were grown in an air-conditioned greenhouse at 70% relative humidity, supplemented with additional light (model SON-T plus 400 W, Philips, Eindhoven, The Netherlands) providing a day length of 16 h light (long day), with light intensity 125 mmol m^{-2} s^{-1}, and maintained at a temperature of 22–25 °C (day) and 18 °C (night). NILs were grown in a randomized complete block design with eight replicates. An experimental plot consisted of a row of 12 plants. At harvest the seeds of eight plants were bulked. Three of the eight replicates were used for the microarray analyses.

T-DNA knock-out lines: Lines were screened for homozygous insertions and grown with the wild type Columbia (Col) under greenhouse conditions using Rock wool supplemented with a Hyponex solution, in a randomized complete block design with four replicates per genotype.

Sample preparation for microarray analyses

Dormant seeds were imbibed for 24 h in continuous light at 22 °C and then stored at −80 °C until RNA isolation. After-ripened seeds were imbibed for 24 h under the same conditions as the dormant seeds as soon as the seeds reached 100% germination in the germination experiment, also these seeds were stored at –80 °C until RNA isolation.

Microarray analysis

Total RNA was prepared from 24-h imbibed seeds using RNAqueous columns with Plant RNA isolation aid (Ambion, Austin, TX, USA) according to the manufacturer's protocol. The RNA was further purified through precipitations with isopropanol and a high salt solution containing 0.24 M sodium citrate and 0.16 M sodium chloride and subsequently with 2 M lithium chloride. RNA was qualitatively assessed and quantified using an Agilent 2100 Bioanalyzer with the RNA 6000 Nano Labchip® kit (Agilent, Santa Clara CA, USA) and Nanodrop1000™ spectrometry (NanoDrop Technologies, Inc., Wilmington, DE, USA). RNA was processed and cRNA synthesized according to the 3′ GeneChips OneCycle kit and hybridized on the ATH1 GeneChip (Affymetrix Inc., Santa Clara, CA, USA). The GeneChip data were analyzed using the R statistical programming environment and the Bioconductor packages [29, 51, 54]. The data was normalized using the RMA algorithm and a linear model was fitted to the data for comparisons of dormant to after-ripened seed within each genotype, the empirical Bayes method was used to reduce the gene wise sample variance [49]. The P values were then adjusted for multiple testing with the Benjamini and Hochberg method to control for false positives [5]. The microarray data were deposited in NCBI's Gene Expression Omnibus (GEO number GSE90162). Microarray quality and reproducibility data is presented in Additional file 2: Figure S1. Dormancy up regulated genes (D-up) represent genes up-regulated ($P > 0.0001$) in the following comparisons, Ler dormant vs Ler after-ripened, NIL$DOG1$ dormant vs NIL$DOG1$ after-ripened, NIL$DOG2$ dormant vs NIL$DOG2$ after-ripened, NIL$DOG3$ dormant vs NIL$DOG3$ after-ripened, NIL$DOG6$ dormant vs NIL$DOG6$ after-ripened and vice versa After-ripening up regulated genes (AR-up) represent genes up-regulated (P > 0.0001) in the following comparisons, Ler after-ripened vs Ler dormant, NIL$DOG1$ after-ripened vs NIL$DOG1$ dormant, NIL$DOG2$ after-ripened vs NIL$DOG2$ dormant, NIL$DOG3$ after-ripened vs NIL$DOG3$ dormant, NIL$DOG6$ after-ripened vs NIL$DOG6$ dormant.

T-DNA knock-out genotype analyses

A quick isolation method modified from [13] was performed to extract genomic DNA from leaves. In short, samples were ground in an extraction buffer containing 2 M NaCl, 200 mM Tris–HCl (pH 8), 70 mM EDTA and 20 mM Na2S2O5. The grinding was conducted with a stainless steel ball at 30 Hz for 1 min (96-well plate shaker, Mo Bio Laboratory). Then samples were incubated at 65 °C for 1 h. Supernatants were collected after centrifugation at maximum speed for 10 min. DNA was precipitated by adding iso-propanol and 10 M NH4Ac with ratio of 1:1/2:1 to the supernatant. This mixture was incubated at room temperature for at least 15 min, then centrifuged for 20 min at maximum speed. The DNA pellet was retrieved and rinsed with 70% ethanol followed by centrifugation for 5 min at maximum speed to recover the pellet. After drying, the DNA pellet was dissolved in distilled water. Homozygous T-DNA insertion lines were screened with gene specific primers (left and right) and insert border primers (Additional file 1: Table S1). T-DNA plants that amplified only the insertion product were consider to be homozygous mutants.

Polymerase chain reactions (PCR) were performed in a 12.5 µL-volume containing approximately 30 ng DNA, 25 µM of each dNTP, 25 ng of forward and reverse primers, 0.05 U of DNA polymerase (Firepol, Solis Bio-Dyne), 312.5 µM of MgCl$_2$. The reaction protocol was as follows; denaturation at 95 °C for 5 min followed by 30s at 95 °C, 30s annealing at 52 to 57 °C and a 45 s to 2 min extension at 72 °C, this cycle was repeated for 35 times, and ended with a final amplification for 10 min at 72 °C. The polymorphism was detected by agarose gel electrophoresis at concentrations from 1.5% and higher (w/v) depending on size of differences.

Germination assays

Germination tests to follow the release of seed dormancy were performed as described by Alonso-Blanco et al. [2] with small adjustments. In short, at several time intervals during seed dry storage until all seed batches reached 100% germination aliquots of 50 to 100 seeds of each genotype were evenly sown on a filter paper soaked with 0.7 ml demineralized water in a 6-cm Petri dish. Petri dishes were placed in moisture chambers consisting of plastic trays containing a filter paper saturated with tap water and closed with transparent lids. Chambers were stored for 1 week in a climate chamber illuminated with 38-W Philips TL84 fluorescent tubes at 8 W m2 in continuous light at 22 °C. After that, the total number and the number of germinating seeds was scored and the percentage of germinating seeds was calculated.

Germination under stress conditions was performed on fully after-ripened seeds. Stress conditions were: osmotic stress (–1 MPa mannitol; Sigma-Aldrich), salt stress (130 mM NaCl; Sigma-Aldrich), ABA stress (0.15 µM ABA; Duchefa Biochemie). ABA was dissolved in 10 mM MES buffer (Sigma-Aldrich) and the pH adjusted to 5.8. To measure seed longevity, an accelerated aging test was performed by incubating seeds above a saturated ZnSO4 solution (40 °C, 85% relative humidity) in a closed tank for 5 days. Then the seeds were taken out and germinated on demineralized water as described before.

Results

Identification of seed dormancy and after-ripening up-regulated genes

Seeds of Ler, NIL*DOG1*-Cvi, NIL*DOG2*-Cvi, NIL*DOG3*-Cvi and NIL*DOG6*-Kas-2 were investigated for their dormancy status. After-ripening was followed by performing germination tests during a time course of dry seed storage (Fig. 1a). After 120 days all genotypes had lost dormancy and showed 100% germination. NIL*DOG2* was less dormant and NIL*DOG3*, NIL*DOG6* and NIL*DOG1* were more dormant when compared to Ler. Both dormant and after-ripened seeds of each genotype were sampled 24 h after sowing (HAS) for microarray analysis, allowing the

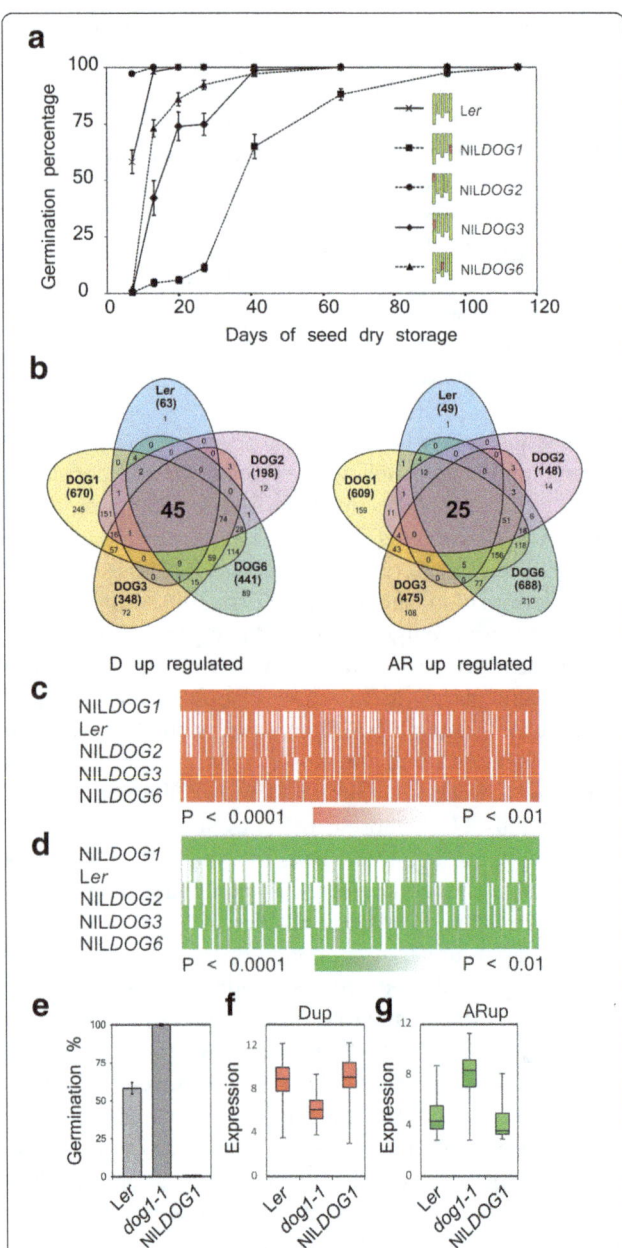

Fig. 1 Microarray analysis of dormant and after-ripened seeds after 24 h of imbibition of five genotypes with differing dormancy levels; Ler, NIL-*DOG1*, NIL*DOG2*, NIL*DOG3* and NIL*DOG6*. **a** After-ripening requirement of the five genotypes. On the right graphical representations of the NILs are depicted showing the 5 chromosomes with the introgressed regions (in red) in an otherwise Ler background (in green). **b** Venn diagrams showing the number of genes that are differentially expressed ($P < 0.0001$) in dormant (D-up) and after ripened (AR-up) 24-h imbibed seeds of different genotypes. For each genotype the total number of differential expressed genes is indicated between brackets. In the intersection of all genotypes the number of genes that are investigated in this study are presented, 46 and 25 for the D-up and AR-up set, respectively. **c** Heat map consisting of 245 NIL*DOG1* D-up genes ($P < 0.0001$). The significance of these genes in the other genotypes is indicated, the white color indicates the genes that are not significantly different in the other genotypes ($P < 0.01$). **d** Heat map consisting of 159 NIL*DOG1* AR-up genes ($P < 0.0001$). The significance of these genes in the other genotypes is indicated, the white color represents the genes that are not significantly different in the other genotypes ($P < 0.01$). **e** Germination behaviour of freshly harvested seeds of Ler, *dog1* and NIL*DOG1*. **f** Box plot showing the expression of the 45 D-up genes in freshly harvested imbibed Ler, *dog1* and NIL*DOG1* seeds (expression data taken from Dekkers et al. [16]). **g** Box plot showing the expression of the 25 D-up genes in freshly harvested imbibed Ler, *dog1* and NIL*DOG1* seeds

comparison of the dormant and after-ripened seed transcriptomes of these five genotypes with varying levels of primary dormancy.

The transcriptome data was investigated to identify genes that are up-regulated in 24 h imbibed dormant (D) Ler, NIL*DOG1*, NIL*DOG2*, NIL*DOG3* and NIL*DOG6* seeds and genes that are up-regulated in 24 h imbibed after-ripened (AR; non-dormant) seeds of the same lines. One thousand eight hundred ninety-six genes ($P < 0.0001$) were differentially expressed when performing within-genotype comparisons for the two stages analysed (dormant versus AR). In dormant seeds 63, 670, 198, 348, and 441 genes were up regulated and 49, 609, 148, 475 and 688 genes were up-regulated in AR seeds for Ler, NIL*DOG1*, NIL*DOG2*, NIL*DOG3* and NIL*DOG6*, respectively (Fig. 1b).

A large proportion of the differentially expressed genes is specific for the genotypes analysed at $P < 0.0001$; however, these genes were differentially expressed in the other genotypes at lower significances. This has been visualised for the genes that are specific for NIL*DOG1* (Fig. 1 c,d). Most of the 245 NIL*DOG1* D-up and 159 NIL*DOG1* AR-up genes are differentially expressed ($P < 0.01$) in the other genotypes. This indicates that the genes that are specifically differentially expressed are based on quantitative expression differences rather than qualitative.

Genes that are important for dormancy and AR are expected to be differentially expressed between these stages in all genotypes tested (intersections in the Venn diagrams of Fig. 1b). This led to the identification of 45 up-regulated genes in all dormant genotypes

(Dormancy-up; D-up; Table 1) and 25 genes that were up-regulated in all after-ripened genotypes (After-ripened–up; AR-up; Table 2). Further investigation of the expression patterns using the Seed EFP browser (http://www.bioinformatics.nl/efp/cgi-bin/efpWeb.cgi) revealed that, in general, all the genes that were up-regulated in dormant seeds at 24 HAI, were highly expressed in dry seeds, remained high during the imbibition of dormant seeds but were down regulated during the germination of AR seeds. Vice versa, genes that were up-regulated in AR seeds, had a low expression in dry seeds that increased with imbibition time (Additional file 2: Figure S2). Furthermore, the relation with dormancy becomes clear when the expression of the individual genes is investigated in imbibed seeds of Ler, *dog1-1* and NIL*DOG1*-Cvi that have very clear dormancy differences (Fig. 1e). D-up genes are highly expressed in dormant Ler and NIL*DOG1*-Cvi

Table 1 Mutants isolated from D-up genes

SALK/SAIL entry	AGI code	Knock out #	Encoded protein
SALK_073011C	AT2G29300	KO 1	NAD(P)-binding Rossmann-fold superfamily protein (RFSP?)
SALK_028749.55.25.x	AT2G31350	KO 2	Mitochondrial glyoxalase 2 (GLX2-5)
SALK_054451.53.45.x	AT2G33830	KO 3	Dormancy/auxin associated family protein(ATDRM2)
SALK_025507C	AT2G38800	KO 4	Plant calmodulin-binding protein-related (PCBP)
SALK_082639C	AT3G14880	KO 5	Transcription factor-related
SALK_150592C	AT5G01670	KO 6	NAD(P)-linked oxidoreductase superfamily protein
SALK_059351	AT5G64210	KO 7	Alternative oxidase2 (AOX2)
SALK_104275C	AT1G01240	KO 8	Unknown protein
SALK_110011C	AT1G05840	KO 9	Eukaryotic aspartyl protease family protein
SALK_027164C	AT1G27990	KO 10	Unknown protein
SALK_036898C	AT2G19900	KO 11	The malic enzyme1(ATNADP-ME1)
SALK_037108.56.00.x	AT1G13640	KO 12	Phosphatidylinositol 3- and 4-kinase family protein
SALK_101144	AT1G56600	KO 13	Galactinol synthase(GOLS2)
SALK_138905.29.65.x	AT2G27940	KO 14	RING/U-box superfamily protein
SALK_094895	AT3G02990	KO 15	Member of Heat Stress Transcription Factor family (HSFA1E)
SALK_025488.38.10	AT3G03310	KO 16	Lecithin:cholesterol acyltransferase 3 (LCAT3)
SALK_038352	AT3G22490	KO 17	Seed maturation protein
SALK_082777C	AT3G53410	KO 18	Paralog of ubiquitin E3 ligase (LUL2)
SALK_090239C	AT3G62090	KO 19	Phytochrome-Interacting Factors (PIF6)
SAIL_512_E03	AT4G19390	KO 20	Uncharacterised protein family
SALK_137617.43.90.x	AT5G02840	KO 21	LHY/CCA1-LIKE 1 (LCL1)
SALK_101433C	AT1G13340	KO 22	Regulator of Vps4 activity in the MVB pathway protein
SALK_025893C	AT1G20650	KO 23	Altered Seed Germination 5 (ASG5)
SALK_087702C	AT1G77450	KO 24	NAC domain-containing protein 32 (NAC032)
SALK_003223C	AT1G79440	KO 25	Succinate-semialdehyde dehydrogenase 1 (SSADH1)
SAIL_563_D10	AT1G80090	KO 26	Cystathionine beta-synthase family protein (CBSX4)
SALK_078702	AT3G50740	KO 27	UDP-glucosyl transferase 72E1 (UGT72E1)
SALK_116062C	AT3G53040	KO 28	Late embryogenesis abundant (LEA)protein
SALK_082087C	AT4G09600	KO 29	Gibberellin-regulated gene family(GASA3)
SALK_112631	AT4G20070	KO 30	Allantoate Amidohydrolase (AtAAH)
SALK_105045	AT4G25580	KO 31	CAP160 protein
SALK_043547C	AT4G36700	KO 32	RmlC-like cupins superfamily protein
SALK_135551C	AT5G65280	KO 33	GCR2-like 1 (GCL1)
SAIL_1256_F11	AT5G58650	KO 34	Plant peptide containing sulfated tyrosine 1(PSY1)

The table includes information about the affected genes (according to TAIR10[a])
[a]TAIR database website: www.arabidopsis.org

Table 2 Mutants isolated from AR-up genes

SALK entry	AGI code	Knock out #	Encoded protein
SALK_043889	AT4G34135	KO 35	UDP-Glucosyltransferase 73B2 (UGT73B2)
SALK_070860C	AT3G26060	KO 36	PEROXIREDOXIN Q (PRXQ)
SALK_094069C	AT3G26570	KO 37	Phosphate transporter 2;1 (PHT2;1)
SALK_091600.51.00.x	AT5G49910	KO 38	Chloroplast heat shock protein 70–2 (CPHSC70-2)
SALK_097487C	AT4G34131	KO 39	UDP-glucosyl transferase 73B3 (UGT73B3)
SALK_086616C	AT3G20210	KO 40	Delta vacuolar processing enzyme (DELTA-VPE)
SAIL_547_D05	AT4G31330	KO 41	Protein of unknown function
SALK_007230.56.00.x	AT5G13400	KO 42	Peptide transporter 5
SALK_017818.55.50.x	AT2G45180	KO 43	Lipid-transfer protein/seed storage 2S albumin superfamily protein
SALK_095678	AT1G07890	KO 44	Ascorbate peroxidase 1 (APX1)
SALK_090550.52.85.x	AT1G47128	KO 45	Responsive to dehydration 21 (RD21)
SALK_015756	AT3G45010	KO 46	Serine carboxypeptidase-like 48 (scpl48)
SALK_132995.40.05.x	AT4G34260	KO 47	Altered Xyloglucan 8 (AXY8)

The table includes information about the affected gene (according to TAIR10[a])[a]
TAIR database website: www.arabidopsis.org

seeds, whereas AR-up genes are highly expressed in the non-dormant *dog1-1* mutant (Fig. 1 f,g).

Among the identified genes there were several that had been related to seed dormancy or germination before, including *PHYTOCHROME-INTERACTING FACTOR 6* (*PIF6, KO19, AT3G62090*)) [47], *GIBBERELLIN 3-OXIDASE 2* (*GA3OX2*) [59] and *ALTERED SEED GERMINATION 5* (*ASG5*, KO23, AT1G20650) [4]. In addition, we found genes encoding for late embryogenesis abundant (LEA) proteins which are known to accumulate during seed desiccation and in response to water deficit induced by drought, low temperature, or salinity [30, 43]. The identified genes cover various GO molecular function categories among which by far the largest proportion is enzyme-related, including transferase activity, kinase activity and hydrolase activity, next to nucleotide binding proteins, including transcription factors.

Isolation of T-DNA mutants for genes involved in seed dormancy and germination

To investigate whether the identified genes indeed affect dormancy and AR we have analysed their T-DNA knock-out lines for seed performance phenotypes. For most of the identified genes, T-DNA mutants could be selected from the SALK and SAIL collections (NASC, http://arabidopsis.info/), but for eight genes no T-DNA insertion mutants were available (Additional file 1: Table S1). In all cases, homozygous lines were generated and confirmed using a PCR-based approach. For 47 genes a homozygous KO mutant could be selected. For nine genes (mostly in the AR-up set) no insertion was found in any of the plants genotyped (described as 'all wild type' in Additional file 1: Table S1). Moreover, for two genes *FRUCTOSE-BISPHOSPHATE ALDOLASE* (*FBA2*;

AT4G38970) and *DELTA-9 DESATURASE1* (*ADS1* AT1G06080) T-DNA insertions were identified, but no homozygous mutants could be selected. Likely, the mutants homozygous for these genes are lethal; therefore siliques of these lines were dissected to investigate possible seed abortion. This confirmed that homozygous mutant seeds of these lines were aborted (around a quarter) at an early stage of seed development (Fig. 2). Complete genotyping information is given in Additional file 1: Table S1.

All the homozygous T-DNA lines were grown together with wild type Columbia (Col) for phenotypic analysis. This revealed normal plant phenotypes for most of the mutants; however, for the *NAD(P)-BINDING ROSSMANN-FOLD SUPER FAMILY PROTEIN* (*NBRSFP*; KO1, AT2G29300) and *SUCCINATE-SEMI-ALDEHYDE DEHYDROGENASE* (*SSADH1*, KO25, At1G79440) mutants the phenotype was dramatically altered (Fig. 2). After seed harvest seeds were tested for their seed performance phenotype.

Altered seed dormancy for knock-out mutants in the dormancy and after-ripened up gene sets

Initially only seed dormancy levels were examined by assessing the number of days of seed dry storage that were required to reach 50% of germination (Days of Seed Dry Storage to reach 50% of germination; DSDS50; Fig. 3a). Thereafter, the fully after-ripened seeds were tested for seed longevity (Fig. 3b) and germination in salt (Fig. 3c), mannitol (Fig. 3d) and ABA (Fig. 3e). Thirteen lines showed a dormancy level (DSDS50) that was significantly distinct from the wild type, of which seven were less dormant (KO11, 14, 16, 17, 20, 36, 41 and 43) and six were more dormant (KO5, 16, 19, 23, 25 and 30)

Fig. 2 Plant phenotypes of T-DNA knock-out lines in comparison with wild type Columbia (Col). **a** Aborted seeds in siliques from heterozygous T-DNA lines with insertions in *FBA2* (AT1G06080) and *ADS1* (AT4G38970). **b** Four-week old plants of the NAD(P)-BINDING ROSSMANN-FOLD SUPER FAMILY PROTEIN (*nbrsfp*; KO1, AT2G29300) and *SUCCINATE-SEMIALDEHYDE DEHYDROGENASE* mutant (*ssadh1*, KO25, At1G79440) (**c**) *nbrsfp*, Col and *ssadh1* 6 weeks after germination

MEMBER (KO43, AT2G455180). *Pif6* displayed a more than two times higher DSDS50 (27.7 days) than the wild type (11.67). *PIF6* has been previously found to negatively regulate seed dormancy [47]. The other mutants were affected for at least one other seed performance trait, as well. Lines with mutations in *MALATE ENZYME 1* (*ME1*, KO11, At2G19900), *ASG5* and an *unknown protein* (*PUF*, KO41, At4G31330) displayed both a dormancy and a seed longevity phenotype. Interestingly, *atnadp-me1* and *puf* had reduced dormancy and longevity and *puf* was also less sensitive to salt stress. *Asg5* showed increased dormancy and longevity. A KO in *ALLANTOATE AMIDOHYDROLASE* (*AAH*; KO30, AT4G20070) displayed a phenotype for all investigated seeds traits except for seed longevity. This gene encodes an enzyme that hydrolyses ureide allantoate to ureidoglycolate, CO_2, and two molecules of ammonium. The *aah* mutant was more dormant, and more sensitive to salt and mannitol, but less sensitive to ABA. A KO in a *U-BOX SUPER FAMILY PROTEIN* (KO14, At2G27940) appeared slightly less dormant than wild type but was far more sensitive to ABA. A KO in the *UNCHARACTERISED PROTEIN FAMILY* gene (*UPF*; At4G19390, KO20) showed a very strong non-dormant phenotype, and was rather insensitive to mannitol and salt.

Other seed performance phenotypes for D-up and AR-up genes

Among the selected mutants were also genotypes that were not affected in their seed dormancy levels but displayed altered phenotypes for other seed performance traits.

Mutants with altered seed longevity phenotype

The *nbrsfp* mutant (KO1, AT2G29300) is besides its reduced seed longevity, also more sensitive to germination in salt and mannitol. Lines mutated in *GIBBERELLIN-REGULATED GENE FAMILY* (*GASA3*, AT4G09600, KO29) and *UDP-GLUCOSYL TRANSFERASE 73B3* (*UGT73B3*, AT4G34131, KO39) showed a longevity phenotype. A role for these genes in seed longevity has not been reported previously.

Mutants with altered response to NaCl and/or osmotic stress

Lines mutated in mitochondrial *GLYOXALASE 2* (*GLX2*; KO2, AT2G31350), *DELTA VACUOLAR PROCESSING ENZYME* (*DELTA-VPE*, KO40, At3G20210) and *SERINE CARBOXYPEPTIDASE-LIKE 48* (*SCPL48*,KO46, AT3G45010) showed reduced germination in salt but tolerated low osmotic potentials caused by high concentrations of mannitol. Although more sensitive to salt, *glx2* was more resistant to ABA than wild type. The KO in *MYB TRANSCRIPTION FACTOR LHY-CCA1-LIKE1* (*LCL1*, KO21, AT5G02840) was more resistant to

than the wild type. Several of the mutants were specifically affected in their seed dormancy levels, so no other seed performance phenotypes were detected. Among these mutants were a transcription factor related gene which is known to respond to karrikin (KO5, At3G14880), *LECITHIN:CHOLESTEROL ACYLTRANSFERASE 3* (*LCAT3*; KO16, At3G03310) which is involved in lipid metabolism, a seed maturation protein (KO17; AT3G22490), *SSADH1*, *PIF6*, the antioxidant gene *PEROXIREDOXIN Q* (*PRXQ*; KO36, AT3G26060), and a *SEED STORAGE 2S ALBUMIN SUPER FAMILY*

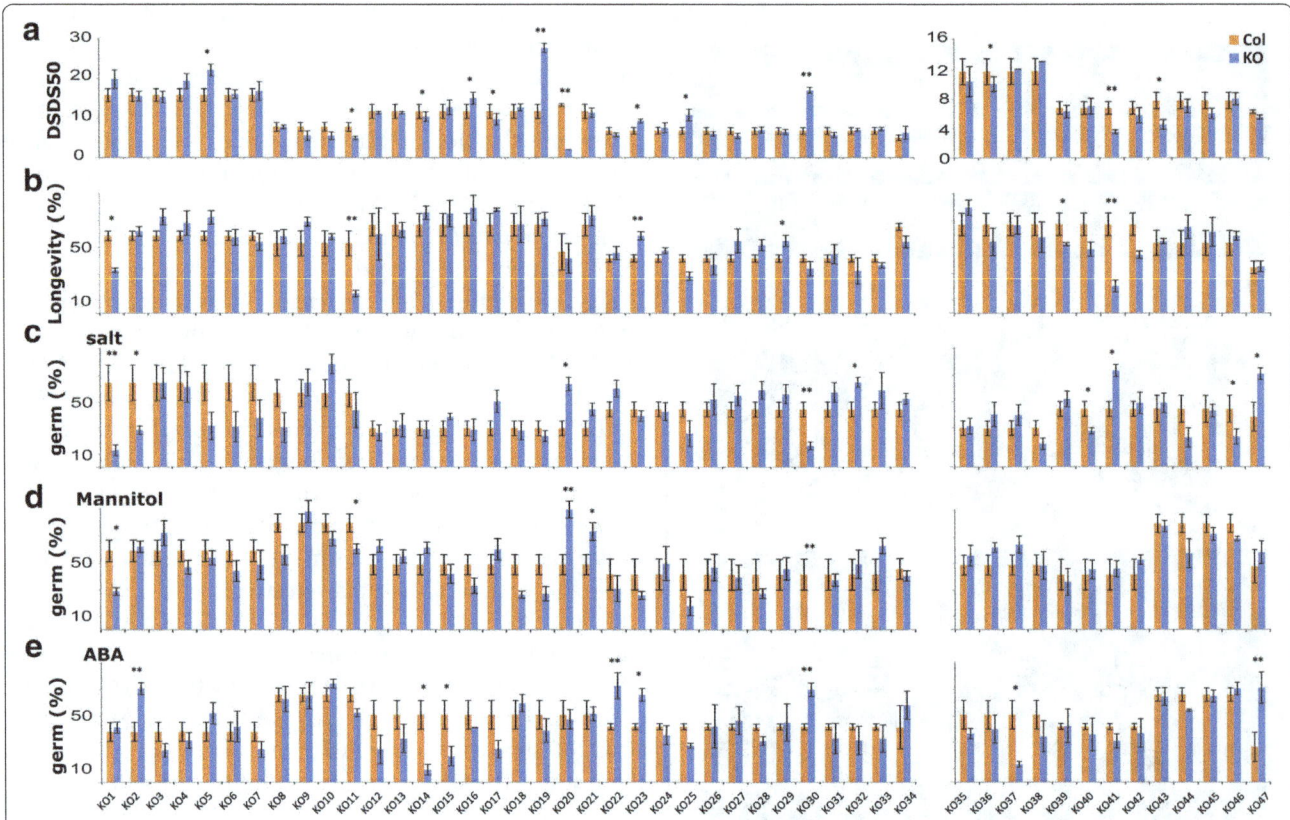

Fig. 3 Germination behaviour of knock-out mutants (KO) in dormancy (left) and after –ripened (right) upregulated genes: (**a**) Average DSDS50 (Days of Seed Dry Storage until 50% germination) values. **b** germination after accelerated aging. **c** germination in salt 130 mM; d) in mannitol (−1 MPa) and e) in ABA(0.15 μM) solutions. Significant differences are indicated (*P < 0.05 and **P < 0.01). There are differences in Col-0 values between the different experiments, however, every knock-out line has been compared to the Col-0 that was grown in the same experiment

germination in mannitol compared to wild type. A similar trend was seen after germination in salt but this effect was not significant. Lines mutated in *CUPINS SUPER FAMILY PROTEIN* (KO32, AT4G36700) and *ALTERED XYLOGLUCAN 8* (AXY8; KO47, AT4G34260) showed a salt resistance phenotype but their germination in mannitol was similar to wild type. Besides lower sensitivity to salt, line *axy8* also showed reduced sensitivity to ABA.

Responses of dormancy related genes to ABA

Several lines that showed an altered response to ABA have already been mentioned above because they had at least also one other phenotype. However, there are three lines that showed a phenotype only for germination in the presence of ABA. Two lines, KOs in *HEAT SHOCK TRANSCRIPTION FACTOR A1E* (*HSFA1E*; KO15, AT3G02990) and *LOW AFFINITY PHOSPHATE TRANSPORTER* (*PHT2;1*, KO37,AT3G26570) were more sensitive, whereas the line mutated in *REGULATOR OF VPS4 ACTIVITY* in the MVB pathway protein (KO22, AT1G13340) was more resistant to ABA.

Discussion

In our search for novel players in the regulation of *Arabidopsis* seed dormancy we employed a comparative transcriptomics approach for 24 h imbibed dormant and after-ripened *DOG*NILs and L*er* seeds. The same genotypes were earlier used to investigate the transcriptome of dormant dry seeds [6], which revealed that seed dormancy in the *DOG*NILs is mainly controlled by different additive genetic and molecular pathways. In dry seeds hardly any differences are found between dormant and after-ripened seeds, however as soon as seeds are being exposed to water, differences in the transcriptomes are evident. Based on these results we hypothesize that dormancy induction in the *DOG*NILs during seed maturation, for which dry seeds are the readout, is largely regulated by distinct molecular pathways, however dormancy maintenance during seed imbibition and the start of germination are likely very conserved processes. This conservation allowed us to identify a robust set of genes which are expressed at 24 h imbibed dormant and AR seeds. The genes identified depend a lot on the time-points chosen. From our earlier work we know that already at early imbibition (3 h after the start of imbibition) the first differences between dormant and after-ripened seeds can be identified [17]. However, we also know that most of changes in gene expression are related to seed rehydration itself and that those changes are similar between dormant and after-ripened seeds. To exclusively identify differences that are related to dormancy maintenance and germination we have chosen to investigate the transcriptome at 24 h after imbibition. This robustness of the identified genes was confirmed by comparison with previously published expression analysis that were performed with seeds of the Cvi accession at a range of physiological states [10, 17]. Of the dormancy and AR-up genes 63% and 32%, respectively, overlapped with genes that were identified by Cadman et al. [10] (Additional file 1: Table S1). Moreover, the D-up genes are also clearly higher expressed in dormant Cvi seeds as compared to AR Cvi seeds, as well as that the AR-up genes are on average higher expressed in AR seeds when compared to dormant seeds (Additional file 2: Figure S2). Some of the identified genes had previously been shown to play a role during germination and/or priming in several plant species. Among these are *GA3OX2* (AT1G80340) a key gene in the gibberellin biosynthesis pathway, *PIF6*, involved in the phytochrome signalling pathway and *ASG5* which is involved in protein and amino acid phosphorylation. The identification of these known dormancy mutants was the incentive to investigate the other dormancy and after-ripening specific genes. We took a reverse genetics approach by using T-DNA insertion lines for the differentially expressed genes and, indeed, we identified genes that had not been related to seed dormancy or germination before. Out of our target list of 66 genes, eight do not currently have any confirmed knock-out line available. This is consistent with a recent report that 12% of Arabidopsis genes do not have insertion lines available in previously generated collections [44]. The fact that a majority of the mutants showed near-wild type dormancy phenotypes, can be explained in several ways. The location of the T-DNA insertion may be decisive, e.g. whether in an intron, an exon or in untranscribed regions, such as promoters. Also, T-DNA–induced mutations do not always result in highly effective mutagenesis. Insertion in the protein-coding region of a gene generates a knockout in 86% of the cases and only 41% of the cases if the insertion is in front of the start codon [56]. Furthermore, gene redundancy may mask any phenotypic difference in plants in which the expression of only one homologue is disrupted [28]. In addition, in our experiments we used mutants with a Columbia-0 background that normally has low dormancy, which consequently does not allow the visualisation of small effects towards a decreasing dormancy level. Seed dormancy can be regulated by either inhibitory or promoting gene expression, considering the fact that in D-up genes we found both mutants that are less dormant (Fig. 3; KO11, 14 and 17) and more

dormant (KO5, 16, 19, 23, 25 and 30). These examples demonstrate the inability to predict phenotypes based on expression pattern alone.

Many seed performance characteristics (i.e. seed desiccation tolerance, seed longevity and seed dormancy) are acquired during seed maturation. If genes affect seed maturation in general, it is likely that pleiotropic effects occur. In our study, some of the mutants showed a phenotype for more than one germination trait. The *aah* mutant for example displayed a phenotype for all investigated seed traits, except longevity. This enzyme degrades allantoate which is required to recycle purine-ring nitrogen in plants. The *aah* T-DNA mutant is unable to grow on allantoin as sole nitrogen source [57]. Furthermore, it is well known that conditions favouring nitrate accumulation in mother plant may lead to lower seed dormancy levels [1]. Since AAH is a key gene in the purine pathway [57], we speculate that defects in this gene block the pathway and, hence, availability of ammonium, resulting in increased primary dormancy and also affecting other seed performance traits of the mutant. *Atnadh-me1*, *asg5* and *ufp* mutants affected both dormancy and longevity and one additional trait. It is known that seed longevity can be a pleotropic effect of genes that regulate other traits, such as seed maturation [53], response to temperature [37], oxidative stress [14] and dormancy [7, 39]. Previous studies, using mutant analysis, have shown that the seed dormancy mutants *dog1* and *rdo4* also have a reduced seed longevity phenotype [7, 39].

Loss of dormancy is expressed as opening of the germination window (permissive range of environments) [19]. It is because of this that germination under stress (i.e. salt or osmotic stress) often correlates with initial seed dormancy levels. We revealed two cases for which reduced dormancy indeed coincided with reduced sensitivity to salt stress (*upf* (KO20) and *puf* (KO41)). For some mutants, *nbrsfp* (KO1), *upf* (KO20) and *aah* (KO30), germination patterns on both NaCl and mannitol correlated positively, probably because both treatments confer osmotic stress. Two of these mutants (*nbrsfp* and *aah*) displayed enhanced sensitivity to salinity and osmotic stress. The NAD(P)-binding Rossmann-fold superfamily protein has oxidoreductase activity, binding, catalytic activity and, based on TAIR annotation, it is located in the endomembrane system. *Upf* was the only mutant to be clearly more insensitive to both salt and osmotic stress, which indicates that this mutant is primarily osmotolerant. Furthermore, for some of the salt-tolerant mutants *RmlC-like cupins super family protein* (KO32), *unknown protein* KO41 and *axy8* (KO47) the germination rates on mannitol were similar to that of wild type. Likely, for the salt-tolerant lines, genes were mutated whose products are elements of stress signalling and inhibit germination under conditions of saline stress. The salt sensitive *glx2* mutant was more tolerant to the application of ABA. The glyoxalase pathway consists of the two enzymes GLX1 and GLX2 and has a vital role in chemical detoxification. In *Arabidopsis thaliana*, GLX2 is required during abiotic stress, as was concluded from the higher sensitive of *glx2-1*to salt stress and anoxia seeds compared to wild type seeds. Moreover, GLX2-1-OE seeds are more resistant to anoxic stress than wild type [18].

Interestingly, *axy8* (KO47) showed both a higher tolerance to ABA and to salt. *AXY8* encodes an α-fucosidase acting on hemicellulose xyloglucan (XyG) that occurs in the primary cell wall of all vascular plants. Due to its high levels in elongating tissues [11] and structural alterations during cell elongation [46], XyG has been proposed to be a major player in extension growth [15]. This was confirmed by the induction of genes involved in XyG metabolism during cell elongation and upon the addition of the growth hormone auxin [50]. Overall these findings emphasize the importance of cell wall remodelling in the germination process, especially in response to stress conditions.

In both the dormancy and after-ripened sets we identified many genes encoding enzymes. This result might be linked to the fact that we looked at 24-h imbibed seeds. At this stage most of the cells in the embryo are potentially metabolically active. This also activates hydrolytic and synthetic enzymes and growth hormones to mobilize nutrients and synthesize ingredients for growth. These include the genes encoding ABA- and GA-biosynthesis- and -deactivation enzymes that play critical roles in determining the ABA-GA balance in seeds, and hence, dormancy and germination [42, 58].

Among the identified genes are also nucleotide binding proteins and transcription factors, such as members of the heat stress transcription factor family (*HSFA1E*, KO15), members of the ERF transcription factor family, transcription factor-related protein (KO5), TRANSCRIPTION FACTOR HOMOLOGOUS TO ABI5 and *PIF6* (KO19). Interestingly all are found in the D-up state and for the ones that mutants were analysed they showed either more dormancy (*transcription factor-related protein* and *pif6*) or more sensitivity to ABA (*hsfa1a*).

Conclusion

We identified seed dormancy and germination phenotypes for genes that had not been associated with seed dormancy before. We tested only one T-DNA allele per line which may not be a definitive prove that the insertional mutation causes the observed phenotype, as many as 50% of the lines may contain additional inserts at unknown loci [22, 44]. We identified germination related phenotypes for nearly 50% of the investigated genes, which is far higher compared to what can be expected from a random selection of genes. It is also far higher

compared to the genotypes that we identified for genes identified in earlier transcriptome analyses. Nevertheless, it is possible that a second locus may cause the phenotype of interest, or may alter the phenotypic effect of a knockout mutation. This work therefor represented an inventory of genes that are likely involved in the control of seed dormancy or germination. However, in depth studies are required to reveal the molecular mechanism by which these genes affect these important seed traits.

Additional files

Additional file 1: Figure S1. Microarray quality and reproducibility. All 28 ATH1 arrays used showed after hybridization similar patterns of intensity (A and B). Slide hybridization patterns were inspected manually without detecting artefacts. The RNAs used as templates for cRNA synthesis were shown to be intact based on Bio-analyzer 2001 analysis of both RNA template and biotinylated cRNA. In agreement with this were the hybridization patterns of control genes on the slide showing a near-identical pattern of hybridization (c). The uniformity of normalized unscaled SE (NUSE) and relative log expression (RLE) indicate high quality and uniformity of the hybridization data (d and e) (1). Raw intensity data were subjected to RMA normalization (2), which kept the uniformity of general levels between the different slides (f and g). Between replicate reproducibility of the experiment was high, exemplified by the high correlation between the data of two biological replicates (h). Array 1 and array 2 are hybridized with cRNA from different replicates of Ler seeds. **Figure S2.** Spatial and temporal expression patterns of the selected dormancy and after-ripening up-regulated genes. Mean relative expression of (a) D-up and (b) AR-up genes across the Arabidopsis germination time course in the micropylar and chalazal endosperm (MCE) and radicle and hypocotyl (RAD) in dry, 1, 3, 7, 12, 16, 20, 25, 31 and 38 hours after imbibition. Data was taken from Seed EFP Browser (http://www.bioinformatics.nl/efp/cgi-bin/efpWeb.cgi). **Figure S3.** Log2 expression differences for the D-up and AR-up genes that are presented in Fig. 1c and d. (a) Heat map showing Log2 expression differences of the 245 NIL*DOG1* D-up genes (P < 0.0001) in NIL*DOG1* and the other genotypes. (b) Log2 expression differences of the 159 NIL*DOG1* AR-up genes (P < 0.0001) in NIL*DOG1* and the other genotypes.

Additional file 2: Table S1. T-DNA selection of the 46 D-up and 25 AR-up genes. Details like, T-DNA identification, genotype, primers used for genotyping, knock-out # in the analyses, where the T-DNA is inserted and whether the genes overlap with the study of Cadman et al. [10] are indicated.

Abbreviations

AAH: ALLANTOATE AMIDOHYDROLASE; ABA: Abscisic acid; ADS1: DELTA-9 DESATURASE1; AR: After-ripened; AR-up: After-ripening up-regulated genes; ASG5: ALTERED SEED GERMINATION 5; AXY8: ALTERED XYLOGLUCAN 8; Col: Columbia; DELTA-VPE: DELTA VACUOLAR PROCESSING ENZYME; *DOG: DELAY OF GERMINATION*; DOGNILs: *DELAY OF GERMINATION* near isogenic lines; DSDS50: Days of seed dry storage to reach 50% of germination; D-up: Dormancy up-regulated genes; FBA2: FRUCTOSE-BISPHOSPHATE ALDOLASE; GA3OX2: GIBBERELLIN 3-OXIDASE 2; GAs: Gibberellins; GASA3: GIBBERELLIN-REGULATED GENE FAMILY; GLX2: GLYOXALASE 2; HAS: Hours after sowing; HSFA1E: HEAT SHOCK TRANSCRIPTION FACTOR A1E; KO: Knock-out; LCAT3: LECITHIN:CHOLESTEROL ACYLTRANSFERASE 3; LCL1: MYB TRANSCRIPTION FACTOR LHY-CCA1-LIKE1; LEA: Late embryogenesis abundant; Ler: Landsberg *erecta*; ME1: MALATE ENZYME; NBRSFP: NAD(P)-BINDING ROSSMANN-FOLD SUPER FAMILY PROTEIN; PHT2;1: LOW AFFINITY PHOSPHATE TRANSPORTER; PIF6: PHYTOCHROME-INTERACTING FACTOR 6; PRXQ: PEROXIREDOXIN Q; QTL: Quantitative trait loci; RIL: Recombinant inbred line; SCPL48: SERINE CARBOXYPEPTIDASE-LIKE 48; SSADH1: SUCCINATE-SEMIALDEHYDE DEHYDROGENASE; UGT73B3: UDP-GLUCOSYL TRANSFERASE 73B3; UPF: UNCHARACTERISED PROTEIN FAMILY

Acknowledgements
Not applicable

Funding
This work was supported by the Dutch Technology Foundation (STW), which is the applied science division of The Netherlands Organization for Scientific Research and the Technology Program of the Ministry of Economic Affairs (to LB) and Bio4Energy, a Strategic Research Environment appointed by the Swedish government (to JH). The funding agencies were not involved in the data collection, analysis, and interpretation neither in writing the manuscript.

Authors' contributions
LB and JH designed, performed and analysed the transcriptome analyses. FY analysed the T-DNA KO lines for their germination performance. FY, HH and LB wrote the manuscript. All authors read and approved the final manuscript.

Competing interests
The authors declare that they have no competing interests.

Author details
[1]Wageningen Seed Laboratory, Laboratory of Plant Physiology, Wageningen University, Droevendaalsesteeg 1, 6708, PB, Wageningen, The Netherlands. [2]Umeå Plant Science Center, Department of Plant Physiology, Umeå University, SE-901 87 Umeå, Sweden. [3]Department of Molecular Plant Physiology, Utrecht University, Padualaan 8, 3584, CH, Utrecht, The Netherlands.

References
1. Alboresi A, Gestin C, Leydecker MT, Bedu M, Meyer C, Truong HN. Nitrate, a signal relieving seed dormancy in Arabidopsis. Plant Cell Environ. 2005;28:500–12.
2. Alonso-Blanco C, Aarts MGM, Bentsink L, Keurentjes JJB, Reymond M, Vreugdenhil D, Koornneef M. What has natural variation taught us about plant development, physiology, and adaptation? Plant Cell. 2009;21:1877–96.
3. Bai B, Novák O, Ljung K Hanson J, Bentsink L. Combined transcriptome and translatome analyses reveal a role for transcriptional inhibition of tryptophan dependent auxin biosynthesis in the control of DOG1 dependent seed dormancy. 2017. Unpublished.
4. Bassel GW, Glaab E, Marquez J, Holdsworth MJ, Bacardit J. Functional network construction in Arabidopsis using rule-based machine learning on large-scale data sets. Plant Cell. 2011;23:3101–16.
5. Benjamini Y, Hochberg Y. Controlling the false discovery rate: a practical and powerful approach to multiple testing. J Roy Statist Soc B Mec. 1995;57: 289–300.
6. Bentsink L, Hanson J, Hanhart CJ, Blankestijn-de Vries H, Coltrane C, Keizer P, El-Lithy M, Alonso-Blanco C, de Andrés MT, Reymond M, van Eeuwijk F, Smeekens S, Koornneef M. Natural variation for seed dormancy in Arabidopsis is regulated by additive genetic and molecular pathways. Proc Natl Acad Sci U S A. 2010;107:4264–9.
7. Bentsink L, Jowett J, Hanhart CJ, Koornneef M. Cloning of DOG1, a quantitative trait locus controlling seed dormancy in Arabidopsis. Proc Natl Acad Sci U S A. 2006;103:17042–7.
8. Bewley JD. Seed germination and dormancy. Plant Cell. 1997;9:1055.
9. Bewley JD, Bradford K, Hilhorst H. Seeds: physiology of development, germination and dormancy. New York: Springer; 2012.
10. Cadman CS, Toorop PE, Hilhorst HW, Finch-Savage WE. Gene expression profiles of Arabidopsis Cvi seeds during dormancy cycling indicate a common underlying dormancy control mechanism. Plant J. 2006;46:805–22.
11. Carpita NC, Gibeaut DM. Structural models of primary cell walls in flowering plants: consistency of molecular structure with the physical properties of the walls during growth. Plant J. 1993;3:1–30.
12. Carrera E, Holman T, Medhurst A, Dietrich D, Footitt S, Theodoulou FL, Holdsworth MJ. Seed after-ripening is a discrete developmental pathway associated with specific gene networks in Arabidopsis. Plant J. 2008;53:214–24.
13. Cheung W, Hubert N, Landry B. A simple and rapid DNA microextraction method for plant, animal, and insect suitable for RAPD and other PCR analyses. Genome Res. 1993;3:69–70.
14. Clerkx EJ, Vries BD, Ruys GJ, Groot SP, Koornneef M. Genetic differences in seed longevity of various Arabidopsis mutants. Physiol Plant. 2004;121: 448–61.
15. Cosgrove DJ. Enzymes and other agents that enhance cell wall extensibility. Annu Rev Plant Biol. 1999;50:391–417.
16. Dekkers BJW, Pearce SP, van Bolderen-Veldkamp RPM, Holdsworth MJ, Bentsink L. Dormant and after-ripened Arabidopsis thaliana seeds are distinguished by early transcriptional differences in the imbibed state. Front Plant Sci. 2016b;7:1323–38.
17. Dekkers BJ, He H, Hanson J, Willems LA, Jamar DC, Cueff G, Rajjou L, Hilhorst HW, Bentsink L. The Arabidopsis DELAY OF GERMINATION 1 gene affects ABSCISIC ACID INSENSITIVE 5 (ABI5) expression and genetically interacts with ABI3 during Arabidopsis seed development. Plant J. 2016a;85:451–65.
18. Devanathan S, Erban A, Perez-Torres R Jr, Kopka J, Makaroff CA. (2014) Arabidopsis thaliana Glyoxalase 2-1 is required during abiotic stress but is not essential under normal plant growth. PLoS One. 2014;9:e95971.
19. Finch-Savage WE, Leubner-Metzger G. Seed dormancy and the control of germination. New Phytol. 2006;171:501–23.
20. Fu Q, Wang B, Jin X, Li H, Han P, Wei K-H, Zhang X-M, Zhu Y. Proteomic analysis and extensive protein identification from dry, germinating Arabidopsis seeds and young seedlings. J Biochem Mol Biol. 2005;38:650.
21. Gallardo K, Job C, Groot SPC, Puype M, Demol H, Vandekerckhove J, Job D. Proteomic analysis of arabidopsis seed germination and priming. Plant Physiol. 2001;126:835–48.
22. Gase K, Weinhold A, Bozorov T, Schuck S, Baldwin IT. Efficient screening of transgenic plant lines for ecological research. Mol Ecol Resour. 2011;11:890–902.
23. Griffiths J, Murase K, Rieu I, Zentella R, Zhang ZL, Powers SJ, Gong F, Phillips AL, Hedden P, Sun TP, Thomas SG. Genetic characterization and functional analysis of the GID1 gibberellin receptors in Arabidopsis. Plant Cell. 2006;18:3399–414.
24. Gubler F, Millar AA, Jacobsen JV. Dormancy release, ABA and pre-harvest sprouting. Curr Opin Plant Biol. 2005;8:183–7.
25. Holdsworth MJ, Bentsink L, Soppe WJJ. Molecular networks regulating Arabidopsis seed maturation, after-ripening, dormancy and germination. New Phytol. 2008a;179:33–54.
26. Holdsworth MJ, Finch-Savage WE, Grappin P, Job D. Post-genomics dissection of seed dormancy and germination. Trends Plant Sci. 2008b;13:7–13.
27. Howell KA, Narsai R, Carroll A, Ivanova A, Lohse M, Usadel B, Millar AH, Whelan J. Mapping metabolic and transcript temporal switches during germination in rice highlights specific transcription factors and the role of RNA instability in the germination process. Plant Physiol. 2009;149: 961–80.
28. Hua J, Meyerowitz EM. Ethylene responses are negatively regulated by a receptor gene family in Arabidopsis thaliana. Cell. 1998;94:261–71.
29. Irizarry RA, Hobbs B, Collin F, Beazer-Barclay YD, Antonellis KJ, Scherf U, Speed TP. Exploration, normalization, and summaries of high density oligonucleotide array probe level data. Biostatistics. 2003;4:249–64.
30. Ismail AM, Hall AE, Close TJ. Allelic variation of a dehydrin gene cosegregates with chilling tolerance during seedling emergence. Proc Natl Acad Sci U S A. 1999;96:13566–70.
31. Iuchi S, Suzuki H, Kim YC, Iuchi A, Kuromori T, Ueguchi-Tanaka M, Asami T, Yamaguchi I, Matsuoka M, Kobayashi M, Nakajima M. Multiple loss-of-function of Arabidopsis gibberellin receptor AtGID1s completely shuts down a gibberellin signal. Plant J. 2007;50:958–66.
32. Joosen RV, Arends D, Li Y, Willems LA, Keurentjes JJ, Ligterink W, Jansen RC, Hilhorst HW. Identifying genotype-by-environment interactions in the metabolism of germinating arabidopsis seeds using generalized genetical genomics. Plant Physiol. 2013;162:553–66.

33. Koornneef M, Van der Veen JH. Induction and analysis of gibberellin-sensitive mutants in *Arabidopsis thaliana* (L.) Heynh. Theor Appl Genet. 1980;58:257–63.

34. Koornneef M, Jorna M, Brinkhorst-Van der Swan D, Karssen C. The isolation of abscisic acid (ABA) deficient mutants by selection of induced revertants in non-germinating gibberellin sensitive lines of *Arabidopsis thaliana* (L.) Heynh. Theor Appl Genet. 1982;61:385–93.

35. Koornneef M, Reuling G, Karssen C. The isolation and characterization of abscisic acid-insensitive mutants of *Arabidopsis thaliana*. Physiol Plant. 1984; 61:377–83.

36. Le BH, Cheng C, Bui AQ, Wagmaister JA, Henry KF, Pelletier J, Kwong L, Belmonte M, Kirkbride R, Horvath S. Global analysis of gene activity during Arabidopsis seed development and identification of seed-specific transcription factors. Proc Natl Acad Sci U S A. 2010;107:8063–70.

37. Lee B-h, Lee H, Xiong L, Zhu J-K. A mitochondrial complex I defect impairs cold-regulated nuclear gene expression. Plant Cell. 2002;14:1235–51.

38. Liu X, Zhang H, Zhao Y, Feng Z, Li Q, Yang H-Q, Luan S, Li J, He Z-H. (2013) Auxin controls seed dormancy through stimulation of abscisic acid signaling by inducing ARF-mediated ABI3 activation in Arabidopsis. Proc Natl Acad Sci U S A. 2013;110:15485–90.

39. Liu Y, Koornneef M, Soppe WJ. (2007) The absence of histone H2B monoubiquitination in the Arabidopsis *hub1* (*rdo4*) mutant reveals a role for chromatin remodeling in seed dormancy. Plant Cell. 2007;19:433–44.

40. Matakiadis T, Albores A, Jikumaru Y, Tatematsu K, Pichon O, Renou J-P, Kamiya Y, Nambara E, Truong H-N. The Arabidopsis abscisic acid catabolic gene CYP707A2 plays a key role in nitrate control of seed dormancy. Plant Physiol. 2009;149:949–60.

41. Nakabayashi K, Okamoto M, Koshiba T, Kamiya Y, Nambara E. Genome-wide profiling of stored mRNA in Arabidopsis Thaliana seed germination: epigenetic and genetic regulation of transcription in seed. Plant J. 2005;41:697–709.

42. Nambara E, Marion-Poll A. Abscisic acid biosynthesis and catabolism. Annu Rev Plant Biol. 2005;56:165–85.

43. Nylander M, Svensson J, Palva ET, Welin BV. (2001) Stress-induced accumulation and tissue-specific localization of dehydrins in *Arabidopsis thaliana*. Plant Mol Biol. 2001;45:263–79.

44. O'Malley RC, Ecker JR. Linking genotype to phenotype using the Arabidopsis unimutant collection. Plant J. 2010;61:928–40.

45. Oh E, Kang H, Yamaguchi S, Park J, Lee D, Kamiya Y, Choi G. Genome-wide analysis of genes targeted by PHYTOCHROME INTERACTING FACTOR 3-LIKE5 during seed germination in Arabidopsis. Plant Cell. 2009;21:403–19.

46. Pauly M, Qin Q, Greene H, Albersheim P, Darvill A, York WS. Changes in the structure of xyloglucan during cell elongation. Planta. 2001;212:842–50.

47. Penfield S, Josse E-M, Halliday KJ. A role for an alternative splice variant of PIF6 in the control of Arabidopsis primary seed dormancy. Plant Mol Biol. 2010;73:89–95.

48. Preston J, Tatematsu K, Kanno Y, Hobo T, Kimura M, Jikumaru Y, Yano R, Kamiya Y, Nambara E. (2009) Temporal expression patterns of hormone metabolism genes during imbibition of *Arabidopsis thaliana* seeds: a comparative study on dormant and non-dormant accessions. Plant Cell Physiol. 2009;50:1786–800.

49. Ritchie ME, Phipson B, Wu D, Hu Y, Law CW, Shi W, Smyth GK. Limma powers differential expression analyses for RNA-sequencing and microarray studies. Nucleic Acids Res. 2015;43:e47.

50. Sánchez M, Gianzo C, Sampedro J, Revilla G, Zarra I. Changes in α-Xylosidase during intact and Auxin-induced growth of pine hypocotyls. Plant Cell Physiol. 2003;44:132–8.

51. Smyth GK. Linear models and empirical bayes methods for assessing differential expression in microarray experiments. Stat Appl Genet Mol. 2004; 3:Article 3.

52. Sreenivasulu N, Usadel B, Winter A, Radchuk V, Scholz U, Stein N, Weschke W, Strickert M, Close TJ, Stitt M. Barley grain maturation and germination: metabolic pathway and regulatory network commonalities and differences highlighted by new MapMan/PageMan profiling tools. Plant Physiol. 2008; 146:1738–58.

53. Sugliani M, Brambilla V, Clerkx EJ, Koornneef M, Soppe WJ. The conserved splicing factor SUA controls alternative splicing of the developmental regulator ABI3 in Arabidopsis. Plant Cell. 2010;22:1936–46.

54. Team, RC R. A language and environment for statistical computing. Vienna, Austria: R Foundation for Statistical Computing; 2012. ISBN 3-900051-07-0

55. Toh S, Imamura A, Watanabe A, Nakabayashi K, Okamoto M, Jikumaru Y, Hanada A, Aso Y, Ishiyama K, Tamura N. High temperature-induced abscisic acid biosynthesis and its role in the inhibition of gibberellin action in Arabidopsis seeds. Plant Physiol. 2008;146:1368–85.

56. Wang YH. How effective is T-DNA insertional mutagenesis in Arabidopsis? J Biochem Technol. 2008;1:11–20.

57. Werner AK, Sparkes IA, Romeis T, Witte C-P. Identification, biochemical characterization, and subcellular localization of allantoate amidohydrolases from Arabidopsis and soybean. Plant Physiol. 2008;46:418–30.

58. Yamaguchi S. Gibberellin metabolism and its regulation. Annu Rev Plant Biol. 2008;59:225–51.

59. Yamaguchi S, Smith MW, Brown RG, Kamiya Y, Sun T-P. Phytochrome regulation and differential expression of gibberellin 3β-hydroxylase genes in germinating Arabidopsis seeds. Plant Cell. 1998;10:2115–26.

60. Yamauchi Y, Ogawa M, Kuwahara A, Hanada A, Kamiya Y, Yamaguchi S. Activation of gibberellin biosynthesis and response pathways by low temperature during imbibition of *Arabidopsis thaliana* seeds. Plant Cell. 2004;16:367–78.

10

Indole-3-butyric acid promotes adventitious rooting in *Arabidopsis thaliana* thin cell layers by conversion into indole-3-acetic acid and stimulation of anthranilate synthase activity

L. Fattorini[†], A. Veloccia[†], F. Della Rovere[†], S. D'Angeli, G. Falasca and M. M. Altamura[*] ⓘ

Abstract

Background: Indole-3-acetic acid (IAA), and its precursor indole-3-butyric acid (IBA), control adventitious root (AR) formation *in planta*. Adventitious roots are also crucial for propagation via cuttings. However, IBA role(s) is/are still far to be elucidated. In *Arabidopsis thaliana* stem cuttings, 10 μM IBA is more AR-inductive than 10 μM IAA, and, in thin cell layers (TCLs), IBA induces ARs when combined with 0.1 μM kinetin (Kin). It is unknown whether arabidopsis TCLs produce ARs under IBA alone (10 μM) or IAA alone (10 μM), and whether they contain endogenous IAA/IBA at culture onset, possibly interfering with the exogenous IBA/IAA input. Moreover, it is unknown whether an IBA-to-IAA conversion is active in TCLs, and positively affects AR formation, possibly through the activity of the nitric oxide (NO) deriving from the conversion process.

Results: Revealed undetectable levels of both auxins at culture onset, showing that arabidopsis TCLs were optimal for investigating AR-formation under the total control of exogenous auxins. The AR-response of TCLs from various ecotypes, transgenic lines and knockout mutants was analyzed under different treatments. It was shown that ARs are better induced by IBA than IAA and IBA + Kin. IBA induced IAA-efflux (*PIN1*) and IAA-influx (*AUX1/LAX3*) genes, IAA-influx carriers activities, and expression of *ANTHRANILATE SYNTHASE -alpha1* (*ASA1*), a gene involved in IAA-biosynthesis. ASA1 and ANTHRANILATE SYNTHASE -beta1 (ASB1), the other subunit of the same enzyme, positively affected AR-formation in the presence of exogenous IBA, because the AR-response in the TCLs of their mutant *wei2wei7* was highly reduced. The AR-response of IBA-treated TCLs from *ech2ibr10* mutant, blocked into IBA-to-IAA-conversion, was also strongly reduced. Nitric oxide, an IAA downstream signal and a by-product of IBA-to-IAA conversion, was early detected in IAA- and IBA-treated TCLs, but at higher levels in the latter explants.

Conclusions: Altogether, results showed that IBA induced AR-formation by conversion into IAA involving NO activity, and by a positive action on IAA-transport and ASA1/ASB1-mediated IAA-biosynthesis. Results are important for applications aimed to overcome rooting recalcitrance in species of economic value, but mainly for helping to understand IBA involvement in the natural process of adventitious rooting.

Keywords: Adventitious roots, Anthranilate synthase genes, *ech2ibr10* mutant, Indole-3-butyric acid, Indole-3-acetic acid, Indole-3-acetic acid influx carriers, Indole-3-acetic acid efflux carriers, In vitro culture, Nitric oxide, Stem thin cell layers

* Correspondence: mariamaddalena.altamura@uniroma1.it
[†]Equal contributors
Dipartimento di Biologia Ambientale, Sapienza Università di Roma, Roma, Italy

Background

Roots of higher plants can be classified into primary roots (PRs), developed from the root pole of the embryo, and post-embryonic lateral and adventitious roots, developed after seed germination. Lateral roots (LRs) are formed by the pericycle of the PR, whereas adventitious roots (ARs) are formed *in planta* by tissues of the PR in secondary vascular structure, and, mainly, by tissues of the aerial organs [1]. Moreover, the formation of ARs is crucial for vegetative propagation via cuttings, and in horticulture and forestry the formation of ARs allows for the cloning of superior genotypes and is an essential part of the breeding programs [2, 3]. In different types of explants and species, indole-3-acetic acid (IAA), and its natural precursor indole-3-butyric acid (IBA) [4], are the main inducers of ARs, when applied exogenously, alone or combined with other phytohormones, e.g. cytokinin and methyl jasmonate (MeJA) [5, 6]. In *Arabidopsis thaliana* dark-grown seedlings, IAA is the endogenous inducer of AR formation from the hypocotyl, with biosynthesis, signalling, and transport involved [7–9], however the exogenous application of IBA (10 μM), combined or not with a cytokinin [kinetin (Kin)], improves AR formation [7, 9]. It has been demonstrated a long time ago that IBA applied at 10 μM in combination with 0.1 μM Kin induces AR formation in tobacco and arabidopsis thin cell layers (TCLs) [10, 11]. By contrast, the role of IBA alone in inducing AR formation has not yet demonstrated in this culture system. However, IBA, at 10 μM, induces AR formation in arabidopsis stem cuttings, and better than IAA at the same concentration [12]. *Arabidopsis thaliana* TCLs consist of stem inflorescence tissues external to the vascular system, i.e., epidermis, cortical parenchyma, endodermis and, occasionally, of one or two layers of fibers [10]. In stem cuttings including the vascular system, it has been hypothesized that the promotion of AR formation by exogenous IBA occurs by an interaction with the endogenous IAA content [12], whereas there is no information about the endogenous IBA and IAA content in the TCLs. The first aim of the research was to determine the endogenous levels of IBA and IAA at the onset of the culture in the arabidopsis TCLs to establish whether IBA (alone or combined with Kin), and IAA (alone), might control AR formation either by an interaction with the endogenous auxin pool or by a total exogenous control.

In planta, IBA is an important component of the auxin pool [13], and in arabidopsis, there is evidence that it is inactive during its cell-to-cell transit, becoming active, by conversion to IAA, in the target cells only [14]. Moreover, by the use of seedlings of the *ech2ibr10* mutant, blocked into IBA-to-IAA-conversion [15], it has been recently shown that the promotion of AR formation by exogenous IBA alone (10 μM) requires conversion into IAA and

interaction with ethylene signalling [9]. However, the possibility that IBA can promote the AR process *in planta* also in a way different from its conversion to IAA has been also hypothesized [9]. Moreover, nitric oxide (NO) is known to be an IAA downstream signal, but is known to derive from IBA-to-IAA conversion ([16], and references therein). NO positively affects AR formation in numerous explant types, e.g. cucumber hypocotyl explants [17], however, its role in adventitious rooting from TCLs is unknown.

Transcriptome analyses of tea cuttings and mung bean seedlings in response to IBA treatments show the existence of a lot of IBA-regulated genes associated with adventitious rooting, including genes coding for proteins involved in auxin signalling and cellular influx and efflux [18, 19]. In accordance, IAA transport via PIN-FORMED (PIN) efflux carriers, e.g. PIN1, and via influx carriers, i.e., AUXIN RESISTANT1 (AUX1) and LIKE AUXIN RESISTANT3 (LAX3), has been demonstrated to be essential for the AR process in arabidopsis seedlings treated without exogenous hormones and with IBA + Kin [7, 8, 20]. The same carriers are active in the IBA + Kin-cultured TCLs, and the AR response strongly declines in TCLs excised from the *aux1* and *lax3aux1* mutants [7, 8]. Moreover, *in planta*, the activity of the promoters of *PIN1* and *LAX3* increases in the wild type (WT), and the AR density decreases in the *lax3aux1* mutant, also in the presence of IBA alone (10 μM) [9]. This suggests that IBA is sufficient to stimulate IAA transport *in planta*, whereas it remains to be determined whether this occurs also in the TCLs.

The *WEAK ETHYLENE-INSENSITIVE2/ANTHRANILATE SYNTHASE alpha1* (*WEI2/ASA1*) and *WEI7/ANTHRANILATE SYNTHASE beta1* (*ASB1*) genes encode, respectively, the α- and β-subunits of anthranilate synthase, a rate-limiting enzyme of an early step of the tryptophan-dependent IAA biosynthesis [21]. In arabidopsis seedlings, by the use of the *wei2wei7* mutant and the *ASA1::GUS* and *ASB1::GUS* lines, it has been shown that the anthranilate synthase is required for AR formation, with the transcriptional induction of the α-anthranilate synthase isoform *(WEI2/ASA1)* mainly involved [9]. However, it is unknown whether the same genes are involved in the AR-formation in TCLs.

The second aim of the research was to understand whether IBA alone was able to induce AR formation in arabidopsis TCLs, in comparison with IBA + Kin, IAA alone and Kin alone, whether the IAA transport by PIN1, LAX3, and AUX1 was affected, whether an IBA conversion into IAA was needed and possibly involved NO formation, and whether an IAA biosynthesis by WEI2/ASA1 and WEI7/ASB1 was also involved.

Nitric oxide is known to activate genes involved in jasmonic acid (JA) biosynthesis [22]. In addition, methyl jasmonate (MeJA) is known to control IAA biosynthesis by enhancing both *ASA1* and *ASB1* expression [23]. In tobacco IBA + Kin-treated TCLs, when applied at 0.1 and 0.01 μM, MeJA is rapidly cleaved to JA, and JA action results into enhanced AR formation [6]. However, MeJA effects on *ASA1* expression in IBA-treated arabidopsis TCLs are presently unknown. It is important to underline that ARs are formed in the TCLs following the same developmental stages that characterize AR formation in entire hypocotyls [7]. This means that the study of AR formation in the TCLs is representative of the natural process occurring *in planta*. In addition, the strict and continuous interrelation between biosynthesis and utilization of IAA and its precursor IBA [13, 14] make it impossible to determine with certainty the role of IBA in the AR process *in planta*. By contrast, it may be determined in the TCLs. In fact, it has been reported that they are unable to produce ARs under hormone free (HF) conditions [10] as a possible consequence of a poor or no endogenous auxin(s) content.

Results show that endogenous IBA and IAA are undetectable at culture onset, and that the AR response is totally dependent on the exogenous auxin, IBA alone in particular. The expression and activities of AUX1 and LAX3 and the expression of *PIN1* occur in the IBA-alone-treated TCLs during AR formation, and the conversion of IBA into IAA, followed by NO formation, is strictly necessary. Exogenous IBA, either alone, or mainly when combined with MeJA, enhances the expression of *WEI2/ASA1*, and ASA1 and ASB1 positively affect AR-formation in the presence of exogenous IBA. Altogether the results demonstrate that the IBA-promotion of adventitious rooting in TCLs involves conversion into IAA and NO production, and promotes IAA biosynthesis and transport.

Methods
Plant material and growth conditions
Seeds of *Arabidopsis thaliana* Col, Col-0 and Col-gl1 ecotypes, of *ASA1::GUS*, *DR5::GUS*, *PIN1::GUS*, *AUX1::-GUS*, and *LAX3::GUS* transgenic lines, and of *wei2-1wei7-1*, *lax3aux1-21* and *ech2-1ibr10-1* double mutants were stratified for 3 days at 4 °C under continuous darkness and sown on a commercial soil. The seeds of the *DR5::GUS* line and the *PIN1::GUS* line were a generous gift of Sabrina Sabatini (Sapienza University Rome) and Stefano Bencivenga (University of Milan), respectively. The seeds of the *AUX1::GUS* and *LAX3::GUS* lines and of the *lax3aux1-21* mutant were kindly provided by Malcom Bennett (University of Nottingham), and those of the *ech2-1ibr10-1* mutant by Bonnie Bartel (Rice University Huston). The seeds of the *wei2-1wei7-1* mutant and of the *ASA1::GUS* transgenic line were bought by NASC (Nottingham Arabidopsis Stock Centre, School of Biosciences, University of Nottingham, UK). The seeds of Col, Col-0 and Col-gl1 ecotypes came from

stocks of our laboratory. The plantlets obtained from these seeds were grown until the reproductive stage (40 days after germination) in the same growth chamber, at 22 ± 2 °C, 70% humidity and long days (white light of 22 Wm^{-2} light intensity).

TCL culture
Superficial TCLs, about 0.5 × 8 mm, composed by six cell layers including the epidermis, were excised from the internodes of the inflorescence stem. The TCLs were cultured, epidermal side up, under continuous darkness, at 22 ± 2 °C, up to day 15 on a medium consisting of MS [24] salts supplemented with 0.55 mM myo-inositol, 0.1 μM thiamine-HCL, 1% (*w/v*) sucrose, 0.8% agar (*w/v*) (pH 5.7) (HF medium). Col-0 TCLs were cultured on this medium with the addition of 10 μM IBA, 10 μM IBA plus 0.1 μM Kin, 0.1 μM Kin, 10 μM IAA, 10 μM IBA plus 0.01 μM MeJA, and under HF as experimental control. TCLs from the *ech2ibr10*, *wei2-1wei7-1*, and *lax3aux1-21* mutants and their WT were cultured with 10 μM IBA, 10 μM IAA or under HF. *ASA1::GUS* TCLs were cultured with either 10 μM IBA, or 10 μM IAA or IBA plus 0.01 μM MeJA. *DR5::GUS*, *PIN1::GUS*, *LAX3::GUS*, *AUX1::GUS* TCLs, and TCLs from Col and Col-gl1 ecotypes, were cultured with 10 μM IBA. One hundred explants per genotype and treatment were used per replicate. The pH was adjusted to 5.7 with 1 M NaOH before autoclaving. For macroscopic analyses, the explants of the WT and mutants were examined under a LEICA MZ8 stereomicroscope at culture end, and the AR response evaluated as the percentage of explants either remaining at the initial stage at culture end or forming macroscopic callus and ARs, and as mean number of ARs (±SE) per rooting explant.

Histochemical analysis of GUS activity
TCLs of *DR5::GUS*, *PIN1::GUS*, *LAX3::GUS*, *AUX1::-GUS*, *ASA1::GUS* lines were harvested at day 8 and day 15 of culture, and processed with the GUS staining as described by Willemsen et al. [25], with minor modifications, as reported by Veloccia et al. [9]. After infiltration for 15 min in a vacuum belljar, the samples were incubated at 37 °C in the dark either for 30 min (*DR5::GUS* and *LAX3::GUS*), or 45 min (*AUX1::GUS*, *ASA1::GUS*), or 2.5 h (*PIN1::GUS*). After GUS assay, the samples were fixed in 70% (*v/v*) ethanol, dehydrated by a graded ethanol series, embedded in Technovit 7100 (Heraeus Kulzer), longitudinally sectioned at 12 μm with a Microm HM 350 SV microtome (Microm, Walldorf, Germany), and observed under light microscopy.

Hormone quantification
TCLs of Col-0 were collected at time 0 (i.e., soon after the excision) and conserved to −80 °C until the analyses. The

extraction of IAA and IBA was performed using aliquots of 50 mg of TCLs according to Veloccia et al. [9]. Quantitative determinations of IAA and IBA were carried out by Rapid Resolution-Reversed Phase-HPLC (RR-RP-HPLC) separation followed by MS/MS detection with a triple quadrupole (QqQ) mass-spectrometer with an ESI-interface (G6420A Agilent Technologies, CA, USA). Pure standards, internal standards, and the quantification of the two auxins were according to Veloccia et al. [9].

Nitric oxide detection

Intracellular NO content in Col-0 TCLs cultured with either 10 μM IBA or 10 μM IAA was quantified using the cell-permeable diacetate derivative diamino-fluorescein-FM (DAF-FMDA; Sigma) under epifluorescence microscopy. TCLs were incubated in 20 mM HEPES/NaOH buffer (pH 7.4) supplemented with 5 μM DAF-FMDA for 20 min [26] at 2, 3 and 6 days of culture, after having verified that no significant epifluorescence signal was detectable with the buffer alone (Additional file 1: Figure S1 a-b). After washing three times with the buffer to remove the excess of the fluorescent probe, TCLs were observed using a Leica DMRB microscope equipped with the specific set of filters (EX 450–490, DM 510, LP 515). The images were acquired with a LEICA DC500 digital camera and analysed with the IM1000 image-analysis software (Leica). Ten observations in each of 20 TCLs per treatment were randomly carried out, and the intensities of the fluorescence signal (in green colour) were quantified using the ImageJ software (National Institute of Health, Bellevue, WA, USA) and expressed in Arbitrary Units (AUs; from 0 to 255). The values were averaged and normalized to the control ones, i.e., to those measured in TCLs incubated in the buffer without the fluorescent probe.

Statistical analysis

Data were expressed as means (±SE). One-way or two-way analysis of variance (ANOVA, $P < 0.05$) was used to compare effects of treatments and/or genotypes, and, if ANOVA showed significant effects, Tukey's post-test was applied (GraphPad Prism 6.0). The significance of the differences between the percentages was statistically evaluated using χ^2 test ($P < 0.05$). All the experiments were repeated three times in two following years, and similar results were obtained (data of the replicate from the second year shown).

Results

IBA induces AR formation in the TCLs independently of the addition of cytokinin, and better than IAA

TCLs of *Arabidopsis thaliana*, ecotype Col-0, were grown under darkness for 15 days in the presence of IBA alone (10 μM), Kin alone (0.1 μM), IBA (10 μM) + Kin (0.1 μM), IAA alone (10 μM), and HF (control). The aim was to evaluate whether AR formation was inducible by IBA without cytokinin, and in the affirmative case, whether IBA was more efficient than IAA.

At culture end, the explants treated with Kin alone did not show any morphogenic/organogenic response, i.e., all remained at the initial stage, similarly to the explants cultured without hormones (Fig. 1a). By contrast, under the other three treatments the percentage of explants remaining at the initial stage at culture end was low, but it was significantly higher under IAA than under IBA + Kin and IBA alone (Fig. 1a). The percentage of explants with ARs followed an inverse trend, being very high under IBA + Kin (90%) and IBA (83%), without significant differences between the two, but under IAA it was significantly lower than IBA and IBA + Kin (Fig.1a). The AR production per explant was significantly higher under IBA alone than under IBA + Kin and IAA, whereas there was no significant difference between the latter two treatments (Fig. 1b). In addition, AR elongation and hair differentiation characterized both the IBA alone-treated explants and the IAA alone-treated ones, but the formation of calli was higher under IAA than IBA (Fig. 2a and b, in comparison). By contrast, ARs at the root primordium (ARP) stage were prevalently observed under IBA + Kin (Fig. 2c, arrow), and callus formation was higher than under IBA alone (Fig. 2a and c, in comparison). Interestingly, IBA alone treatment highly increased also the formation of LRs from the ARs in comparison with IBA + Kin (Additional file 1: Figure S1c-d). No HF-treated explant formed either ARs or macroscopic callus (Fig. 1a and Fig. 2a, inset).

To exclude possible differences in the AR response due to the genotype, the AR response of TCLs from different genotypes, i.e., Col-0, Col, and Col-gl1 was compared under IBA alone. No significant difference occurred in the mean number of ARs per explant (Additional file 1: Figure S1e). Thus, results from Figs. 1 and 2 show that 10 μM IBA was the best treatment for inducing AR formation in arabidopsis TCLs.

Despite the absence of auxin at the excision time, IAA is active in the IBA-treated TCLs during AR formation

The steady state levels of endogenous IAA and IBA were measured in the TCLs soon after excision from the inflorescence stem. Neither IAA nor IBA were detected in the explants, showing that, in accordance with previous results in tobacco TCLs [6], the arabidopsis TCLs did not contain any endogenous auxin content at the excision time.

The *DR5::GUS* line is a well-known reporter of the localization of IAA-induced gene expression in the AR-forming explants [7, 8, 27]. In entire portions of arabidopsis stem, the signal is occasional in the epidermal

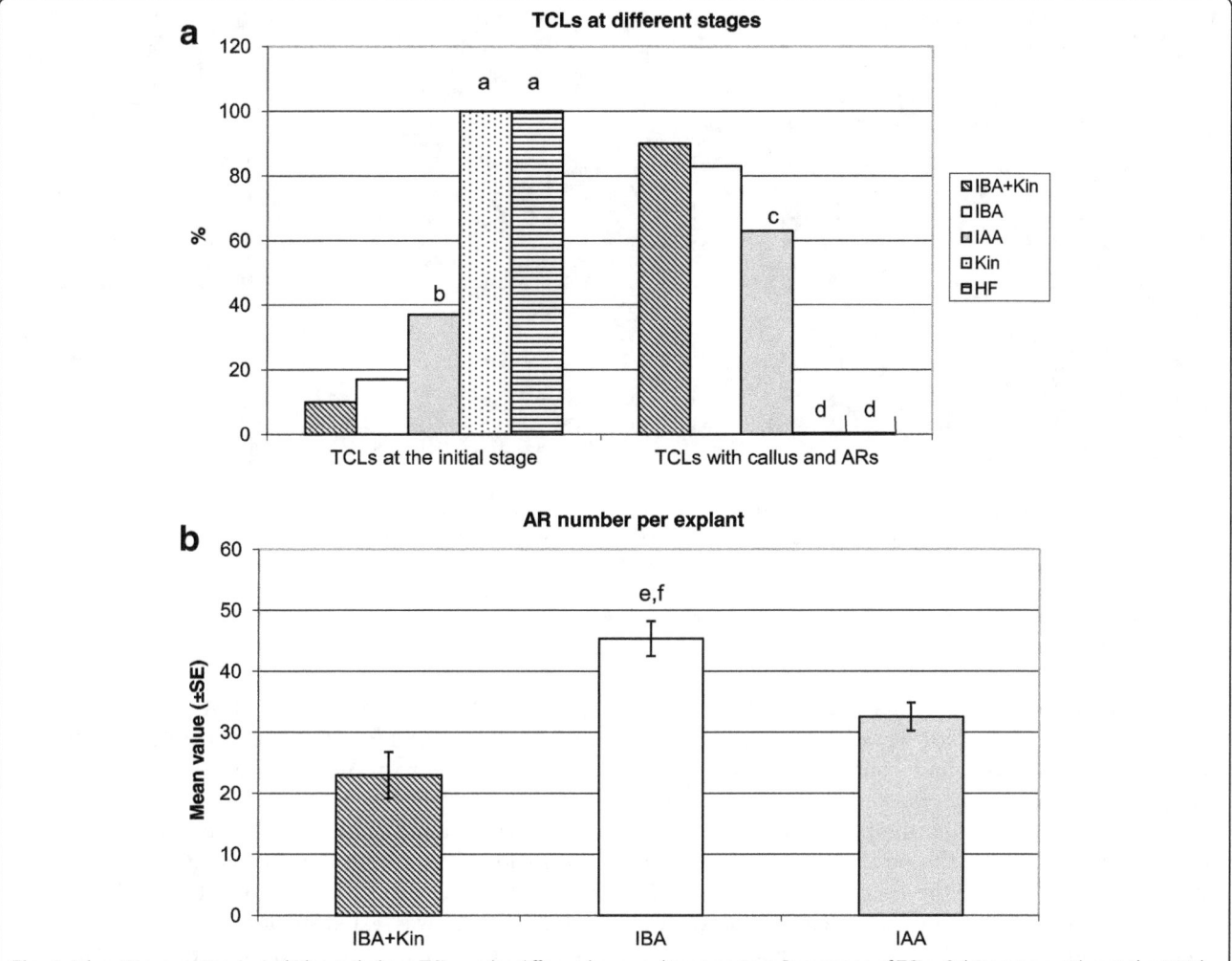

Fig. 1 Adventitious rooting in *Arabidopsis thaliana* TCLs under different hormonal treatments. **a** Percentage of TCLs, Col-0 ecotype, either at the initial stage or with macroscopic callus and ARs, after 15 days of culture without hormones (HF) or with IBA (10 μM) + Kin (0.1 μM), IBA (10 μM), IAA (10 μM) or Kin (0.1 μM). **b** Productivity of AR-forming TCLs evaluated as mean number (±SE) of ARs per TCL under either IBA (10 μM) + Kin (0.1 μM), or IBA (10 μM), or IAA (10 μM). [a,d] $P < 0.01$ difference with respect to the other treatments within the same developmental stage. [b,c] $P < 0.01$ difference with respect to IBA + Kin and IBA within the same developmental stage. [e] $P < 0.01$ difference with respect to IAA. [f] $P < 0.001$ difference with respect to IBA + Kin. Columns with no letter or the same letter within the same developmental stage are not significantly different. $N = 100$

and cortical cells before the culture [8, 27]. By contrast, under the same plant growth and culture conditions presently used, in IBA + Kin TCLs the signal has been reported to appear in the meristematic cell clusters formed by the stem endodermis, and to continue in the meristemoids and tips of ARPs and ARs [7]. Because there is no information about the *DR5*-driven IAA signal in the presence of IBA alone in the TCLs, it was monitored histologically. A slight IAA-signal was observed in the meristematic cell clusters produced by the stem endodermis at day 8, and in the meristemoids (Fig. 3a) and ARPs (Fig. 3b). At day 15, the signal was reinforced in the tips of the elongating and mature ARs, marking the quiescent centre and some initial and cap cells around (Fig. 3c-d). Coupled with the initial absence of any endogenous IAA content in the TCLs, the results

support that the IAA activity necessary for the AR process was totally dependent on the exogenous IBA input, sufficient per se to induce the IAA-signal specific for the AR process.

The AR-response of *ech2ibr10* TCLs demonstrates that exogenous IBA is converted into IAA to induce AR formation

Mutations in genes encoding enzymes specific for IBA-to-IAA conversion confer IBA-resistance without altering IAA-response [28]. Among these enzymes, coded by genes with single alleles, there are the enoyl-CoA hydratase IBR10 and the ENOYL-COA HYDRATASE2 (ECH2). The *ibr10-1* mutant is resistant to the inhibitory effects of IBA on the elongation of light-grown roots [29], and dark-grown hypocotyls [15]. Moreover, *ibr10-1*

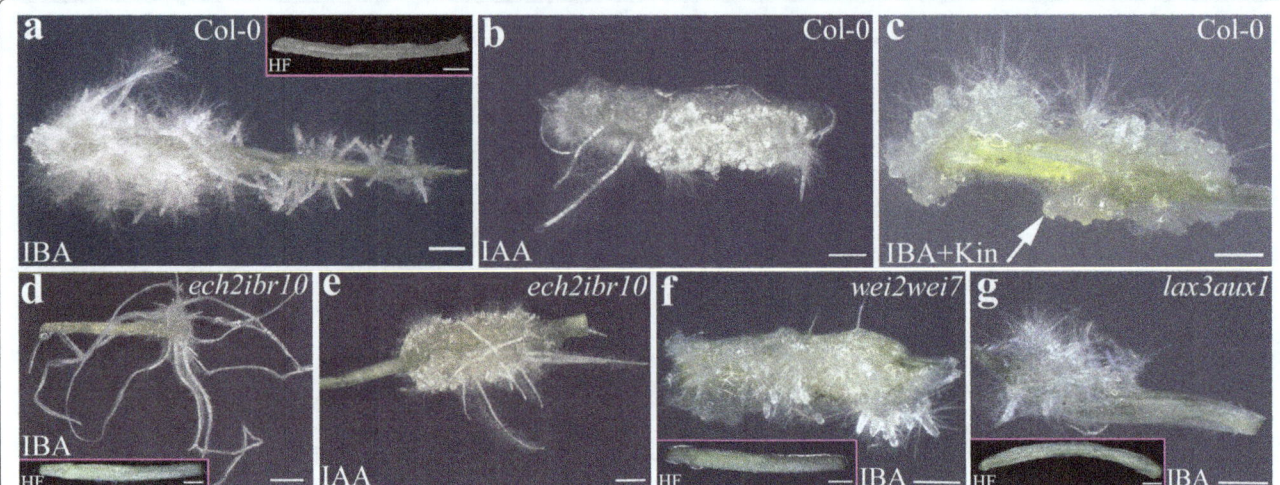

Fig. 2 Macroscopic adventitious rooting response on TCLs from various genotypes under different hormonal treatments. **a–g** Images under the stereomicroscope at the end of the in vitro culture (day 15) with IBA (10 μM) (**a**, **d**, **f**, **g**), IBA (10 μM) + Kin (0.1 μM) (**c**), IAA (10 μM) (**b**, **e**) or under HF (Insets in **a**, **d**, **f**, **g**). **a** Col-0 TCLs showing a poor callus formation and a lot of elongated ARs with hairs. **b** Col-0 TCLs with elongated hairy ARs, and high callus formation. **c** Col-0 TCLs with ARPs (*arrow*) and no elongated AR, and callus. **d** *ech2ibr10* TCLs with a poor number of highly elongated ARs, and a very reduced callus formation. **e** *ech2ibr10* TCLs with callus and elongated ARs. **f–g** *wei2wei7* (**f**) and *lax3aux1* (**g**) TCLs with a very few number of ARs which were not elongated. Insets in **a**, **d**, **f**, and **g** show the absence of AR formation in the HF-treated control explants. Bars = 1 mm (**a-c**, **e-g**, and insets in **a**, **d**, **f**, **g**), 2 mm (**d**)

produces dramatically fewer LRs than the WT in response to IBA [29]. Like *ibr10-1*, *eich2-1* mutant displays IBA-resistant hypocotyls and roots, and resistance is greatly enhanced when the two mutations are combined [15]. The synergism between the phenotypes of the two mutants is supported by the observation that, in comparison with the WT and the single mutants, *ech2ibr10* is unable to produce LRs in the absence of auxin exogenous treatments, and displays decreased auxin reporter activity [15]. About AR formation, it has been recently shown that *ech2ibr10* seedlings also show a reduced AR number in comparison with the WT in the presence of 10 μM IBA alone [9], whereas there is no information about *ech2ibr10* TCLs.

In the absence of hormones, the *ech2ibr10* TCLs did not form ARs (Fig. 2d, inset), as the WT TCLs. The mean number of ARs per IBA-cultured TCL was about 5-folds lower than in the WT ($P < 0.0001$, Fig. 4a, and Fig. 2d and a, in comparison). This AR reduction was similar to that occurring in *ech2ibr10* seedlings grown under the same IBA concentration and experimental conditions [9]. Mutant and WT TCLs treated with IAA (Fig. 2e and b) showed the same high number of ARs (Fig. 4a). Taken together, data show that exogenous IBA needs to be converted into IAA to exhibit its action on AR formation in TCLs, as *in planta*.

Both exogenous IAA and exogenous IBA induce the expression of the *ASA1* IAA-biosynthetic gene

To understand whether IBA had an effect on the IAA synthesis induced by the α subunit of the anthranilate synthase (WEI2/ASA1) [21], *ASA1* expression was evaluated in *ASA1::GUS* TCLs cultured with IBA alone, and the pattern compared with that obtained under IAA alone.

The expression pattern under either IAA alone or IBA alone showed that both the exogenous auxins were able to induce *ASA1* expression up to culture end (Fig. 3e-f), but the signal was extended to a wider portion of the callus under IAA in comparison with IBA (Fig. 3e-f), whereas the AR tips presented the same pattern of expression in both treatments (insets in Fig. 3e and f).

The histological analysis at day 8 of the IAA-treated TCLs showed that the expression started in the endodermis derivatives initiating both the AR process and the xylogenesis (Fig. 3h), and continued in the apical part of the forming ARPs (Fig. 3i). At day 15, the signal was diffused in the ARPs entrapped in the callus (Fig. 3j), but was also shown by the tips of the maturing ARs which were frequently fused (Fig. 3k). Moreover, the meristematic cells of the xylogenic nodules showed a faint expression (Fig. 3l).

In the presence of IBA, the expression pattern did not change in comparison with IAA alone (Fig. 3 m-p), however, at day 8, the signal in the endodermis derivatives was higher than with IAA alone (Fig. 3m and h, in comparison). In the elongating ARPs and in the ARs of day 15, the signal characterized the initial cells of the niche and the protodermis of the apical meristem, but faintly the columella (Fig. 3o-p). Xylogenesis sporadically occurred under IBA, but the forming xylary cells showed the same expression pattern as under IAA (data not shown).

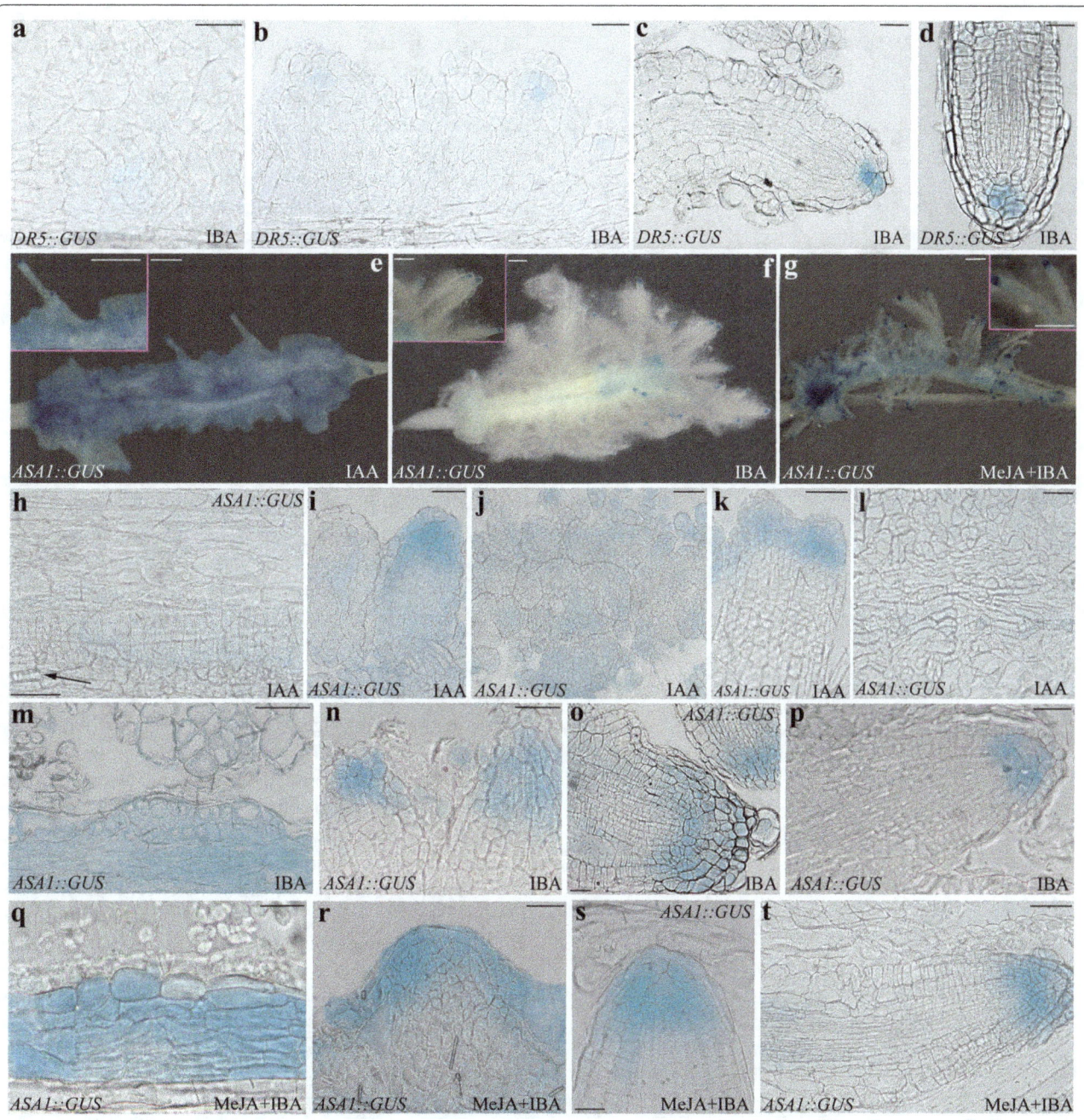

Fig. 3 IAA-driven *DR5::GUS* expression in IBA-cultured TCLs, and *ASA1* expression under IAA, IBA, and MeJA + IBA. **a–d** Expression in TCLs treated with IBA (10 µM) for 8 (**a-b**) and 15 days (**c-d**). **a-b** Beginning of the signal in early meristemoids (**a**) and in ARPs (**b**). **c-d** *DR5*-signal in the quiescent centre and some initial and cap cells in the apex of elongating (**c**) and mature (**d**) ARs. **e–t** *ASA1::GUS* expression. **e–g** Expression in TCLs observed under the stereomicroscope at day 15 under IAA (10 µM) (**e**), IBA (10 µM) (**f**), or MeJA (0.01 µM) + IBA (10 µM) (**g**), showing differences in signal intensity and localization among the treatments at explant level, but not in the AR apex (Insets). **h–l** Histological analysis of the expression at 8 (**h-i**) and 15 (**j-l**) day in IAA-treated TCLs. **h–i** Expression in the endodermis derivatives (**h**) and in de novo formed xylary cells (*arrow*), and in the apex of the developing ARPs (**i**). **j–k** Widespread expression in the ARPs entrapped in the callus (**j**), and in the apices of the frequently fused ARs (**k**). **l** Faint expression in the meristematic cells of the xylogenic nodules. **m–p** Histological analysis of the expression at 8 (**m-n**) and 15 (**o-p**) days in IBA-treated TCLs. **m–n** High expression in the endodermis derivatives (**m**), and in the apical part of the forming ARPs (**n**). **o–p** Signal in the initial cells of the niche and in the protodermis of the apex of the elongating ARPs (**o**) and ARs (**p**), with faint expression in the columella in both cases. **q–t** Histological analysis of the expression at 8 (**q-r**) and 15 (**s-t**) days in MeJA + IBA-treated TCLs. **q–r** Strong signal in the endodermis derivatives (**q**), and in the apical part of the forming ARPs (**r**). **s–t** High signal in the initial cells of the niche and protodermis of the apex of the elongating ARPs (**s**), and ARs (**t**), with a lower expression in the columella in both cases. Bars = 20 µm (**r**, **s**), 40 µm (**a-c**, **h**, **i**, **k-p**, **q**, **t**), 50 µm (**d**, **j**, **o**), 500 µm (**e-g** and Insets)

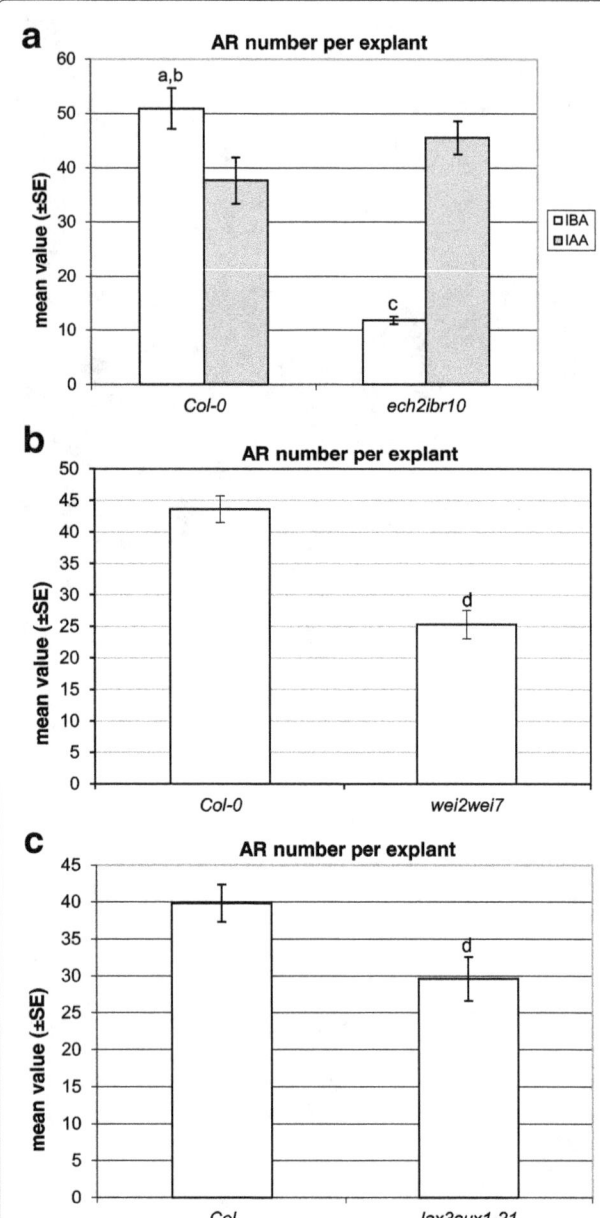

Fig. 4 Adventitious rooting on TCLs from various genotypes cultured with IBA (10 μM) or IAA (10 μM). **a** Mean number (±SE) of ARs per IBA- and IAA-cultured TCL of Col-0 and *ech2ibr10* at day 15. **b** Mean number (±SE) of ARs per IBA-cultured TCL of Col-0 and *wei2wei7* at day 15. **c** Mean number (±SE) of ARs per IBA-cultured TCL of Col and *lax3aux1* at day 15. [a], $P < 0.0001$ difference with respect to *ech2ibr10* within the same treatment; [b], $P < 0.05$ and [c], $P < 0.0001$ difference with respect to IAA within the same genotype; [d], $P < 0.01$ difference with respect to the WT (Col-0 in **b**, Col in **c**). Columns with no letter are not significantly different. $N = 100$

The induction of AR formation by exogenous IBA is reduced in the *wei2wei7* TCLs, supporting that ASA1/ASB activity is required for IBA-induced AR formation

For a deep insight into the action of exogenous IBA on AR formation from TCLs through the activity of WEI2/ASA1, but also of its isoform WEI7/ASB1, explants of

the *wei2wei7* mutant, blocked at the level of both genes [21], were treated with 10 μM IBA, and the response compared with WT TCLs.

The mean number of ARs per IBA-treated TCL was significantly reduced in the double mutant in comparison with the WT (Fig. 4b and Fig. 2f and a, in comparison).

Coupled with the localization and timing of IBA-induced *ASA1* expression (Fig. 3m-p) and the *eich2ibr10* response (Fig. 4a), results demonstrate that exogenous IBA enhances AR formation in the TCLs through its conversion into IAA, with this event related to a stimulation of IAA biosynthesis by ASA1/ASB1.

Exogenous IBA induces AUX1- and LAX3-mediated IAA-influx, and PIN1-mediated IAA-efflux in the AR-forming TCLs

The expression of *PIN1*, and of *AUX1* and *LAX3* genes, was analysed in the IBA-cultured TCLs.

PIN1 was expressed in a wide population of the endodermis derivatives (Fig. 5a), and at the base and along the procambium in the early ARPs (Fig. 5b). Moreover, in the elongating ARPs, *PIN1* was expressed in the differentiating vascular system and in the central part of the apical meristem (Fig. 5c and d). In the mature ARs, *PIN1::GUS* signal was present in the vasculature, and faintly at the apex (Fig. 5e and inset). Altogether the expression pattern of *PIN1* under IBA alone repeated that previously observed under IBA + Kin under experimentally comparable conditions [7], suggesting that IBA does not need Kin for causing IAA cellular efflux by PIN1 during AR formation in arabidopsis TCLs.

LAX3 expression started in the meristematic cell clusters initiating the AR process (Fig. 5f-g), and continued at the base of the developing ARPs (Fig. 5h). In the elongated ARPs, the signal was present in the forming procambium and in a few apical cells (Fig. 5i). The expression signal remained in the vasculature of the maturing ARs (Fig. 5j). The expression pattern of *LAX3* did not differ from that previously observed under IBA + Kin with the same experimental conditions [7, 8].

The expression of *AUX1* started in the meristematic cell clusters, and was uniformly observed in all the cells of the early primordia and developing ARPs (Fig. 5k-l). The expression pattern changed in the elongated ARPs, because the signal was localized in the cap, protodermis, and developing procambium at ARP base (Fig. 5m). In the mature ARs, *AUX1* expression characterized the cap, the protodermis, and, faintly, some niche cells (Fig. 5n). The signal reappeared in the procambium of the differentiation zone (Fig. 5n, arrow), and, mainly, in the vascular parenchyma of the primary structure zone (Fig. 5o). Also in the case of *AUX1* there was no substantial difference in the expression pattern between the IBA-

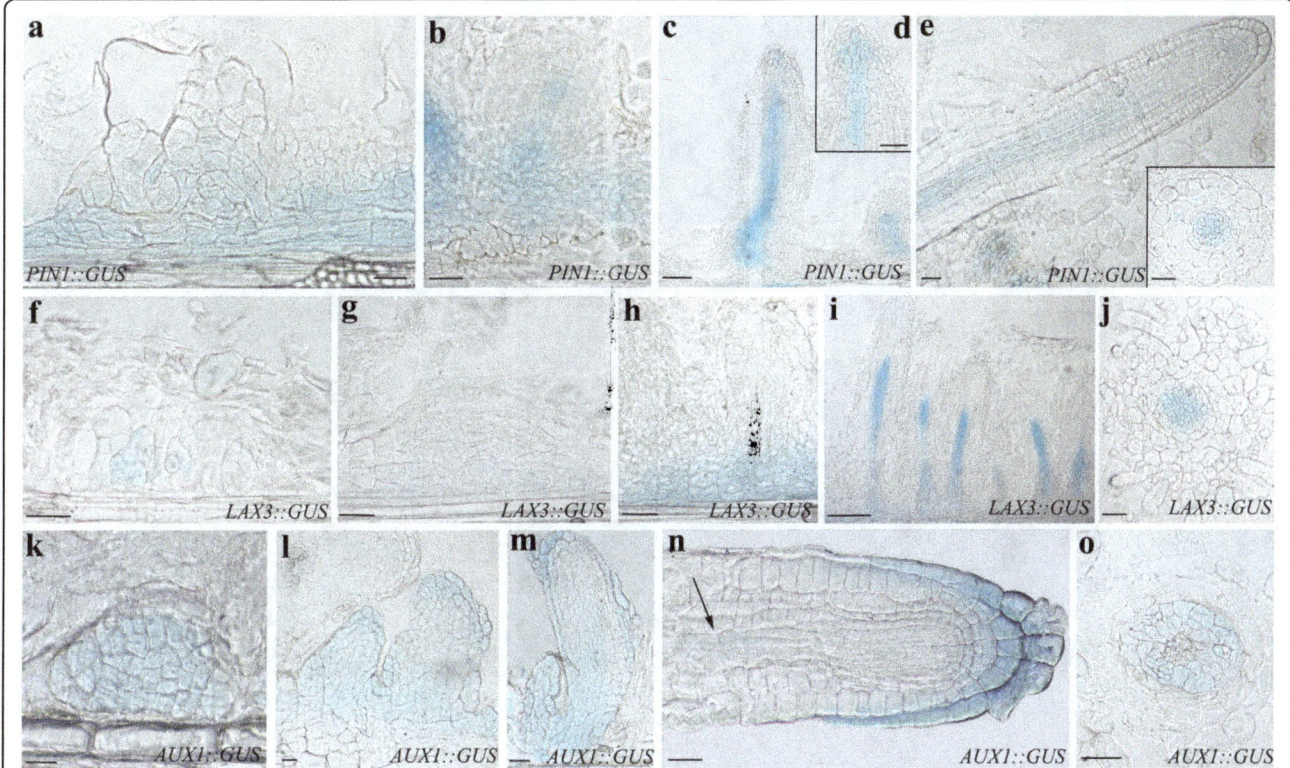

Fig. 5 Expression pattern of *PIN1*, *LAX3*, and *AUX1* during AR formation in IBA-cultured TCLs at day15. **a–e** *PIN1::GUS* expression. **a** Signal in a wide population of endodermis derivatives and in meristemoids. **b–c** Signal in the basal part of young ARPs (**b**), all along the developing vasculature (**c**), and in the central cells of the apex of elongating ARPs (**d**). **e** Expression in the vasculature (Inset), and faintly in the apex of mature ARs. **f–j** *LAX3::GUS* expression. **f–g** Onset of expression in the meristematic cell clusters formed by the endodermis. **h** Expression at the base of the differentiating ARPs (**h**). **i–j** Strong *LAX3* signal in the procambium (**i**), and vasculature of the maturing ARs (**j**). **k–o** *AUX1::GUS* expression. **k–l** Uniform signal in the meristematic cell clusters formed by the endodermis (**k**), and in early primordia (**l**). **m** Signal in the cap, protodermis, and developing procambium in the elongating ARPs. **n** *AUX1* expression in the cap, protodermis, and faintly in the niche and procambium (*arrow*) in a mature AR. **o** Expression in the vascular parenchyma of the AR primary structure zone. **a–e**, **f-i**, **k-n**, longitudinal sections, **j**, **o**, and Inset in **e**, transections. Bars = 40 μm (**a**, **b**, **f-k**, **n**, **o**), 50 μm (**c-e**, **l**, **m**, and Inset in **e**)

alone-treated TCLs and the previously examined IBA + Kin-cultured ones ([8], and present results).

To further support the importance of an IAA-influx mediated by the exogenous IBA in the target cells of the AR process, the response of the TCLs from *lax3aux1* double mutant was investigated, and compared with that occurring under IAA alone.

The IBA-treated TCLs of this double mutant showed an AR response significantly reduced in comparison with the WT both as percentage of AR-forming explants (40%, $P < 0.01$ difference with the WT) and as mean number of macroscopic ARs per explant (Figs. 2g and 4c).

The results support that exogenous IBA activates the IAA influx by AUX1 and LAX3 in the AR forming WT-TCLs. It cannot be excluded that the IBA-promotion of the activity of these IAA-transporters involved NO formation.

Nitric oxide and methyl jasmonate enhance AR-formation in IBA-cultured TCLs

Nitric oxide is an IAA downstream signal, and an early by-product of IBA-to-IAA conversion [16]. Moreover, it

is known to be positively involved in AR formation [17]. At 48 h, NO presence under IBA treatment was detected in a wider number of cells than under IAA (Fig. 6a-b and c-d, in comparison), but with a similar localization (cells of the deepest explant layers). At day 3, the difference of the epifluorescence signal became more evident between the auxin treatments. In fact, a lot of the endodermis derivatives showed the signal in the presence of IBA (Fig. 6e), whereas epifluorescence remained in scattered cells in the presence of IAA (Fig. 6f). Interestingly, at day 6, in the presence of exogenous IBA, entire layers of derivatives of the stem endodermis appeared green fluorescent, the same as the first formed ARPs (Fig. 6g-h). By contrast, only scattered cells or thin-layered endodermis derivatives were epifluorescent in the presence of exogenous IAA (Fig. 6i-j).

The intensity of green fluorescence was also quantified, and significant ($P < 0.0001$) increases occurred in the presence of IBA alone in comparison with IAA alone at both day 2 and day 3 of culture (Fig. 6k), supporting the microscopic observations.

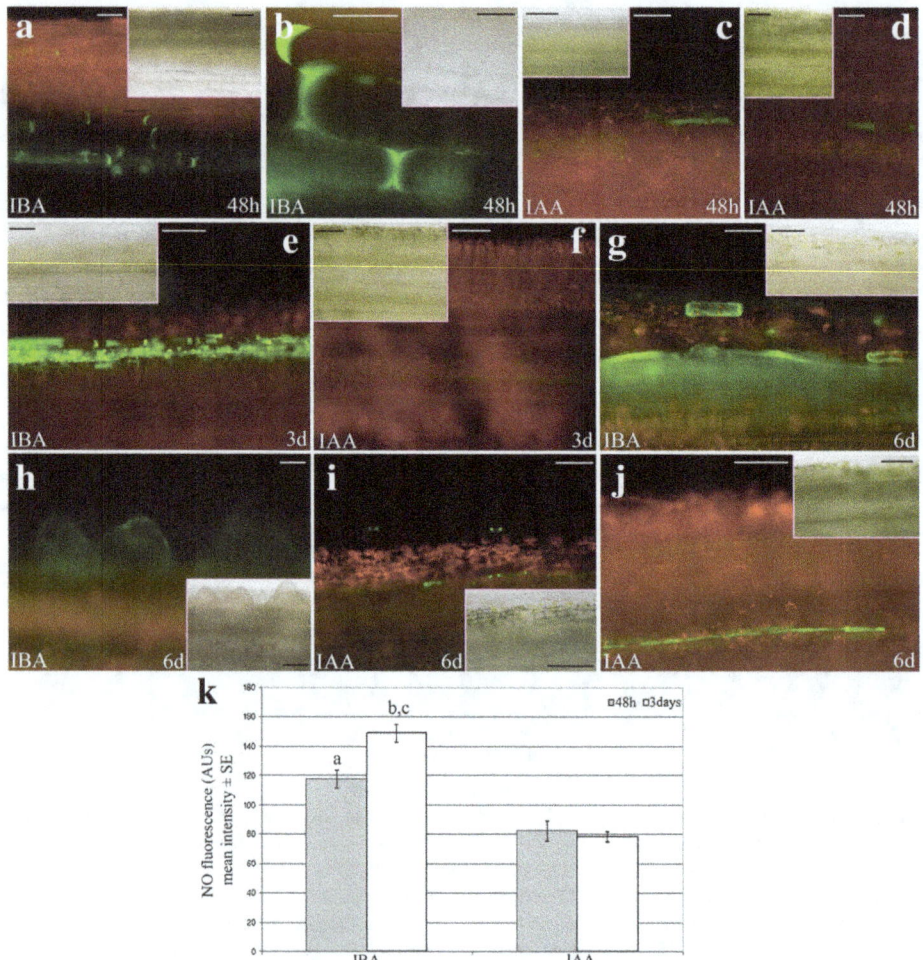

Fig. 6 Detection and quantification of the epifluorescence signal caused by NO in IBA- or IAA-cultured TCLs. **a–d** Presence of the epifluorescence signal (*green colour*) at 48 h in cells of the deepest layers of TCLs cultured with IBA (10 µM) (**a-b**), or IAA (10 µM) (**c-d**). **e** Numerous endodermis derivative cells showing the NO green signal in TCLs cultured with IBA for 3 days. **f** Rare cells with a faint signal in the deepest layers of the explant in the presence of IAA at day 3. **g** Detail of the numerous layers of the endodermis derivatives showing the green epifluorescence at day 6 (IBA treatment). **h** Presence of the green signal in the first formed ARPs (day 6, IBA treatment). **i–j** Very faint signal in scattered cells (**i**), and in thin-layered endodermis derivatives (**j**) of the explant at day 6 (IAA treatment). TCL longitudinal views. The same images under light microscopy are shown in the Insets. **k** Mean intensity (±SE) of NO fluorescence (AUs) in TCLs cultured with either IBA (10 µM) or IAA (10 µM) for 48 h and 3 days. [a,b], $P < 0.0001$ difference with IAA within the same culture time. [c], $P < 0.001$ difference with the other culture time within the same treatment. Columns with no letter are not significantly different. $N = 200$. Bars = 50 µm (**b, c, e, g–j** and Insets in **b, f, g, i**), 70 µm (**a, d, f**, and Insets in **c, e, h, j**), 100 µm (Insets in **a** and **d**)

Nitric oxide is known to activate genes of the JA biosynthesis [22], and MeJA treatments are known to enhance *ASA1* and *ASB1* expression [23].

To obtain information about NO downstream signals affecting AR formation in the TCLs, MeJA was applied at 0.01 µM [6] in combination with IBA (10 µM). At day 15, the treatment resulted into a significant ($P < 0.01$) increase in the mean number of ARs per TCL in comparison with IBA alone, i.e., 50 ± 2.5 and 40 ± 2.4, respectively. Moreover, *ASA1* expression signal increased in the TCLs under 0.01 µM MeJA + IBA in comparison with IBA alone (Fig. 3 f and g), without changing the expression pattern during the entire AR process (Fig. 3q-t and m-p, in comparison).

Altogether, results support a positive involvement of jasmonates, possibly formed downstream to NO, on ASA1/ASB1 expression/activity during IBA-mediated AR formation in TCLs.

Discussion

Results showed that exogenous IBA alone induced AR-formation in arabidopsis TCLs. The AR-process was totally under the control of this exogenous auxin because the TCLs were devoid of IAA and IBA at culture onset. However, IBA needed to be converted into IAA to give AR formation, and favoured IAA-transport by PIN1, AUX1 and LAX3, and ASA1/ASB1-mediated IAA-biosynthesis. The latter two roles seemed to involve the action of the NO formed during the conversion process.

The IAA-precursor IBA is the main player of the AR process positively affecting the biosynthesis of IAA, but its action is indirect

IBA is present in numerous plant species, in which it represents a variable percentage of the total pool of auxins. In arabidopsis seedlings, IBA levels are low, and differ depending on growth conditions and detection methodologies [30, 31]. However, IBA levels increase *in planta* when the AR process occurs. In fact, in arabidopsis AR forming hypocotyls of dark-grown seedlings, both IAA and IBA are present, with IBA levels about 10% of IAA levels. When exogenous IBA is applied at 10 μM, the endogenous IBA level triplicates, whereas endogenous IAA doubles [9]. This result shows that, in arabidopsis, as in other plants/culture systems, exogenous IBA is converted into endogenous IAA, and acts as source of IAA [12, 32, 33].

It has been suggested that the ASA1/ASB1-system functions when the auxin biosynthetic pathway is hyperactive [21]. In accordance, the endogenous IAA deficiency at the excision time presently observed in the TCLs might trigger a feedback loop, with exogenous IAA rapidly inducing its own biosynthesis via ASA1, as confirmed by the observed *ASA1::GUS* signal (Fig. 3h-l). This biosynthetic activity might cause an initial rise in the endogenous IAA, which might be useful for rapidly inducing both AR-formation and xylogenesis. In fact, also the latter program is auxin-inducible [34], and uses auxin to form xylary cells in competition with AR formation ([3], and references therein). The xylogenic response, frequently observed in the TCLs treated with IAA (Fig.3 h,l), but occasional in those treated with IBA, might explain the final reduction in AR formation occurring in the former treatment in comparison with the latter (Fig. 1).

By the use of the *DR5::GUS* system, it has been demonstrated that in arabidopsis stem cuttings exposed to IBA, the GUS signal appears, and is mainly associated with the root initiation sites [27]. Also in arabidopsis TCLs cultured with IBA + Kin for 14 days under the same experimental conditions presently used, the *GUS* signal characterizes the cells initiating the AR process, but is also observed in the de novo formed ARPs and ARs [7, 8]. Present results show that this pattern also occurs in the TCLs cultured with IBA alone, supporting that this exogenous auxin is able per se to induce IAA biosynthesis in the explants.

Recent transcriptome analyses of IBA-induced AR formation in *Camellia sinensis* cuttings and mung bean seedlings have allowed the identification of a lot of differentially expressed genes, and mainly genes involved in auxin homeostasis and signalling [18, 19]. However, no *ASA1* expression has been shown to be activated by IBA. By contrast, by the analysis of the expression pattern of

ASA1 under IBA alone (Fig. 3m-p), and the highly-reduced AR response of *wei2wei7* TCLs under the same treatment (Fig. 4b), it is presently shown that exogenous IBA induces *ASA1*, and the rooting promotion by IBA requires ASA1/ASB1. IBA conversion to IAA is catalysed by the action of peroxisomal ß-oxidation enzymes, e.g., IBR10 and ECH2 [15]. The highly-reduced AR-response observed in the *eich2ibr10* TCLs (Fig. 4 a) supports that the peroxisomal IBA-to-IAA conversion occurs in the IBA-treated TCLs. This conversion occurs also during AR formation in arabidopsis *in planta* [9], and during LR formation in arabidopsis and *Zea mays* [16]. In the latter study, it has been demonstrated that the conversion of IBA into IAA is followed by peroxisomal NO formation, and that the spatially and temporally coordinated release of NO and IAA from peroxisomes is the causative agent of the promotion of LR formation [16]. Nitric oxide also mediates the auxin response leading to AR formation [17, 35]. Not only IBA, but also IAA uses NO as downstream signal for LR formation, however, peroxisomes accumulate more NO under IBA treatment than under IAA ([16], and references therein). Present results show that this is also the case for AR formation from TCLs. In fact, an earlier and enhanced detection of NO occurred in the IBA-cultured explants in comparison with the IAA-cultured ones (Fig. 6a-d, and k), supporting an important involvement of the NO coming from IBA-to-IAA conversion. Taken together, NO might be a messenger in IBA-induced AR formation.

An interaction of NO with auxin synthesis and transport has been reported [36]. Moreover, NO is known to activate *Allene oxide synthase* (*AOS*) and *lypoxygenase2* (*LOX2*) genes involved in JA biosynthesis [22]. Jasmonates induce AR formation in arabidopsis (present results) and tobacco TCLs [6], when combined with exogenous IBA and IBA + Kin, respectively. ASA1 is required for the JA-induced IAA biosynthesis necessary to LR formation in arabidopsis [23]. Present results showed that MeJA, combined with IBA, enhanced *ASA1* expression in comparison with IBA alone without changing the expression pattern of the gene during the AR process (Fig. 3 q-t, and m-p, in comparison). Consequently, ASA1 might be an interaction node through which jasmonate integrates its action with auxin to regulate AR formation. In our hypothesis, the NO formed during the IBA-to-IAA peroxisomal conversion might induce *AOS* and *LOX2*, involved in JA biosynthesis, and the produced JA might induce ASA1/ASB1 expression/activity, increasing the IAA content coming from conversion, leading to the endogenous IAA pool necessary for AR formation (Fig. 7).

Only hypotheses may be advanced about the auxin signalling and perception necessary to successful AR formation in TCLs. It is known that genes that are up-

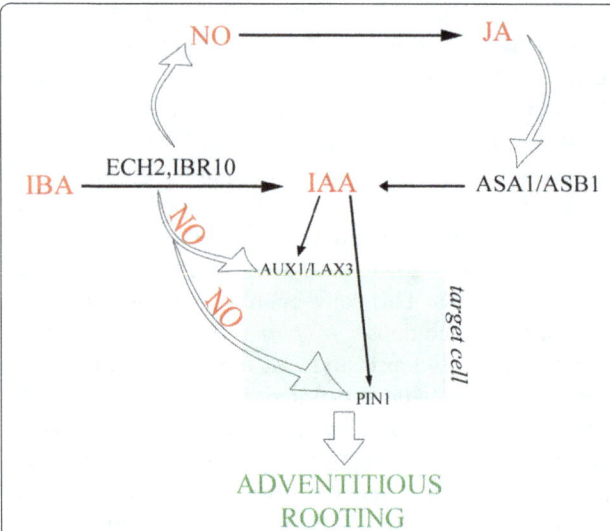

Fig. 7 Model explaining the promotion by exogenous IBA (10 μM) of AR formation in arabidopsis TCLs. Nitric oxide (NO) formed during the exogenous IBA-to-IAA conversion by ECH2/IBR10 induces the synthesis of JA, which, in turn, induces ASA1/ASB1 activity. The IAA, coming from IBA conversion and biosynthesis by ASA1/ASB1, is transported into the target cells of the rhizogenic process by the efflux carrier PIN1 and the influx carriers AUX1 and LAX3. NO might also positively affect PIN1 and AUX1, enhancing the endogenous IAA transport required for adventitious rooting. (See the text for further explanations)

regulated or down-regulated by auxin contain auxin response elements (AuxREs, 5′ tgtctc 3′) in their promoters, which bind transcription factors of the auxin response factor (ARF) family [37]. At high auxin levels, the ARFs become active because released by their repressors, the Aux/IAAs proteins, when the latter are degraded after interaction with the SCF$^{TIR1/AFB}$ complex [38]. In intact hypocotyls of de-etiolated arabidopsis seedlings, AR initiation is controlled by a balance between the negative AR regulator ARF17 and the positive AR regulators ARF6 and ARF8, with these three ARFs controlling each other's expression at both transcriptional and post-transcriptional level [39]. In contrast to ARF17, ARF6 and ARF8 positively affect the auxin-inducible genes GH3.3, GH3.5, and GH3.6, required for fine-tuning AR initiation by modulating JA homeostasis [40]. These results suggest a regulatory pathway at the crosstalk of IAA and JA, in which ARF6, ARF8, and ARF17 and their downstream targets GH3.3, GH3.5, and GH3.6 are involved. The same pathway might be active in the dark-grown AR-forming TCLs, with perhaps also NO involved. Nitric oxide might mediate auxin signalling via modification of the TIR1/AFB-Aux/IAA-ARF interaction. In fact, NO is known to enhance auxin signalling via S-nitrosylation of the auxin receptor protein TIR1, thereby facilitating Aux/IAA degradation [41]. Our preliminary results about the expression patterns of these *ARF* and *GH3* genes in dark-grown arabidopsis

TCLs cultured with IBA + Kin support the hypothesis (Fattorini et al., unpublished results), which however needs to be confirmed by the response of the null mutants.

The IAA-precursor IBA is the main player of the AR process positively affecting IAA transport, but its action is indirect

It has been suggested that AUX1 recognizes endogenous IAA and not IBA, whereas IBA may be a substrate of LAX3, at least in arabidopsis hypocotyls [42]. During AR formation in entire dark-grown seedlings, *AUX1* expression appears at the onset of the AR process, and continues during ARP formation, and in the ARs, but the pattern does not change under HF condition in comparison with IBA + Kin treatment, and remains the same also in IBA + Kin-treated TCLs [8]. *LAX3* expression is enhanced in the seedlings by IBA + Kin in comparison with the HF treatment, but also in this case there is no change in the IBA + Kin-induced pattern *in planta* in comparison with TCLs [7]. Present data support that the exogenous Kin does not affect the expression pattern of both these IAA-influx carriers in the arabidopsis TCLs, because IBA alone (Fig. 5f-o) did not induce any significant change in the expression pattern in comparison with previous results with IBA + Kin under comparable conditions [7, 8]. In tobacco TCLs [11], Kin had been supposed to act synergistically with IBA to induce the mitotic activity necessary for callus formation and meristemoid growth, and for this reason it had been then used for arabidopsis TCL culture [10]. By contrast, present results show that Kin is not necessary to arabidopsis TCLs, because the AR meristemoids are formed with IBA alone.

Moreover, present data also show a post-transcriptional role of exogenous IBA when applied alone, because the knockout of both *AUX1* and *LAX3* IAA-carrier genes caused a reduced AR response in comparison with the WT (Fig. 4c). Previous data show that TCLs from *aux1* mutant treated with IBA + Kin also have a reduced AR response in comparison with the WT, whereas this does not occur in *lax3* ones [8, 20], collectively suggesting a pivotal role for AUX1, independent/partially dependent on the exogenous hormonal treatment, in early controlling the AR process.

PIN proteins are encoded by a multigene family, with high homology among species. In *Medicago truncatula*, IAA treatments increase *MtPIN1* and *MtPIN2* expression, up-regulate most of the *PINs* in rice [43, 44], and positively affect *PIN1* promoter activity in arabidopsis PR [45]. Coupled with the inhibition of AR formation reported for *pin1* de-rooted seedlings [46], and the reduced AR response of *lax3/aux1* IBA-alone-treated seedlings, and IBA + Kin-treated TCLs [8, 9], the present results about *PIN1* and *AUX1* expression (Fig.

5a-e, and k-o) and *lax3aux1* response (Fig. 4c) suggest that both AUX1 and PIN1 are activated by exogenous IBA in arabidopsis AR formation. However, IBA action would occur by the IAA coming from the IBA-to-IAA conversion. Following this hypothesis, in addition to the IAA formed by conversion, a by-product of the same conversion process, e.g., NO, would be another regulator of the action of AUX1 and PIN1. Post-translational modification, such as protein phosphorylation, is crucial for many aspects of functional biology of plant proteins, with NO as an important regulator ([47], and references therein). By a quantitative phosphoproteomics analysis of NO responsive phosphoproteins in cotton leaf, it has been recently demonstrated that both PIN1 and AUX1 are activated by NO-mediated phosphorylation [47]. Taken together, in the AR-forming IBA-cultured TCLs, the NO formed during the IBA-to-IAA conversion might not only affect the jasmonate-induced ASA1 expression/activity, but also enhance the endogenous IAA transport by phophorylation of both PIN1 and AUX1 (Fig. 7).

Conclusions

IBA is the main player of adventitious rooting in arabidopsis TCLs, and possibly in many other culture systems and species characterized by very low endogenous auxin contents. IBA acts by conversion into IAA, and by enhancing IAA biosynthesis and transport. The nodal point of its action is the regulation of the endogenous IAA pool. IBA-regulation of IAA homeostasis involves the activity of other compounds downstream to its peroxisomal conversion, NO and jasmonates. The relationship of IBA with NO and jasmonates, and the downstream auxin signalling and perception, needs further investigation. Even if useful for planning experiments to overcome the rooting recalcitrance of species of economic value, the main implication of the findings is to help in understanding the mechanism by which IBA controls the natural process of adventitious rooting,

Additional file

Additional file 1: Figure S1. NO-buffer controls, lateral-rooting on ARs of IBA- and IBA + Kin-cultured TCLs, and AR-formation on IBA-treated TCLs from different ecotypes. Description of data: (**a-b**) Autofluorescence signal in the explants after buffer (HEPES/NaOH) alone treatment at 48 h of culture and at day 3, under IBA (10 μM). The same images under light microscopy are shown in the Insets. (**c**) Percentage of ARs with lateral root (LR) formation on *Arabidopsis thaliana* TCLs, Col-0 ecotype, at the end (day 15) of culture on IBA (10 μM) + Kin (0.1 μM) or IBA (10 μM) alone. [a], $P < 0.01$ difference with IBA + Kin, $N = 260$. (**d**) Detail of a TCL from Col-0 ecotype at day 15 of culture on IBA, showing an AR with two LRs (*arrows*). (**e**) Mean number (±SE) of ARs per TCL from Col-0, Col, and Col-gl1 ecotypes after 15 days of culture with IBA (10 μM). $N = 100$. Columns with no letter are not significantly different. Longitudinal views (**a-b**), and stereomicroscope image (**d**). Bars = 20 μm (**a** and Inset in **a**), 40 μm (**b** and Inset in **b**), 200 μm (**d**).

Abbreviations

AOS: Allene Oxide Synthase; AR: Adventitious root; ARF: Auxin response factor; ASA1: ANTHRANILATE SYNTHASE-alpha1; ASB: ANTHRANILATE SYNTHASE-beta1; AUX1: AUXIN RESISTANT1; *ECH2*: *ENOYL-COA HYDRATASE2*; HF: Hormone free; IAA: Indole-3-acetic acid; IBA: Indole-3-butyric acid; *IBR10*: *INDOLE-3-BUTYRIC ACID RESPONSE10*; JA: Jasmonic acid; Kin: Kinetin; LAX3: LIKE AUXIN RESISTANT3; LOX2: Lypoxygenase 2; LR: Lateral root; MeJA: Methyl jasmonate; NO: Nitric oxide; PIN1: PIN-FORMED1; PR: Primary root; TCLs: Thin cell layers; WEI2: WEAK ETHYLENE-INSENSITIVE2; WEI7: WEAK ETHYLENE-INSENSITIVE7; WT: Wild type

Acknowledgements

We thank the colleagues who provided the seeds of the transgenic lines/mutants, namely Sabrina Sabatini (Sapienza University Rome), Stefano Bencivenga (University of Milan), Malcom Bennett (University of Nottingham), and Bonnie Bartel (Rice University Huston).

Funding

Sapienza Università di Roma, grant number C26H157ANK and RP116154C3D60B9D to MMA. Funds were used for the experimental design, the analysis of data, and the open access payment.

Authors' contributions

AV carried out the TCL experiments. LF contributed to NO detection and carried out the quantification and the statistical evaluation of the results. FDR performed the histological and histochemical examination of the samples. SDA performed the hormone quantification. GF analyzed and interpreted the data. MMA planned the experiments, interpreted the data, and was a major contributor in writing the manuscript. All authors read and approved the final manuscript.

Competing interests

The authors declare that they have no competing interests.

References

1. Fahn A. Plant Anatomy. 4th ed. Oxford: Pergamon Press; 1990.
2. Hartmann HT, Kester DE, Davies FT. Plant propagation: principles and practices. 5th ed. Englewood Cliffs: Prentice-Hall; 1990.
3. Ricci A, Rolli E, Brunoni F, Dramis L, Sacco E, Fattorini L, et al. 1,3-di(benzo[d]oxazol-5-yl) urea acts as either adventitious rooting adjuvant or xylogenesis enhancer in carob and pine microcuttings depending on the presence/absence of exogenous indole-3-butyric acid. PCTOC. 2016;126:411–27.
4. Simon S, Petrášek J. Why plants need more than one type of auxin. Plant Sci. 2011;180:454–60.
5. Altamura MM, Falasca G. Adventitious rooting in model plants and in *in vitro* systems: an integrated molecular and cytohistological approach. In: Niemi K, Scagel C, editors. Adventitious root formation of forest trees and horticultural plants-from genes to applications. Kerala: Research Signpost; 2009. p. 123–44.
6. Fattorini L, Falasca G, Kevers C, Mainero Rocca L, Zadra C, Altamura MM. Adventitious rooting is enhanced by methyl jasmonate in tobacco thin cell layers. Planta. 2009;231:155–68.

7.   Della Rovere F, Fattorini L, D'Angeli S, Veloccia A, Falasca G, Altamura MM. Auxin and cytokinin control formation of the quiescent centre in the adventitious root apex of Arabidopsis. Ann Bot. 2013;112:1395–407.

8.   Della Rovere F, Fattorini L, D'Angeli S, Veloccia A, Del Duca S, Cai G, et al. Arabidopsis SHR and SCR transcription factors and AUX1 auxin influx carrier control the switch between adventitious rooting and xylogenesis in planta and in in vitro cultured thin cell layers. Ann Bot. 2015;115:617–28.

9.   Veloccia A, Fattorini L, Della Rovere F, Sofo A, D'Angeli S, Betti C, et al. Ethylene and auxin interaction in the control of adventitious rooting in Arabidopsis thaliana. J Exp Bot. 2016;67:6445–58.

10. Falasca G, Zaghi D, Possenti M, Altamura MM. Adventitious root formation in Arabidopsis thaliana thin cell layers. Plant Cell Rep. 2004;23:17–25.

11. Tran Thanh Van M, Dien NT, Chlyah A. Regulation of organogenesis in small explants of superficial tissue of Nicotiana tabacum L. Planta. 1974;119:149–59.

12. Ludwig-Müller J, Vertocnik A, Town CD. Analysis of indole-3-butyric acid-induced adventitious root formation on Arabidopsis stem segments. J Exp Bot. 2005;56:2095–105.

13. Korasick DA, Enders TA, Strader LC. Auxin biosynthesis and storage forms. J Exp Bot. 2013;64:2541–55.

14. Sauer M, Robert S, Kleine-Vehn J. Auxin: simply complicated. J Exp Bot. 2013;64:2565–77.

15. Strader LC, Wheeler DL, Christensen SE, Berens JC, Cohen JD, Rampey RA. Multiple facets of Arabidopsis seedling development require indole-3-butyric acid-derived auxin. Plant Cell. 2011;23:984–99.

16. Schlicht M, Ludwig-Müller J, Burbach C, Volkmann D, Baluska F. Indole-3-butyric acid induces lateral root formation via peroxisome-derived indole-3-acetic acid and nitric oxide. New Phytol. 2013;200:473–82.

17. Pagnussat GC, Simontacchi M, Puntarulo S, Lamattina L. Nitric oxide is required for root organogenesis. Plant Physiol. 2002;129:954–6.

18. Wei K, Wang L-Y, Wu L-Y, Zhang C-C, Li H-L, Tan L-Q, et al. Transcriptome analysis of indole-3-butyric acid-induced adventitious root formation in nodal cuttings of Camellia sinensis (L.). PLoS One. 2014;9(9):e107201.

19. Li S-W, Shi R-F, Leng Y, Zhou Y. Transcriptomic analysis reveals the gene expression profile that specifically responds to IBA during adventitious rooting in mung bean seedlings. BMC Genomics. 2016;17:43.

20. Della Rovere F, Fattorini L, Ronzan M, Falasca G, Altamura MM. The quiescent centre and the stem cell niche in the adventitious roots of Arabidopsis thaliana. Plant Signal Behav. 2016;11(5):e1176660.

21. Stepanova AN, Hoyt JM, Hamilton AA, Alonso JM. A link between ethylene and auxin uncovered by the characterization of two root-specific ethylene-insensitive mutants in Arabidopsis. Plant Cell. 2005;17:2230–42.

22. Mira MM, Wally OSD, Elhiti M, El-Shanshory A, Reddy DS, Hill RD, et al. Jasmonic acid is a downstream component in the modulation of somatic embryogenesis by Arabidopsis class 2 phytoglobin. J Exp Bot. 2016;67:2231–46.

23. Sun J, Xu Y, Ye S, Jiang H, Chen Q, Liu F, et al. Arabidopsis ASA1 is important for jasmonate-mediated regulation of auxin biosynthesis and transport during lateral root formation. Plant Cell. 2009;21:1495–511.

24. Murashige T, Skoog F. A revised medium for rapid growth and bio assays with tobacco tissue cultures. Physiol Plant. 1962;15:473–97.

25. Willemsen V, Wolkenfelt H, de Vrieze G, Weisbeek P, Scheres B. The HOBBIT gene is required for formation of the root meristem in the Arabidopsis embryo. Development. 1998;125:521–31.

26. Chen WW, Yang JL, Qin C, Jin CW, Mo JH, Ye T, et al. Nitric oxide acts downstream of auxin to trigger root ferric-chelate reductase activity in response to iron deficiency in Arabidopsis. Plant Physiol. 2010;154:810–9.

27. Welander M, Geier T, Smolka A, Ahlman A, Fan J, Zhu L-H. Origin, timing, and gene expression profile of adventitious rooting in Arabidopsis hypocotyls and stems. Am J Bot. 2014;101:255–66.

28. Strader LC, Bartel B. Transport and metabolism of the endogenous auxin precursor indole-3-butyric acid. Mol Plant. 2011;4:477–86.

29. Zolman BK, Martinez N, Millius A, Adham AR, Bartel B. Identification and characterization of Arabidopsis indole-3-butyric acid response mutants defective in novel peroxisomal enzymes. Genetics. 2008;180:237–51.

30. Ludwig-Müller J, Sass S, Sutter EG, Wodner M, Epstein E. Indole-3-butyric acid in Arabidopsis thaliana I. Identification and quantification. Plant Growth Regul. 1993;13:179–87.

31. Novák O, Hényková E, Sairanen I, Kowalczyk M, Pospíšil T, Ljung K. Tissue-specific profiling of the Arabidopsis thaliana auxin metabolome. Plant J. 2012;72:523–36.

32. Pacurar DI, Perrone I, Bellini C. Auxin is a central player in the hormone cross-talks that control adventitious rooting. Physiol Plant. 2014;151:83–96.

33. Rout GR. Effect of auxins on adventitious root development from single node cuttings of Camellia sinensis (L.) Kuntze and associated biochemical changes. Plant Growth Regul. 2006;48:111–7.

34. Faivre-Rampant O, D'Angeli S, Falasca G, Dommes J, Gaspar T, Altamura MM. Rooting blockage in the tobacco rac mutant occurs at the initiation phase, and induces diversion to xylem. Plant Biosyst. 2003;137:163–74.

35. Yadav S, David A, Baluška F, Bhatla SC. Rapid auxin-induced nitric oxide accumulation and subsequent tyrosine nitration of proteins during adventitious root formation in sunflower hypocotyls. Plant Signal Behav. 2013;8(3):e23196.

36. Sanz L, Albertos P, Mateos I, Sánchez-Vicente I, Lechón T, Fernández-Marcos M, et al. Nitric oxide (NO) and phytohormones crosstalk during early plant development. J Exp Bot. 2015;66:2857–68.

37. Guilfoyle TJ, Hagen G. Auxin response factors. Curr Opin Plant Biol. 2007;10:453–60.

38. Wang R, Estelle M. Diversity and specificity: auxin perception and signaling through the TIR1/AFB pathway. Curr Opin Plant Biol. 2014;21:51–8.

39. Gutierrez L, Bussell JD, Pacurar DI, Schwambach J, Pacurar M, Bellini C. Phenotypic plasticity of adventitious rooting in Arabidopsis is controlled by complex regulation of AUXIN RESPONSE FACTOR transcripts and microRNA abundance. Plant Cell. 2009;21:3119–32.

40. Gutierrez L, Mongelard G, Flokova K, Pacurar DI, Novak O, Staswick P, et al. Auxin controls Arabidopsis adventitious root initiation by regulating jasmonic acid homeostasis. Plant Cell. 2012;24:2515–27.

41. Terrile MC, Paris R, Calderon-Villalobos LIA, Iglesias MJ, Lamattina L, Estelle M, et al. Nitric oxide influences auxin signaling through S-nitrosylation of the Arabidopsis TRANSPORT INHIBITOR RESPONSE 1 auxin receptor. Plant J. 2012;70:492–500.

42. Liu X, Barkawi L, Gardner G, Cohen JD. Transport of indole-3-butyric acid and indole-3-acetic acid in Arabidopsis hypocotyls using stable isotope labeling. Plant Physiol. 2012;158:1988–2000.

43. Pii Y, Crimi M, Cremonese G, Spena A, Pandolfini T. Auxin and nitric oxide control indeterminate nodule formation. BMC Plant Biol. 2007;7:21.

44. Wang JR, Hu H, Wang GH, Li J, Chen JY, Wu P. Expression of PIN genes in rice (Oryza sativa L.): tissue specificity and regulation by hormones. Mol Plant. 2009;2:823–31.

45. Omelyanchuk NA, Kovrizhnykh VV, Oshchepkova EA, Pasternak T, Palme K, Mironova VV. A detailed expression map of the PIN1 auxin transporter in Arabidopsis thaliana root. BMC Plant Biol. 2016;16:5.

46. Sukumar P, Maloney GS, Muday GK. Localized induction of the ATP-binding cassette B19 auxin transporter enhances adventitious root formation in Arabidopsis. Plant Physiol. 2013;162:1392–405.

47. Fan S, Meng Y, Song M, Pang C, Wei H, Liu J, et al. Quantitative phosphoproteomics analysis of nitric oxide-responsive phosphoproteins in cotton leaf. PLoS One. 2014;9(4):e94261.

11

Sinapic acid or its derivatives interfere with abscisic acid homeostasis during *Arabidopsis thaliana* seed germination

Baodi Bi[†], Jingliang Tang[†], Shuang Han, Jinggong Guo and Yuchen Miao[*]

Abstract

Background: Sinapic acid and its esters have broad functions in different stages of seed germination and plant development and are thought to play a role in protecting against ultraviolet irradiation. To better understand the interactions between sinapic acid esters and seed germination processes in response to various stresses, we analyzed the role of the plant hormone abscisic acid (ABA) in the regulation of sinapic acid esters involved in seed germination and early seedling growth.

Results: We found that exogenous sinapic acid promotes seed germination in a dose-dependent manner in *Arabidopsis thaliana*. High-performance liquid chromatography mass spectrometry analysis showed that exogenous sinapic acid increased the sinapoylcholine content of imbibed seeds. Furthermore, sinapic acid affected ABA catabolism, resulting in reduced ABA levels and increased levels of the ABA-glucose ester. Using mutants deficient in the synthesis of sinapate esters, we showed that the germination of mutant *sinapoylglucose accumulator 2* (*sng2*) and *bright trichomes 1* (*brt1*) seeds was more sensitive to ABA than the wild-type. Moreover, *Arabidopsis* mutants deficient in either *abscisic acid deficient 2* (*ABA2*) or *abscisic acid insensitive 3* (*ABI3*) displayed increased expression of the *sinapoylglucose:choline sinapoyltransferase* (*SCT*) and *sinapoylcholine esterase* (*SCE*) genes with sinapic acid treatment. This treatment also affected the accumulation of sinapoylcholine and free choline during seed germination.

Conclusions: We demonstrated that sinapoylcholine, which constitutes the major phenolic component in seeds among various minor sinapate esters, affected ABA homeostasis during seed germination and early seedling growth in *Arabidopsis*. Our findings provide insights into the role of sinapic acid and its esters in regulating ABA-mediated inhibition of *Arabidopsis* seed germination in response to drought stress.

Keywords: Sinapic acid, Sinapic acid esters, Abscisic acid homeostasis, Seed germination, *Arabidopsis thaliana*

Background

Phenylpropanoid metabolism leads to a diverse group of compounds that are derived from the carbon skeleton of phenylalanine and are involved in plant defense, structural support, and survival [1]. Sinapic acid is a small, naturally occurring member of the phenylpropanoid family that serves as a common precursor for soluble secondary metabolites [2]. In brassicaceous plants, including *Arabidopsis thaliana*, sinapic acid is converted into a broad spectrum of *O*-ester conjugates. These abundant soluble sinapic acid esters reflect a well-known metabolic network and are produced at different stages of plant development. The accumulation of these sinapic acid esters and soluble phenylpropanoids also provides protection against ultraviolet (UV)-B stress and functions in the defense response to pathogens such as *Verticillium longisporum* in *Arabidopsis* [3, 4].

It has been reported that three major sinapic acid esters, sinapoylglucose, sinapoylmalate, and sinapoylcholine, accumulate in *Arabidopsis* and other members of the *Brassicaceae* family (Fig. 1) [5, 6]. Sinapoylglucose, which is the immediate precursor of sinapoylcholine and sinapoylmalate that accumulate in seeds and leaves, is

* Correspondence: miaoych@henu.edu.cn

[†]Equal contributors

Institute of Plant Stress Biology, State Key Laboratory of Cotton Biology, Department of Biology, Henan University, 85 Minglun Street, Kaifeng 475001, China

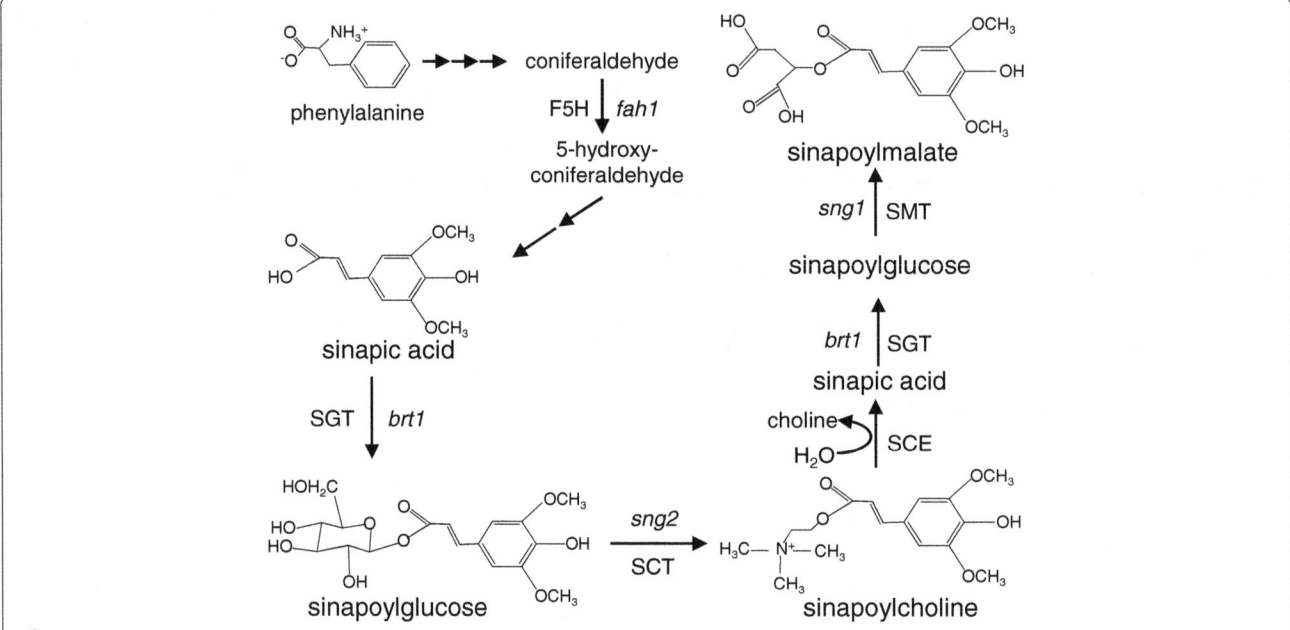

Fig. 1 Schema of sinapate ester metabolism in *Arabidopsis*. F5H, ferulate 5-hydroxylase; SGT, UDP-glucose:sinapate glucosyltransferase; SCT, sinapoylglucose:choline sinapoyltransferase; SCE, sinapine esterase; SMT, sinapoylglucose:malate sinapoyltransferase. *Arabidopsis* mutant names are written in italics: *brt 1*, bright trichomes 1; *sng 1, 2*, sinapoylglucose accumulator 1, 2; *fah 1*, ferulate hydroxylase 1

produced by a UDP-glucose: sinapic acid glucosyltransferase (SGT) that transfers the glucose moiety of UDP-glucose to the carboxyl group of sinapic acid [5]. The resulting 1-*O*-sinapoylglucose is a β-acetal ester that acts as an energy-rich acyl donor [7]; this donor provides the necessary free energy for the transacylation reaction catalyzed by sinapoylglucose:malate sinapoyltransferase (SMT) that generates sinapoylmalate in vegetative tissues [8]. In developing seeds, sinapoylglucose is also converted to sinapoylcholine by sinapoylglucose:choline sinapoyltransferase (SCT) [9–11]. Sinapoylcholine plays an important role during seed germination [12]. When the seeds start to germinate, the sinapoylcholine is hydrolyzed by sinapoylcholine esterase (SCE) to liberate sinapic acid and choline [12], and subsequently, the choline is oxidized to glycine betaine that functions as a stress-protecting agent by stabilizing proteins and membranes [13]. Thus, sinapoylcholine may serve as an important seed storage form of choline for the subsequent synthesis of phosphatidylcholine in developing seedlings. Currently, several *Arabidopsis* mutants have been identified to dissect the effects of sinapic acid ester accumulation at various stages of development on plant growth and yield (Fig. 1). For example, the *bright trichomes 1* (*brt1*) mutant, defective in *SGT*, showed a reduced epidermal fluorescence phenotype [9, 14]. The *Arabidopsis* mutant *sinapoylglucose accumulator 1* (*sng1*), defective in the gene encoding SMT, accumulated sinapoylglucose instead of sinapoylmalate in its leaves. Interestingly, the *Arabidopsis* loss-of-function mutant for *SCT* (*sng2*) was

found to have increased sinapoylglucose content in the seeds and a corresponding decrease in the level of sinapoylcholine [5, 15, 16]. Additionally, the *FERULIC ACID HYDROXYLASE1* (*fah1*) and *REDUCED EPIDERMAL FLUORESCENCE1* (*ref1*) mutations lack sinapic acid and sinapate ester synthesis, respectively [9, 17–19].

It is well known that various stresses trigger the activation of the phenylpropanoid pathway in plants [20–22]. The phytohormone abscisic acid (ABA) is a major regulator of plant development and stress responses, including seed dormancy, germination, and drought resistance responses [23–25]. It is generally believed that both phenolic compounds and ABA act as inhibitors of plant growth and development, and they are involved in plant growth regulation [26]. In addition to their individual inhibitory actions, phenolic compounds have been demonstrated to antagonize some effects of ABA, for instance, reversing ABA-induced abscission, hypocotyl growth, and seed germination [27–29]. Earlier studies revealed that phenolic compounds such as *t*-cinnamic acid and *p*-coumaric acid reverse ABA-induced stomatal closure [30]. Furthermore, ABA also causes an increase in stomatal diffusive resistance that is recovered by *t*-cinnamic acid and *p*-coumaric acid [31]. Moreover, some phenolic compounds such as scopoletin and umbelliferone were found to be associated with substantial retention of K⁺ in guard cells, antagonizing the effect of ABA and ABA-mediated increases in epidermal diffusive resistance [32]. Conversely, hydrophenolic compounds were completely inactive [32]. Recently, the use of microarrays and quantitative proteomics has found that treatment with ABA alters the transcript levels of phenylpropanoid pathway genes in *Arabidopsis* suspension cells [20]. ABA also can activate the expression of *MYB10*, a transcription factor that plays a major role in the regulation of flavonoid/phenylpropanoid metabolism during ripening in *Fragaria x ananassa* fruit [33]. Together, these studies suggest that phenylpropanoid metabolism plays an important role in the response to ABA. It is possible, therefore, that phenolics affect plant growth and development by inhibiting ABA synthesis and signaling processes. However, direct biochemical and genetic evidence for this is lacking.

In this study, we investigated the roles of sinapic acid during seed germination in *Arabidopsis*. Our results show that sinapic acid is involved in regulating ABA-mediated inhibition of seed germination. To test the contribution of sinapic acid esters to the interaction with ABA, sinapic acid ester-accumulating *Arabidopsis* mutants were analyzed. Our findings suggest a novel model for the involvement of sinapic acid esters in ABA homeostasis during seed germination.

Results

Effects of sinapic acid on seed germination and early seedling growth

As previously reported, sinapic acid esters are involved in protection against UV radiation, seed germination, and seedling development in brassicaceous plants [34]; it is, however, unclear how sinapic acid esters regulate seed germination. We, therefore, examined the role of sinapic acid in plant seed germination and early seedling development. First, we compared the germination rates of wild-type *Arabidopsis* seeds on Murashige and Skoog (MS) [35] medium containing different concentrations of sinapic acid. As shown in Fig. 2a, wild-type seed germination was promoted by sinapic acid concentrations ranging from 0.1 to 1 mM, with the germination rate of wild-type seeds in MS medium containing 0.5 mM sinapic acid increased by ~9.2% compared with the control (Fig. 2a, b). Next, we observed the effect of sinapic acid on root growth and early seedling development (Fig. 2c, d). Sinapic acid promoted root growth, causing an ~44% increase in root length compared with the mock (dimethyl sulfoxide was added, as the same volume of sinapic acid) treatment at 8 d after seed imbibition (Fig. 2c, e). Treatment with 0.5 mM sinapic acid for 20 d increased fresh seedling weight by ~20% compared with the mock treatment (Fig. 2f). To remove the effects of any trace of chemicals in the MS medium that could interfere with seed germination we also performed the germination assay on water medium. Freshly harvested *Arabidopsis* seeds were used in this study. The rate of seed germination increased by ~9%

with 0.5 mM sinapic acid compared with the control, indicating that sinapic acid significantly promoted early seedling growth (Additional file 1: Figure S1). These findings are consistent with the seed germination results using MS medium. Together, these data suggest that sinapic acid is involved in seed germination and early plant development.

To test whether exogenous sinapic acid is converted into sinapic acid esters during seed germination in *Arabidopsis*, we analyzed sinapoylglucose and sinapoylcholine production using high-performance liquid chromatography (HPLC) mass spectrometry. *Arabidopsis* seeds were germinated on MS medium containing 0.5 mM sinapic acid. As shown in Fig. 2g, wild-type seeds, after imbibition with 0.5 mM sinapic acid for 2 d, accumulated ~36 μmol g^{-1} dry weight (DW) sinapoylcholine, while the control accumulated only ~17 μmol g^{-1} DW. When soluble phenolic compounds were extracted from seedlings after 20 d of growth, the levels of sinapoylglucose and sinapoylmalate following 0.5 mM sinapic acid treatment were two to three times higher than in the mock-treated seedlings (Fig. 2h). These results suggest that exogenous sinapic acid may be channeled into the phenylpropanoid pathway where it is subsequently converted into the corresponding sinapic acid esters by sinapoyltransferase. This could potentially support seed germination and increase the development of young seedlings.

Effect of sinapic acid on ABA homeostasis

ABA is the key hormone associated with seed germination and vegetative growth and is known to interact with phenolic compound synthesis during plant development [20, 23–25]. To examine whether exogenous sinapic acid affects ABA homeostasis during seed germination, we first analyzed the expression of genes encoding UDP-glycosyltransferases (UGTs) and hydroxylases using real-time quantitative reverse transcriptase PCR (qRT-PCR). Previously reported UGTs UGT71C5, UGT71B6, UGT71B7, and UGT71B8, which displayed in vitro glucosylation activity toward ABA, belong to UGT subfamily 1 in *Arabidopsis* [36–38]. Under sinapic acid treatment, the transcript levels of *UGT71C5*, *UGT71B6*, *UGT71B7*, and *UGT71B8* were upregulated compared with the control (Fig. 3a). *UGT71B7* expression was upregulated more than 3-fold in seeds pretreated with sinapic acid for 12 h compared with the mock treatment. To further study whether sinapic acid affects ABA catabolism, we analyzed the expression of members of the *Arabidopsis* cytochrome P450 *CYP707A* gene family that encode proteins with ABA 8'-hydroxylase activity [39, 40]. There were no significant differences in the expression of *CYP707A1, 2, 3,* and *4* (Additional file 2: Figure S2a).

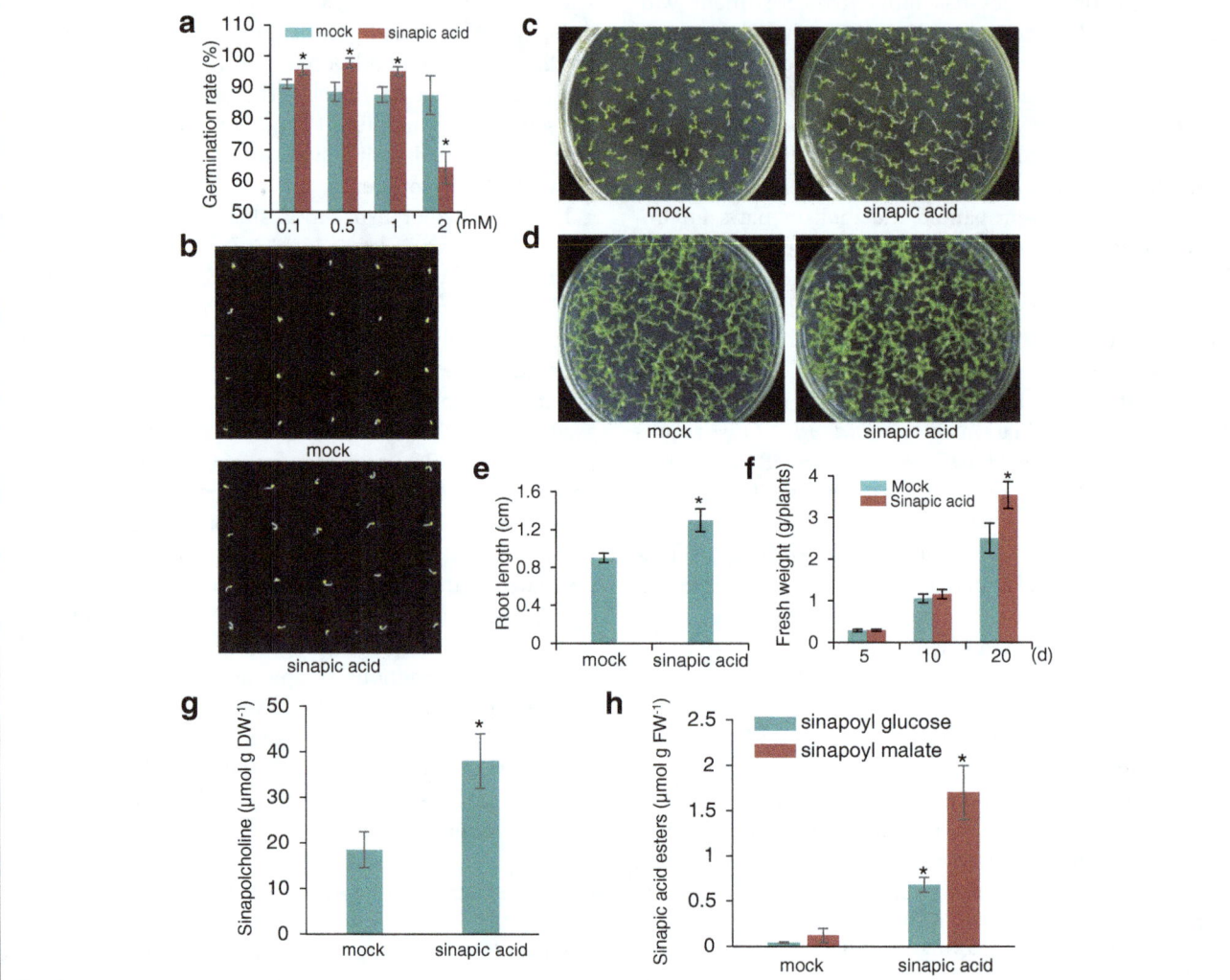

Fig. 2 Sinapic acid promotes seed germination and early seedling growth in *Arabidopsis*. **a** Comparison of the germination rates of *Arabidopsis* seeds after exposure to different concentrations of sinapic acid for 2 d. Seeds were germinated and grown on MS medium with sinapic acid or mock-treatment (dimethyl sulfoxide, DMSO) as a control. **b–d** Seeds were germinated and grown on MS medium with 0.5 mM sinapic acid. The photographs were taken at 2 d (**b**), 8 d (**c**), and 20 d (**d**) after seed imbibition. **e** Quantitative analysis of root length after 0.5 mM sinapic acid treatment for 8 d. **f** Fresh weight (FW) biomass per five plants at different times for seedlings grown on MS medium with 0.5 mM sinapic acid. **g** Quantification of sinapoylcholine released from the wild-type seeds. After 2 d of treatment with 0.5 mM sinapic acid, sinapic acid esters were extracted from germinating seeds and sinapoylcholine was quantified by HPLC; DW, dry weight. **h** Sinapoylglucose and sinapoylmalate analyses in seedlings grown for 10 d after exogenous 0.5 mM sinapic acid treatment. Values are means ± SD of three independent experiments. Asterisks indicate significant changes compared with the mock ($P < 0.05$) calculated using Student's *t*-test

In addition to de novo ABA biosynthesis, it has been shown that the β-glucosidase (BG) homologs in *Arabidopsis*, β-glucosidase 1 (BG1) and BG2, generate ABA from ABA- glucose ester (GE) in the endoplasmic reticulum and vacuole, respectively [41, 42]. To determine whether sinapic acid affects hydroxylation of ABA-GE to ABA by the BGs, we analyzed the expression of *BG1* and *2* under sinapic acid treatment. As shown in Additional file 2: Figure S2B, there was no significant difference in the expression of *BG1* and *2* with sinapic acid, suggesting that exogenous sinapic acid mainly participates in the regulation of the endogenous level of free

ABA by glucosylation, but does not affect ABA catabolism or hydroxylation of ABA-GE during seed germination.

To further demonstrate that sinapic acid affects ABA glucosylation, we examined the levels of free ABA and ABA-GE using reversed-phase HPLC in *Arabidopsis* seeds pretreated with sinapic acid for 48 h. We found that the level of ABA in the imbibed seeds was ~560 pmol g^{-1} after pretreatment with sinapic acid, compared with ~620 pmol g^{-1} after mock pretreatment. However, the level of ABA-GE in the seeds with sinapic acid pretreatment was ~190 pmol g^{-1}, compared with ~60 pmol g^{-1} in the mock-treated seeds (Fig. 3b).

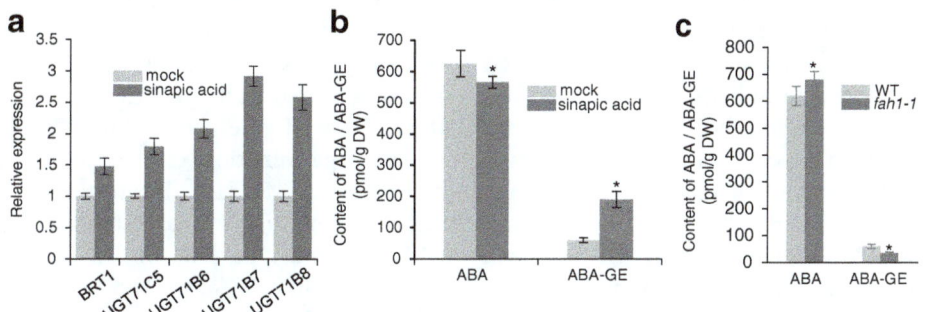

Fig. 3 Analysis of the endogenous ABA and ABA- glucose ester (GE) levels following sinapic acid pretreatment of seeds in *Arabidopsis*. **a** qRT-PCR analyses of ABA UDP-glucosyltransferase genes after pretreatment of seeds with sinapic acid or DMSO (mock-treated control). The seeds were incubated with 0.5 mM sinapic acid or DMSO for 36 h at 4 °C in the dark and then germinated on MS agar medium for 1 d. Total RNAs were isolated from the treated and mock-treated seeds; *Actin2* primers were used in PCR as an internal control. **b** Analysis of endogenous ABA and ABA-GE levels using HPLC-mass spectrometry. Seeds were immersed in 0.5 mM sinapic acid or DMSO (mock) for 36 h at 4 °C in the dark and then transferred to MS medium for 1 d of germination. Data are based on three independent replicates (± SD). **c** Analysis of endogenous ABA and ABA-GE levels in *fah1–1 Arabidopsis* plants. Seeds were immersed in water for 36 h at 4 °C in the dark and then transferred to MS medium for 1 d of germination. Data are based on three independent replicates (± SD). Asterisks indicate significant changes compared with the control ($P < 0.05$) calculated using Student's *t*-test

In *Arabidopsis*, *FAH1* encodes ferulate-5-hydroxylase, an enzyme in the phenylpropanoid pathway responsible for sinapic acid ester synthesis. The *fah1* mutant fails to accumulate sinapic acid esters [9]. To further confirm the degree of sinapic acid regulation of endogenous ABA and ABA-GE, free ABA and ABA-GE was measured in the null *fah1–1* mutant with decreased sinapic acid levels (Additional file 3: Figure S3). Consistently, the level of ABA reached ~680 pmol g^{-1} in *fah1–1* mutant seeds without sinapic acid treatment, compared with ~620 pmol g^{-1} in the wild-type seeds. Additionally, the level of ABA-GE was ~30 pmol g^{-1} in *fah1–1* compared with ~36 pmol g^{-1} in wild-type (Fig. 3c). Hence, our work provides genetic evidence that sinapic acid functions in ABA glucosylation during seed germination in *Arabidopsis*. Taken together, these findings suggest that sinapic acid influences ABA homeostasis by regulating ABA glucosylation in plants.

Sinapic acid partly reversed ABA-mediated inhibition of *Arabidopsis* seed germination

To further investigate the function of sinapic acid in the regulation of ABA homeostasis, we sowed *Arabidopsis* seeds onto a plate and soaked them in mock (dimethyl sulfoxide was added, as the same volume of sinapic acid) or 0.5 mM sinapic acid solutions at 4 °C in the dark. After 36 h incubation, the seeds were germinated in MS medium with 0.2 μM ABA. As shown in Fig. 4, a greater percentage of the seeds pretreated with sinapic acid germinated in MS medium containing ABA compared with the control; and 2 d after sowing, the germination percentage of seeds soaked in 0.5 mM sinapic acid was ~65%, while only ~40.5% of

the mock-pretreated seeds had germinated. At 4 d, no significant differences were observed in germination percentage between sinapic acid-incubated seeds and mock-treated seeds (Fig. 4). These findings are consistent with sinapic acid affecting the balance between ABA and ABA-GE during seed germination, and further confirm that exogenous sinapic acid might be the primary phenolic compound that is channeled into seeds and converted into sinapic acid esters to regulate ABA homeostasis.

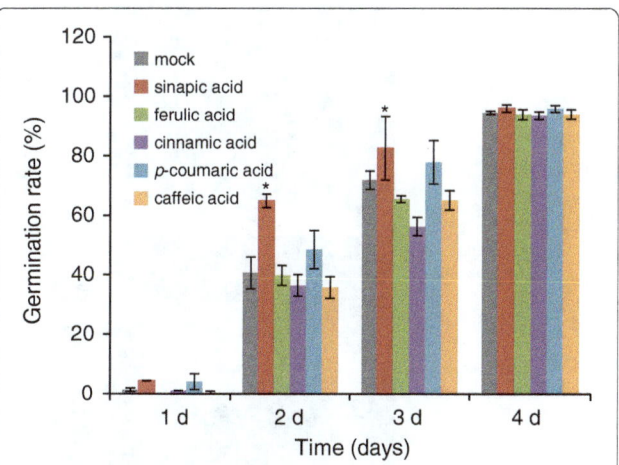

Fig. 4 Sinapic acid decreased ABA-induced inhibition of seed germination in *Arabidopsis*. Wild-type *Arabidopsis* seeds were incubated with 0.5 mM sinapic acid, ferulic acid, cinnamic acid, *p*-coumaric acid, or caffeic acid for 36 h at 4 °C in the dark, and then germinated on MS agar medium with 0.2 μM ABA. Values are means ± SD of three independent experiments (>100 seeds per data point). Asterisks indicate significant changes compared with the control ($P < 0.05$) calculated using Student's *t*-test

To evaluate whether the accumulation of phenyl-propanoids has the same function as sinapic acid on seed germination in response to ABA, we examined the effects of cinnamic acid and its hydroxylated derivatives (caffeic acid, ferulic acid, p-coumaric acid, and *t*-cinnamic acid) on seed germination. As shown in Fig. 4, the pretreatment of seeds with caffeic acid, ferulic acid, or cinnamic acid did not result in antagonistic effects upon ABA-mediated inhibition of seed germination; conversely, these compounds slightly enhanced ABA inhibition of seed germination compared with the mock treatment. However, p-coumaric acid slightly increased seed germination upon ABA treatment.

Loss of sinapic acid esters enhances susceptibility to ABA

To provide further genetic evidence that sinapic acid interacts with ABA during seed germination and early plant growth, we analyzed mutants with impaired function of different enzymes involved in sinapic acid ester biosynthesis. First, the seed germination rate was determined in *sng2*, the null mutant for *SCT* [16]. Compared with the wild-type, seed germination of the *sng2* mutant was more sensitive to ABA, and its germination rate was reduced by 50% at 2 d (Fig. 5a, b). Additionally, the number of green cotyledons for *sng2* was reduced 4-fold compared with the wild-type at 8 d (Fig. 5a, c). *brt1-1*

mutants, which are defective in SGT production, are responsible for the production of the sinapoyl donor sinapoylglucose [14]. Consistent with the role of *sng2*, *brt1-1* also reduced the rate of seed germination and the number of green cotyledons (Fig. 5). We also tested the seed germination rate in *sng1-1* mutants that are defective in *SMT* [15]. Interestingly, the seed germination rate of *sng1-1* was insensitive to ABA compared with the wild-type (Fig. 5b); moreover, there were no significant differences in the number of green cotyledons in *sng1-1* when compared with wild-type after 8 d of seedling growth (Fig. 5c).

To further investigate the relationship between sinapic acid ester biosynthesis and the ABA response during seed germination, we examined the sinapoylcholine level in *sng1-1*, *sng2*, and *brt1-1* mutants under ABA treatment. As shown in Fig. 5d, the level of sinapoylcholine in *sng1-1* and wild-type seeds was significantly higher following exposure to ABA than with the mock treatment. Almost no sinapoylcholine was found in *sng2*, however, and there were no significant differences between *brt1-1* and the wild-type. These findings suggest that the loss of *SCT* and *SGT* function resulted in a decrease in the levels of sinapoylcholine that might enhance susceptibility to ABA during seed germination.

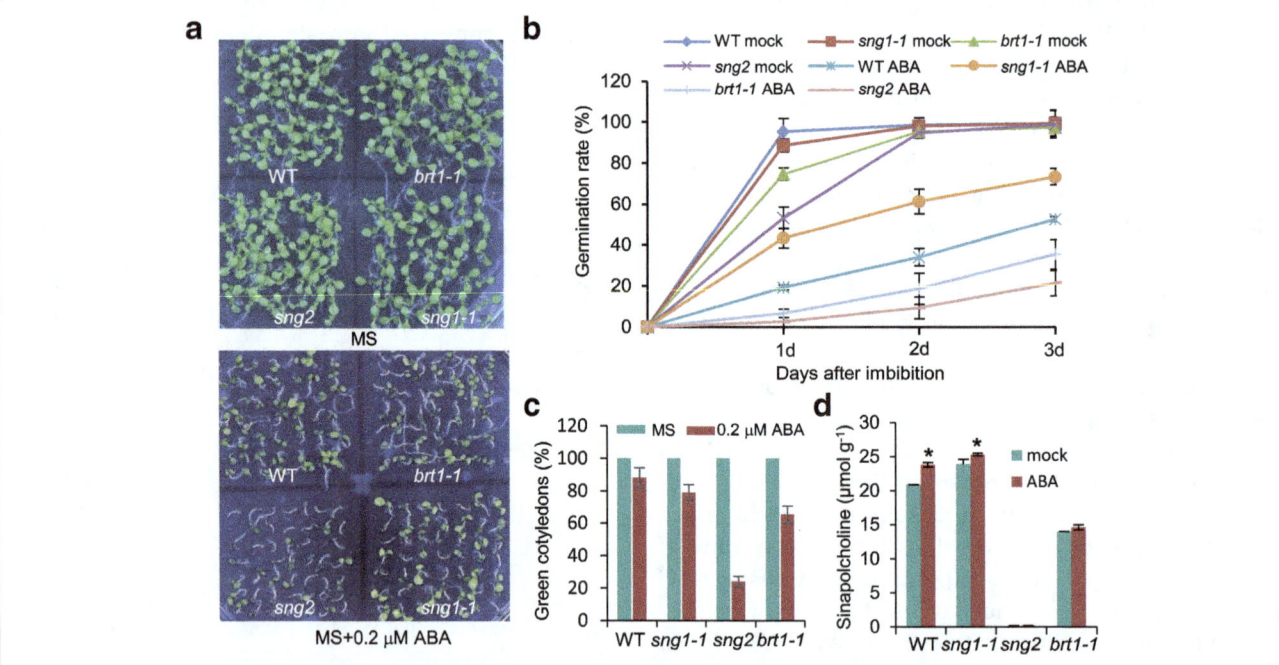

Fig. 5 Effect of ABA on seed germination and early seedling growth in sinapic acid ester-accumulating mutants of *Arabidopsis*. **a** Eight-day-old wild-type, *sng1-1*, *sng2*, and *brt1-1* seedlings were grown in MS medium with or without 0.2 μM ABA. **b** Seed germination rates of wild-type (WT), *sng1-1*, *sng2*, and *brt1-1* in the presence of 0.2 μM ABA. **c** The number of green cotyledons was recorded for plants grown as in (**a**). **d** Quantification of sinapoylcholine released from 0.2 μM ABA pretreated seeds of wild-type and sinapic acid ester-accumulating mutants of *Arabidopsis*. Values are means ± SD of three independent experiments (>120 seeds per data point)

Sinapoylcholine might be a key sinapic acid ester in antagonizing the effect of ABA during seed germination

Mutants defective in ABA biosynthesis or catabolism have been shown to have reduced or increased seed dormancy. *ABSCISIC ACID DEFICIENT2* (*ABA2*) encodes a short-chain dehydrogenase/reductase family member in *Arabidopsis* that plays a unique and specific role in ABA biosynthesis [43]. The *ABA2*-deficient mutant *aba2-1* [44] was used to determine the link between sinapic acid and the ABA pathway during seed germination. The germination rate of *aba2-1* on MS medium lacking sinapic acid was ~27% at 1 d after imbibition. However, when the seeds were placed on MS medium containing 0.5 mM sinapic acid, the germination rate of *aba2-1* increased by ~15% compared with the mock treatment. For wild-type seeds, almost no germination had occurred at 1 d (Fig. 6a). To determine whether sinapic acid interferes with seed-specific ABA signal transduction, the role of *ABSCISIC ACID INSENSITIVE3* (*ABI3*) [45], a major regulator of seed maturation in *Arabidopsis*, was analyzed. As shown in Fig. 6a, the seed germination rate of the *Arabidopsis* loss of function mutant of *abi3-1* was also increased with 0.5 mM sinapic acid, compared with the mock treatment. These data are consistent with the greater insensitivity to sinapic acid shown by the *aba2-1* mutant (Fig. 6a).

To test whether sinapoylcholine is involved in ABA-mediated inhibition of seed germination, we performed qRT-PCR to examine *SCT* expression. With sinapic acid treatment, *SCT* expression was higher in the *aba2-1* and *abi3-1* mutants than in the wild-type (Fig. 6b). Moreover, SCT enzymatic activity, measured using extracts of sinapic acid-treated wild-type, *aba2-1*, and *abi3-1* seeds, was 27.02 ± 1.35, 34.22 ± 1.65, and 35.2 ± 2.1 pKat mg^{-1} protein, respectively (Fig. 6c). These data are consistent with the increased expression of *SCT* in *aba2-1* and *abi3-1* mutants.

Having found that sinapic acid led to the production of sinapoylcholine in pretreated seeds (Fig. 1g), we determined the sinapoylcholine content of the *aba2-1 and abi3-1* mutants. When the seeds of wild-type, *aba2-1*, and *abi3-1* were pretreated in 0.5 mM sinapic acid for 36 h before being germinated on MS medium for 48 h, the sinapoylcholine content of *aba2-1* and *abi3-1* had increased by ~2.58 and ~3.4 µmol g^{-1}, respectively, compared with the wild-type (Fig. 6d). Therefore, we conclude that sinapoylcholine is responsible for ABA-mediated inhibition of seed germination.

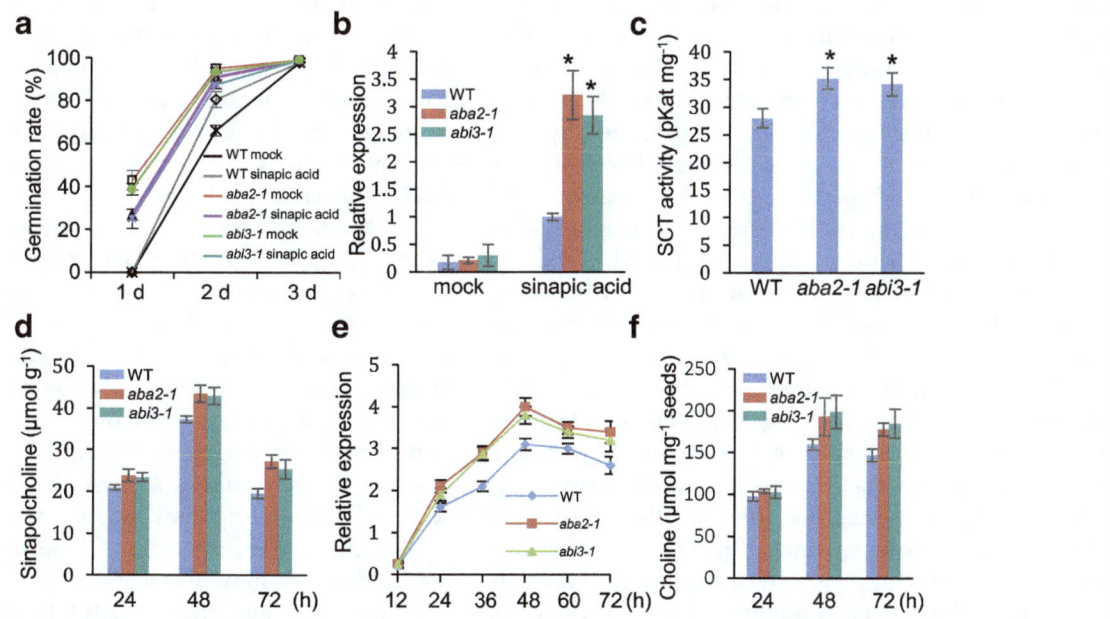

Fig. 6 Effect of the *ABA2* and *ABI3* mutations on sinapic acid-mediated seed germination in *Arabidopsis*. **a** Germination of *aba2-1* and *abi3-1* seeds in the presence of 0.5 mM sinapic acid; WT, wild-type. **b** qRT-PCR analysis of *SCT* expression in the *aba2-1* and *abi3-1* mutant plants in response to sinapic acid. The seeds were incubated with 0.5 mM sinapic acid for 36 h at 4 °C in the dark and then germinated on MS medium for 1 d. Total RNAs were isolated from treated and untreated seedlings; *Actin2* primers were used in PCR as an internal control. **c** Analysis of SCT activity in *aba2-1* and *abi3-1* crude seed extracts. SCT activity was determined in the presence of sinapoylglucose and choline chloride as substrates. **d** Quantification of sinapoylcholine released from *aba2-1* and *abi3-1* seeds pretreated with 0.5 mM sinapic acid. **e** Expression of *SCE* in *aba2-1* and *abi3-1* mutant plants in response to 0.5 mM sinapic acid. **f** Quantification assay of free choline in wild-type, *aba2-1*, and *abi3-1* plants treated with 0.5 mM sinapic acid. Values are means ± SD of three independent experiments. Asterisks indicate significant changes compared with the control ($P < 0.05$) calculated using Student's *t*-test

To further investigate the physiological relevance of the interaction between sinapoylcholine and ABA, the expression of *SCE* was assayed in *aba2-1* and *abi3-1* mutants. As shown in Fig. 6e, *SCE* expression increased ~29 and ~23% in *aba2-1* and *abi3-1*, respectively, compared with wild-type. Similarly, free choline accumulation in germinating seeds also increased in the *aba2-1* and *abi3-1* mutants upon sinapic acid treatment (Fig. 6f). Next, we tested whether choline chloride increased seed germination. Similar to sinapic acid, choline chloride also increased seed germination (Additional file 4: Figure S4). These data suggest that sinapoylcholine metabolism might regulate ABA-mediated inhibition of *Arabidopsis* seed germination.

Discussion

In this study, we investigated the role of sinapic acid in *Arabidopsis* seed germination and seedling growth using the *sng2* and *brt1-1* mutants. Our results showed that the sinapic acid ester metabolic pathways are involved in regulating ABA-mediated inhibition of seed germination and early seedling growth in *Arabidopsis*.

As mentioned above, phenylpropanoids suppress seed germination, cause root growth disorders, and inhibit plant growth [46]. Several known phenolic compounds such as cinnamic acid, flavonoids, and coumarins have been classified as natural inhibitors of plant growth regulation [26]. Sinapoylcholine is a member or derivative of the phenylpropanoid family that specifically accumulates in the seeds of cruciferous plants [34]. During seed germination, sinapoylcholine is hydrolyzed by SCE activity [47–50]. However, the physiological role of sinapoylcholine as a seed-specific ester is still unknown. In this study, we have shown that exogenous sinapic acid at concentrations of between ~0.1 and 1 mM could increase seed germination and the development of young seedlings (Fig. 2). A higher concentration of sinapic acid suppressed seed germination (Fig. 2a). When exogenous sinapic acid (0.5 mM) was applied to the medium, wild-type imbibed seeds and seedlings contained two to three times more of the sinapic acid esters sinapoylcholine and sinapoylglucose than did mock-treated control seedlings (Fig. 2g, h). Hence, we conclude that exogenous sinapic acid might be channeled into seeds where it is converted into sinapic acid esters, leading to the accumulation of these compounds in imbibed *Arabidopsis* seeds. This differs from the mechanisms of other phenolic compounds involved in the regulation of seed germination, root growth, and early seedling growth.

A number of phenolic compounds are known to antagonize the effects of ABA by, for instance, reversing ABA-induced abscission, hypocotyl growth, and seed germination [27, 29]. Some phenolic compounds, such as vanillic acid, gallic acid, salicylic acid, cinnamic acid,

p-coumaric acid, ferulic acid, coumarin, chlorogenic acid, rutin, and morin antagonize ABA-induced stomatal closure [51]. Interestingly, all the benzoic acids, including sinapic acid, resulted in the recovery of ABA-induced stomatal closure [28]. These results suggest that phenolic compounds might be involved in ABA metabolism or ABA signaling in response to stress. This hypothesis was confirmed in assays examining the germination of wild-type seeds after pretreatment with sinapic acid that found that sinapic acid partly reversed the ABA-mediated inhibition of *Arabidopsis* seed germination (Fig. 4). However, pretreatment with several simple phenolic compounds such as *p*-coumaric acid, caffeic acid, and ferulic acid did not recover seed germination following ABA exposure (Fig. 4). Our data, therefore, reveal an important role for sinapic acid in regulating seed germination together with ABA, though other phenolic compounds do not mirror this relationship. These results strongly support the presence of a correlation between the accumulation of sinapic acid esters and ABA homeostasis during seed germination. Therefore, sinapic acid not only enables the recovery of ABA-induced stomatal closure [28] but may also antagonize ABA-mediated inhibition of seed germination.

As the metabolism of sinapic acid esters in seeds is controlled by the sinapoylglucose-dependent sinapoyltransferase UGT enzyme family that includes BRT1 (UGT84A2), it is possible that sinapic acid ester metabolism might be involved in ABA glucosylation. Indeed, expression of *BRT1* was induced by sinapic acid (Fig. 3a). Interestingly, the transcript levels of *UGT71B5*, *UGT71B6*, *UGT71B7*, and *UGT71B8* were apparently upregulated by sinapic acid compared with mock-treated samples (Fig. 3a). Conversely, the expression of *CYP707A* genes and *BG* genes were less affected by sinapic acid (Additional file 2: Figure S2). The endogenous ABA and ABA-GE levels determined using LC-mass spectrometry in this study confirm the hypothesis that sinapic acid plays a major role in ABA glucosylation. Exogenous sinapic acid treatment led to dynamic changes in endogenous ABA/ABA-GE concentrations (Fig. 3b), suggesting that sinapic acid may influence ABA homeostasis in germinating seeds. Accordingly, a loss of function mutation in *FAH1* also led to increased ABA-GE levels and reduced ABA levels (Fig. 3C). The genetic analysis showed that seed germination decreased the sensitivity to sinapic acid in *aba2* and *abi3* mutants compared with wild-type seeds (Fig. 6a) and that sinapic acid increased *SCT* gene expression and enzyme activity (Fig. 6b, c). Importantly, sinapic acid enhanced *SCE* gene expression and sinapoylcholine level. When exogenous sinapic acid was added, the level free choline increased along with seed germination (Fig. 6f).

Hence, it is possible that exogenous sinapic acid is channeled via sinapoylglucose (1-*O*-sinapoyl-glucose) to sinapoylcholine, simultaneously enhancing free choline levels in the germinating seeds of *aba2-1* and *abi3-1* mutants, and antagonizing some effects of ABA-mediated inhibition of seed germination.

The involvement of sinapic acid or its derivatives in ABA-mediated inhibition of seed germination is also confirmed by *Arabidopsis* lines with mutations in the *SGT*, *SCT*, and *SMT* genes. As illustrated in Fig. 1, the *brt1-1* mutation impairs the gross metabolic flux toward sinapoylmalate in leaves or sinapoylcholine in seeds [52]. The *sng2* mutant accumulated a high level of sinapoylglucose in its seeds, corresponding to a decreased level of sinapoylcholine [5, 15, 16]. The fact that *sng2* and *brt1-1* seed germination exhibited greater sensitivity to ABA than seed germination in wild-type suggests that sinapoylcholine plays an important role in ABA-regulated seed germination (Fig. 5). Furthermore, when the seeds were planted on MS medium containing 0.2 μM ABA, the *sng1-1* mutant was found to be less sensitive to ABA than the wild-type, *sng2*, or *brt1-1* (Fig. 5). One possibility is that the *Arabidopsis sng1-1* mutant accumulated sinapoylglucose instead of sinapoylmalate in the imbibed seeds, corresponding to an increased level of sinapoylcholine and resulting in a blocking of ABA synthesis and breaking of seed dormancy. These results were also partly confirmed by an assay of the sinapoylcholine content of the *sng1-1* mutant (Fig. 5d). Consistently, defects in the ABA pathway increased sinapic acid-induced seed germination and *SCT* expression (Fig. 6a, c).

Overall, the accumulation of sinapoylcholine in the seed is a typical characteristic of many members of the *Brassicaceae* family. Once seeds begin to germinate, sinapoylcholine is mobilized by SCE hydrolysis to liberate sinapic acid and choline for germinating seedlings. Therefore, it is possible that the accumulation of sinapic acid and choline disturbs ABA homeostasis during seed germination. However, the link between sinapic acid-induced free choline accumulation and ABA-mediated inhibition of seed germination is further strengthened by our observations.

Conclusion

We demonstrate that sinapic acid esters might regulate ABA-mediated inhibition of dormancy breakage, germination, and growth in *Arabidopsis*. Hence, our investigation highlights the importance of sinapic acid metabolism in the conjugation cycle of ABA in ABA homeostasis during seed germination. Further research is needed to ascertain how sinapic acid esters regulate seed germination through a negative feedback loop modulating ABA homeostasis.

Methods

Plant material and growth conditions

Arabidopsis thaliana ecotype Columbia was used in this study. The *sng1-1* [15], *sng2* [16], and *brt1-1* [14] homozygous mutants (Col-0) were generously provided by Clint Chapple (Purdue University, USA). *fah1–2* (CS6172) and *abi3-1* (CS24) were obtained from the *Arabidopsis* Biological Resource Center (Ohio State University). *aba2-1/eas1–1* was screened by chlorophyll fluorescence imaging in our lab [44]. For seed germination, all seeds were sterilized and kept for 2 d at 4 °C in the dark to break dormancy. The seeds were then placed on 0.6% agar-containing MS medium (PhytoTech) with different levels of phenolic compounds or ABA as indicated, or the seeds were placed on wet filter paper (water medium) with sinapic acid or water alone as a control. The plates were incubated at 22 ± 2 °C with a 16-h-light photoperiod.

Pretreatment of *Arabidopsis* seeds

Freshly harvested *Arabidopsis* seeds were surface-sterilized with 0.1% mercuric chloride for 3 min, washed three times with sterile water before sowing, and then incubated in 0.5 mM sinapic acid (50 mM sinapic acid stock solution in 100% dimethyl sulfoxide) or the corresponding control for 36 h in the dark at room temperature. For the germination assay, the pre-treated seeds were then sown on solid MS medium with or without 0.2 μM ABA at 4 °C for 2 d in the dark before being transferred into a growth chamber with a 16/8 h (24/18 °C) day/night cycle with a light intensity of 150 μmol m^{-2} s^{-1}. The number of germinated seeds was recorded daily over 5 d. The experiment was carried out with three replicates, each with a group of 100 seeds per treatment.

RNA extraction and qRT-PCR

Total RNA was extracted using the RNeasy plant mini kit (Qiagen) according to the manufacturer's instructions. RNA samples were quantified with a Nanodrop spectrophotometer (ND-1000; Labtech). Reverse transcription was performed with 3 μg of total RNA and M-MLV (Promega) according to the manufacturer's instructions. All gene expression experiments were repeated at least three times (2 × SYBR Green Realtime Master Mix, Novoprotein). PCR primer sequences are presented in Additional file 5: Table S1.

Determination of ABA and ABA-GE content

Samples of imbibed seeds before solid-phase extraction were treated as described previously [37]. Assays of ABA and ABA-GE were performed essentially as described by Liu [38]. Briefly, 1 ml of pretreated sample with 20 ml of internal standard solution (chloromycetin, 14.426 ng ml^{-1}) was loaded into a Kinetex 2.6 μm C18 Column

(50 × 2.1 mm). Samples were eluted with 2 ml of buffer containing methanol: distilled deionized water: acetic acid (80:19:1, $v/v/v$). The eluent was dried at 40 °C under a gentle stream of nitrogen. The residue was reconstituted by the addition of 200 ml of methanol: water: acetic acid (45:54:1, $v/v/v$); 10 ml aliquots of supernatant were analyzed by LC-mass spectrometry (AB Sciex API QTRAP4000).

Analysis of sinapic acid esters

Rosette leaves and pretreated seeds of *Arabidopsis* were ground into powder in liquid nitrogen and extracted overnight at 4 °C with 80% methanol containing 25 μM chrysin as an internal standard. Samples were ground briefly and then centrifuged at 13,000×g for 10 min. The imbibed seed extracts were prepared from 1 mg of seeds suspended in 0.1 ml of 80% methanol. Sinapic acid ester contents were determined by HPLC (Agilent). The sample was resolved on a Kinetex 2.6 μm C18 Column (50 × 2.1 mm) in 0.2% acetic acid (A) with an increasing concentration gradient of acetonitrile containing 0.2% acetic acid (B) at a constant rate of 0.8 ml min^{-1}: 0–20 min, 30% B; 20–25 min, 100% B. UV absorption was monitored at 330 nm using a multiple-wavelength photodiode array detector (Agilent). Peaks were identified and quantified using commercially available standard substances.

Determination of SCT activity

Enzyme extraction and assay conditions were based on those used previously to purify and assay SCT from *Arabidopsis* [53]. To assay the crude seed extracts, wild-type, *aba2-1*, and *abi3-1* seeds were frozen in liquid nitrogen and each ground to a fine powder. This powder was stirred for 20 min in five volumes of 100 mM potassium phosphate buffer (pH 6.8) containing 20 mM NaCl and 4% w/v insoluble polyvinylpolypyrrolidone. The samples were filtered through Miracloth (Calbiochem, La Jolla, CA, USA) and centrifuged for 20 min at 14,000×g. The supernatant was added to 0.1% w/v protamine sulfate, stirred for 20 min, and centrifuged for 20 min at 14,000×g. The supernatant was again filtered through Miracloth (Calbiochem), and the protein was precipitated by adding ammonium sulfate to 85% saturation, followed by centrifugation at 14,000×g for 20 min. The pellet was resuspended in 100 mM potassium phosphate buffer (pH 7.0) containing 50 mM NaCl, desalted into 100 mM potassium phosphate buffer (pH 7.0) using PD-10 Sephadex G-25 M columns (Supelco, Bellefonte, PA, USA), and used for the determination of SCT activity. Each assay contained 50 ml of 2.5 mM sinapoylglucose, 10 ml of 100 mM choline chloride, and 30 ml of protein extract. The assays were incubated for 60 min at 30 °C, stopped by the addition of 400 ml of cold 50% methanol, and analyzed by HPLC (Agilent).

Choline assay

Free choline from germinating seeds was assayed using a Choline Assay Kit (Abnova) in accordance with the manufacturer's instructions. For an assay of crude seed extracts, wild-type, *aba2-1*, and *abi3-1* seeds were frozen in liquid nitrogen and each ground to a fine powder. This powder was stirred in cold PBS buffer (pH 6.8) for 1 h. The samples were filtered through Miracloth (Calbiochem) and centrifuged for 5 min at 14,000×g. For each sample, 300 μl of supernatant was transferred to a clean tube and neutralized with 50 μl 6 M NaOH. The neutralized supernatant was then assayed by spectrophotometer (Thermo Fisher).

Additional files

Additional file 1: Figure S1. Sinapic acid promotes seed germination and early seedling growth in *Arabidopsis*. **a** Comparison of germination rates of *Arabidopsis* after exposure to different concentrations of sinapic acid for 2 d. Fresh seeds were germinated and grown on wet filter paper with 0.5 mM sinapic acid or water alone as a control. Values are means ± SD of three independent experiments. Asterisks indicate significant changes compared with the mock ($P < 0.05$) calculated using Student's t-test. **b** Photographs taken 7 d after seed imbibition. Seeds were germinated and grown on water alone, or water with 0.5 mM sinapic acid.

Additional file 2: Figure S2. Expression of ABA metabolism genes in response to sinapic acid. **a** Quantitative real-time RT-PCR (qRT-PCR) to examine the expression of ABA catabolism genes (*CYP707A1*, *CYP707A2*, *CYP707A3*, and *CYP707A4*) with sinapic acid. **b** qRT-PCR to examine the expression of de novo ABA biosynthesis genes (*BG1* and *BG2*) with sinapic acid. The seeds were incubated with 0.5 mM sinapic acid or dimethyl sulfoxide (DMSO) for 36 h at 4 °C in the dark and then germinated on MS agar medium for 1 d. Total RNA was isolated from the treated and mock-treated seeds; *Actin2* primers were used as an internal control.

Additional file 3: Figure S3. Quantification of sinapoylcholine released from wild-type and *fah1-1* seeds pretreated with 0.5 mM sinapic acid. After 2 d of treatment with 0.5 mM sinapic acid, sinapic acid esters were extracted from seeds and sinapoylcholine was quantified by HPLC; DW, dry weight.

Additional file 4: Figure S4. Germination of *aba2-1* and *abi3-1* seeds in the presence of 400 mg l^{-1} choline chloride.

Additional file 5: Table S1. List of primers used in this study.

Acknowledgment
We thank Clint Chapple of Purdue University (USA) for kindly providing the *sng1-1*, *sng2*, and *brt1-1* mutants.

Funding
This work was supported by the National Natural Science Foundation of China (31370332), the Doctoral Program Foundation of Institutions of Higher Education of China (20134103110001), and the Plan for Scientific Innovation Talent of Henan Province (144200510017).

Authors' contributions
YM conceived and designed the experiments. BB, JT, SH, and JG performed the experiments. YM, BB, and JT analyzed the data. YM, BB, and JT contributed reagents/materials/analytical tools. YM wrote the paper. All authors read and approved the final manuscript.

Competing interests
The authors declare that they have no competing interests.

Accession numbers

Sequence data from this report can be found in the *Arabidopsis* Genome Initiative or the GenBank/EMBL database under the following accession numbers: *SGT*, At3g50310; *SCT*, At5g09640; *SMT*, At2g22990; *UGT71C5*, At1g07240; *UGT71B6*, At3g21780; *UGT71B7*, At3g21790; *UGT71B8*, At3G21800; *CYP707A2*, At2g29090; *ABA2*, At1g52340; *ABI3*, At3g24650 and *ACTIN2*, At3g18780.

References

1. Vogt T. Phenylpropanoid biosynthesis. Mol Plant. 2010;3:2–20.
2. Shahidi F, Naczk M. Cereals, legumes, and nuts. In phenolics in food and nutraceuticals, Ch, vol. 2. Boca Raton: CRC press; 2004. p. 17–166.
3. König S, Feussner K, Kaever A, Landesfeind M, Thurow C, Karlovsky P, et al. Soluble phenylpropanoids are involved in the defense response of *Arabidopsis* against *Verticillium longisporum*. New Phytol. 2014;202:823–37.
4. Landry LG, Chapple CCS, Last RL. *Arabidopsis* mutants lacking phenolic sunscreens exhibit enhanced ultraviolet-B injury and oxidative damage. Plant Physiol. 1995;109:1159–66.
5. Lorenzen M, Racicot V, Strack D, Chapple C. Sinapic acid ester metabolism in wild type and a sinapoylglucose-accumulating mutant of *Arabidopsis*. Plant Physiol. 1996;112:1625–30.
6. Mock H-P, Vogt T, Strack D. Sinapoylglucose:malate sinapoyltransferase activity in *Arabidopsis thaliana* and *Brassica napus*. Z. Naturforsch. 1992;47c:680–2.
7. Mock H-P, Strack D. Energetics of the uridine 5-diphosphoglucose: hydroxycinnamic acid acyl-glucosyltransferase reaction. Phytochemistry. 1993;32:575–9.
8. Sharma V, Strack D. Vacuolar localization of 1-sinapolglucose:l-malate sinapoyltransferase in protoplasts from cotyledons of *Raphanus sativus*. Planta. 1985;163:563–8.
9. Chapple CCS, Vogt T, Ellis BE, Somerville CR. An *Arabidopsis* mutant defective in the general phenylpropanoid pathway. Plant Cell. 1992;4:1413–24.
10. Shirley AM, Chapple C. Biochemical characterization of sinapoylglucose: choline sinapoyltransferase, a serine carboxypeptidase-like protein that functions as an acyltransferase in plant secondary metabolism. J Biol Chem. 2003;278:19870–7.
11. Strack D, Knogge W, Dahlbender B. Enzymatic synthesis of sinapine from 1-O-sinapoyl-β-D-glucose and choline by a cell-free system from developing seeds of red radish (*Raphanus sativus* L. var. *sativus*). Z Naturforsch. 1983;38c:21–7.
12. Strack D. Sinapine as a supply of choline for the biosynthesis of phosphatidylcholine in *Raphanus sativus* seedlings. Z Naturforsch. 1981;36c:215–21.
13. Sakamoto A, Murata N. Genetic engineering of glycinebetaine synthesis in plants: current status and implications for enhancement of stress tolerance. J Exp Bot. 2000;51:81–8.
14. Sinlapadech T, Stout J, Ruegger MO, Deak M, Chapple C. The hyper-fluorescent trichome phenotype of the *brt1* mutant of *Arabidopsis* the result of a defect in a sinapic acid: UDPG glucosyltransferase. Plant J. 2007;49:655–68.
15. Lehfeldt C, Shirley AM, Meyer K, Ruegger MO, Cusumano JC, Viitanen PV, et al. Cloning of the *SNG1* gene of *Arabidopsis* reveals a role for a serine carboxypeptidase-like protein as an acyltransferase in secondary metabolism. Plant Cell. 2000;12:1295–306.
16. Shirley AM, McMichael CM, Chapple C. The *sng2* mutant of *Arabidopsis* is defective in the gene encoding the serine carboxypeptidase-like protein sinapoylglucose:choline sinapoyltransferase. Plant J. 2001;28:83–94.
17. Meyer K, Cusumano JC, Somerville C, Chapple CCS. Ferulate-5-hydroxylase from *Arabidopsis thaliana* defines a new family of cytochrome P450-dependent monooxygenases. Proc Natl Acad Sci U S A. 1996;93:6869–74.
18. Nair RB, Bastress KL, Ruegger MO, Denault JW, Chapple C. The *Arabidopsis thaliana REDUCED EPIDERMAL FLUORESCENCE1* gene encodes an aldehyde dehydrogenase involved in ferulic acid and sinapic acid biosynthesis. Plant Cell. 2004;16:544–54.
19. Ruegger M, Meyer K, Cusumano JC, Chapple C. Regulation of ferulate-5-hydroxylase expression in *Arabidopsis* in the context of sinapate ester biosynthesis. Plant Physiol. 1999;119:101–10.
20. Bohmer M, Schroeder JI. Quantitative transcriptomic analysis of abscisic acid-induced and reactive oxygen species-dependent expression changes and proteomic profiling in *Arabidopsis* suspension cells. Plant J. 2011;67:105–18.
21. Hura T, Grzesiak S, Hura K, Thiemt E, Tokarz K, Wedzony M. Physiological and biochemical tools useful in drought-tolerance detection in genotypes of winter triticale: accumulation of ferulic acid correlates with drought tolerance. Ann Bot. 2007;100:767–75.
22. Santos-Filho PR, Vitor SC, Frungillo L, Saviani EE, Oliveira HC, Salgado I. Nitrate reductase- and nitric oxide-dependent activation of sinapoylglucose: malate sinapoyltransferase in leaves of *Arabidopsis thaliana*. Plant Cell Physiol. 2012;53:1607–16.
23. Cutler SR, Rodriguez PL, Finkelstein RR, Abrams SR. Abscisic acid: emergence of a core signaling network. Annu Rev Plant Biol. 2010;61:651–79.
24. Kim TH, Böhmer M, Hu H, Nishimura N, Schroeder JI. Guard cell signal transduction network: advances in understanding abscisic acid, CO_2, and Ca^{2+} signalling. Annu Rev Plant Biol. 2010;61:561–91.
25. Raghavendra AS, Gonugunta VK, Christmann A, Grill E. ABA perception and signalling. Trends Plant Sci. 2010;15:395–401.
26. Kefeli VI, Kadyrov CC. Natural growth inhibitors, their chemical and physiological properties. Annu Rev Plant Biol. 1971;22:185–96.
27. Apte PV, Laloraya MM. Inhibitory action of phenolic compounds on abscisic acid-induced abscission. J Exp Bot. 1982;33:826–30.
28. Purohit S, Laloraya MM, Bharti S. Effect of phenolic compounds on abscisic acid-induced stomatal movement: structure activity relationship. Physiol Plant. 1991;81:79–82.
29. Ray SD, Guruprasad KN, Laloraya MM. Antagonistic action of phenolic compounds on abscisic acid-induced inhibition of hypocotyl growth. J Exp Bot. 1980;31:1651–6.
30. Laloraya MM, Nozzolillo C, Purohit S, Lisa S. Reversal of abscisic acid-induced closure by *trans*-cinnamic and *p*-coumaric acid. Plant Physiol. 1986;81:253–8.
31. Mittelheuser CJ, Van Steveninck RFM. Rapid action of abscisic acid on photosynthesis and stomatal resistance. Planta. 1971;97:83–6.
32. Purohit S, Laloraya MM, Bharti S, Nozzolillo C. Effect of phenolic compounds on ABA-induced changes in K^+ concentration of guard cells and in epidermal diffusive resistance. J Exp Bot. 1992;43:103–10.
33. Medina-Puche L, Cumplido-Laso G, Amil-Ruiz F, et al. MYB10 plays a major role in the regulation of flavonoid/phenylpropanoid metabolism during ripening of *Fragaria x ananassa* fruits. J Exp Bot. 2014;65:401–17.
34. Milkowski C, Strack D. Sinapate esters in brassicaceous plants: biochemistry, molecular biology, evolution and metabolic engineering. Planta. 2010;232:19–35.
35. Murashige T, Skoog F. A revised medium for rapid growth and bioassays with tobacco tissue culture. Physiol Plant. 1962;15:473–97.
36. Lim EK, Doucet CJ, Hou B, Jackson RG, Abrams SR, Bowles DJ. Resolution of (+)-abscisic acid using an *Arabidopsis* glycosyltransferase. Tetrahedron Asymmetry. 2005;16:143–7.
37. Priest DM, Ambrose SJ, Vaistij FE, Elias L, Higgins GS, Ross AR, et al. Use of the glucosyltransferase UGT71B6 to disturb abscisic acid homeostasis in *Arabidopsis thaliana*. Plant J. 2006;46:492–502.
38. Liu Z, Yan JP, Li DK, Luo Q, Yan Q, Liu ZB, et al. UDP-glucosyltransferase71C5, a major glucosyltransferase, mediates abscisic acid homeostasis in *Arabidopsis*. Plant Physiol. 2015;167:1659–70.
39. Kushiro T, Okamoto M, Nakabayashi K, Yamagishi K, Kitamura S, Asami T, Hirai N, Koshiba T, Kamiya Y, Nambara E. The *Arabidopsis* cytochrome P450 CYP707A encodes ABA 8'-hydroxylases: key enzymes in ABA catabolism. EMBO J. 2004;23:1647–56.
40. Saito S, Hirai N, Matsumoto C, Ohigashi H, Ohta D, Sakata K, Mizutani M. *Arabidopsis* CYP707As encode (+)-abscisic acid 8'-hydroxylase, a key enzyme in the oxidative catabolism of abscisic acid. Plant Physiol. 2004;134:1439–49.
41. Lee KH, Piao HL, Kim HY, Choi SM, Jiang F, Hartung W, Hwang I, Kwak JM, Lee IJ, Hwang I. Activation of glucosidase via stress-induced polymerization rapidly increases active pools of abscisic acid. Cell. 2006;126:1109–20.
42. Xu ZY, Lee KH, Dong T, Jeong JC, Jin JB, Kanno Y, Kim DH, Kim SY, Seo M, Bressan RA, et al. A vacuolar β-glucosidase homolog that possesses glucose-conjugated abscisic acid hydrolyzing activity plays an important role in osmotic stress responses in *Arabidopsis*. Plant Cell. 2012;24:2184–99.

43. Cheng WH, Endo A, Zhou L, Penney J, Chen HC, Arroyo A, Leon P, Nambara
     E, Asami T, Seo M, Koshiba T, Sheen J. A unique short-chain
     dehydrogenase/reductase in *Arabidopsis* glucose signaling and abscisic acid
     biosynthesis and functions. Plant Cell. 2002;14:2723–43.

44. Song Y, Xiang F, Zhang G, Miao Y, Miao C, Song C-P. Abscisic acid as an
     internal integrator of multiple physiological processes modulates leaf
     senescence onset in *Arabidopsis thaliana*. Front. Plant Sci. 2016;7:181.

45. Giraudat J1, Hauge BM, Valon C, Smalle J, Parcy F, Goodman HM. Isolation of
     the *Arabidopsis ABI3* gene by positional cloning. Plant Cell. 1992;4:1251–61.

46. Weir TL, Park S-W, Vivanco JM. Biochemical and physiological mechanisms
     mediated by allelochemicals. Curr Opin Plant Biol. 2004;7:472–9.

47. Nurmann G, Strack D. Sinapine esterase. Part I. Characterization of
     sinapine esterase from cotyledons of *Raphanus sativus*. Z Naturforsch.
     1979;34c:715–20.

48. Strack D. Enzymatic synthesis of 1-sinapoylglucose from free sinapic acid
     and UDP-glucose by a cell-free system from *Raphanus sativus* seedlings.
     Z Naturforsch. 1980;35c:204–8.

49. Tzagolo VA. Metabolism of sinapine in mustard plants. I. Degradation of
     sinapine into sinapic acid and choline. Plant Physiol. 1963;38:202–6.

50. Tzagolo VA. Metabolism of sinapine in mustard plants. II. Purification and
     some properties of sinapine esterase. Plant Physiol. 1963;38:207–13.

51. Rai VK, Sharma SS, Sharma S. Reversal of ABA-induced stomatal closure by
     phenolic compounds. J Exp Bot. 1986;37:129–34.

52. Meißner D, Albert A, Böttcher C, Strack D, Milkowski C. The role of UDP-
     glucose:hydroxycinnamate glucosyltransferases in phenylpropanoid
     metabolism and the response to UV-B radiation in *Arabidopsis thaliana*.
     Planta. 2008;228:663–74.

53. Vogt T, Aebershold R, Ellis B. Purification and characterization of sinapine
     synthase from seeds of *Brassica napus*. Arch Biochem Biophys. 1993;300:622–8.

12

Genome-wide analysis of WOX genes in upland cotton and their expression pattern under different stresses

Zhaoen Yang[1,2†], Qian Gong[2†], Wenqiang Qin[1,2†], Zuoren Yang[2], Yuan Cheng[2], Lili Lu[2], Xiaoyang Ge[2], Chaojun Zhang[2], Zhixia Wu[2] and Fuguang Li[1,2*] (iD)

Abstract

Background: WUSCHEL-related homeobox (WOX) family members play significant roles in plant growth and development, such as in embryo patterning, stem-cell maintenance, and lateral organ formation. The recently published cotton genome sequences allow us to perform comprehensive genome-wide analysis and characterization of WOX genes in cotton.

Results: In this study, we identified 21, 20, and 38 WOX genes in *Gossypium arboreum* (2n = 26, A_2), *G. raimondii* (2n = 26, D_5), and *G. hirsutum* (2n = 4x = 52, $(AD)_t$), respectively. Sequence logos showed that homeobox domains were significantly conserved among the WOX genes in cotton, *Arabidopsis*, and rice. A total of 168 genes from three typical monocots and six dicots were naturally divided into three clades, which were further classified into nine sub-clades. A good collinearity was observed in the synteny analysis of the orthologs from At and Dt (t represents tetraploid) sub-genomes. Whole genome duplication (WGD) and segmental duplication within At and Dt sub-genomes played significant roles in the expansion of WOX genes, and segmental duplication mainly generated the WUS clade. Copia and Gypsy were the two major types of transposable elements distributed upstream or downstream of WOX genes. Furthermore, through comparison, we found that the exon/intron pattern was highly conserved between *Arabidopsis* and cotton, and the homeobox domain loci were also conserved between them. In addition, the expression pattern in different tissues indicated that the duplicated genes in cotton might have acquired new functions as a result of sub-functionalization or neo-functionalization. The expression pattern of WOX genes under different stress treatments showed that the different genes were induced by different stresses.

Conclusion: In present work, WOX genes, classified into three clades, were identified in the upland cotton genome. Whole genome and segmental duplication were determined to be the two major impetuses for the expansion of gene numbers during the evolution. Moreover, the expression patterns suggested that the duplicated genes might have experienced a functional divergence. Together, these results shed light on the evolution of the WOX gene family, and would be helpful in future research.

Keywords: *Gossypium hirsutum*, WUSCHEL-related homeobox, Segmental duplication, Transcript factor, Embryogenesis

* Correspondence: aylifug@163.com
†Equal contributors
[1]Xinjiang Research Base, State Key Laboratory of Cotton Biology, Xinjiang Agricultural University, Urumqi 830052, China
[2]Institute of Cotton Research, Chinese Academy of Agricultural Sciences, Anyang 455000, China

Background

WUSCHEL-related homeobox (WOX), one of the sub-clades of homeodomain (HD) superfamily, is a plant-specific homeobox (HB) transcription factor, characterized by a short stretch (60–66 residues) of amino acids that form a DNA-binding domain named as homeodomain [1]. Previous reports have shown that WOXs assay a wide variety of important roles in development and growth process of plants, such as in embryonic patterning, stem cell maintenance, and organ formation [2, 3]. In *Arabidopsis thaliana*, which is a model plant, the evolution and function of WOXs have been well studied and characterized. As of date, 16 WOXs have been discovered by whole genome identification [4]; these have been clustered into three clades, namely, the ancient, WUS, and intermediate clades, by phylogenetic analysis [5]. WOXs of lower plants, such as those of green algae and non-vascular moss, *Physcomitrella patens*, belong only to the ancient clade, whereas those of higher plants, such as *Arabidopsis*, sorghum, maize, and rice, are present in all the three clades [1, 2]. *AtWUS* expressed in shoot apical meristem, ovule, and anther has been proven to play important roles in stem cell maintenance [6]. Ectopic expression of *AtWUS* in upland cotton was demonstrated to promote callus dedifferentiation resulting in the formation of embryo callus [7]. *AtWOX3*, as well as its orthologs in maize, *NS1(narrow sheath1)* and *NS2(narrow sheath2)*, perform a highly conserved function during the recruitment of founder cells to form lateral domains of vegetative and floral organs [8]. *AtWOX1* was demonstrated to express in the initiating vascular primordium of the cotyledons during heart and torpedo stages, and its overexpression resulted in defects in the development of meristem and dwarf phenotype [9]. *AtWOX4* was determined to be an essential regulator in auxin-dependent cambium stimulation that regulates lateral plant growth [10]. *AtWOX5*, induced by turanose and auxin, was shown to play vital roles in a correct pattern of root-formation through mediation of auxin homeostasis and by maximizing the auxin content in a restricted area [11]. *AtWOX6/PSF2* was revealed to play important roles in ovule development via the regulation of cell proliferation of the maternal integuments and through differentiation of the megaspore mother cell [12]. *WOX6/HOS9* was also shown to be an essential component, mediating cold tolerance in *Arabidopsis* through a CBF-independent pathway [13]. *AtWOX2* and *AtWOX8* were demonstrated to be essential for determination of the boundary between the cotyledons through the activation of three *CUPSHAPED COTYLEDON* (*CUC*) genes [14]. *STIMPY/AtWOX9* was found to be crucial for the growth of vegetative shoot apical meristem through maintenance of cell division and prevention of premature differentiation [15]. Also, *STIMPY/AtWOX9* determined

the fate of meristem by activation of the cytokinin signaling pathway in meristematic tissue [16]. *WOX11*, being a key regulator of shoot-borne crown development, was associated with the activation of crown root emergence and growth via directly repressing *PR2* to regulate cell proliferation during crown root development [17]. *AtWOX11* directly responds to an auxin maximum, induced by wounding in and around the procambium, and like *AtWOX12*, it positively upregulates *LATERAL ORGAN BOUNDARIES DOMAIN 16* and *29* genes, resulting in the initiation of a leaf procambium or the transition of its neighboring parenchyma cells to root founder cells [18]. *AtWOX13* mainly expresses in meristematic tissues including replum and promotes replum during the development of the *Arabidopsis thaliana* fruits [19]. *AtWOX14* and *WOX4*, regulatory elements downstream of *PHLOEM INTERCALATED WITH XYLEM* (*PXY*), regulate vascular cell division instead of vascular organization, playing crucial roles in stem formation [20].

Cotton is an important fiber crop, which provides the natural renew fiber for textile industry [21]. The roles of WOXs have been well-documented in embryogenesis in *Arabidopsis*; however, the functions of WOXs in cotton, especially in somatic embryogenesis, are largely unknown, thus far. The completion of genome sequencing in cotton allows comprehensive identification and analysis of WOXs in cotton [22–26]. We, therefore, conducted a thorough investigation on WOXs in cotton; the study included the identification of gene families, phylogenetic tree analysis, as well as the analyses of segmental duplication, gene structure, chromosome location, and expression pattern.

Methods

Sequence identification

The complete genome sequence data of three cotton species, *Gossypium arboreum* (BJI, version 1.0), *G. raimondii*, (JGI, version 2.0), and *G. hirsutum* (NAU, version 1.1, BJI, version 1.0), available from COTTONGEN (http://www.cottongen.org) [27] were used. The rice (version 7.0), sorghum (version 2.1), cacao (version 1.1), poplar (version 1.1), and maize (version 1.1) genome sequence data were retrieved from JGI (https://phytozome.jgi.doe.gov/pz/portal.html). The amino acid sequences of WOXs from *Arabidopsis thaliana* were acquired from TAIR 10 (http://www.arabidopsis.org); these were used as query sequences to search the *G. arboreum* protein database for candidate sequences employing blastp program. Thereafter, Interproscan 56.0 [28] was used to search for the HB domain (IPR001356) in the candidate sequences, and eventually the WOX sequences were identified. WOXs in rice, sorghum, cacao, *G. hirsutum*, and *G. raimondii* were identified using the same method as employed for *G. arboreum*.

Conserved sequence and phylogenetic analysis

Multiple sequence alignment was performed using Clustal X 2.0. For sequence logo analysis, the conserved HB domain sequences of WOXs from rice, *Arabidopsis*, and upland cotton were aligned, and the multiple alignment result was submitted to an online tool, WEBLOG [29], for generating the logos. For phylogenetic analysis, the full-length WOX sequences from *Arabidopsis*, rice, sorghum, cacao, *G. arboreum*, *G. raimondii*, and *G. hirsutum* were aligned, and MEGA 7.0 [30] was used to construct a neighbor-joining (NJ) tree. Bootstrap method was used to test the tree with 1000 replicates. Substitution was evaluated by Poisson model using the default parameters. To validate the phylogenetic tree, constructed using the NJ method, the minimum-evolution method was also used. The bootstrap method was used to test the tree with 1000 replicates.

Chromosome location and collinearity analysis

The gene loci of WOXs were extracted from the annotation gff3-file. Mapchart was then used to obtain the chromosome location [31]. All the protein sequences of upland cotton were included in a local database using Basic Local Alignment Search Tool (BLAST). The entire protein sequences were used as queries to search the above-mentioned database with an e-value of 1e-5. The blastp result was analyzed by MCSCAN to produce the collinearity blocks across the whole genome. The collinearity pairs belonging to WOX family were extracted to draw a collinearity map within WOXs by CIRCOS software [32].

Calculation of Ka/Ks values

The amino acid sequences from segmentally duplicated pairs and orthologous pairs were first aligned using Clustal X 2.0; thereafter, the aligned sequences were converted to the original cDNA sequences using the PAL2-NAL program [33] (http://www.bork.embl.de/pal2nal/). The CODEML program of the PAML package [34] was used to estimate the synonymous (Ks) and nonsynonymous (Ka) substitution rates.

Annotation and analysis of transposable elements

The *de novo* prediction and homolog search method based on Repbase [35] were used in the present study to identify the repeat content. For the *de novo* analysis, PILER-DF, RepeatModeler, and LTR_FINDER [36, 37] were used to predict the transposable elements (TEs) in the genome. For the analysis using the homology-based approach, the known TE library was used; the TEs were identified at the DNA level with RepeatMasker [38] using Repbase TE. To analyze the function of TEs in the expansion of the WOX family, we identified the TEs located 10,000 and 2,000 bp upstream and downstream of

the gene and made statistics of the different types of TEs (mutator-like transposable element (MULE), hAT, CACTA, helitron, retrogenes, and retrotransposons) present.

Gene structure analysis

Arabidopsis and *G. hirsutum* sequences were aligned with Clustal × 2.0, and MEGA 7.0 [30] was used to construct an NJ tree using the method and parameters as described above. The exon positions were acquired from the bed-file, and they were displayed by an online tool, GSDS 2.0 [39].

Transcriptome data analysis and gene expression heatmap

The raw data of RNA-seq was downloaded from the NCBI Sequence Read Archive (SRA: PRJNA248163). Tophat and cufflinks were used to analyze the RNA-seq expression, and the gene expressions were uniformed in fragments per kilobase million (FPKM) [40]. The expression of WOXs was extracted from the total expression data. Heatmap was drawn by Genesis software [41].

Real-time PCR

Cotton seeds of TM-1 were obtained from the Institute of Cotton Research of the Chinese Academy of Agricultural Sciences. The cotton (TM-1) seeds were germinated on a wet filter paper for 3 days at 28 °C, and then transferred to a liquid culture medium [42]. At the 3-leaf stage, the seedlings were treated with 10% PEG 6000 and 300 NaCl; the true leaves were sampled at 0, 1, 3, 6, and 12 h of the treatment and were immediately frozen in liquid nitrogen and stored at -80 °C. The total RNA was extracted from the seedlings using RNAprep Pure Plant Kit (TIANGEN, Beijing, China), as per the manufacturer's instructions. The first strand cDNA was synthesized using a PrimeScript® RT reagent kit (Takara, Dalian, China). SYBR Premix Ex TaqTM II (Takara) was used for PCR amplifications. The cotton histone 3 (GenBank accession no. AF024716) was used as an internal control.

Results

Gene identification and homeobox domain retrieval

We used WOXs from *Arabidopsis thaliana* as queries for searching the rice, sorghum, poplar, maize, cacao, *G. arboreum*, *G. raimondii*, and *G. hirsutum* databases using blastp program and hits with e-values of 1e-5 were considered significant. In preliminary analysis, we recognized 14 candidates in rice, 14 in sorghum, 18 in poplar, 20 in maize, 14 in cacao, 26 in *G. arboreum*, 31 in *G. raimondii*, 50 in *G. hirsutum* (NAU), and 33 in *G. hirsutum* (BJI). Thereafter, PROSITE (http://prosite.expasy.org/) and InterProscan 56.0 (http://www.ebi.ac.uk/interpro/) were used to search for the HB domain in the obtained sequences and 14, 12, 12, 19, 11, 21, 20, 38, and 33 genes were confirmed as WOX family members in rice, sorghum, poplar, maize, cacao, *G. arboreum*, *G. raimondii*, *G. hirsutum* (NAU), and *G. hirsutum* (BJI), respectively (Additional file 1:Table S1). According to newly sequenced A genome database (Unpublished) by PacBio RS II [43], we found the previous annotations of Cotton_A_11936 and Cotton_A_11937 were not accurate, and the total WOX genes in *G. arboreum* should be 20 , so we used *GaWOX1* to represent them for further study. Comparing the genes from the two *G. hirsutum* genomes (NAU and BJI), we found that they were highly similar, and the genes from NAU contained all the genes from BJI; therefore, we took genes from NAU for most of the analyses (Additional file 2: Table S2). The total number of WOX genes identified in the two diploid species, *G. arboreum* (AA) and *G. raimondii* (DD), was higher than that in the tetraploid *G. hirsutum* (AADD), which is derived from hybridization of progenitors resembling *G. arboreum* and *G. raimondii*. The number of WOXs in the two diploid cotton species (*G. arboreum* and *G. raimondii*) was found to be much higher than that in cacao [11], poplar [12], *Arabidopsis* [16], sorghum [12], and rice [14], indicating that the WOX family in cotton has undergone enlargement during the evolution.

Conserved amino residues within homeobox domains

The WOX gene family is a plant-specific clade of HD-containing superfamily, typically characterized by the

presence of a conserved HB domain in the full-length sequence. To investigate the presence of homologous domain sequences and the degree of conservation of each residue in the HB domains, we performed multiple sequence alignment to generate sequence logos of the HB domain in *G. hirsutum*, *Arabidopsis*, and rice. The sequence logos revealed that the residue distribution in the HB domains was highly similar in these three plants (Fig. 1a-c). Some amino acid residues in the HB domain, for instance, Q, L, and E in helix 1, P, I, and L in helix 2, and I, N, V, F, V, W, F, Q, N, R, and R in helix 3, were highly conserved. In contrast, the amino acid residues in the loop and turn were more variable; for example, the twentieth amino acid was blank due to the insertion of an extra amino acid in some atypical HB domains (Fig. 1a-c). Therefore, our results demonstrate that the HB domain sequences from WOXs are highly conserved among the typical dicot and monocots species.

Phylogenetic analysis and nomenclature of WOX genes

To determine the evolutionary relationship of WOX genes among cotton (*G. hirsutum*, *G. arboreum*, and *G. raimondii*) and other species, we constructed a phylogenetic tree by MEGA 7.0 using the NJ method. The cotton WOX genes were named based on the phylogenetic analysis. Ga, Gr, Gh, and At were used as prefixes before the names of WOX genes from *G.*

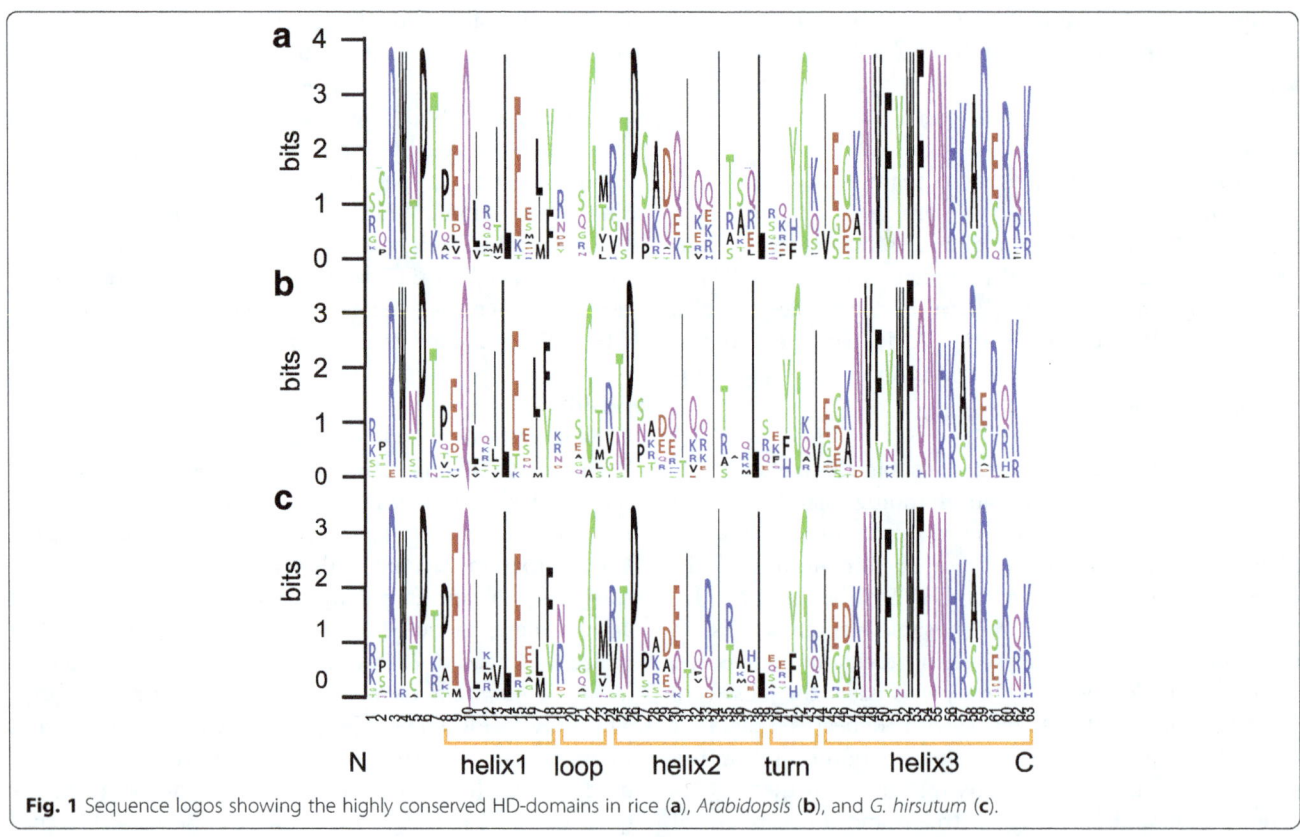

Fig. 1 Sequence logos showing the highly conserved HD-domains in rice (**a**), *Arabidopsis* (**b**), and *G. hirsutum* (**c**).

arboreum, *G. raimondii*, *G. hirsutum*, and *Arabidopsis*, respectively. Moreover, following rules were also considered for the nomenclature: 1) cotton WOXs were named after their orthologs in *Arabidopsis*; 2) if there were no orthologous counterparts in *Arabidopsis*, the cotton WOXs were named based on their homologs in the same clades; and 3) "a" and "b" were appended to the gene names to distinguish the relatively recent paralogs in a particular lineage. The phylogenetic tree showed that WOX genes could be naturally classified into three clades, namely, the ancient clade, WUS clade, and intermediate clade (Fig. 2). To validate the phylogenetic tree constructed using the NJ method, we also used the minimum-evolution method to construct a tree. As show in Additional file 3: Figure. S1, WOXs were divided into three clades as shown in Fig. 2. Although, there were differences between the topologies of the two trees, the

member within the subclades and the topology within the subclades were relative stable, which indicated that the NJ tree could be used for further analysis.

Compared to those of WUS and intermediate clades, the WOX genes of ancient clade were probably the earliest to diverge because the ancient clade separated the WUS and intermediate clades in the midpoint root. We found that the number of genes in the WUS clade (95) was greater than the sum of genes in ancient (28) and intermediate clades (45). To further investigate the evolutionary relationship and to predict the gene functions, we divided WOXs into nine sub-clades, named α through ι; six sub-clades were included in the WUS clade, two in the intermediate clade, and one in the ancient clade (Fig. 2; Additional file 3: Figure S1). Furthermore, we found that all the sub-clades comprised of dicot and monocot species. It is noteworthy that the

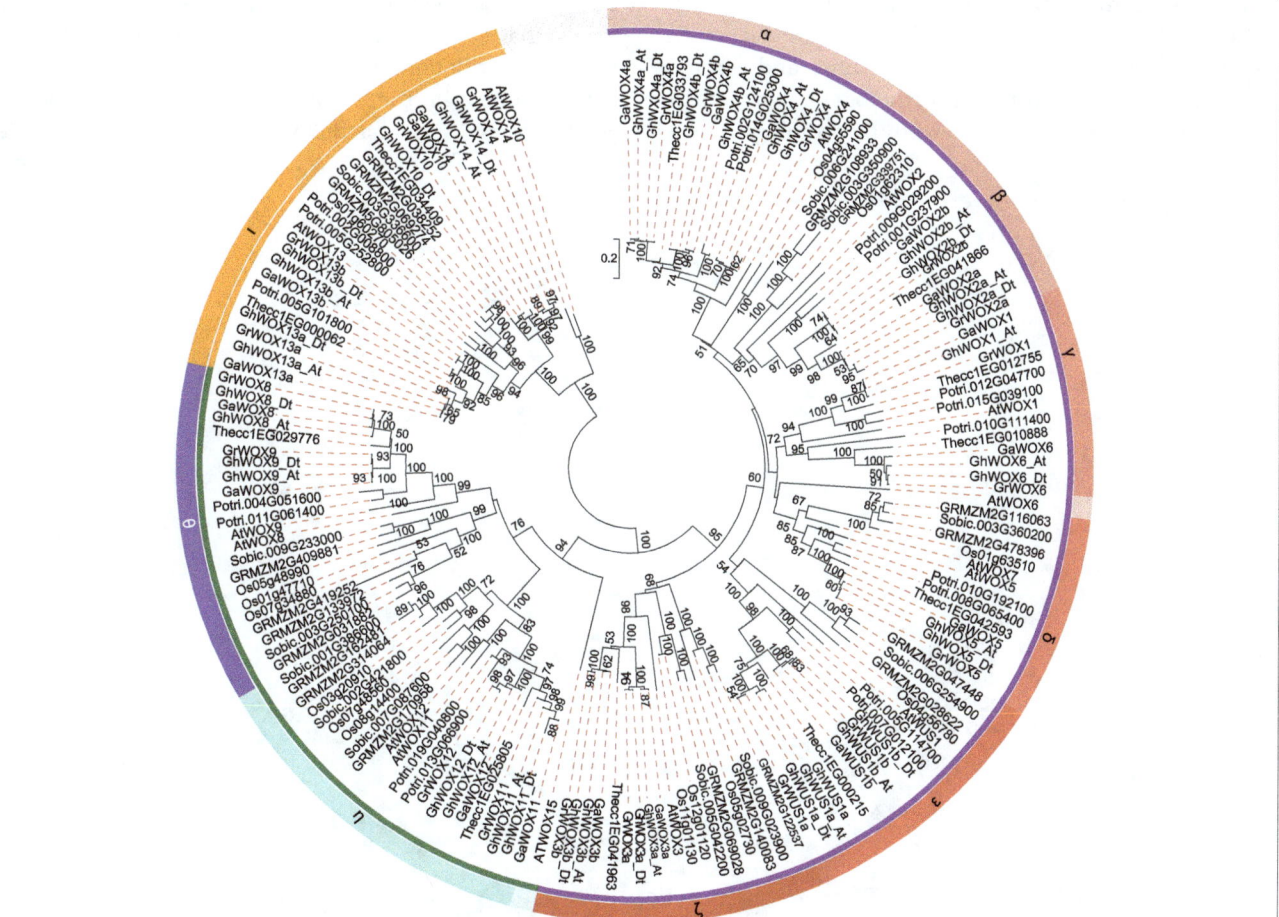

Fig. 2 Phylogenetic tree of WOX genes indicating that WOX genes could be divided into three clades. MEGA 7.0 was used for constructing the neighbor joining (NJ) tree. The inner circle is marked in purple, green, and orange representing the WUS, intermediate, and ancient clades, respectively. Each clade was classified into sub-clades, marked in different colors on the outer circle. α to η represent the sub-clades in the WUS clade, θ and ι represent the sub-clades in the intermediate clade, and κ represents the ancient clade. The prefixes Ga, Gr, Gh, Potri, At, Os, GRMZM, Sobic, and Thecc stand for *G. arboreum*, *G. raimondii*, *G. hirsutum*, *Populus trichocarpa*, *Arabidopsis thaliana*, *Oryza sativa*, *Zea mays*, *Sorghum bicolor*, and *Theobroma cacao*, respectively. The appendices At and Dt in the upland cotton indicate the A- and D-subgenome, respectively. The bootstrap values are shown near the nodes, and only those values greater than 50 are displayed.

genes within the sub-clades clustered with a dicot- or monocot-specific pattern. The number of WOXs in each species was variable within the sub-clades. For example, in the sub-clade ε, only one member was present in *Arabidopsis*, sorghum, rice, and cacao, each, but each of the other four species contained two members.

Compared to other species, WOX genes in cotton showed a closer relationship with that in cacao because they always clustered closely to each other in the phylogenetic tree. However, their gene number were not similar within the sub-clades, and in most cases one cacao gene corresponded to two homologous genes of *G. arboreum* and *G. raimondii*, whereas in some sub-clades, one cacao gene only had one corresponding gene in *G. arboreum* and *G. raimondii*. For example, in sub-clades θ, *Thecc1EG029776* had two orthologs in *G. arboreum* and *G. raimondii* each, whereas in sub-clade γ, *Thecc1EG012755* and *Thecc1EG010888* had only one ortholog in *G. arboreum* and *G. raimondii*, respectively.

As shown in Fig. 2 and Additional file 3: Figure S1, almost all the orthologs from A genome and At subgenome tended to form an orthologous pair at the branch end; same was the case with the orthologs from D

genome and D_t sub-genome, indicating that the orthologs from At-A or Dt-D had a more closer relationship.

Gene enlargement and synteny analysis

G. hirsutum, the typical allotetraploid, is an ideal material for studying the effect of naturally occurring polyploidy [26]. To study the locus relationship of orthologs between the At and Dt genomes, we investigated the gene locus on chromosome and performed synteny analysis. The synteny analysis revealed that most of the WOX loci were highly conserved between the At and Dt sub-genomes (Fig. 3). We also found that 38 WOX genes in *G. hirsutum* were located on 20 chromosomes, and their distribution was uneven; for example, there were no WOX genes on chromosomes A04, D04, A06, D06, A09, and D09. Except for A02-D02, A03-D03, A11-D11, and A12-D12, the number of genes located on the chromosome in At sub-genome was the same as that on its homologous chromosome in Dt sub-genome, indicating that some genes might have been lost during the evolution or incomplete sequencing of genome might have resulted in the identification of less number of genes than were actually present; for example,

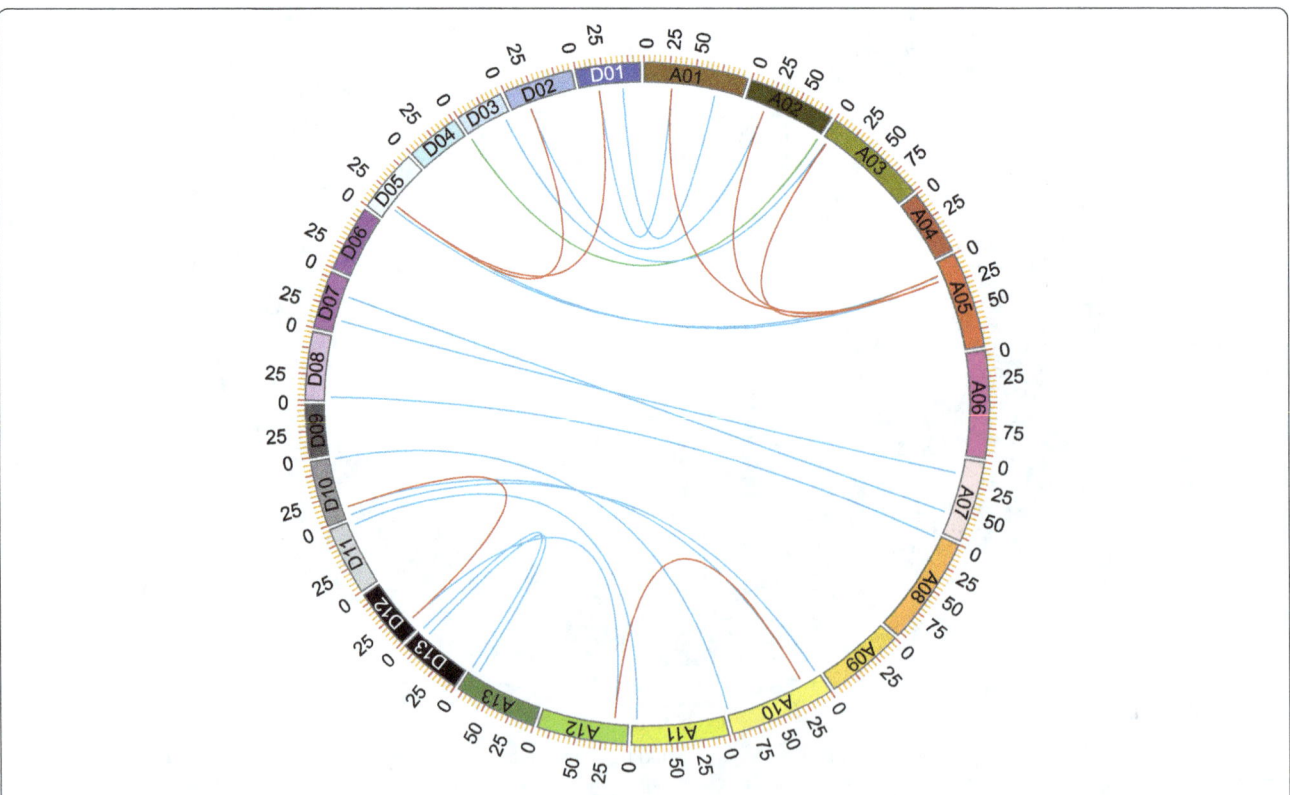

Fig. 3 Collinearity analyses within *G. hirsutum*. The ends of blue lines link two genes from the homologous chromosome pairs in At and Dt sub-genomes, respectively, showing the orthologous pairs that diverged from a same ancestor. The green line links two duplicated genes locating in non-homologous chromosome pair in At and Dt sub-genomes. The ends of the red lines link the homologous pairs formed by segmental duplication within the At- and Dt-subgenomes. A01 to A13 represent the chromosomes in At sub-genome and D01 to D13 represent the chromosomes in Dt sub-genome.

GhWOX1_At was located on A12, but no corresponding ortholog was found in the Dt sub-genome. We further checked the genome sequence released by BJI [24]; unfortunately, we identified only one un-oriented orthologous gene (*CotAD_21819*), which was located on scaffold506.1 (Additional file 2: Table S2). Therefore, we took eleven genes in the collinearity block around *GhWOX1_At* (from *Gh_A12G2428* to *Gh_A12G2439*) to align with the cotton genome and coding sequencing database to determine its collinearity block in the Dt sub-genome. As shown in Additional file 4: Figure S2a, we found that there was a corresponding collinearity block on D12 containing 11 genes, among which *Gh_D12G2554* corresponded to *GhWOX1_At*. Thereafter, the sequences from the promoter and genomic regions (from 58,423,052 to 58,426,551 bp) were extracted from the genome to align with the A12 chromosome. As shown in Additional file 4: Figure S2b, we found that there was a deletion (998 bps) between the promoter and genomic regions on D12, which resulted in *Gh_D12G2554* losing the typical HB domain of the WOX genes (Additional file 4: Figure S2c). We also found that *GhWOX13b_At* was located on A02, but its corresponding ortholog, *GhWOX13b_Dt*, was located on D03. To confirm this, we checked *GhWOX13b* in another genome (BJI) and found two corresponding genes (*CotAD_49496* and *CotAD_38760*); however, both of them were located on the unanchored scaffold (Additional file 2: Table S2). Therefore, we used the protein sequence of *GhWOX13b* to align with the newly sequenced A genome protein database (Unpublished) by PacBio RS II [43], and found that *WOX13b* was also located on Chr02, which is the homologous chromosome of A02 (Additional file 5). We found that the collinearity around the *GhWOX1 3b_Dt* locus was conserved between the Dt03 and D03 genomes of *G. raimondii*; therefore, we presumed that a chromosomal translocation has been accomplished between Chr02 and Chr03 before cotton polyploidization forming an allotetraploid (Fig. 3; Additional file 6: Figure S3). MCSCAN was used to identify the duplicate gene types. Almost half of the WOX genes were determined to be singletons, whereas a total of 12 genes were observed to have undergone segmental duplication. No tandem and proximal duplications (in the nearby chromosomal region but not in the adjacent region) were found among WOXs. Eight WOX genes were present as dispersed genes (duplications other than segmental, tandem, and proximal) (Additional file 7: Table S3). *GhWOX4b_At* and *GhWOX4b_Dt* seem to have been more active during evolution because both of these formed two collinearity gene pairs within At and Dt sub-genome, respectively.

During the long evolutionary history, duplicated genes might have experienced functional divergence, including nonfunctionalization (loss of original functions), neofunctionalization (acquisition of novel functions), or subfunctionalization (partition of original functions) [44, 45]. To investigate whether Darwinian positive selection was associated with the WOX gene divergence after duplication, the non-synonymous divergence levels (Ka) versus synonymous divergence levels were calculated for 25 homologous pairs. Based on the Ka/Ks ratio, we can presume the selection pressure for the duplicated genes. It is generally accepted that a value of Ka/Ks = 1 indicates that the genes are pseudogenes with neutral selection, Ka/Ks < 1 indicates that the duplicated genes have a tendency to purify, and Ka/Ks > 1 shows an accelerated evolution with positive selection. In our study, we found that the Ka/Ks ratios from 15 gene pairs were smaller than 0.5 and those from seven gene pairs were between 0.5 and 1.0. Only three gene pairs (*GhWOX13a_At-GhWOX13a*, *GhWOX13b_At-GhWOX13b_Dt*, and *GhWUS1b_At-GhWUS1b_Dt*) had Ka/Ks larger than 1, and these gene pairs might have experienced relatively rapid evolution following duplication (Table 1). Because most of the Ka/Ks values were smaller than 1.0, we presumed that the Cotton WOX gene family has undergone strong purifying selection pressure with limited functional divergence that occurred after segmental duplications and whole genome duplication (WGD).

The TEs are spread throughout the genome, and many of them are located in the vicinity of host genes [46]. Under abiotic or biotic stress, the TEs can be activated or repressed [47]. To explore whether TEs were involved in the WOX family expansion, we used *de novo* prediction and homolog search methods to identify the TEs in the whole genome, and the TEs close to the WOX genes were taken out. As show in Table 2, when checking the 2,000-bp region around the gene locus of WOX, only two retroelements (L1 and Caulimovirus) were found (Additional file 8: Table S4). We then broadened the scanning region to 10,000 bp upstream and downstream of the genes, respectively, and thirty one TEs were identified including five DNA transposons and 26 retroelements. All the five DNA transposons belonged to CMC-EnSpm family, whereas the retroelements were made up of Caulimovirus (1), Copia (13), and Gypsy (12) (Additional file 9: Table S5). Upon further inspection of the distribution of TEs within the 2,000-bp region, only one L1 was identified downstream of *GhWOX4_Dt* and one LTR/Copia was found to be located downstream of *GhWOX2a_At*. Within the 10,000-bp region around the gene locus, four CMC-EnSpm TEs were found to be located upstream of *GhWOX4a_Dt*, one CMC-EnSpm was located downstream of *GhWOX8_At*, three Copia elements were located downstream of *GhWOX13a_Dt* and *GhWOX2a_Dt*, two Copia TEs were located upstream of *GhWOX4_At* and *GhWOX6_At*, one Copia was located

Table 1 The Ka and Ks values for homologous pairs

Paralogous pairs	Ka	Ks	Ka/Ks
GhWOX11_At-GhWOX11_Dt	0.007	0.036	0.192
GhWOX12_At-GhWOX12_Dt	0.004	0.031	0.119
GhWOX13a_At-GhWOX13a_Dt	0.008	0.006	1.456
GhWOX13b_At-GhWOX13b_Dt	0.008	0.006	1.323
GhWOX14_At-GhWOX14_Dt	0.031	0.044	0.716
GhWOX2a_At-GhWOX2a_Dt	0.019	0.036	0.515
GhWOX2b_At-GhWOX2b_Dt	0.007	0.072	0.101
GhWOX3a_At-GhWOX3a_Dt	0.023	0.026	0.900
GhWOX3a_At-GhWOX3b_At	0.189	0.490	0.386
GhWOX3b_At-GhWOX3b_At	0.011	0.031	0.346
GhWOX4_At-GhWOX4_Dt	0.010	0.034	0.285
GhWOX4a_At-GhWOX4a_Dt	0.011	0.026	0.424
GhWOX4b_At-GhWOX4_At	0.144	0.361	0.400
GhWOX4b_At-GhWOX4a_At	0.124	0.463	0.267
GhWOX4b_At-GhWOX4b_Dt	0.006	0.012	0.520
GhWOX4b_Dt-GhWOX4_Dt	0.133	0.366	0.363
GhWOX4b_Dt-GhWOX4a_Dt	0.097	0.456	0.213
GhWOX5_At-GhWOX5_Dt	0.002	0.042	0.052
GhWOX6_At-GhWOX6_Dt	0.020	0.025	0.803
GhWOX8_At-GhWOX8_Dt	0.012	0.049	0.243
GhWOX9_At-GhWOX9_Dt	0.009	0.014	0.619
GhWUS1a_At-GhWUS1b_At	0.144	0.438	0.329
GhWUS1a_At-GhWUS1a_Dt	0.017	0.020	0.834
GhWUS1a_Dt-GhWUS1b_Dt	0.141	0.454	0.310
GhWUS1b_At-GhWUS1b_Dt	0.003	0.000	99.042

downstream of *GhWOX3a_At*, *GhWUS1b* and *GhWOX3 a_Dt*, five Gypsy TEs were located upstream and downstream of *GhWOX13a_Dt*, three Gypsy TEs were located downstream of *GhWOX4_At*, two Gypsy TEs were located downstream of *GhWOX4_At*, and one Gypsy element was located downstream of *GhWOX13a_At* and *GhWOX3b_At*. We noticed that most of the TEs correlated with the presence of duplicated genes, which suggested that TEs, especially the retroelements, played important roles in the WOX family expansion. Compared to TEs, simple repeat sequences are more abundant, most of which locate downstream or upstream of the genes, and only 39 were located in the genomic region. The length of simple repeat sequence was very variable, which might play important roles in the divergence of gene function after duplication.

Analysis of gene structure and homeodomain location

Gene structure is closely related to its function and, together with phylogenetic analysis, it can reflect the phylogenetic relation among the WOX genes. To further study the phylogenetic relationship between *Arabidopsis* and *G. hirsutum*, an NJ tree was generated with MEGA 7.0 using *Arabidopsis* and cotton WOX protein sequences (Fig. 4a), and positions of exons and introns in their genes were determined (Fig. 4b). Because the number of genes used for generating the phylogenetic tree described earlier (Fig. 2) was different from the one shown in Fig. 4a, the topologies of the two trees were different; however, the gene members within the sub-clades were mostly the same. As shown in Fig. 4a, WOX genes from cotton were grouped into nine sub-clades; *AtWOX15* could not be divided into any of the sub-clades. As evident from Fig. 4a, WOXs in α, ε, and ζ sub-clades of cotton might have undergone duplication because one *Arabidopsis* WOX in these sub-clades matched more than two orthologs from cotton (Fig. 4a) and this speculation was confirmed through collinearity analysis within At and Dt sub-genomes (Fig. 3). The exon number in *Arabidopsis* ranged from one to four, with the average being 2.9375, and in cotton, it ranged from two to four, with an average of 2.7297. In general, the gene structures among most of the sub-clades were conserved; for instance, in ζ, δ, and β sub-clades, the members in each sub-clade consisted of two exons. On comparing the exons in the orthologs from the same position in *Arabidopsis* and cotton, we found that the intron lengths were more divergent between the two; for example, the intron length of *AtWOX5* was much shorter than that of its ortholog in cotton. In the θ sub-clade, 6-bp-long exons were found at the fourth exon locus of *AtWOX8* and *AtWOX9* but they were not found in *GhWOX8_At/Dt* and *GhWOX9_At/Dt*, which might be due to an intron insertion in the third exon of the ancestor of *AtWOX8* and *AtWOX9* or these exons might have been lost in cotton during the evolution. We noticed that the intron lengths of *GhWOX5_At* and *GhWOX5_Dt* were different. Therefore, *GhWOX5* genomic sequences were extracted from the chromosome sequences to perform multiple alignments. We found many indels in the genomic sequence and statistically analyzed the indels more than 2 bps in length. The result showed that the introns of *GhWOX5* in the At and Dt sub-genomes were less conserved than the exons. This was because of the fact that of the eight indels present in the genomic sequences, seven were located in the introns and only one was present in the exons (Additional file 10: Figure S4). The typical conserved domains (HB domain) are marked in the exons with orange color. The lengths of HB domain ranged from 177 to 183 bp, and were considerably conserved. The HB domain locus in the same sub-clade was significantly conserved; for example, in α, γ, θ, and ι sub-clades, the HB domains were located in the second exon, and in δ, ε, ζ, β, and η sub-clades, they were located in the first exon (Fig. 4b). Only

Table 2 The TEs around the WOX gene locus

Type	Number of elments	Length occupied	Percentage of sequence (%)	Number of elments	Length occupied	Percentage of sequence (%)
	10,000 bp region			2000 bp region		
DNA transposons	5	1264 bp	0.16	0	0 bp	0.00
CMC-EnSpm	5	1264 bp	0.16	0	0 bp	0.00
MULE-MuDR	0	0 bp	0.00	0	0 bp	0.00
PIF-Harbinger	0	0 bp	0.00	0	0 bp	0.00
TcMar-Pogo	0	0 bp	0.00	0	0 bp	0.00
hAT	0	0 bp	0.00	0	0 bp	0.00
hAT-Ac	0	0 bp	0.00	0	0 bp	0.00
hAT-Charlie	0	0 bp	0.00	0	0 bp	0.00
hAT-Tag1	0	0 bp	0.00	0	0 bp	0.00
hAT-Tip100	0	0 bp	0.00	0	0 bp	0.00
Retroelements	31	15,657 bp	1.92	2	1400 bp	0.69
LINE:	5	1501 bp	0.18	1	373 bp	0.18
L1	5	1501 bp	0.18	1	373 bp	0.18
LTR:	26	14,156 bp	1.74	1	1027 bp	0.51
Caulimovirus	1	126 bp	0.02	0	0 bp	0
Copia	13	8097 bp	0.99	1	1027 bp	0.51
Gypsy	12	5933 bp	0.73	0	0 bp	0.00
RC:	0	0 bp	0.00	0	0 bp	0.00
Helitron	0	0 bp	0.00	0	0 bp	0.00
Low_complexity	6	257	0.03	0	0 bp	0.00
Simple_repeat	559	23,669	2.91	173	6793	3.35
Unspecified	68	24,602	3.02	7	1825	0.90

GhWOX10_Dt was used in our study, and its ortholog in At (*Gh_A11G2876*) was observed to have lost the third exon and part of the second exon, which resulted in the HB domains being incomplete. Further investigation revealed a point mutation at 1433 bp, turning 'CAG' to 'TAG', which lead to a premature termination of *Gh_A11G2876* protein (Additional file 11: Figure S5). To validate the above results, we searched the BJI genome; however, no corresponding orthologs were identified. Therefore, we used the RNA-seq aligned data (Bam file) to verify the result. As shown in Additional file 12: Figure S6, we found the clean reads mapped to the reference genome very well, and according to the mapping rate, we clearly found that there were two introns in *GhWOX10_Dt*. However, there was only one intron in *Gh_A11G2876*. We enlarged the mapping results near the mutation site and clearly saw that *Gh_A11G2876* acquired a termination codon at 92,69,533 bp. The point mutant might have affected the *Gh_A11G2876* gene structure, indicating that *Gh_A11G2876* might have a function different from that of *GhWOX10_Dt* or might have lost its biological function.

Gene expression pattern in different tissues and under multiple stresses

Because gene expression is associated with the biological function, we inspected the expression patterns of different WOX genes. RNA-seq data were downloaded from NCBI and analyzed. As shown in Fig. 5a, we found that WOXs were widely expressed in the vegetative (root, stem, and leaf) and reproductive (torus, petal, stamen, pistil, calycle, and -3, -1, 0, 1, 3, 5, 10, 20, 25 and 35 days post-anthesis (DPA) ovule) tissues as well as in the fiber (5, 10, 20, and 25 DPA), indicating that WOXs have diverse biological functions and work in different tissues. We found that some WOXs did not express in the vegetative tissues and had a very low expression levels in the reproductive tissues. For instance, we could not detect *GhWOX2_At/Dt*, *GhWOX3a_At/Dt*, and *GhWOX3b_At/Dt* expression in root, stem, and leaf, and very low expression levels were detected in some of the reproductive tissues. On comparing the expression patterns of the orthologs between At and Dt, we found that the expression patterns and the levels of expression of the two were not always the same; for example, *GhWOX8_Dt*

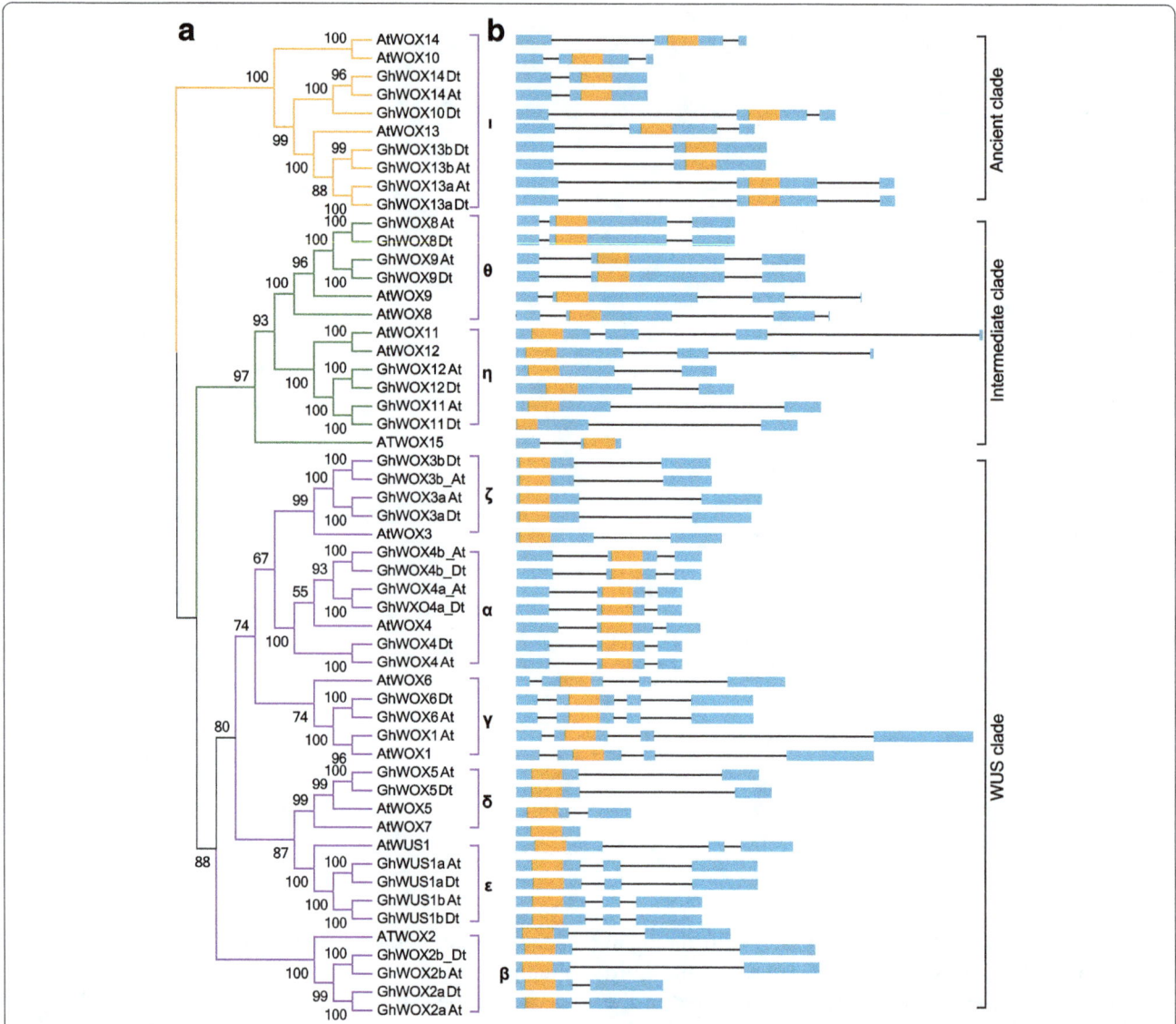

Fig. 4 Comparison of the gene structures between *A. thaliana* and *G. hirsutum*. **a** NJ tree analysis of *A. thaliana* and *G. hirsutum*. Orange, green, and purple in the phylogenetic tree represent the ancient, intermediate, and WUS clades. **b** The number, length, and position of exons and introns within WOX genes. Boxes indicate the exons and black lines indicate the introns. The orange boxes represent the HB domains.

was expressed in root and leaf but *GhWOX8_At* was not. *GhWOX8_Dt* had higher expression levels in 20, 25, and 35 DPA ovule and 10 DPA fiber compared to that in *GhWOX8_At*. *GhWOX13a_At/Dt* and *GhWOX13b_At/Dt* were not only close to each other in the phylogenetic tree (Fig. 2 and Fig. 4a), but also had similar expression patterns (Fig. 5), suggesting that they have a similar biological function.

Cotton faces multiple abiotic stresses during its growth and development. Therefore, a comprehensive analysis of the expression pattern of WOXs was performed in the present study. No obvious changes in the expression levels were observed for more than half of the WOXs under hot, cold, salt, and PEG 6000 conditions.

GhWOX10_Dt responded to multiple stress treatments. The expression of *GhWOX13a_At/Dt* and *GhWOX13-b_At/Dt* was strongly induced by multiple stresses, indicating that they might take part in response to stress and their expression might be regulated by stress.

Examination of the expression of WOX genes by qPCR

To verify the expression profiles of the WOX genes obtained by RNA-seq data, qPCR was performed using the leaves from plants treated with PEG6000 and NaCl. Eight genes, including three orthologous pairs, that were presumed to be highly expressed under PEG6000 or NaCl treatment based on the RNA-seq data, were selected for qPCR, (Fig. 5b; Fig.6). It was difficult to

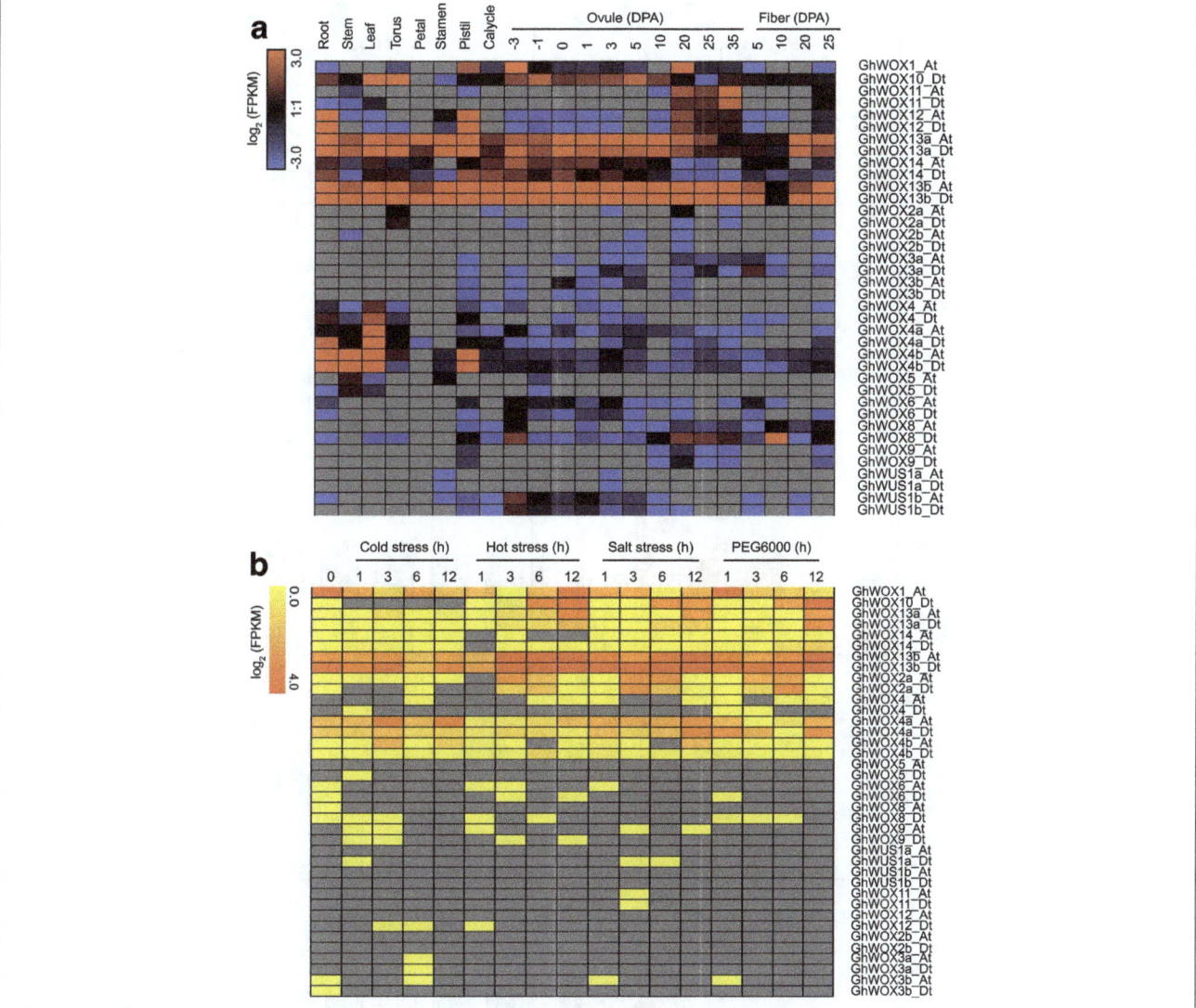

Fig. 5 Pattern of gene expression in different tissues (**a**) and under different stresses (**b**). In (**b**) true leaves were harvested at 0, 1, 3, 6, and 12 h after the treatment.

distinguish between the orthologous pairs by qPCR because the sequences of the orthologous genes (*GhWOX13-a_At-GhWOX13a_Dt*, *GhWOX13b_At-GhWOX13b_Dt*, and *GhWOX4a_At-GhWOX4a_Dt*) were highly similar. Therefore, we designed primers to amplify the orthologous pairs together (Additional file 13: Table S6). The pattern of gene expression assessed by qPCR showed a similar tendency with that detected using the RNA-seq data, as shown in Fig. 5b and Fig. 6. The expression of *GhWOX1_At* was down-regulated under PEG 6000 and NaCl treatment. In contrast, the expression levels of *GhWOX10_Dt*, *GhWOX13b_At/Dt* and *GhWOX13a_At/Dt* were increased under both PEG 6000 and NaCl treatment, indicating that they might play vital roles in the stress response and could be the candidate genes for further study on cotton stress biology.

Discussion

Previous analysis of the WOX gene family has been performed in rice, sorghum, maize, *Arabidopsis*, and poplar [2]. However, there were no reports on the analysis of this family in cotton. In this study, we performed a comprehensive identification of WOX genes in *G. hirsutum*, *G. arboreum*, and *G. raimondii*, mainly focusing on the allotetraploid cotton *G. hirsutum*, with the aim of understanding the roles of this gene family in cotton, in future studies.

Cotton WOX family underwent enlargement during the evolution

Upland cotton is one of the most important cash crops worldwide, providing more than 90% of the natural renew fiber for textile industry. It is also an ideal material

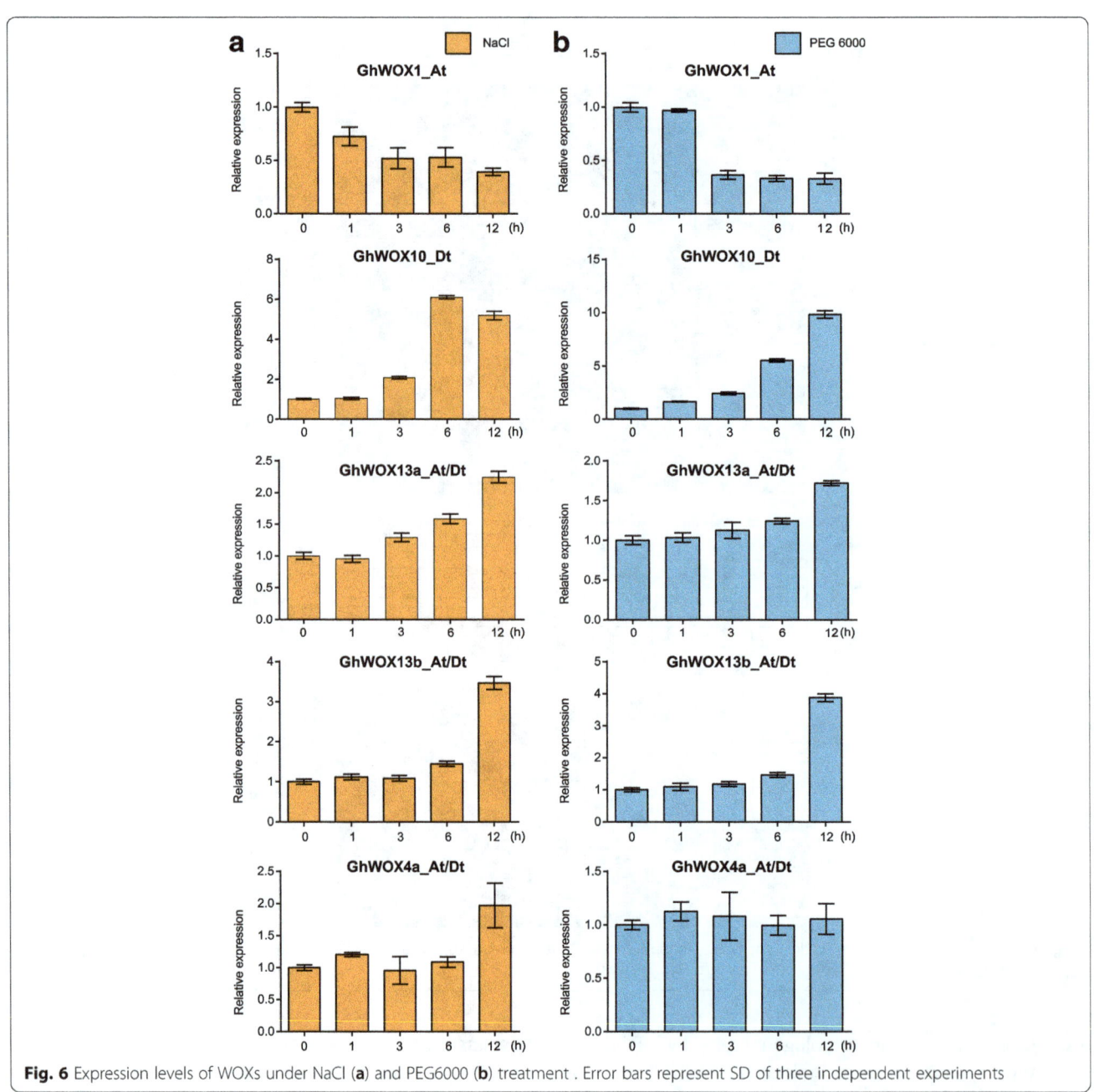

Fig. 6 Expression levels of WOXs under NaCl (**a**) and PEG6000 (**b**) treatment . Error bars represent SD of three independent experiments

for studying the effects of natural polyploidy [48]. A-genome and D-genome diploid cotton were native to Africa and Mexico, respectively, and they diverged about 5–10 million years ago. About 1–2 million years ago, A-genome cotton resembling *G. arboreum* and *G. herbaceum* hybridized with D-genome cotton resembling *G. raimondii*, which was followed by chromosome doubling, and the eventual formation of nascent AtDt allopolyploid, including the upland cotton [22, 48].

A total of 38 WOX genes were identified in the upland cotton, which is more than the number of this gene reported in most of the other species whose genome has

been sequenced [4]. One of the most important features was that the upland cotton experienced polyploidization, resulting in WGD. The At and Dt donors of the upland cotton are close relatives, with almost equal number of orthologs; therefore, the nascent duplication resulted in a number of WOX doublings [49]. Polyploidization is an important event during the evolution of flowering plants that might play important roles in the adaptation of plants to new environment [50]. Although the total number of upland cotton dramatically increased, gene loss also happened during the evolution of upland cotton, as evident from the comparison of the number of genes in At or Dt genomes with that in A- (*G. arboreum*) or D-genome (*G. raimondii*). Gene loss always happened during the rapid arrangement of genomic sequences after hybridization and following chromosome doubling during polyploidization [51]. Compared to the paleopolyploid maize and *Brassica*, cotton displays fewer changes in their genomic sequences [52, 53].

Although polyploidy was the main contributor in duplication, segmental duplication has also been responsible for the increase in the WOX gene number. Segmental duplication is one of the most important impetuses for evolution; it occurs most frequently in plants since most plants are diploidized polyploids and retain large amount of duplicated chromosomal blocks within their genomes [54]. The WRKY genes in soybean underwent large scale segmental duplication, with the duplicated genes exhibiting differential expression and functional divergence [55]. Heat shock proteins play important roles in response to drought stress in sesame; evolutionary analysis showed that segmental duplication was the primary force promoting the expansion of heat shock promoting genes [56]. In our study, 12 out of 38 genes were associated with segmental duplication, indicating that segmental duplication played significant roles in the expansion of WOX genes, and differential expression suggested that the duplicated genes might have experienced functional divergence.

Although we did not find tandem duplication in our study, it is also a basic contributor to gene expansion, which arises from unequal crossing over, chromosomal anomalies, transposon insertions, and other reverse-transcriptase-mediated processes [57]. After duplication, the new gene would be redundant with the previously existing one. The redundancy has been considered a driving force for evolutionary innovation [57]. Some models have been proposed for understanding the gene duplication; these include the neofunctionalization, DDC-subfunctionalization, escape from adaptive conflict (EAC), and dosage-balance models, which provide a theoretical framework for further studying the process [57].

Cotton WOXs have been highly conserved during the evolution

WOX is a plant-specific sub-clade with few members of HD-containing superfamily, which consists of a leucine zipper (HD-ZIP), plant homeodomain associated to a finger domain (PHD finger), the distinctive Bell domain (Bell), zinc finger associated to a homeodomain (ZF-HD), knotted-related homeobox (KNOX), and WUSCHEL-related homeobox (WOX) [58]. Previous studies have shown that HDs (HB-domain) might have diverged before the separation of the branches forming plants, animals, and fungi [59]. The members of HD-containing superfamily were different not only in the sequence of HB domain, but also in their size, HB location, other related domains

as well as in their structures [58]. WOX predominantly has HB domain, and can be distinguished by the phylogenetic relatedness of its homeodomains [1]. It has been reported that HB domain contains some conserved amino acids located in the helices; for example, it contains Q, L, and Y in helix 1 and I, V, W, F, N, K, R, and R in helix 3 [2]. Our results demonstrate that the above amino acids were also conserved in cotton. Previous studies have indicated that the amino acids in loops were less conserved [2], which is in consonance with our results. Previous studies have also shown that WOXs could be divided into three clades [1, 2, 60], as was observed in the present study for cotton (Fig. 2). In a study by Zhang et al., WOX family was further classified into nine sub-clades, as was the case in the present study [2]. We observed that *AtWOX6* was not divided into any sub-clades (Fig. 2) because it had a bootstrap value lower than 50. We found that in the study by Zhang et al., *AtWOX6* was clustered into the same clade as *AtWOX1* despite the bootstrap value not supporting this classification. However, as shown in Fig. 4, we found that *AtWOX6* and *AtWOX1* might have originated from the same ancestor, a fact supported by a high bootstrap value. Therefore, we presumed that the increasing number of sequences in Fig. 2 would make the total number of sequences more divergent, resulting in *AtWOX6* being not stable between Fig. 2 and Fig. 4. Because the annotation is gradually improving, *AtWOX15* (*AT5G46010.1*) was identified as a novel member that was not reported in previous studies [1, 2]. We found that it belonged to the intermediate clade, but it could not be divided into any sub-clade because as per our definition in the present study, a new sub-clade should have at least two members. The monocot- and dicot-specific cluster pattern suggested that the main functions of WOX genes had been determined before the monocots and dicots split. Previous studies have shown that in ζ (NS/WOX3) sub-clade, the genes from monocots formed two separate branches [2], which was consistent with our results. However, we found that there were two pairs of duplicated genes (*GhWOX3a_At-GhWOX3b_At* and *GhWOX3a_Dt-GhWOX3b_Dt*) in cotton, with no orthologs in poplar, which indicated that poplar might have lost these genes during evolution or this might have resulted from an incomplete sequencing result. In the report of Zhang et al., it was thought that the subgroup B (corresponding to sub-clade α in our study) might be an ancient subgroup, because only one member from each species was present in this clade [2]. However, our data did not support this opinion because we found that *GhWOX4_At/Dt*, *GhWOX4a_At/Dt*, and *GhWOX4b_At/Dt* were duplicated genes (Fig. 3). Cacao and cotton diverged from a common ancestor about 18–58 million years ago, and both of them had undergone the ancient duplication event together; cotton had subsequently experienced a nascent duplication event again [25], our data also indicated that cacao and cotton were close relative and probably derived from the same ancestor. One cacao WOX gene should theoretically correspond to two orthologs in the diploid cotton, but in fact some genes in *G. arboreum* and *G. raimondii* might have been lost during the evolution [51]. It is generally believed that *G. hirsutum* (2n = 4x = 52, (AD)$_t$) was reunited by hybridization of an A-genome species resembling *G. arboreum* (2n = 26, A$_2$) with a D-genome material resembling *G. raimondii* (2n = 26, D$_5$), followed by chromosome doubling [24]. Our data suggested that WOXs in A genome and At sub-genome had a common ancestor, and those in D genome and Dt sub-genome had a common ancestor, which was consistent with the above hypothesis, indicating that cotton WOXs were highly conserved during the evolution.

Expression of duplicated WOX genes

Previous studies on the expression and function of WOX genes have indicated that the WOX family members play crucial roles in key developmental processes of plants, such as in embryonic patterning, stem-cell maintenance, and organ formation [1]. Our data (Fig. 5a) suggest that most of the WOX genes had very low expression levels or they did not express in selected tissues, the reason for which might be that WOX family members mainly function in the process of embryogenesis because of which their expression is restricted. Despite the expression levels and the number of expressed genes being different among the different tissues (Fig. 5a), except for *GhWOX8_At/Dt*, the orthologous gene pairs did not show A- or D-ortholog bias, which suggests that they may have conserved functions. Segmental duplication was one of the most important impetuses for increasing the diversity at the molecular level. After duplication, the coding regions could have acquired new regulatory context through the acquisition/deletion of tissue-specific enhancers and repressors, causing the spatial and/or temporal change in the expression pattern of the duplicated gene, promoting diversification of gene functions, like subfunctionalization (acquisition of part of the function of a pre-existing gene) or neofunctionalization (acquisition of a new function) [61]. *AtWOX4* works in procambium development, which is associated with vascular patterning and leaf complexity [62]. In upland cotton, we observed the expansion of *WOX4* through WGD and segmental duplication, and the duplicated genes showed different expression levels in root and stem, which indicated that these genes might have acquired subfunctionalization or a novel function [57].

In *Arabidopsis thaliana*, WOX8 protein not only positively regulates early embryonic growth, but also interacts with *AtCLE8* to promote seed growth and the overall seed size [63]. In the present study, *GhWOX8_Dt*, the ortholog

of *AtWOX8* in upland cotton, was observed to express abundantly during the seed development, indicating that *GhWOX8_Dt* might have similar function in seed development as *AtWOX8* has in *Arabidopsis*. However, *GhWOX8_At* might have a divergent function in cotton seed development, because it had a different expression pattern in ovule compared to that of *GhWOX8_Dt*, suggesting that *GhWOX8s* might have experienced functional divergence after duplication and *GhWOX8_Dt* could be a key candidate gene in cotton seed development that should be studied further.

AtWOX11 interacts with *WOX12*; these two genes are involved in the first-step of cell fate transition during the *de novo* root organogenesis [18]. *GhWOX12* was expressed specifically in root, which suggested that *GhWOX12* might play important roles in root development; however, confirmation as to whether it works together with *GhWOX11* in regulating the *de novo* root organogenesis would need future investigation. *GhWOX11* and *GhWOX12* had similar expression patterns as *GhWOX8_Dt* had in ovule; however, whether these genes have similar functions in seed development would need to be ascertained in further studies.

AtWOX13 is a key component in the regulation of floral transition and root development, and has a high expression level during primary and lateral root initiation and development, in gynoecium, and during embryo development [64]. We found that its orthologs in cotton, *GhWOX13a_At/Dt* and *GhWOX13b_At/Dt*, were dramatically expressed in the root, in reproductive organs, as well as during the embryo development, indicating that *WOX13* genes in cotton and *Arabidopsis* have a conserved expression pattern, and might have a similar biological function. We also noticed that the other homologs of *WOX13* in cotton from the same sub-clade also expressed in all the assessed tissues, which suggested that the genes in the ancient clades might be more active during cotton growth and development. Although most of the reports on WOXs determined their roles in embryogenesis, they might play important roles in stress response. The mutant *hos9-1*, which was generated by a T-DNA insertion into the fourth exon of *AtWOX6*, was reported to be hypersensitive to freezing treatment [13]. Three WOXs from paper mulberry were induced by cold exposure [65]. A previous study showed that *Os08g14400* was up-regulated under desiccation and salt stress, *Os03g20910* was up-regulated under salt stress, and *Os01g63510* was down-regulated under salt stress [61]. In our study, some of the WOX genes were also induced by different stress, indicating that they may mediate the stress response. We found that the duplicated genes *GhWOX4_At/Dt*, *GhWOX4a_At/Dt*, and *GhWOX4b_At*/Dt responded differently when exposed to different stresses, suggesting that they have different functions under different stresses. *GhWOX13b_At/Dt* was induced by multiple stresses, indicating that *GhWOX13b_At/Dt* might be the node for multiple stress regulation. Although WOX genes were shown to have different expression levels under multiple stresses, there are no reports on the validation of the function of cotton WOXs in stress. Therefore, future studies should focus on determining the function of WOX genes in stress response.

Conclusion

Previous studies have illustrated that members of the WOX gene family play significant roles in the regulation of plant growth and development by determining the cell fate. The results of the present study indicate that WOX genes are highly conserved among cotton and other plant species. Furthermore, whole genome and segmental duplication have probably been the two major ways for gene amplification during the expansion of the WOX family in upland cotton. Moreover, the duplicated genes in cotton seem to have experienced functional divergence because the duplicated genes showed different expression patterns in different tissues and organs. In addition, some WOX gene members are likely to be involved in the mediation of stress response. Our results will not only deepen the understanding of the evolutionary processes of cotton WOX genes, but would also be helpful in formulating further functional genomic studies of WOX gene family in cotton.

Additional files

Additional file 1: Table S1. Information on the gene exons used in this study.

Additional file 2: Table S2. The correspondence between WOX genes from two versions of the *G. hirsutum* genome.

Additional file 3: Figure S1. Phylogenetic tree of WOX genes indicating that WOX genes could be divided into three clades. MEGA 7.0 was used for constructing the tree using the minimum-evolution method. The inner circle is marked in purple, green, and orange representing the WUS, intermediate, and ancient clades, respectively. The bootstrap values are shown near the nodes, and only those values greater than 50 are displayed.

Additional file 4: Figure S2. Comparative analysis of *GhWOX1_At* and *Gh_D12G2554*. (a) The collinearity analysis between chromosome A12 from 86,759,600 bp to 86,860,416 bp and D12. (b) An indel on D12 resulted in coding sequence that is divergent in *GhWOX1_At* and *Gh_D12G2554*. (c) Amino sequence alignment of *GhWOX1_At* and *Gh_D12G2554*).

Additional file 5: Blastn result showing the ortholog of *WOX10* in *G. arboreum* located on Chr02.

Additional file 6: Figure S3. Location of cotton WOX genes on chromosomes. The red dotted lines link the orthologs located on At and Dt.

Additional file 7: Table S3. Information on duplicated genes.

Additional file 8: Table S4. Identification of repeat sequences in the region 2,000 bp upstream to 2,000 bp downstream of WOX genes.

Additional file 9: Table S5. Identification of repeat sequences in the region 10,000 bp upstream to 10,000 bp downstream of WOX genes.

Additional file 10: Figure S4. Multiple sequence alignment of *GhWOX5_At* and *GhWOX5_Dt*.

Additional file 11: Figure S5. Multiple sequence alignment of *Gh_A11G2876* and *Gh_WOX10_Dt*.

Additional file 12: Figure S6. Mapping reads around *GhWOX10_Dt* (a, b) and *Gh_A11G2876* (c, d).

Additional file 13: Table S6. The primers used in present study for qPCR.

Abbreviations

BLAST: Basic Local Alignment Search Tool; CUC: Cupshaped cotyledon; DPA: Days post-anthesis; EAC: Escape from adaptive conflict; FPKM: Fragments per kilobase million; *G. arboreum*: *Gossypium arboreum*; *G. hirsutum*: *Gossypium hirsutum*; *G. raimondii*: *Gossypium raimondii*; HB: Homeobox; HD: Homeodomain; NJ: Neighbor-joining; PXY: Phloem intercalated with xylem; TE: Transposable elements; WGD: Whole genome duplication; WOX: WUSCHEL-related homeobox

Acknowledgements

Not Applicable.

Funding

This work was supported by The Major Program of Joint Funds (Sinkiang) of the National Natural Science Foundation of China (Grant No. U1303282) and a grant from Henan Province (No. 162300410160). The funding body supported the funds for research and was not involved in the design, data collection, and analysis, decision to publish, or preparation of the manuscript.

Authors' contributions

FL and Z-E.Y conceived and designed the study; XG, YC. LL and Z-R.Y carried out experiments; Z-E.Y, QG and WQ prepared Figures; YC, LL and Z-R.Y analyzed and interpreted the data; Z-E.Y, QG and WQ prepared the manuscript; XG, CZ and ZW participated in the design of the experiments, wrote part of the manuscript and critical review for intellectual content of it. All the authors have read, edited, and approved the current version of the manuscript.

Competing interests

The authors declare that they have no competing interests.

References

1. van der Graaff E, Laux T, Rensing SA. The WUS homeobox-containing (WOX) protein family. Genome Biology. 2009;10(248):1–9.
2. Zhang X, Zong J, Liu J, Yin J, Zhang D. Genome-Wide Analysis of WOX Gene Family in Rice, Sorghom, Maize, Arabidopsis and Poplar. Journal of integrative plant biology. 2010;52(11):1016–26.

3.   Dolzblasz A, Nardmann J, Clerici E, Causier B, van der Graaff E, Chen J, Davies B, Werr W, Laux T. Stem Cell Regulation by Arabidopsis WOX Genes. Molecular plant. 2016;9(7):1028–39.

4.   Jin J, Tian F, Yang D-C, Meng Y-Q, Kong L, Luo J, Gao G. PlantTFDB 4.0: toward a central hub for transcription factors and regulatory interactions in plants. Nucleic acids research. 2016; doi: 10.1093/nar/gkw982.

5.   Nardmann J, Reisewitz P, Werr W. Discrete Shoot and Root Stem Cell-Promoting WUS/WOX5 Functions Are an Evolutionary Innovation of Angiosperms. Molecular biology and evolution. 2009;26(8):1745–55.

6.   Laux T, Mayer KFX, Berger J, Jürgens G. The WUSCHEL gene is required for shoot and floral meristem integrity in Arabidopsis. Development. 1996;122:87–96.

7.   Wu Z, Xueyan Z, Zuoren Y, Jiahe W, Fenglian L, Fuguang L. AtWuschel Promotes Formation of the Embryogenic Callus in Gossypium hirsutum. PloS one. 2014;9(1):e87502.

8.   Shimizu R, Ji J, Kelsey E, Ohtsu K, Schnable PS, Scanlon MJ. Tissue Specificity and Evolution of Meristematic WOX3 Function. Plant physiology. 2009;149(2):841–50.

9.   Zhang Y, Wu R, Qin G, Chen Z, Gu H, Qu L-J. Over-expression of WOX1 Leads to Defects in Meristem Development and Polyamine Homeostasis in ArabidopsisF. Journal of integrative plant biology. 2011;53(6):493–506.

10. Suer S, Agusti J, Sanchez P, Schwarz M, Greb T. WOX4 Imparts Auxin Responsiveness to Cambium Cells in Arabidopsis. The Plant Cell Online. 2011;23(9):3247–59.

11. Gonzali S, Novi G, Loreti E, Paolicchi F, Poggi A, Alpi A, Perata P. A turanose-insensitive mutant suggests a role for WOX5 in auxin homeostasis in Arabidopsis thaliana. The Plant Journal. 2005;44(4):633–45.

12. Park SO. The PRETTY FEW SEEDS2 gene encodes an Arabidopsis homeodomain protein that regulates ovule development. Development. 2005;132(4):841–9.

13. Zhu J, Shi H, Lee B, Damsz B, Cheng S, Stirm V, Zhu JK, Hasegawa PM, Bressan RA. An Arabidopsis homeodomain transcription factor gene, HOS9, mediates cold tolerance through a CBF-independent pathway. Proceedings of the National Academy of Sciences. 2004;101(26):9873–8.

14. Lie C, Kelsom C, Wu X. WOX2andSTIMPY-LIKE/WOX8promote cotyledon boundary formation in Arabidopsis. The Plant Journal. 2012;72(4):674–82.

15. Wu X, Dabi T, Weigel D. Requirement of Homeobox Gene STIMPY/WOX9 for Arabidopsis Meristem Growth and Maintenance. Current Biology. 2005;15(5):436–40.

16. Skylar A, Hong F, Chory J, Weigel D, Wu X. STIMPY mediates cytokinin signaling during shoot meristem establishment in Arabidopsis seedlings. Development. 2010;137(4):541–9.

17. Zhao Y, Hu Y, Dai M, Huang L, Zhou DX. The WUSCHEL-Related Homeobox Gene WOX11 Is Required to Activate Shoot-Borne Crown Root Development in Rice. The Plant Cell Online. 2009;21(3):736–48.

18. Liu J, Sheng L, Xu Y, Li J, Yang Z, Huang H, Xu L. WOX11 and 12 are involved in the first-step cell fate transition during de novo root organogenesis in Arabidopsis. The Plant cell. 2014;26(3):1081–93.

19. Romera-Branchat M, Ripoll JJ, Yanofsky MF, Pelaz S. TheWOX13homeobox gene promotes replum formation in theArabidopsis thalianafruit. The Plant Journal. 2013;73(1):37–49.

20. Etchells JP, Provost CM, Mishra L, Turner SR. WOX4 and WOX14 act downstream of the PXY receptor kinase to regulate plant vascular proliferation independently of any role in vascular organisation. Development. 2013;140(10):2224–34.

21. Yang ZR, Zhang CJ, Yang XJ, Liu K, Wu ZX, Zhang XY, Zheng W, Xun QQ, Liu CL, Lu LL, et al. PAG1, a cotton brassinosteroid catabolism gene, modulates fiber elongation. New Phytologist. 2014;203(2):437–48.

22. Li F, Fan G, Lu C, Xiao G, Zou C, Kohel RJ, Ma Z, Shang H, Ma X, Wu J, et al. Genome sequence of cultivated Upland cotton (Gossypium hirsutum TM-1) provides insights into genome evolution. Nature biotechnology. 2015;33(5):524–30.

23. Zhang T, Hu Y, Jiang W, Fang L, Guan X, Chen J, Zhang J, Saski CA, Scheffler BE, Stelly DM, et al. Sequencing of allotetraploid cotton (Gossypium hirsutum L. acc. TM-1) provides a resource for fiber improvement. Nature biotechnology. 2015;33(5):531–7.

24. Wang K, Wang Z, Li F, Ye W, Wang J, Song G, Yue Z, Cong L, Shang H, Zhu S, et al. The draft genome of a diploid cotton Gossypium raimondii. Nature genetics. 2012;44(10):1098–103.

25. Li F, Fan G, Wang K, Sun F, Yuan Y, Song G, Li Q, Ma Z, Lu C, Zou C, et al. Genome sequence of the cultivated cotton Gossypium arboreum. Nature genetics. 2014;46(6):567–72.

26. Paterson AH, Wendel JF, Gundlach H, Guo H, Jenkins J, Jin D, Llewellyn D, Showmaker KC, Shu S, Udall J, et al. Repeated polyploidization of Gossypium genomes and the evolution of spinnable cotton fibres. Nature. 2012;492(7429):423–7.

27. Yu J, Jung S, Cheng CH, Ficklin SP, Lee T, Zheng P, Jones D, Percy RG, Main D. CottonGen: a genomics, genetics and breeding database for cotton research. Nucleic acids research. 2014;42(Database issue):D1229–36.

28. Jones P, Binns D, Chang HY, Fraser M, Li W, McAnulla C, McWilliam H, Maslen J, Mitchell A, Nuka G, et al. InterProScan 5: genome-scale protein function classification. Bioinformatics. 2014;30(9):1236–40.

29. Crooks GE, Hon G, Chandonia J-M, Brenner SE. WebLogo: A Sequence Logo Generator. Genome Research. 2004;14:1188–90.

30. Kumar S, Stecher G, Tamura K. MEGA7: Molecular Evolutionary Genetics Analysis Version 7.0 for Bigger Datasets. Molecular Biology and Evolution. 2016;33(7):1870–4.

31. Voorrips RE. MapChart: Software for the Graphical Presentation of Linkage Maps and QTLs. Journal of Heredity. 2001;93(1):77–8.

32. Krzywinski M, Schein J, Birol I, Connors J, Gascoyne R, Horsman D, Jones SJ, Marra MA. Circos: An information aesthetic for comparative genomics. Genome Research. 2009;19(9):1639–45.

33. Suyama M, Torrents D, Bork P. PAL2NAL: robust conversion of protein sequence alignments into the corresponding codon alignments. Nucleic acids research. 2006;34(Web Server issue):W609–12.

34. Yang Z. PAML 4: phylogenetic analysis by maximum likelihood. Molecular biology and evolution. 2007;24(8):1586–91.

35. Jurka J. Repbase Update a database and an electronic journal of repetitive elements. Trends in Genetics. 2000;16(9):418–20.

36. Myers RCEEW. PILER: identification and classification of genomic repeats. Bioinformatics. 2003;1(1):1–7.

37. Xu Z, Wang H. LTR_FINDER: an efficient tool for the prediction of full-length LTR retrotransposons. Nucleic acids research. 2007;35(Web Server issue):265–8.

38. Chen N. Using Repeat Masker to Identify Repetitive Elements in Genomic Sequences. Current Protocols in Bioinformatics. 2004;4(4):4.10.11–14.10.14.

39. Hu B, Jin J, Guo AY, Zhang H, Luo J, Gao G. GSDS 2.0: an upgraded gene feature visualization server. Bioinformatics. 2014;31(8):1296–7.

40. Trapnell C, Roberts A, Goff O, Pertea G, Kim D, Kelley DR, Pimentel H, Salzberg SL, Rinn JL, Pachter L. Differential gene and transcript expression analysis of RNA-seq experiments with TopHat and Cufflinks. Nat Protoc. 2013;7(3):562–78.

41. Sturn A, Quackenbush J, Trajanoski Z. Genesis: cluster analysis of microarray data. Bioinformatics applications note. 2002;18(1):207–8.

42. Yang Z, Zhang C, Yang X, Liu K, Wu Z, Zhang X, Zheng W, Xun Q, Liu C, Lu L, et al. PAG1, a cotton brassinosteroid catabolism gene, modulates fiber elongation. The New phytologist. 2014;203(2):437–48.

43. Nan Du YS. Improve homology search sensitivity of PacBio data by correcting frameshifts. Bioinformatics. 2016;32(17):529–37.

44. Prince VE, Pickett FB. Splitting pairs: the diverging fates of duplicated genes. Nature reviews Genetics. 2002;3(11):827–37.

45. Vandepoele K. Evidence That Rice and Other Cereals Are Ancient Aneuploids. The Plant Cell Online. 2003;15(9):2192–202.

46. Oliver KR, McComb JA, Greene WK. Transposable elements: powerful contributors to angiosperm evolution and diversity. Genome biology and evolution. 2013;5(10):1886–901.

47. Bennetzen JL, Wang H. The contributions of transposable elements to the structure, function, and evolution of plant genomes. Annual review of plant biology. 2014;65:505–30.

48. Wendel JF, Cronn RC. Polyploidy and the evolutionary history of cotton. Advances in Agronomy. 2003;78:139-186.

49. Otto SP. The evolutionary consequences of polyploidy. Cell. 2007;131(3):452–62.

50. Ramsey J, Schemske DW, Soltis PS. Soltis DE. Pathways, mechanisms, and rates of polyploid formation in flowering plants. 1998;60:467–501.

51. Paterson AH, Bowers JE, Chapman BA. Ancient polyploidization predating divergence of the cereals, and its consequences for comparative genomics. Proceedings of the National Academy of Sciences. 2004;101(26):9903–8.

52. Woodhouse MR, Schnable JC, Pedersen BS, Lyons E, Lisch D, Subramaniam S, Freeling M. Following tetraploidy in maize, a short deletion mechanism removed genes preferentially from one of the two homologs. PLoS biology. 2010;8(6):e1000409.

53. Gaeta RT, Pires JC, Iniguez-Luy F, Leon E, Osborn TC. Genomic changes in resynthesized Brassica napus and their effect on gene expression and phenotype. The Plant cell. 2007;19(11):3403–17.

54. Cannon SB, Mitra A, Baumgarten A, Young ND, May G. The roles of segmental and tandem gene duplication in the evolution of large gene families in Arabidopsis thaliana. BMC plant biology. 2004;4:10.

55. Yin G, Xu H, Xiao S, Qin Y, Li Y, Yan Y, Hu Y. The large soybean (Glycine max) WRKY TF family expanded by segmental duplication events and subsequent divergent selection among subgroups. BMC plant biology. 2013;13:148.

56. Dossa K, Diouf D, Cisse N. Genome-Wide Investigation of Hsf Genes in Sesame Reveals Their Segmental Duplication Expansion and Their Active Role in Drought Stress Response. Frontiers in plant science. 2016;7:1522.

57. Flagel LE, Wendel JF. Gene duplication and evolutionary novelty in plants. The New phytologist. 2009;183(3):557–64.

58. Ariel FD, Manavella PA, Dezar CA, Chan RL. The true story of the HD-Zip family. Trends in plant science. 2007;12(9):419–26.

59. Bharathan G, Janssen B-J, Kellogg EA, Sinha N. Did homeodomain proteins duplicate before the origin of angiosperms, fungi, and metazoa? Proceedings of the National Academy Science of USA. 1997;94:13749–53.

60. Ge Y, Liu J, Zeng M, He J, Qin P, Huang H, Xu L. Identification of WOX Family Genes in Selaginella kraussiana for Studies on Stem Cells and Regeneration in Lycophytes. Frontiers in plant science. 2016;7:93.

61. Jain M, Tyagi AK, Khurana JP. Genome-wide identification, classification, evolutionary expansion and expression analyses of homeobox genes in rice. The FEBS journal. 2008;275(11):2845–61.

62. Ji J, Shimizu R, Sinha N, Scanlon MJ. Analyses of WOX4 transgenics provide further evidence for the evolution of the WOX gene family during the regulation of diverse stem cell functions. Plant signaling & behavior. 2010;5(7):916–20.

63. Fiume E, Fletcher JC. Regulation of Arabidopsis embryo and endosperm development by the polypeptide signaling molecule CLE8. The Plant cell. 2012;24(3):1000–12.

64. Deveaux Y, Toffano-Nioche C, Claisse G, Thareau V, Morin H, Laufs P, Moreau H, Kreis M, Lecharny A. Genes of the most conserved WOX clade in plants affect root and flower development in Arabidopsis. BMC Evolutionary Biology. 2008;8(1):291.

65. Peng X, Wu Q, Teng L, Tang F, Pi Z, Shen S. Transcriptional regulation of the paper mulberry under cold stress as revealed by a comprehensive analysis of transcription factors. BMC plant biology. 2015;15:108.

13

Combined biotic stresses trigger similar transcriptomic responses but contrasting resistance against a chewing herbivore in *Brassica nigra*

Christelle Bonnet[1], Steve Lassueur[1], Camille Ponzio[2], Rieta Gols[2], Marcel Dicke[2] and Philippe Reymond[1]* (iD)

Abstract

Background: In nature, plants are frequently exposed to simultaneous biotic stresses that activate distinct and often antagonistic defense signaling pathways. How plants integrate this information and whether they prioritize one stress over the other is not well understood.

Results: We investigated the transcriptome signature of the wild annual crucifer, *Brassica nigra*, in response to eggs and caterpillars of *Pieris brassicae* butterflies, *Brevicoryne brassicae* aphids and the bacterial phytopathogen *Xanthomonas campestris* pv. *raphani* (*Xcr*). Pretreatment with egg extract, aphids, or *Xcr* had a weak impact on the subsequent transcriptome profile of plants challenged with caterpillars, suggesting that the second stress dominates the transcriptional response. Nevertheless, *P. brassicae* larval performance was strongly affected by egg extract or *Xcr* pretreatment and depended on the site where the initial stress was applied. Although egg extract and *Xcr* pretreatments inhibited insect-induced defense gene expression, suggesting salicylic acid (SA)/jasmonic acid (JA) pathway cross talk, this was not strictly correlated with larval performance.

Conclusion: These results emphasize the need to better integrate plant responses at different levels of biological organization and to consider localized effects in order to predict the consequence of multiple stresses on plant resistance.

Keywords: *Brassica nigra*, *Brevicoryne brassicae*, Combined stresses, Herbivory, *Pieris brassicae*, Transcriptome, *Xanthomonas campestris* pv. *raphani* (*Xcr*)

Background

Biotic and abiotic stresses impose a strong pressure on plants in nature. When combined, stresses such as heat, drought or high light intensity have profound effects on crop performance and yields [1]. Plants have developed specific mechanisms to precisely detect environmental changes and respond to complex stress conditions to minimize damage and conserve sufficient resources for growth and reproduction. Over the years, research has focused mainly on responses to a single stress in several model plants including Arabidopsis [2–5]. However, there is a growing recognition for the need to consider the effects of multiple stresses at the molecular level and at higher levels of biological organization [6–9]. Such an approach is crucial as we need to know how plants adapt to novel environmental factors in the context of co-occurring stresses [10].

Insect herbivory is a major biotic stress under natural conditions. Therefore, plants have evolved sophisticated constitutive and inducible defenses to resist or reduce the effects of insect attack [11]. Several studies have shown that plants subjected to abiotic stress or nutritional limitation differentially affect the performance and behavior of insects [12–17]. In addition to insects, plant

---
* Correspondence: Philippe.Reymond@unil.ch
[1]Department of Plant Molecular Biology, University of Lausanne, Biophore Building, 1015 Lausanne, Switzerland
Full list of author information is available at the end of the article

pathogens are a major threat to plant growth and survival, but also impact on the colonization by and performance of herbivores feeding on pathogen-infected plants [18, 19]. As biotrophic and necrotrophic phytopathogens exhibit distinct infection pathways, they induce different plant responses [20]. Their effects on plants may influence the phytochemical environment of the insect attacker in different ways. Moreover, pathogen effects on plant resistance to insects will depend on the biology of the herbivore, e.g., whether it is a phloem feeder or a chewing larva [19]. For example, the necrotrophic pathogen *Botrytis cinerea* inhibited the development, fecundity and survival rate of the aphid *Aphis fabae* in *Vicia faba*, whereas the biotrophic fungus *Uromyces viciae-fabae* enhanced aphid performance [21]. Most interestingly, the effect of combined *B. cinerea* and *U. viciae-fabae* stress on aphid performance seemed to fluctuate depending on the order of infection [21]. In contrast, *B. cinerea* pretreatment had no significant effect on further performance of *Pieris rapae* caterpillars in Arabidopsis [22]. Tomato plants challenged by *Pseudomonas syringae* reduced *Spodoptera exigua* growth, whereas tomato mosaic virus increased caterpillar performance but decreased aphid colonization [23]. *Manduca sexta* larvae feeding on *Nicotiana attenuata* plants treated with the bacterial quorum-sensing *N*-acylhomoserine lactone were significantly heavier than on untreated plants. This effect was attributed to an inhibition of plant defenses against herbivores [24]. Oviposition by *Pieris brassicae* inhibited growth of *P. syringae* strains in Arabidopsis [25]. Furthermore, *P. brassicae* larvae showed a poor performance on *P. syringae*-infected Arabidopsis plants, suggesting that insect eggs inhibit plant defenses for the benefit of their progeny [25]. In summary, the outcome of a biotic pretreatment on herbivore performance is difficult to predict and depends primarily on the severity and duration of the infection, attack strategies of the pathogens and herbivores involved, and the plant species that is attacked.

Induced defenses are controlled by phytohormones. Biotrophic pathogens, which obtain nutrients from living tissues activate mainly the salicylic acid (SA) pathway, while necrotrophs obtaining nutrients from dead host tissues and chewing herbivores activate especially the jasmonic acid (JA) and ethylene (ET) pathways [26]. These pathways regulate the expression of defense genes that provide specific resistance to the attacker. The existence of antagonism between the SA and JA pathways is well established [26]. It is thought to modulate prioritization of defense allocation towards different attackers [27, 28] but is also the target of defense manipulation by plant pathogens and insect herbivores [26, 29–33]. Stimulation of the SA pathway attenuated plant response to generalist herbivores, i.e. herbivores feeding on plant species in different plant families, but had no effect on the specialist *P. brassicae*, which primarily feeds on plant species within the Brassicaceae family [34, 35]. At the molecular level, treatments with SA or pathogens that enhance SA levels reduced the expression of the anti-herbivore *VSP2* in Arabidopsis [36, 37]. Pathway cross talk may thus represent a crucial component of plant responses to combined stresses.

Transcriptome analyses have been conducted to better understand plant responses to multiple stresses. For instance, one study analyzed transcriptomic differences in ten ecotypes of Arabidopsis challenged by single or dual (a) biotic stress combinations. The authors concluded that the majority of changes in gene regulation in response to combined stresses were not predictable using expression profiles from single treatments [38]. Drought or flooding pretreatment significantly modified the transcriptome signature of *Solanum dulcamara* plants infested with *S. exigua* [16]. Simultaneous attack by sap-feeding and chewing herbivores in *N. attenuata* triggered a transcriptional response that was distinct from those in response to single attackers [39]. An overview of 33 different combined stresses revealed that each treatment seems to generate a unique response, reflecting the plant's ability to specifically adapt to a changing and complex environment [8]. The same conclusion was reached for the impact of combined stresses at the metabolomic and proteomic level, with several unique metabolites and proteins accumulating after multiple stresses but not after single stresses [8, 40]. However, Arabidopsis plants challenged by both nematodes and drought responded primarily to drought [41].

Thus, plant responses to multiple stresses are interconnected and result in complicated and unpredictable outcomes. More studies on plant responses to combined stress conditions are critical to understand the effects of these interactions. This requires analysis at multiple levels, transcriptional and hormonal responses, defense compound accumulation and ecological consequences, using different plant species. Here, we investigated the effects of combined biotics stresses on plant transcriptomic changes, changes in plant hormones and metabolites and insect performance, using and ecologically relevant system. We selected the wild annual crucifer, *Brassica nigra*, subjected to feeding by a naturally associated lepidopteran pest, caterpillars of the large cabbage white *P. brassicae*, alone or in combination with a second stress. Combined stresses consisted of a pretreatment with *P. brassicae* egg extract, the cabbage aphid *Brevicoryne brassicae*, or the necrotrophic bacterial phytopathogen *Xanthomonas campestris* pv. *raphani* (*Xcr*), followed by caterpillar herbivory. All stresses used here occur naturally on *B. nigra* in the field [42, 43]. Whereas plant defenses against chewing larvae are primarily regulated by the JA pathway, eggs and aphids activate primarily the SA pathway [35, 44–46], and defense against *Xcr* is mediated by SA, JA and ET [47]. Given the known mutual antagonistic actions of these signaling pathways, we were expecting significant effects of a primary stress on the responses to *P. brassicae* larvae. Interactions of *B. nigra*-attacker interactions are well-investigated at the ecological level under field conditions where multiple attackers occur [43, 48, 49]. However, much less is known about the mechanistic aspects of the responses of *B. nigra* plants to single as compared to combined stresses. This is the topic of the present study.

Results

Effect of combined stresses on transcriptional responses to herbivory

We used whole-genome Arabidopsis CATMA microarrays [50, 51] to assess gene expression changes in *B. nigra*. Previous studies have shown that Arabidopsis microarrays can be successfully used to study transcriptional responses of *Brassica oleracea* or *B. nigra* [52, 53]. After 1 day of feeding by *P. brassicae* larvae on *B. nigra*, 218 genes were significantly upregulated ($\log_2 > 0.585$, $P < 0.05$) and 49 genes were significantly downregulated ($\log2 > -0.585$, $P < 0.05$) (Additional file 1: Table S1). Gene ontology (GO) search of the upregulated genes revealed a highly significant enrichment of terms including response to wounding (GO:0009611), response to stress (GO:0006950), response to jasmonic acid stimulus (GO:0009753), response to biotic stimulus (GO:0009607), response to chitin (GO:0010200), defense response (GO:0006952), jasmonic acid biosynthesis (GO:0009695), oxylipin biosynthesis (GO:0031408), secondary metabolic process (GO:0019748). Downregulated genes were enriched in terms like photosynthesis (GO:0015979), cellular metabolic process (GO:0044237), nitrogen metabolism (GO:0034631, GO:0044271), chloroplast (GO:0005907, GO:0044434) (Additional file 2: Table S2). This transcriptional signature confirms results from previous studies on the response to caterpillar herbivory in other plant species, which have identified a crucial role for the jasmonate pathway in inducing anti-insect defense genes and observed a downregulation of photosynthesis-related genes [45, 54–59]. Because we used Arabidopsis microarrays, some more distantly related *B. nigra* defense genes may have been missed in the hybridization procedure. A more exhaustive list of insect-responsive genes will await transcriptome analyses by RNA sequencing once a *B. nigra* reference genome is available.

Then, to investigate how a biotic pretreatment may affect *P. brassicae*-induced transcriptome changes, we challenged *B. nigra* plants with *P. brassicae* egg extract, the bacterial pathogen *Xcr*, or *B. brassicae* aphids before

adding *P. brassicae* larvae for 24 h. As control experiments, we subjected *B. nigra* plants to each single stress. Strikingly, an expression-based clustering analysis of all experiments showed that transcriptomes from the three combined stress treatments were grouped with the transcriptome of *P. brassicae* larval treatment, whereas transcriptomes from egg extract, the bacterial pathogen, or aphid single treatments were clearly separated (Fig. 1a). Indeed, from the list of 218 uregulated and 49 downregulated genes after herbivory alone, 206 (94%), repectively 43 (88%), were still similarly regulated after egg extract pretreatment, 155 (72%), respectively 38 (78%), after pathogen pretreatment, and 201 (92%), respectively 46 (94%), after aphid pretreatment, indicating that the biotic pretreatments applied had a weak effect on the subsequent transcriptional response to herbivory (Fig. 1b, c). Analysis of the 50 top up- and downregulated genes after single treatment with caterpillars showed that expression of 48, respectively 42 genes, did not differ significantly between single or combined stress with egg extract. Similarly, 36 upregulated and 46 downregulated genes were not expressed differently between herbivory and combined stress with aphids (Additional files 3, 4: Figures S1, S2). However, in the case of pathogen pretreatment, 22 of the top-50 genes showed a significantly reduced induction, including known JA-regulated genes like *LOX3*, *CORI3*, and *OPR3*, suggesting that bacterial infection inhibits defense against herbivory (Additional file 3: Figure S1).

A combination of stresses may activate genes that are normally not regulated during single stresses. To identify a specific signature of a combined stress, we searched for genes that were significantly induced or repressed only in the three dual-stress treatments (egg extract/caterpillars, pathogen/caterpillars/, or aphids/caterpillars). There were respectively 7, 23, and 52 upregulated genes and 16, 15 and 13 downregulated genes meeting these criteria. Strikingly, a comparison of these combined-stress-specific genes indicated that only one gene was commonly regulated in egg extract/caterpillars and pathogen/caterpillars while other genes were specifically regulated by each combination of stresses (Fig. 2). A GO search of the combined-stress responsive genes did not reveal enrichment of any particular or conserved biological process (Additional files 5, 6: Figures S3, S4). These results indicate that there is no typical transcriptional signature of a combined stress but that each combination activates a relatively small number of additional genes.

Effect of combined stresses on larval performance and plant defense compounds

Performance of *P. brassicae* caterpillars was also measured in terms of weight gain on plants pretreated with egg extract or the pathogen and on untreated plants.

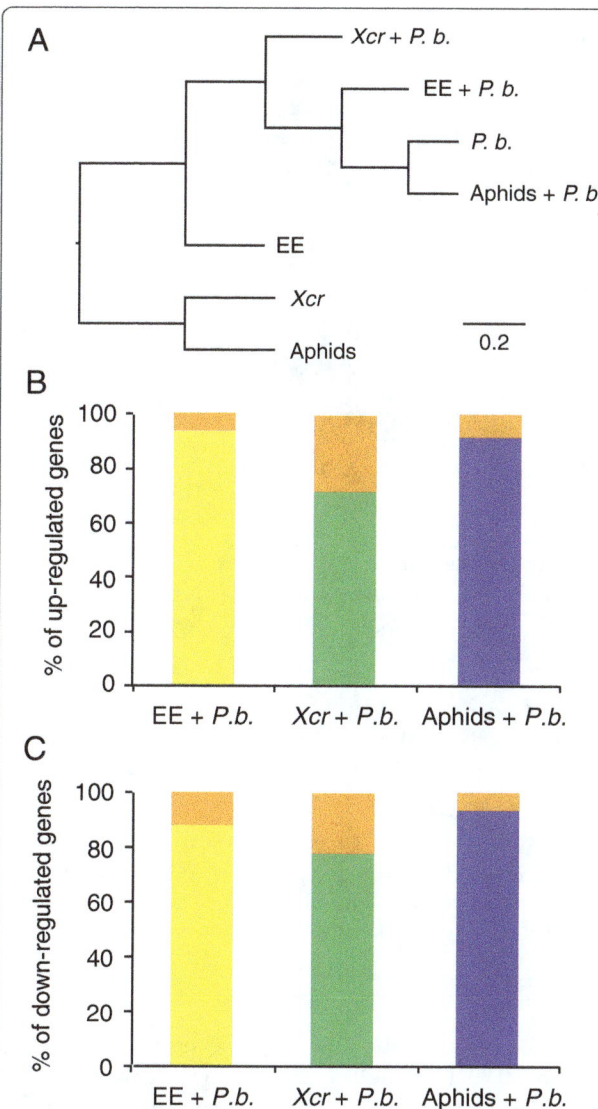

Fig. 1 Expression profiles in response to single and combined stresses in *Brassica nigra* plants. **a** Correspondence analysis of expression profiles including all induced or repressed genes in at least one experiment (−0.585 < log₂ ratio > 0.585, P < 0.05, n = 961). Clustering and node length calculations were performed with MultiExperiment Viewer 4.8.1 using Pearson's correlation. **b** Proportion of *P. brassicae*-upregulated genes that are also upregulated during a combined stress. Each bar segment (*yellow, green, blue*) represents a different combined stress. The proportion of genes specifically induced by *P. brassicae* is shown in orange. The number of genes regulated by herbivory (upregulated, n = 218; downregulated, n = 49) is set to 100%. (**c**) Proportion of *P. brassicae*-downregulated genes also downregulated during a combined stress. *P.b.*, *P. brassicae* larvae; EE, *P. brassicae* egg extract; *Xcr*, *Xanthomonas campestris* pv. *raphani*; Aphids, *Brevicoryne brassicae*

When caterpillars were feeding freely on entire leaves pre-treated with egg extract or the pathogen, their weight gain was significantly reduced compared to that on control plants (linear mixed model (LMM), $P_{[egg\ extract]}$ = 2*10⁻¹⁶ and pathogen, $P_{[pathogen]}$ = 2*10⁻⁸) (Fig. 3a).

To test whether altered insect performance on pretreated plants correlated with changes in defense signals and metabolites, we quantified SA, JA, and glucosinolate (GS) concentrations. GS are potent defense compounds in brassicaceous plants, effective against generalist insects, that accumulate in response to herbivory [60–64]. However, results from whole-leaf analyses showed that concentrations of JA and total GS were not significantly different in leaves that were pretreated with egg extract or the pathogen followed by caterpillar feeding and in leaves exposed to caterpillar feeding alone (Two-way ANOVA, $P_{[JA,\ egg\ extract]}$ = 0.62, $P_{[JA,\ pathogen]}$ = 0.17, $P_{[GS,\ egg\ extract]}$ = 0.46, $P_{[GS,\ pathogen]}$ = 0.41) (Additional file 7: Figure S5). For SA, the presence or absence of caterpillar feeding did not alter the significant accumulation in response to egg extract or pathogen treatment (Two-way ANOVA, $P_{[SA,\ egg\ extract]}$ = 0.20, $P_{[SA,\ pathogen]}$ = 0.49) (Additional file 7: Figure S5).

Since the difference in larval performance between untreated and pretreated plants was not easily explained by transcriptomic data, by changes in defense hormonal signaling, or by GS accumulation in whole *B. nigra* leaves, we decided to study the effect of pretreatment relative to the feeding site of the caterpillars. The rationale was that locally induced changes within the leaf may account for the observed effects. Noteworthy, a recent study on maize has reported statistically significant higher concentration of defense metabolites, i.e. 1,4-benzoxazin-3-ones, in young leaves compared to old ones and that this was negatively correlated with insect performance [65]. We modified the experimental design by constraining larvae in clip cages locally or distally, relative to the pretreatment site. Caterpillar weight gain was significantly reduced after egg extract pretreatment, but only when caterpillars were forced to feed on the pretreatment site (LMM, $P_{[local]}$ = 0.029; $P_{[distal]}$ = 0.37) (Fig. 3b). The effect was similar to the whole-leaf response (Fig. 3a). In contrast, caterpillars performed significantly better on pathogen-pretreated site than on control leaves (LMM, P = 9*10⁻⁷) but their weight was not different when forced to feed distally from the pretreatment site (LMM, P = 0.22), suggesting for instance a local suppression of defenses by bacterial effectors (Fig. 3b). This result was different from the result of the whole-leaf experiment, where caterpillar performance was reduced on pathogen-pretreated leaves (Fig. 3a). Thus, the respective localization of pathogen pretreatment and caterpillar feeding site clearly impacted the effect of the pathogen on insect performance.

To further correlate insect performance and site of treatment with defense compound and signaling hormone accumulation, we quantified SA, JA, and GS concentrations in leaf tissues collected from untreated plants, and from pretreated plants at the site where the

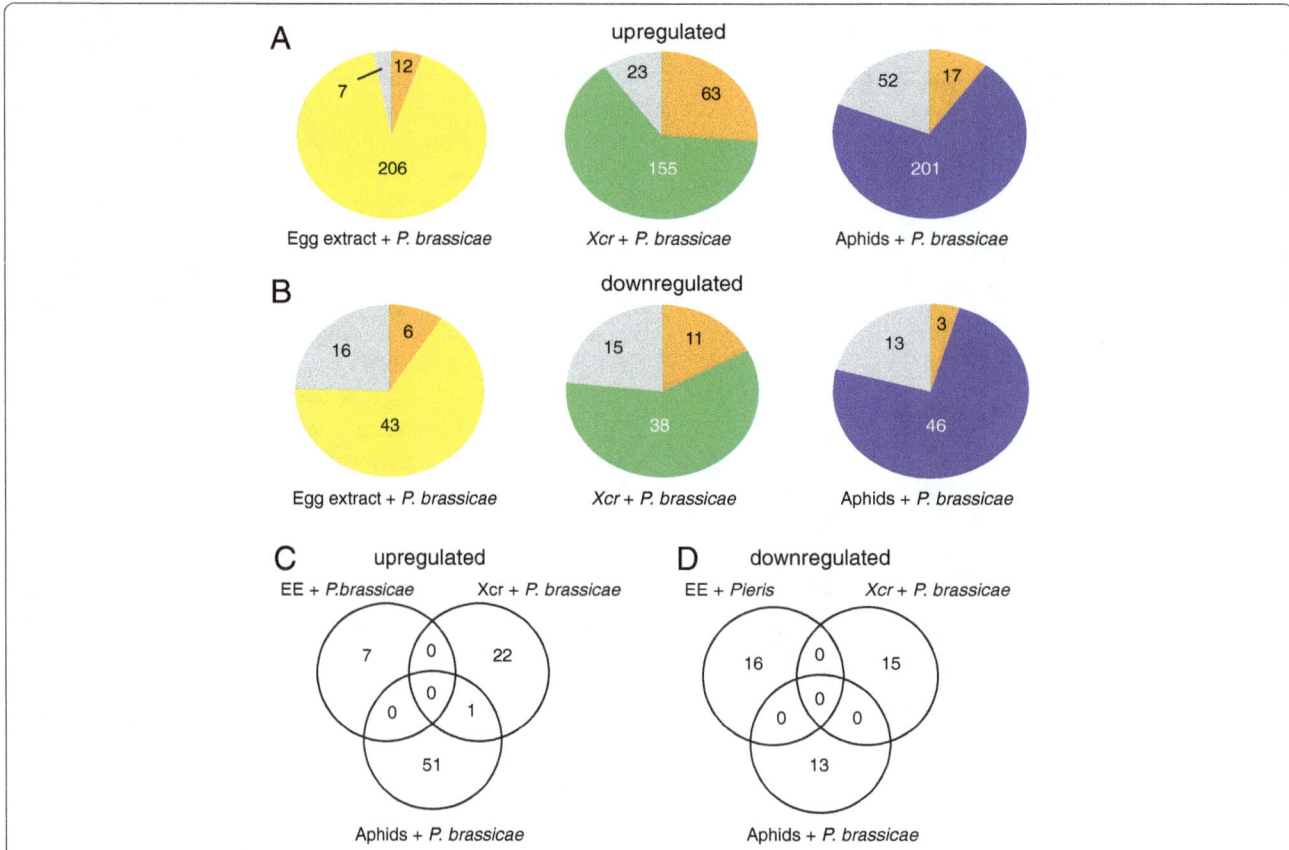

Fig. 2 Number of **a** upregulated and **b** downregulated genes in response to single and combined stresses. Number of genes differentially regulated after combined *P. brassicae* feeding and *P. brassicae* egg extract (*yellow*), *Xanthomonas campestris* pv. *raphani* (*green*) and *Brevicoryne brassicae* (*blue*) are indicated. Number of genes specifically regulated by *P. brassicae* feeding (*orange*) and specifically regulated by a combined stress (*grey*) is also indicated. **c, d** Distribution of genes specifically regulated by a combined stress shows very little overlap

pretreatment stress was applied. In plants that were also exposed to caterpillar feeding, leaf tissues were collected from the areas where the caterpillars were constrained, so at the site of pretreatment or at a site distal from the pretreatment application (Fig. 4a). In the experiment with egg extract, JA accumulated only in response to herbivory. Egg-extract pretreatment itself did not cause JA accumulation and did not alter caterpillar-induced JA concentration (ANOVA, $F = 13,31$, $P = 0.0001$) (Fig. 4b). In contrast, pretreatment with the pathogen resulted in a 10-fold increase of JA, whereas JA concentrations in leaf tissues exposed to caterpillar feeding alone were not significantly different from those found in control plants (ANOVA, $F = 68.85$, $P = 5*10^{-7}$) (Fig. 4b). JA was induced equally (5-fold) when caterpillars were feeding distally from the site where the pathogen was applied, and when caterpillars were feeding at the same site (Fig. 4b). Egg extract and pathogen pretreatments significantly induced SA, only at the treatment site. Moreover, there was no change in SA concentrations in response to herbivory, and caterpillars did not affect egg extract- or pathogen-induced SA concentrations (ANOVA, $F_{[egg\ extract]} = 18.37$,

$P_{[egg\ extract]} = 2*10^{-5}$, $F_{[pathogen]} = 66.18$, $P_{[pathogen]} = 5*10^{-8}$) (Fig. 4c). Thus, *B. nigra* leaves respond locally to biotic challenges by accumulating distinct JA or SA concentrations depending on the biotic stress. Furthermore, the SA response to combined stresses did not differ from the hormonal response to single stresses, whereas it did for JA.

As with whole-leaf experiments, total GS concentrations did not change significantly between control and treated plants (ANOVA, $F_{[egg\ extract]} = 2.71$, $P_{[egg\ extract]} = 0.058$, $F_{[pathogen]} = 0.66$, $P_{[pathogen]} = 0.68$) (Fig. 4d). Among the 13 glucosinolates that were identified and quantified (Additional file 8: Table S3), sinigrin contributed 91% to 96% of the total GS content in different treatments. Consistent with a lack of GS accumulation after biotic stress in *B. nigra*, we observed that expression of 22 out of 27 GS biosynthesis genes was not significantly enhanced in response to single or combined stresses (Additional files 1, 9: Table S1, Figure S6). This contrasts with the coordinated induction of genes involved in all steps of GS biosynthesis in Arabidopsis after herbivory [64].

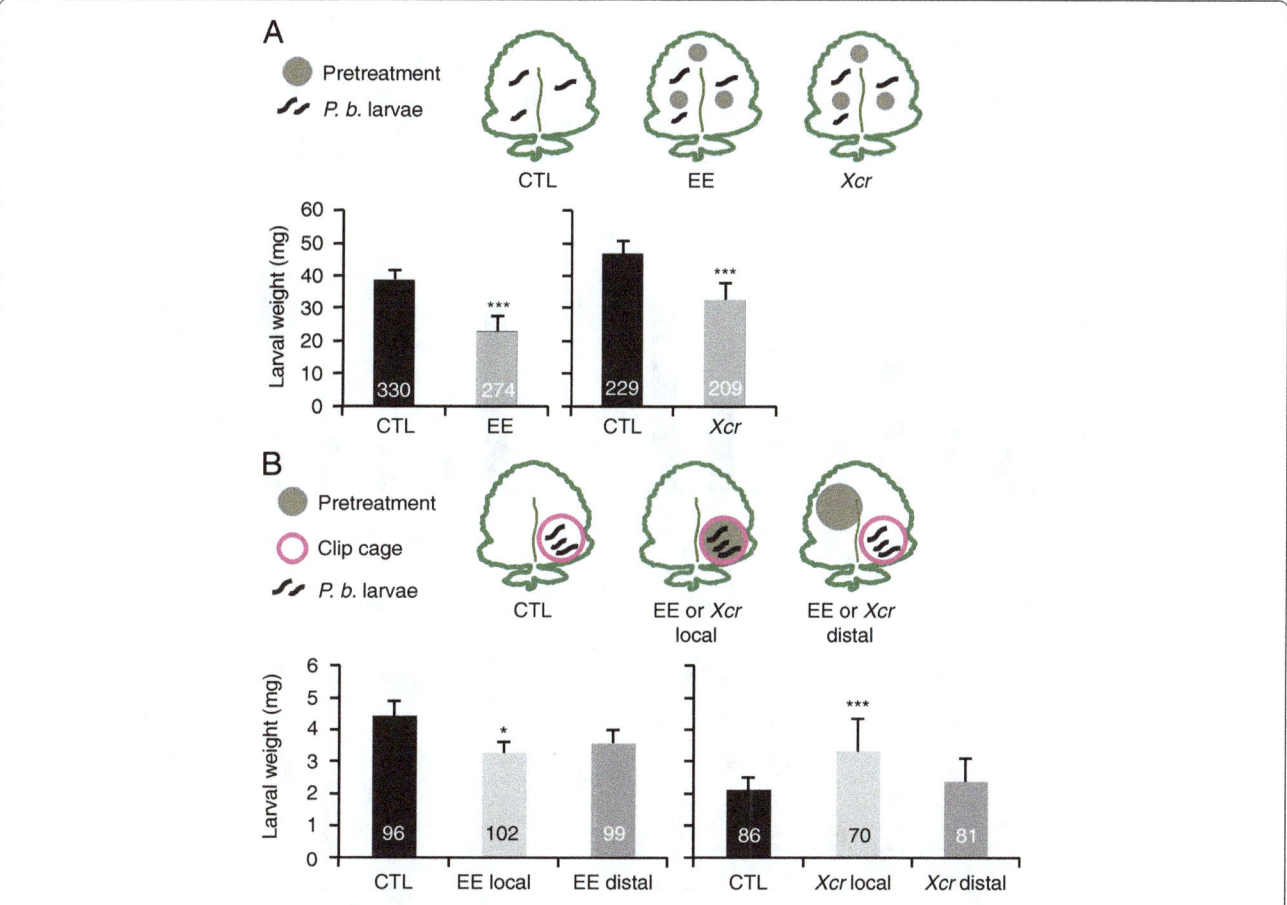

Fig. 3 Impact of pretreatment on insect performance. **a** Larval weight of *P. brassicae* feeding on 5-week-old *B. nigra* plants pretreated for 3 days with *P. brassicae* egg extract or *Xanthomonas campestris* pv. *raphani* (*Xcr*) was measured after 7 days of feeding. Values (± SE) are the mean of independent experiments (*P. brassicae*: CTL/EE, *n* = 5; CTL/*Xcr*, *n* = 4). The total number of larvae is indicated in each column. Significant differences between control and pretreatment are indicated (linear mixed model, ***P < 0.001). **b** Larval weight of *P. brassicae* feeding on 5-week-old *B. nigra* plants pretreated for 3 days with *P. brassicae* egg extract (EE) or *Xanthomonas campestris* pv. *raphani* (*Xcr*) was measured after 4 days. Larvae placed in clip cages were feeding on the treated site (local) or adjacent to the treated site (distal). Values (± SE) are the mean of four (CTL/EE) or three (CTL/*Xcr*) independent experiments. The total number of larvae is indicated in each column. Significant differences between control and pretreatment are indicated (linear mixed model, ***P < 0.001, **P < 0.01, *P < 0.05)

Effect of combined stresses on SA/JA cross talk

Since exposure to egg extract, pathogen treatment and caterpillar feeding triggered SA and JA accumulation to different extents, we decided to investigate the known SA/JA antagonism in response to combined stresses. We designed QPCR primers for *B. nigra* sequences related to *VSP2* and *MYC2*, which are JA- and herbivory-regulated genes [56], and for *PR2* and *SAG13*, which are SA- and egg-regulated genes [66]. In single stress treatments, *BnVSP2* and *BnMYC2* expression were significantly upregulated in tissues exposed to caterpillar feeding, but not in tissues treated with egg extract or the pathogen (Two-way ANOVA, $P_{[VSP2, \text{ egg extract}]}$ = 0.01, $P_{[VSP2, \text{ pathogen}]}$ < 0.0001, $P_{[MYC2, \text{ egg extract}]}$ = 0.04, $P_{[MYC2, \text{ pathogen}]}$ < 0.0001) (Fig. 5). Contrastingly, *BnPR2* and *BnSAG13* were significantly upregulated in tissues treated with egg extract or the pathogen, but not in

tissues exposed to caterpillar feeding (Two-way ANOVA, $P_{[PR2, \text{ egg extract}]}$ < 0.0001, $P_{[PR2, \text{ pathogen}]}$ < 0.0001, $P_{[SAG13, \text{ egg extract}]}$ < 0.0001, $P_{[SAG13, \text{ pathogen}]}$ < 0.0001) (Fig. 5). Interestingly, for combined stresses we found that both egg extract and pathogen pretreatments led to a significantly reduced induction of insect-responsive *BnVSP2* and *BnMYC2*. Combined stresses also reduced the induction of egg extract- or pathogen-responsive *BnSAG13*, whereas *BnPR2* induction was only inhibited by egg extract pretreatment (Fig. 5). These results suggest that under dual-stress conditions a combined accumulation of SA and JA in response to either egg extract or pathogen pretreatment followed by herbivory negatively affects specific JA-and SA-responsive genes. We thus observed a consistent and reciprocal SA/JA cross talk in *B. nigra*, in response to treatment with egg extract or the pathogen followed by caterpillar feeding.

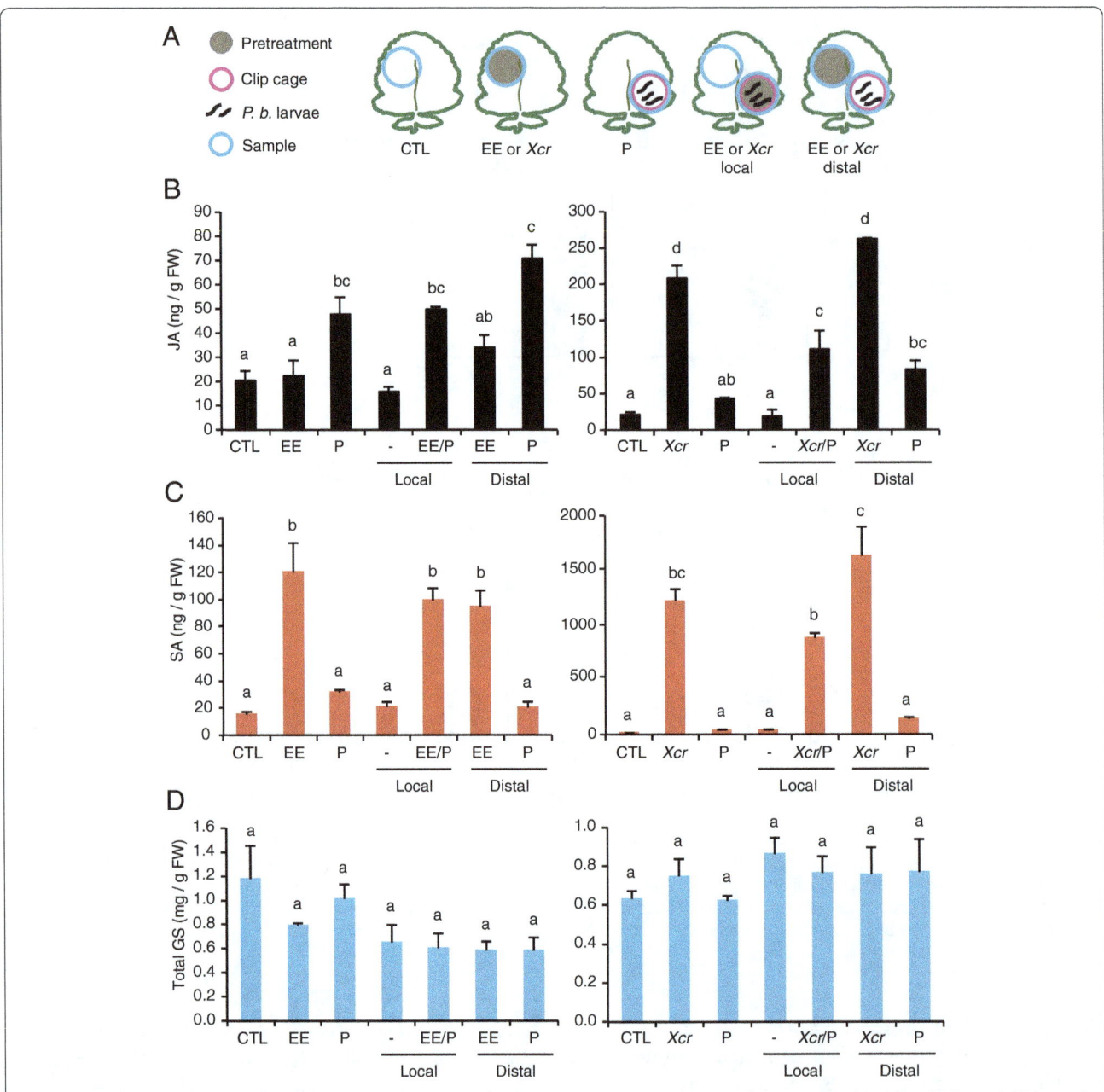

Fig. 4 Quantification of defense signals and glucosinolates. **a** Experimental design. Quantification of jasmonic acid (JA) **b**, salicylic acid (SA) **c** and total glucosinolates (GS) **d** in single and combined stress. Leaf discs (blue circle) were collected on 5-week-old *B. nigra* plants pretreated for 3 days with *P. brassicae* egg extract (EE) or *Xanthomonas campestris* pv. *raphani* (*Xcr*) and further challenged with *P. brassicae* (P) larvae feeding for 24 h at the site or distal to the site of treatment. Controls consisted of untreated plants (CTL) or plants exposed to a single treatment. -, sample distal from the combined treatment. Values (± SE) are the mean of three independent experiments. Letters indicate significant difference between treatments (ANOVA followed by Tukey's honest significant difference test, $P < 0.05$)

Discussion

Exposure to two or more biotic stresses can either be more detrimental than a single stress or, conversely, have an attenuating effect. The ability of plants to recognize and respond to combined and specific stresses appears thus to be important, especially if stresses, such as pathogens and herbivores, trigger different plant defense pathways. Few studies have been conducted on whole-genome responses under multiple stress conditions, and gene expression studies often focused on the plant model Arabidopsis. Here, *B. nigra*, which is also a brassicaceous plant species, was used to investigate how plants respond to combined stresses. Surprisingly, transcriptomic responses of *B. nigra* to different pretreatments followed by *P. brassicae* herbivory revealed that the first stress has only a weak impact on transcriptional responses

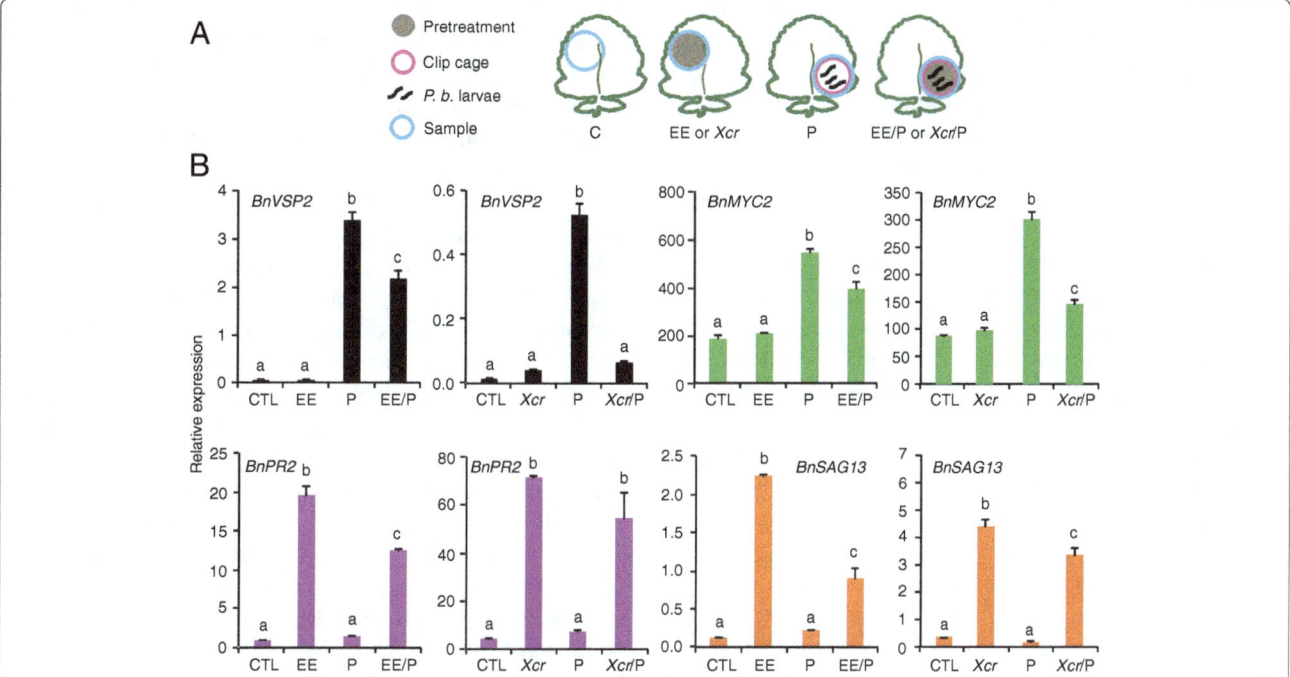

Fig. 5 Expression of JA- and SA-related genes. **a** Experimental design. **b** Expression of *B. nigra* genes was measured by QPCR and normalized to the housekeeping gene *BnSAND*. Leaf discs (blue circle) were collected on 5-week-old *B. nigra* plants pretreated for 3 days with *P. brassicae* egg extract (EE) or *Xanthomonas campestris* pv. *raphani* (*Xcr*) and further challenged with *P. brassicae* (P) feeding for 24 h at the site of treatment. Untreated plants (CTL) or plants exposed to a single stress were included. Means (± SE) of three technical replicates are shown. These experiments were repeated at least twice with similar results. Different letters indicate significant differences (two-way ANOVA followed by Tukey's honest significant difference test, $P < 0.05$)

to the second stress. In addition, no genes common to a combined stress could be identified. It was recently found that the effect of a previous exposure to *B. cinerea* or to drought only slightly changed Arabidopsis transcriptional response to *P. rapae* feeding, suggesting that plants prioritize a response to the second stress [22]. Similarly, Arabidopsis transcriptome after *P. brassicae* feeding was not affected by pre-exposure to *P. brassicae* eggs, although it was impacted by a cold pretreatment [67]. However, a detailed time-course analysis revealed that pre-exposure shifts the timing of caterpillar-induced responses. Plants responded faster to *P. rapae* if they were preceded by a drought or *B. cinerea* treatment [59], indicating that timing of the response needs to be considered. Atkinson and coworkers [41] postulated that during multiple attack, plants respond preferentially to the most damaging stress (see also [68]). In Arabidopsis challenged by drought and/or nematodes, 96% of differentially regulated genes were shared between the combined stress and water stress, whereas only 2% overlapped with nematode feeding [41]. We hypothesize that *B. nigra* prioritized a response to caterpillar feeding rather than to aphids, eggs or bacteria for the benefit of its own fitness, as *P. brassicae* caterpillars are known to be voracious feeders on brassicaceous plants; these caterpillars are florivorous when reaching the second and subsequent

instars, thus reducing fitness directly [69]. Also at the level of the metabolome, changes in *B. nigra* plants exposed to *B. brassicae* aphids and/or *P. brassicae* caterpillars were the strongest in response to feeding by caterpillars both when feeding alone or together with the aphids on the same leaf [68]. It would be interesting to perform reciprocal experiments to see if the transcriptome of *P. brassicae*-pretreated plants is dominated by the signature of a second biotic stress or if plants prioritize the response to herbivory over other stresses. A recent transcriptome study on Arabidopsis plants infected by *Botrytis cinerea* with or without prior herbivory suggests that the first hypothesis is more likely [59]. Another testable hypothesis is that plants respond to the most severe stress, irrespective of the order of attack.

We observed that pretreatment with the biotrophic pathogen *Xcr* had a measurable effect on the *P. brassicae* transcriptome. Indeed, ca. 30% of insect-induced genes, including JA-regulated genes, were significantly less induced in response to the combined stress, but at the same time weight-gain of *P. brassicae* caterpillars was reduced on pathogen-infected plants when feeding on the entire leaf or when restricted to feed distally from the site where the pathogen-pretreatment was applied. Since *Xcr* single treatment triggered SA accumulation, we hypothesize that

SA/JA cross talk was responsible for this attenuation of gene expression. Our targeted analysis of the JA markers *BnVSP2* and *BnMYC2* confirmed this observation, but the insect performance assay also indicated that attenuation of these genes does not have negative consequences for plant defense against *P. brassicae*. Caterpillar weight gain was also reduced on plants pretreated with an egg extract. SA-responsive genes, including *PR1*, were clearly up-regulated by egg extract treatment indicating that the SA pathway was activated by eggs, like in Arabidopsis [35, 66]. This apparent absence of SA/JA cross talk at the whole-genome level may be explained by a relatively less strong response to *P. brassicae* eggs. Indeed, we found a much higher SA accumulation after *Xcr* pathogen than after *P. brassicae* egg extract treatment. In addition, whole-leaf analysis may have diluted a localized response since we detected a localized suppression of *BnVSP2* and *BnMYC2* expression after egg extract treatment. Cross talk between defense signaling pathways is known to strongly modulate the outcome of combined biotic stresses [26]. Here, we also showed that pathogen and egg extract pretreatments inhibited both induction of JA- and SA-regulated genes in response to additional feeding by *P. brassicae* caterpillars, suggesting that SA/JA cross talk reduced the transcription of some genes considered important in plant defense against insects.

Although single *Xcr* infection led to a strong SA and JA accumulation, it is noteworthy that only SA-regulated *BnPR2* and *BnSAG13* were induced but not JA-regulated *BnVSP2* and *BnMYC2*. Moreover, SA/JA cross talk effects were most pronounced on *BnVSP2* and *BnMYC2* transcript levels in response to combined *Xcr*/herbivory stresses, indicating that SA strongly influenced the JA pathway in this particular context. Depending on the hormonal context, accumulation of a defense signal is thus not necessarily correlated with the induction of downstream genes. Conversely, pathogen- and egg-extract induction of SA-signaling related genes *BnPR2* and *BnSAG13* was inhibited by caterpillar feeding, suggesting that the reciprocal JA/SA cross talk was also operating. The consequence of such cross talk on susceptibility to *Xcr* infection was not tested but would be an interesting topic for future research. Thus, JA and SA activation and their mutualistic antagonistic effects may depend on the strength or the nature of the treatment. For example, a study in Arabidopsis reported a synergistic or antagonistic effect on JA- and SA-induced genes if plants were treated with low or high concentrations of each hormone, respectively [70]. In addition, plant responses to herbivory are known to be dynamic and may depend on the sampling time [59, 71]. It will thus be interesting in future experiments to see if our observations

on SA/JA cross talk at a single time point robustly underlie the outcome of the combined interactions. In conclusion, although the emerging picture is that of a domination of the most recent stress on the transcriptional response [22, 59], it would be interesting to confirm this hypothesis by extending the range of reciprocal combinations of biotic and abiotic stresses, including time-course analyses.

We found that insect performance differed between treatments, suggesting that plant resistance status after combined stresses is difficult to predict based on transcriptome, defense hormone profiles or defense pathway cross talk. Indeed, we observed that insect performance after *P. brassicae* egg-extract application was decreased. In Arabidopsis, we previously showed that *P. brassicae* egg deposition had no effect on performance of the specialist *P. brassicae* [35]. Other studies with Arabidopsis and *B. nigra* revealed that *P. brassicae* performed less well or equally in the presence of eggs, depending on the species identity of the egg donor [67, 72–74]. Similar results were found for *P. brassicae* feeding on other wild brassicaceous species, i.e. *Brassica oleracea*, *Moricandia moricandioides* and *Sinapsis arvensis* exposed to *P. brassicae* eggs [43]. On the contrary, the generalist *S. littoralis* performed better on plants already treated with *P. brassicae* egg extract or after natural oviposition [35, 75] but no effect was found for eggs of the generalist *Mamestra brassicae* on subsequent *M. brassicae* larval performance [72]. Thus, whether insect eggs induce plant defenses is context-specific. Similarly, whether there is an effect of exposure to a biotrophic pathogen is context specific [25, 76] and may also depend on the virulence level of the pathogen [77].

Furthermore, *P. brassicae* caterpillars feeding freely on *Xcr*-pretreated leaves gained less weight than caterpillars feeding on an untreated leaf. Thus, at the whole-leaf level, *Xcr* pretreatment impacted plant defense responses similarly to the egg extract pretreatment, although the underlying mechanism might have been different. A study on *Capsicum annuum* L. reported an enhanced performance of *S. exigua* larvae on plants infected with *X. campestris* pv. *vesicatoria* [76]. In contrast, performance of *P. brassicae* larvae was reduced on Arabidopsis infected with *Pseudomonas syringae* pv. tomato [25]. Again, insect performance on plants infected with phytopathogens seems to be variable.

Surprisingly, compared to whole-leaf bioassays, insect performance assays yielded somewhat contrasting conclusions using clip cages to restrict feeding by caterpillars on specific sites on a leaf. Whereas egg-extract treatment impacted larvae similarly regardless whether they were feeding freely on the whole leaf or locally in a clip cage, *P. brassicae* caterpillars feeding on *Xcr*-infiltrated leaf area were larger than those constrained to feed on a non-infected zone or on an untreated leaf. This observation could be explained by our finding of a local inhibition of JA-dependent defense gene expression after *Xcr* pretreatment, although a restricted feeding on egg-extract pretreated tissues did not result in enhanced insect performance. Hence, other factors likely contribute to a localized effect. *Xcr* infection triggered a local accumulation of both SA and JA while *P. brassicae* egg-extract treatment only triggered SA accumulation. Insect herbivores tend to avoid defended leaf areas [78], which they could not do under the constrained clip-cage conditions. Indeed, when given the choice we noticed that *P. brassicae* larvae avoided egg-treated and *Xcr*-infected zones (Additional file 10: Figure S7). This is intriguing with regard to the opposite performance of larvae when feeding on egg-treated or *Xcr*-infected zones. This finding of larval selective feeding deserves further investigation.

Differential activation of the SA and JA signaling pathways may affect metabolite composition. However, we did not observe a differential GS accumulation between treatments. A similar finding was observed after combined ozone and *P. brassicae* treatment in *B. nigra*, although larvae grew less well on ozone-pretreated plants [79]. Contrasting results were reported in the study by Ponzio et al. [68] where the total GS concentration significantly increased in response to feeding by *P. brassicae* caterpillars. Furthermore, induction of GS under dual stress conditions with caterpillars and *B. brassicae* aphids depended on the density of the aphids. The difference in caterpillar densities per leaf, i.e. Thirty in the Ponzio et al. study [68] and 10 here, may explain this discrepancy. As postulated previously [80], other defense compounds may play a crucial role in influencing *P. brassicae* performance on Brassicaceous plant species. The recent identification of flavonoid compounds that negatively impact *P. brassicae* caterpillar performance in Arabidopsis supports this conclusion [81, 82]. In addition, since ET is an important regulator associated with *Xcr* in Arabidopsis [47], local ET signaling may be involved in the local effect of *Xcr* on *P. brassicae* performance. Moreover, plant nutritional quality at the treatment site could also play a role. It may negatively correlate with insect performance on egg-treated sites but positively on *Xcr*-treated sites. For instance, leaf carbohydrate content was found to be controlled by JA and mediated plant susceptibility to an adapted herbivore in *Nicotiana attenuata* [83].

Conclusions

Our transcriptome analysis of *B. nigra* in response to combined stress treatments revealed that the second stress dominates the transcript signature, although pretreatments clearly impacted how plants resisted an herbivore attack. Measurement of defense-signaling hormones and transcript levels of defense marker genes in response to multiple attack by different stresses do not necessarily predict the plant's defense response in a straightforward fashion. Future studies should include more marker genes representing different steps along the molecular sequence of events. Our results show that under conditions of multiple stress the plant responds highly specifically to each stress combination. Contrasting responses strongly suggest that we need to better integrate responses at different levels of biological organization, to consider local versus distant plant responses within a leaf, and to measure the accumulation of a range of (defense) metabolites determining nutritional quality when trying to correlate plant traits with insect performance.

Methods

Biological material

Seeds of *Brassica nigra* were collected from a wild population in Wageningen (The Netherlands) [43]. Plants were grown in soil in growth chambers (16 h light, at 25 °C day, 22 °C night, 60% relative humidity) under white fluorescent light (170 μmol m^{-2} s^{-1}). Seeds were stratified for 3 days at 4 °C after sowing. The soil contained 65% humus, 10% sand, 15% perlite and 10% silt. Growth conditions were the same in the different bioassays described below.

Xanthomonas campestris pv. *raphani* (*Xcr*) (formerly classified as *X. campestris* pv. *armoraciae*) was obtained from the Plant-Microbe Interactions group of Utrecht University (The Netherlands) and was originally acquired from the Department of Plant Pathology at Ohio State University (USA). The pathovar identity was confirmed by pathogenicity assays and PCR. Bacteria were grown in 10 ml of liquid King B culture medium (20 g / l peptone (Sigma-Aldrich), 1.5 g / l dipotassium hydrogen phosphate, 1.5 g / l magnesium sulfate heptahydrate, 12 g /l agar, at a final pH of 7.2) supplemented with rifampicin (25 μg/ ml) and grown in a shaker at 28 °C, 200 rpm, during 48 h. *Xcr* culture was centrifuged at 7000 rpm during 2 min. The supernatant was discarded and the pellet was washed and re-suspended in 10 mM MgCl$_2$ and centrifuged again at 7000 rpm during 2 min. The supernatant was discarded and the pellet diluted in 10 mM MgCl$_2$ and adjusted to an OD 600 of 0.07 to obtain a concentration of 10^7 cfu/ml.

Pieris brassicae was reared on Brussels sprout plants (*Brassica oleracea* var. *gemmifera*) in 1 m^3 cages in a greenhouse (25 ± 5 °C, 60 ± 5% RH, 16/8 h light-dark cycle) at Lausanne University (Switzerland). Eggs were removed manually from the plants and crushed with a pestle in Eppendorf tubes. After centrifugation (15'000 g, 3 min), the supernatant (egg extract) was stored at −20 °C.

Brevicoryne brassicae aphids were reared on *B. oleracea* var. *gemmifera* in a greenhouse (22 ± 3 °C, 65 ± 5%

RH, 16/8 h light-dark cycle) at Wageningen University (The Netherlands), where all experiments with aphids were also performed. *B. nigra* plants were grown in peat soil (Lentse potgrond no. 4, Lent, The Netherlands).

The pest species *P. brassicae* and *B. brassicae*, and *B. nigra* plants were collected in the wild in The Netherlands. This complies with national legislation as The Netherlands allows free access to its biodiversity under the Nagoya Protocol. Correct identification of *B. nigra* was confirmed by Dr. E. H. Poelman (Department of Plant Sciences, Wageningen University, The Netherlands). Seeds of *B. oleracea* var. *gemmifera* were obtained commercially from Semences Zollinger (1897 Les Evouettes, Switzerland) or Syngenta Seeds (2678 LV De Lier, The Netherlands).

Plant treatments

The overall experimental design is summarized in Table S4 (Additional file 11: Table S4). Plants were 5 weeks old when exposed to the various treatments. Pretreatments with egg extract and the pathogen were applied to the three youngest fully developed leaves of three plants. Aphids were applied to a single leaf, i.e. the first fully developed leaf (nine plants in total), according to a design that has been used in previous experiments with the same study system [84, 85]. For *P. brassicae* egg-extract treatment, 12×2 μl of egg extract were added to each of the three leaves and incubated for 72 h. This treatment was equivalent to treatments previously applied to Arabidopsis and corresponds to approximately 10–12 egg batches per leaf, each batch consisting of 20–30 eggs [35, 66]. Leaves of untreated plants were used as controls.

For infection with the bacterial pathogen, *X. campestris* pv. *raphani*, each of the three treatment leaves was subjected to three infiltrations of 10^7 cfu/ml using a 1 ml needleless syringe and incubated for 72 h. Each infiltration zone represented a circle of 1.5 cm^2. In control plants, the same number of 10 mM $MgCl_2$ infiltrations was performed.

For treatment with *B. brassicae* aphids, 100 nymphs were placed on the youngest fully developed leaf on each of nine plants, which were incubated for 48 h. Aphids were not constrained but remained on the leaf on which they had been introduced.

Treatment with caterpillars consisted of the introduction of 10 neonate caterpillars on the three leaves that had received a pretreatment (combined stresses) or on three leaves similar in development of clean plants. Thirty neonate caterpillars were introduced on the single aphid-treated leaf or a single leaf of clean plant. Caterpillars were allowed to feed for 24 h.

All experiments were repeated independently five or more times at intervals of several weeks.

Insect performance assays on plants pretreated with egg extract or pathogen

Five-week-old *B. nigra* plants were placed in $60 \times 60 \times 60$ cm plastic tents (Bugdorm company) in a growth chamber (20 ± 1 °C, $65 \pm 10\%$ relative humidity, 10/14 h light-dark cycle, 100 μmol m^{-2} s^{-1}). For insect bioassays performed on entire leaves, ten neonate caterpillars were placed on each of the three pretreated leaves or on three leaves of clean plants with a total of 30 caterpillars per plant. Caterpillar weight was measured after 7 days of feeding. For bioassays investigating local vs. distal effects of pretreatment, five neonate caterpillars were placed in a clip cage ($36.5 \times 25.4 \times 9.5$ mm, Bio-Quip Products, USA) on each of three pretreated leaves with a total of 15 larvae per plant either at the same site or a site distal from where the pretreatment was applied. Plants were pretreated as described above with egg extract or the pathogen and incubated for 3 days (see Fig. 4a for experimental design). Caterpillar weight was measured after 4 days. For all experiments, insect recovery was similar between treatments.

Each treatment was done on three different plants for each biological replicate. All experiments were repeated independently three or more times at intervals of several weeks.

Hormone and glucosinolate analysis

Leaf tissues that were sampled for hormone (SA and JA) and GS analysis were exposed to egg extract, the pathogen, and/or caterpillar feeding as described above. Entire leaves (experiments with no constraint on caterpillar feeding) or 2.4 cm leaf discs (experiments with constrained caterpillar feeding) were harvested and frozen in liquid nitrogen. Extraction, UHPLC-QTOFMS measurement and data analysis were conducted as described earlier [86, 87]. Three independent biological replicates were analyzed for each treatment.

Transcriptome analyses

Following treatment, entire leaves were harvested, flash-frozen in liquid nitrogen and stored at –80 °C. RNA extraction, probe labeling, hybridization onto Arabidopsis CATMAv4 microarrays, and data analyses have been published previously [51, 56, 88]. For data analysis, we used an expression threshold of $\log_2 > 0.585$ and < -0.585, and an unadjusted *P*-value of 0.05. FDR values are shown in supplementary data for further evaluation. GO enrichment analysis was performed with AgriGO singular enrichment analysis using hypergeometric test [89].

Quantitative PCR

Relative gene expression was measured according to previously published procedures [35, 90]. Briefly, 500 nanograms of total RNA were transcribed to cDNA using M-MLV reverse transcriptase (Invitrogen) and oligo dT primers according to commercial instructions. cDNA synthesis was done in triplicates. QPCR analysis was performed in a final volume of 25 µl according to the Brilliant III Fast SYBR Green instruction manual (Agilent). *B. nigra* primers (Additional file 12: Table S5) were designed on conserved sequences identified by multiple alignments of genes from different species of the *Brassica* family. Sequences were obtained from the *Brassica* database (http://brassica.nbi.ac.uk/BrassicaDB/). Each primer has a Tm of 60 °C and gives an amplicons length between 100 and 250 bp in the conserved part of the cDNA strand. Primer efficiencies were evaluated by five-step dilution regression. Each amplicon produced a single band and was confirmed by Sanger sequencing. For normalization, the *BnSAND* gene was used as housekeeping gene. Similar to Arabidopsis *SAND* gene [91], its expression was stable across experiments.

Additional files

Additional file 1: Table S1. Gene expression ratios (log_2) for all biological replicates.

Additional file 2: Table S2. GO analysis of *P. brassicae*-regulated genes.

Additional file 3: Figure S1. Expression of the top-50 upregulated genes in response to *P. brassicae* feeding and combined stresses. The highest significantly upregulated genes ($log_2 > 0.585$, $P < 0.05$) were extracted from microarray data (orange bars) and plotted with values from combined stresses. (A) Egg extract/*P. brassicae* larvae (yellow bars), (B) *Xanthomonas campestris* pv. *raphani*/*P. brassicae* larvae (green bars), and (C) *Brevicoryne brassicae*/*P. brassicae* larvae (blue bars). Significant differences between single and combined stress are indicated (Student's *t*-test, ***$P < 0.001$, **$P < 0.01$, *$P < 0.05$).

Additional file 4: Figure S2. Expression of the top-50 downregulated genes in response to *P. brassicae* feeding and combined stresses. The highest significantly downregulated genes ($log_2 < -0.585$, $P < 0.05$) were extracted from microarray data (orange bars) and plotted with values from combined stresses. (A) Egg extract/*P. brassicae* larvae (yellow bars), (B) *Xanthomonas campestris* pv. *raphani*/*P. brassicae* larvae (green bars), and (C) *Brevicoryne brassicae*/*P. brassicae* larvae (blue bars). Significant differences between single and combined stress are indicated (Student's *t*-test, ***$P < 0.001$, **$P < 0.01$, *$P < 0.05$).

Additional file 5: Figure S3. GO analysis of genes specifically upregulated by combined stress. GO terms significantly enriched with each combined stress are shown separately. Length of the bars shows the percentage of regulated genes in the respective GO categories.

Additional file 6: Figure S4. GO analysis of genes specifically downregulated by combined stress. GO analysis of genes specifically downregulated by combined stress. GO terms significantly enriched with each combined stress are shown separately. Length of the bars shows the percentage of regulated genes in the respective GO categories.

Additional file 7: Figure S5. Quantification of defense signals and glucosinolates. (A) Experimental design. (B) Quantification of jasmonic acid (JA), salicylic acid (SA) and total glucosinolates (GS) in single and combined stress in whole 5-week-old *B. nigra* leaves. Plants were pretreated for 3 days with *P. brassicae* egg extract (EE) or *Xanthomonas campestris* pv. *raphani* (Xcr) and further challenged with *P. brassicae* (P) larvae for 24 h. Controls (CTL) consisted of untreated plants or plants exposed to a single treatment. Values (± SE) are the mean of three independent experiments. Letters indicate significant difference between treatments (two-way ANOVA followed by Tukey's honest significant difference test, $P < 0.05$).

Additional file 8: Table S3. Glucosinolate content in *B. nigra* leaves.

Additional file 9: Figure S6. Expression of glucosinolate biosynthesis genes in response to *P. brassicae* feeding and combined stresses. Values were extracted from microarray data. *P. brassicae* larvae (orange bars), egg extract/*P. brassicae* larvae (yellow bars), *Xanthomonas campestris* pv. *raphani*/*P. brassicae* (green bars), *Brevicoryne brassicae*/*P. brassicae* (blue bars). Significant differences between single and combined stress are indicated (Student's *t*-test, ***$P < 0.001$, **$P < 0.01$, *$P < 0.05$).

Additional file 10: Figure S7. Feeding behavior of *P. brassicae* larvae in response to combined stresses. Neonate larvae were allowed to feed freely for 2 days (*P. brassicae*) on 5-week-old *B. nigra* plants pretreated for 3 days with egg extract (A) or *Xanthomonas campestris* pv. *raphani* (B). Representative images from three biological replicates are shown. Scale bar = 1 cm.

Additional file 11: Table S4. Overall experimental design. (PDF 48 kb)

Additional file 12: Table S5. List of primers used for QPCR.

Acknowledgements
We thank Blaise Tissot for maintenance of the plants and Johann Weber for help with microarray analyses.

Funding
The Swiss National Science Foundation (grant 31003A_149286 and EUROCORES program EuroVOL grant 31VL30_134414 to PR) and the Earth and Life Sciences Council (ALW) of the Netherlands Organisation for Scientific Research (NWO) (EUROCORES program EuroVOL grant 855.01.171 to MD) supported this work.

Authors' contributions
CB, CP, RG, MD and PR designed the experiments. CB, SL and CP performed the experiments. CB and PR analyzed the data. CB, MD, RG and PR wrote the paper. All authors have read and approved this manuscript.

Competing interests
The authors declare that they have no competing interests.

Author details
[1]Department of Plant Molecular Biology, University of Lausanne, Biophore Building, 1015 Lausanne, Switzerland. [2]Laboratory of Entomology, Wageningen University, P.O. Box 16, 6700 AA Wageningen, The Netherlands.

References
1. Chew IH, Halliday KJ. A stress-free walk form Arabidopsis to crops. Curr Opin Biotechnol. 2011;22:281–6.
2. Bray EA. Plant responses to water deficit. Trends Plant Sci. 1997;2:48–54.

3. Howe GA, Jander G. Plant immunity to insect herbivores. Annu Rev Plant Biol. 2008;59:41–66.

4. Dean R, van Kan JAL, Pretorius ZA, Hammond-Kosack KE, et al. The top 10 fungal pathogens in molecular plant pathology. Mol Plant Pathol. 2012;13:414–30.

5. Xin X-F, He SY. Pseudomonas syringae pv. tomato DC3000: a model pathogen for probing disease susceptibility and hormone signaling in plants. Annu Rev Phytopathol. 2013;51:473–98.

6. Atkinson NJ, Urwin PE. The interaction of plant biotic and abiotic stresses: from genes to the field. J Exp Bot. 2012;63:3523–43.

7. Rejeb IB, Pastor V, Mauch-Mani B. Plant responses to simultaneous biotic and abiotic stress: molecular mechanisms. Plants. 2014;3:458–75.

8. Suzuki N, Rivero RM, Shulaev V, Blumwald E, Mittler R. Abiotic and biotic stress combinations. New Phytol. 2014;203:32–43.

9. Ramegowda V, Senthil-Kumar M. The interactive effects of simultaneous biotic and abiotic stresses on plants: mechanistic understanding from drought and pathogen combination. J Plant Physiol. 2015;176:47–54.

10. Ahuja I, de Vos RCH, Bones AM, Hall RD. Plant molecular stress responses face climate change. Trends Plant Sci. 2010;15:664–74.

11. Schoonhoven LM, van Loon JJA, Dicke M. Insect-plant biology. USA: Oxford University Press; 2005.

12. Huberty AF, Denno RF. Plant water stress and its consequences for herbivorous insects: a new synthesis. Ecology. 2004;85:1383–98.

13. Mewis I, Khan MAM, Glawischnig E, Schreiner M, Ulrichs C. Water stress and aphid feeding differentially influence metabolite composition in Arabidopsis thaliana L. PLoS One. 2012;7:e48661.

14. Chakraborty S, Whitehill J, Hill AL, Opiyo SO, Cipollini D, Herms DA, Bonello P. Effects of water availability on emerald ash borer larval performance and phloem phenolics of Manchurian and black ash. Plant Cell Environ. 2014;37:1009–21.

15. Rivas-Ubach A, Gargallo-Garriga A, Sardans J, Oravec M, Mateu-Castell L, Pérez-Trujillo M, Parella T, Ogaya R, Urban O, Peñuelas J. Drought enhances folivory by shifting foliar metabolomes in Quercus ilex trees. New Phytol. 2014;202:874–85.

16. Nguyen D, D'Agostino N, Tytgat TOG, et al. Drought and flooding have distinct effects on herbivore-induced responses and resistance in Solanum dulcamara. Plant Cell Environ. 2016;39:1485–99.

17. Khan GA, Vogiatzaki E, Glauser G, Poirier Y. Phosphate deficiency induces the jasmonate pathway and enhances resistance to insect herbivory. Plant Physiol. 2016;171:632–44.

18. Tack AJM, Dicke M. Plant pathogens structure arthropod communities across multiple spatial and temporal scales. Funct Ecol. 2013;27:633–45.

19. Lazebnik J, Frago E, Dicke M, van Loon JJA. Phytohormone mediation of interactions between herbivores and plant pathogens. J Chem Ecol. 2014;40:730–41.

20. Stout MJ, Thaler JS, Thomma BPHJ. Plant-mediated interactions between pathogenic microorganisms and herbivorous arthropods. Annu Rev Entomol. 2006;51:663–89.

21. Naemi Al F, Hatcher PE. Contrasting effects of necrotrophic and biotrophic plant pathogens on the aphid Aphis fabae. Entomol Exp Appli. 2013;148:234–45.

22. Davila Olivas NH, Coolen S, Huang P, et al. Effect of prior drought and pathogen stress on Arabidopsis transcriptome changes to caterpillar herbivory. New Phytol. 2016;210:1344–56.

23. Thaler JS, Agrawal AA, Halitschke R. Salicylate-mediated interactions between pathogens and herbivores. Ecology. 2010;91:1075–82.

24. Heidel AJ, Barazani O, Baldwin IT. Interaction between herbivore defense and microbial signaling: bacterial quorum-sensing compounds weaken JA-mediated herbivore resistance in Nicotiana attenuata. Chemoecol. 2010;20:149–54.

25. Hilfiker O, Groux R, Bruessow F, Kiefer K, Zeier J, Reymond P. Insect eggs induce a systemic acquired resistance in Arabidopsis. Plant J. 2014;80:1085–94.

26. Pieterse CMJ, der Does VD, Zamioudis C, Leon-Reyes A, Van Wees SCM. Hormonal modulation of plant immunity. Annu Rev Cell Dev Biol. 2012;28:489–521.

27. Reymond P, Farmer EE. Jasmonate and salicylate as global signals for defense gene expression. Curr Opin Plant Biol. 1998;1:404–11.

28. Spoel SH, Dong X. Making sense of hormone crosstalk during plant immune responses. Cell Host Microbe. 2008;3:348–51.

29. Zarate SI, Kempema LA, Walling LL. Silverleaf whitefly induces salicylic acid defenses and suppresses effectual jasmonic acid defenses. Plant Physiol. 2007;143:866–75.

30. Koornneef A, Pieterse CMJ. Cross talk in defense signaling. Plant Physiol. 2008;146:839–44.

31. Zhang PJ, Zheng SJ, van Loon JJA, Boland W, David A, Mumm R, Dicke M. Whiteflies interfere with indirect plant defense against spider mites in lima bean. Proc Nat Acad Sci USA. 2009;106:21202–7.

32. Zhang PJ, Broekgaarden C, Zheng SJ, Snoeren TAL, van Loon JJA, Gols R, Dicke M. Jasmonate and ethylene signaling mediate whitefly-induced interference with indirect plant defense in Arabidopsis thaliana. New Phytol. 2013;197:1291–9.

33. Caarls L, Pieterse CMJ, Van Wees SCM. How salicylic acid takes transcriptional control over jasmonic acid signaling. Frontiers Plant Sci. 2015;6:170.

34. Cipollini DD, Enright SS, Traw MBM, Bergelson JJ. Salicylic acid inhibits jasmonic acid-induced resistance of Arabidopsis thaliana to Spodoptera exigua. Mol Ecol. 2004;13:1643–53.

35. Bruessow F, Gouhier-Darimont C, Buchala A, Metraux J-P, Reymond P. Insect eggs suppress plant defence against chewing herbivores. Plant J. 2010;62:876–85.

36. Koornneef A, Leon-Reyes A, Ritsema T, Verhage A, Otter Den FC, Van Loon LC, Pieterse CMJ. Kinetics of salicylate-mediated suppression of jasmonate signaling reveal a role for redox modulation. Plant Physiol. 2008;147:1358–68.

37. Leon-Reyes A, Spoel SH, De Lange ES, Abe H, Kobayashi M, Tsuda S, Millenaar FF, Welschen RAM, Ritsema T, Pieterse CMJ. Ethylene modulates the role of NONEXPRESSOR OF PATHOGENESIS-RELATED GENES1 in cross talk between salicylate and jasmonate signaling. Plant Physiol. 2009;149:1797–809.

38. Rasmussen S, Barah P, Suarez-Rodriguez MC, Bressendorff S, Friis P, Costantino P, Bones AM, Nielsen HB, Mundy J. Transcriptome responses to combinations of stresses in Arabidopsis. Plant Physiol. 2013;161:1783–94.

39. Voelckel C, Baldwin IT. Generalist and specialist lepidopteran larvae elicit different transcriptional responses in Nicotiana attenuata, which correlate with larval FAC profiles. Ecol Lett. 2004;7:770–5.

40. Rizhsky L, Liang H, Shuman J, Shulaev V, Davletova S, Mittler R. When defense pathways collide. The response of Arabidopsis to a combination of drought and heat stress. Plant Physiol. 2004;134:1683–96.

41. Atkinson NJ, Lilley CJ, Urwin PE. Identification of genes involved in the response of Arabidopsis to simultaneous biotic and abiotic stresses. Plant Physiol. 2013;162:2028–41.

42. Westman AL, Kresovich S, Dickson MH. Regional variation in Brassica nigra and other weedy crucifers for disease reaction to Alternaria brassicicola and Xanthomonas campestris pv. campestris. Euphytica. 1999;106:253–9.

43. Pashalidou FG, Fatouros NE, van Loon JJA, Dicke M, Gols R. Plant-mediated effects of butterfly egg deposition on subsequent caterpillar and pupal development, across different species of wild Brassicaceae. Ecol Entomol. 2015;40:444–50.

44. Moran PJ, Thompson GA. Molecular responses to aphid feeding in Arabidopsis in relation to plant defense pathways. Plant Physiol. 2001;125:1074–85.

45. de Vos M, Van Oosten VR, Van Poecke RMP, et al. Signal signature and transcriptome changes of Arabidopsis during pathogen and insect attack. Mol Plant-Microbe Interact. 2005;18:923–37.

46. Gouhier-Darimont C, Schmiesing A, Bonnet C, Lassueur S, Reymond P. Signalling of Arabidopsis thaliana response to Pieris brassicae eggs shares similarities with PAMP-triggered immunity. J Exp Bot. 2013;64:665–74.

47. Ton J, Van Pelt JA, Van Loon LC, Pieterse CMJ. Differential effectiveness of salicylate-dependent and jasmonate/ethylene-dependent induced resistance in Arabidopsis. Mol Plant-Microbe Interact. 2002;15:27–34.

48. Lucas-Barbosa D, van Loon JJA, Gols R, van Beek TA, Dicke M. Reproductive escape: annual plant responds to butterfly eggs by accelerating seed production. Funct Ecol. 2013;27:245–54.

49. Lucas-Barbosa D, Dicke M, Kranenburg T, Aartsma Y, Huigens ME, van Beek TA, van Loon JJA. Endure and call for help: strategies of black mustard plants to deal with a specialised caterpillar. Funct Ecol. 2017;31:325–33.

50. Hilson P, Allemeersch J, Altmann T, et al. Versatile gene-specific sequence tags for Arabidopsis functional genomics: transcript profiling and reverse genetics applications. Genome Res. 2004;14:2176–89.

51. Sclep G, Allemeersch J, Liechti R, et al. CATMA, a comprehensive genome-scale resource for silencing and transcript profiling of Arabidopsis genes. BMC Bioinfo. 2007;8:400.

52. Broekgaarden C, Poelman EH, Steenhuis G, Voorrips RE, Dicke M, Vosman B. Responses of Brassica oleracea cultivars to infestation by the aphid Brevicoryne brassicae: an ecological and molecular approach. Plant Cell Environ. 2008;31:1592–605.

53. Broekgaarden C, Voorrips RE, Dicke M, Vosman B. Transcriptional responses of *Brassica nigra* to feeding by specialist insects of different feeding guilds. Insect Sci. 2011;18:259–72.

54. Reymond P, Weber H, Damond M, Farmer EE. Differential gene expression in response to mechanical wounding and insect feeding in Arabidopsis. Plant Cell. 2000;12:707–20.

55. Voelckel C, Baldwin IT. Herbivore-induced plant vaccination. Part II. Array-studies reveal the transience of herbivore-specific transcriptional imprints and a distinct imprint from stress combinations. Plant J. 2004;38:650–63.

56. Reymond P, Bodenhausen N, Van Poecke RMP, Krishnamurthy V, Dicke M, Farmer EE. A conserved transcript pattern in response to a specialist and a generalist herbivore. Plant Cell. 2004;16:3132–47.

57. Major IT, Constabel CP. Molecular analysis of poplar defense against herbivory: comparison of wound- and insect elicitor-induced gene expression. New Phytol. 2006;172:617–35.

58. Broekgaarden C, Poelman EH, Steenhuis G, Voorrips RE, Dicke M, Vosman B. Genotypic variation in genome-wide transcription profiles induced by insect feeding: *Brassica oleracea-Pieris rapae* interactions. BMC Genomics. 2007;8:239.

59. Coolen S, Proietti S, Hickman R, et al. Transcriptome dynamics of Arabidopsis during sequential biotic and abiotic stresses. Plant J. 2016;86:249–67.

60. Kliebenstein DJ, Kroymann J, Mitchell-Olds T. The glucosinolate-myrosinase system in an ecological and evolutionary context. Curr Opin Plant Biol. 2005;8:264–71.

61. Halkier BA, Gershenzon J. Biology and biochemistry of glucosinolates. Annu Rev Plant Biol. 2006;57:303–33.

62. Gigolashvili T, Yatusevich R, Berger B, Müller C, Flügge U-I. The R2R3-MYB transcription factor HAG1/MYB28 is a regulator of methionine-derived glucosinolate biosynthesis in *Arabidopsis thaliana*. Plant J. 2007;51:247–61.

63. Kos M, Houshyani B, Wietsma R, Kabouw P, Vet LEM, van Loon JJA, Dicke M. Effects of glucosinolates on a generalist and specialist leaf-chewing herbivore and an associated parasitoid. Phytochemistry. 2012;77:162–70.

64. Schweizer F, Fernández-Calvo P, Zander M, Diez-Diaz M, Fonseca S, Glauser G, Lewsey MG, Ecker JR, Solano R, Reymond P. Arabidopsis basic helix-loop-helix transcription factors MYC2, MYC3, and MYC4 regulate glucosinolate biosynthesis, insect performance, and feeding behavior. Plant Cell. 2013;25:3117–32.

65. Köhler A, Maag D, Veyrat N, Glauser G, Wolfender J-L, Turlings TCJ, Erb M. Within-plant distribution of 1,4-benzoxazin-3-ones contributes to herbivore niche differentiation in maize. Plant Cell Environ. 2015;38:1081–93.

66. Little D, Gouhier-Darimont C, Bruessow F, Reymond P. Oviposition by pierid butterflies triggers defense responses in Arabidopsis. Plant Physiol. 2007;143:784–800.

67. Firtzlaff V, Oberländer J, Geiselhardt S, Hilker M, Kunze R. Pre-exposure of Arabidopsis to the abiotic or biotic environmental stimuli "chilling" or "insect eggs" exhibits different transcriptomic responses to herbivory. Sci Rep. 2016;6:28544.

68. Ponzio C, Papazian S, Albrectsen BR, Dicke M, Gols R. Dual herbivore attack and herbivore density affect metabolic profiles of *Brassica nigra* leaves, Plant Cell Environ., in press. doi:10.1111/pce.12926.

69. Smallegange RC, van Loon JJA, Blatt SE, Harvey JA, Agerbirk N, Dicke M. Flower vs. leaf feeding by *Pieris brassicae*: glucosinolate-rich flower tissues are preferred and sustain higher growth rate. J Chem Ecol. 2007;33:1831–44.

70. Mur LAJ, Kenton P, Atzorn R, Miersch O, Wasternack C. The outcomes of concentration-specific interactions between salicylate and jasmonate signaling include synergy, antagonism, and oxidative stress leading to cell death. Plant Physiol. 2006;140:249–62.

71. Appel HM, Fescemyer H, Ehlting J, Weston D, Rehrig E, Joshi T, Xu D, Bohlmann J, Schultz J. Transcriptional responses of *Arabidopsis thaliana* to chewing and sucking insect herbivores. Front Plant Sci. 2014;5:565.

72. Pashalidou FG, Lucas-Barbosa D, van Loon J, Dicke M, Fatouros N. Phenotypic plasticity of plant response to herbivore eggs: effects on resistance to caterpillars and plant development. Ecology. 2012;94:702–13.

73. Geiselhardt S, Yoneya K, Blenn B, Drechsler N, Gershenzon J, Kunze R, Hilker M. Egg laying of cabbage white butterfly. (*Pieris brassicae*) on *Arabidopsis thaliana* affects subsequent performance of the larvae. PLoS One. 2013;8:e59661.

74. Pashalidou FG, Frago E, Griese E, Poelman EH, van Loon JJA, Dicke M, Fatouros NE. Early herbivore alert matters: plant-mediated effects of egg deposition on higher trophic levels benefit plant fitness. Ecol Lett. 2015;18:927–36.

75. Schmiesing A, Emonet A, Gouhier-Darimont C, Reymond P. Arabidopsis MYC transcription factors are the target of hormonal SA/JA cross talk in response to *Pieris brassicae* egg extract. Plant Physiol. 2016;170:2432–43.

76. Cardoza YJ, Tumlinson JH. Compatible and incompatible *Xanthomonas* infections differentially affect herbivore-induced volatile emission by pepper plants. J Chem Ecol. 2006;32:1755–68.

77. Ponzio C, Weldegergis BT, Dicke M, Gols R. Compatible and incompatible pathogen-plant interactions differentially affect plant volatile emissions and the attraction of parasitoid wasps. Funct Ecol. 2016;30:1779–89.

78. Perkins LE, Cribb BW, Brewer PB, Hanan J, Grant M, de Torres M, Zalucki MP. Generalist insects behave in a jasmonate-dependent manner on their host plants, leaving induced areas quickly and staying longer on distant parts. Proc Royal Soc London B. 2013;280:20122646.

79. Khaling E, Papazian S, Poelman EH, Holopainen JK, Albrectsen BR, Blande JD. Ozone affects growth and development of *Pieris brassicae* on the wild host plant *Brassica nigra*. Environ Poll. 2015;199:119–29.

80. Poelman EH, Galiart RJFH, Raaijmakers CE, van Loon JJA, van Dam NM. Performance of specialist and generalist herbivores feeding on cabbage cultivars is not explained by glucosinolate profiles. Entomol Exp Appli. 2008;127:218–28.

81. Onkokesung N, Reichelt M, van Doorn A, Schuurink RC, van Loon JJA, Dicke M. Modulation of flavonoid metabolites in *Arabidopsis thaliana* through overexpression of the MYB75 transcription factor: role of kaempferol-3,7-dirhamnoside in resistance to the specialist insect herbivore Pieris Brassicae. J Exp Bot. 2014;65:2203–17.

82. Onkokesung N, Reichelt M, van Doorn A, Schuurink RC, Dicke M. Differential costs of two distinct resistance mechanisms induced by different herbivore species in Arabidopsis. Plant Physiol. 2016;170:891–906.

83. Machado RAR, Arce CCM, Ferrieri AP, Baldwin IT, Erb M. Jasmonate-dependent depletion of soluble sugars compromises plant resistance to *Manduca sexta*. New Phytol. 2015;207:91–105.

84. Ponzio C, Gols R, Weldegergis BT, Dicke M. Caterpillar-induced plant volatiles remain a reliable signal for foraging wasps during dual attack with a plant pathogen or non-host insect herbivore. Plant Cell Environ. 2014;37:1924–35.

85. Ponzio C, Cascone P, Cusumano A, Weldegergis BT, Fatouros NE, Guerrieri E, Dicke M, Gols R. Volatile-mediated foraging behaviour of three parasitoid species under conditions of dual insect herbivore attack. Animal Behav. 2016;111:197–206.

86. Glauser G, Schweizer F, Turlings TCJ, Reymond P. Rapid profiling of intact glucosinolates in Arabidopsis leaves by UHPLC-QTOFMS using a charged surface hybrid column. Phytochem Anal. 2012;23:520–8.

87. Glauser G, Vallat A, Balmer D. 2014. Hormone profiling. Methods Mol Biol 2014;1062:597–608.

88. Papazian S, Khaling E, Bonnet C, Lassueur S, Reymond P, Moritz T, Blande JD, Albrectsen BR. Central metabolic responses to ozone and herbivory affect photosynthesis and stomatal closure. Plant Physiol. 2016;172:2057–78.

89. Du Z, Zhou X, Ling Y, Zhang Z, Su Z. agriGO: a GO analysis toolkit for the agricultural community. Nucl Acids Res. 2010;38:W64–70.

90. Consales F, Schweizer F, Erb M, Gouhier-Darimont C, Bodenhausen N, Bruessow F, Sohby I, Reymond P. Insect oral secretions suppress wound-induced responses in *Arabidopsis*. J Exp Bot. 2012;63:727–37.

91. Czechowski T, Stitt M, Altmann T, Udvardi MK, Scheible WR. Genome-wide identification and testing of superior reference genes for transcript normalization in Arabidopsis. Plant Physiol. 2005;139:5–17.

14

Evaluating the accuracy of genomic prediction of growth and wood traits in two Eucalyptus species and their F$_1$ hybrids

Biyue Tan[1,2], Dario Grattapaglia[3,4], Gustavo Salgado Martins[5], Karina Zamprogno Ferreira[5], Björn Sundberg[2] and Pär K. Ingvarsson[1,6*]

Abstract

Background: Genomic prediction is a genomics assisted breeding methodology that can increase genetic gains by accelerating the breeding cycle and potentially improving the accuracy of breeding values. In this study, we use 41,304 informative SNPs genotyped in a *Eucalyptus* breeding population involving 90 *E.grandis* and 78 *E.urophylla* parents and their 949 F$_1$ hybrids to develop genomic prediction models for eight phenotypic traits - basic density and pulp yield, circumference at breast height and height and tree volume scored at age three and six years. We assessed the impact of different genomic prediction methods, the composition and size of the training and validation set and the number and genomic location of SNPs on the predictive ability (PA).

Results: Heritabilities estimated using the realized genomic relationship matrix (GRM) were considerably higher than estimates based on the expected pedigree, mainly due to inconsistencies in the expected pedigree that were readily corrected by the GRM. Moreover, the GRM more precisely capture Mendelian sampling among related individuals, such that the genetic covariance was based on the true proportion of the genome shared between individuals. PA improved considerably when increasing the size of the training set and by enhancing relatedness to the validation set. Prediction models trained on pure species parents could not predict well in F$_1$ hybrids, indicating that model training has to be carried out in hybrid populations if one is to predict in hybrid selection candidates. The different genomic prediction methods provided similar results for all traits, therefore either GBLUP or rrBLUP represents better compromises between computational time and prediction efficiency. Only slight improvement was observed in PA when more than 5000 SNPs were used for all traits. Using SNPs in intergenic regions provided slightly better PA than using SNPs sampled exclusively in genic regions.

Conclusions: The size and composition of the training set and number of SNPs used are the two most important factors for model prediction, compared to the statistical methods and the genomic location of SNPs. Furthermore, training the prediction model based on pure parental species only provide limited ability to predict traits in interspecific hybrids. Our results provide additional promising perspectives for the implementation of genomic prediction in *Eucalyptus* breeding programs by the selection of interspecific hybrids.

Keywords: Genomic relationship, Genomic heritability, Two-generation, Genome annotation, High-density SNP-chip, Bayesian LASSO, GBLUP, rrBLUP

---

* Correspondence: par.ingvarsson@slu.se
[1]Umeå Plant Science Centre, Department of Ecology and Environmental Science, Umeå University, Umeå SE-90187, Sweden
[6]Present address: Department of Plant Biology, Uppsala BioCenter, Swedish University of Agricultural Sciences, Uppsala SE-75007, Sweden
Full list of author information is available at the end of the article

Background

Eucalyptus species and their hybrids are the most widely planted hardwoods in tropical, subtropical and temperate regions, due to their fast growth, short rotation times, wide environmental adaptability and suitability for commercial pulp and paper production [1, 2]. Interspecific hybrids of *E.grandis* and *E.urophylla*, in particular, are generally superior to their parents in growth, wood quality and biotic and abiotic stresses resistance, by inheriting both the fast growth and good rooting abilities of *E.grandis* while by maintaining disease tolerance and wide adaptability of *E.urophylla* [3]. A conventional breeding cycle toward clonal selection in hybrid populations involves mating, progeny trials, a small-scale clonal trial and a second expanded clonal trial, that together typically take between 12 and 18 years [1, 4]. To accelerate the genetic gain per unit time, new methods that can help shorten the breeding cycles are greatly needed.

Genomic prediction or genomic selection (GS) is one of the most recent developments in genomics-assisted methods that are aimed at improving breeding efficiency and genetic gains. Genomic prediction provides a genome-wide paradigm for marker-assisted selection (MAS) [5, 6]. In GS all genome-wide markers are fitted simultaneously in a model that relies on the principle of linkage disequilibrium (LD) to capture most of the relevant variation throughout the genome, whereas MAS focuses on discrete quantitative trait loci (QTLs) that have previously been detected, usually in underpowered experiments, and thus leaving most of the phenotypic variation unaccounted for [7]. GS are generally performed in three steps: (1) genotyping and phenotyping a 'reference' or 'training population' combined with the development of genomic prediction models that allow for prediction of phenotypes from genotypes; (2) validation of the predictive models in a 'validation population', i.e. a set of individuals that did not participate in model training; (3) application of the models to predict the genomic estimated breeding values (GEBVs) of unphenotyped individuals which are then selected according to their GEBVs [6]. GS has been successfully implemented in the breeding of livestock [7, 8] and crops [9, 10] and several recent papers have also exemplified its great potential in forest tree breeding [11, 12].

The accuracies of genomic prediction models vary depending on the statistical methods employed. Several methods have been developed for GS, including ridge-regression best linear unbiased prediction (rrBLUP), genomic best linear unbiased prediction (GBLUP), BayesA, BayesB, Bayesian LASSO, BayesR and reproducing kernel Hilbert space (RKHS) regression [7, 13]. These methods mainly differ in the assumptions of the distribution and variances of marker effects. For rrBLUP all loci are a priori assumed to explain an equal amount of variance and thus assumes that marker effects follow a normal distribution where all effects are shrunk to a similar and small size. [6, 14] In Bayesian methods (BayesA, BayesB, Bayesian LASSO and BayesR) the genetic variance explained by the ith locus, V_{gi}, is assumed to themselves follow a prior distribution, $p(V_{gi})$. Therefore, the variance can vary across loci, and combining the information from the prior distribution with that of the data yields an estimate of V_{gi} [6, 15]. For instance, BayesA assumes that the genetic variance follow an inverted chi-square distribution whereas Bayesian LASSO assume the genetic variance follow a double exponential distribution. The GBLUP method computes the additive genetic merits from a genomic relationship matrix and is equivalent to rrBLUP under conditions that are generally met in practice [16]. The RKHS regression model is a linear combination of the basic function provided by the reproducing kernel [17]. Recent studies have indicated that the selection of suitable statistical methods relies on the actual data at hand and the pattern of phenotypic variation in the traits of interest and with reference population used [9, 18].

Beside statistical methods, other factors are known to influence the accuracy of genomic prediction models, such as the size of the training population, number of markers employed, and relatedness between the training and validation population and, by extension, to the future selection candidates. Hayes et al. [19] found that for a given effective population size (N_e), increasing the size of reference population leads to improved accuracy of genomic predictions. Closer relationship between training population and selection candidates has also been reported to lead to a higher accuracy of genomic predictions, while enlarging the genetic diversity of the training population resulted in lower accuracy [20]. A number of simulation and empirical studies have shown that increasing the number of markers may improve the predictive accuracy as N_e also increased [9, 21–23]. However, increasing the number of markers in small N_e populations provides little or no improvement on predictive accuracy [24, 25].

Going one step further from previous studies in forest trees, where individuals of the same breeding generation were allocated to training and validation sets for the evaluation of genomic prediction models, in this study we used both the parental and progeny generations of *E. grandis*, *E. urophylla* and their F$_1$ hybrids to build prediction models using different subsets of parents and progeny for training and validation sets. A multi-species single-nucleotide polymorphism (SNP) chip containing 60,904 SNPs [26] were used to provide high-density genotyping of the two generations. Based on these data, we developed genomic prediction models for height, circumference at breast height (CBH), volume, wood basic density and pulp yield, using a number of statistical methods and compared their performance to the traditional pedigree-based prediction. Furthermore, we evaluated the impact of varying the number of SNPs and the composition and size of training and validation sets on the predictive ability (PA) of genomic prediction.

Methods

Breeding population

The breeding population in this study was obtained through controlled crossings of 86 *E. urophylla* and 95 *E. grandis* trees (G0 population) following a incomplete diallel mating design, resulting in 16,660 progeny individuals (G1 population) comprising 476 full-sib families with 35 individuals per family. In 2009, the progenies were deployed in a field trial under a randomized complete block design with single-tree plots and 35 replicates per family in Belmonte (Brazil, 39.19 W, 16.06 S, 210 m above the sea level) at Veracel Celulose S.A. (Eunápolis, BA, Brazil). Our experimental population consists of 168 parents (78 of *E.urophylla* and 90 of *E.grandis*) (G0), as not all parents were still alive at the time of the study, and 958 progeny individuals (G1) sampled across 338 full-sib families by avoiding low performing trees. The number of individuals in each full-sib family ranged from one to 13 with an average of 2.8 individuals per family.

Phenotyping

For the 958 G1 samples, height, volume, and circumference at breast height (CBH) were measured at age three and six years, respectively, and wood traits (basic density and pulp yield) were measured at age five years. For the 168 G0 parents, the same traits had been measured at age seven years for *E. grandis* and at age five years for *E. urophylla*. Briefly, height was measured using a Suunto hypsometer/height meter (PM-5/1520 series) and CBH was measured with a centimetre tape at 130 cm above ground. Wood properties were estimated by employing near-infrared reflectance spectra of sawdust samples collected at breast height using a FOSS NIRSystem 5000-M and applying calibration models developed earlier by Veracel S.A..

A mixed linear model was applied to minimize the impacts of environmental and age differences on each trait.

$$Y = X\beta + Zu + Wb + e$$

where Y is a vector of observations of a single trait; β is a vector of fixed effects, including overall mean, experimental sites and age differences; u is a vector of random additive genetic effect of individuals with a normal distribution, $u \sim N(0, A\ \sigma_u^2)$, A is a matrix of additive genetic relationships among individuals; b is a vector of random incomplete block effect nested in each experimental site; and e is a heterogeneous random residual effect in each experimental site. X, Z and W are incidence matrices for β, u and b, respectively. The phenotypes of each trait were then corrected by subtracting variation of sites, ages and blocks effects for all individuals, and were referred to adjusted phenotypes. The adjusted phenotypic traits were used for calculating the heritability of traits and for building genomic prediction models.

Genotyping and quality control

The 168 G0 and 958 G1 populations were genotyped using the Illumina Infinium EuCHIP60K [24] that contains probes for 60,904 SNPs. EUChip60K intensity data (.idat files) were obtained through GENESEEK (Lincoln, NE, USA). SNP genotypes were called using GenomeStudio (Illumina Inc., San Diego, CA, USA) following standard genotyping and quality control procedures with no manual editing of clusters as described earlier [26]. Further quality control of the genotyped samples was performed using PLINK [27]. Nine G1 individuals with sample call rate less than 70% or inbreeding coefficients greater than one were removed for further analyses. 10,240 SNPs were excluded due to low call rates (less than 70%) and 9243 SNPs were filtered out due to monomorphism or by having minor allele frequency (MAF) less than 0.01. Finally 117 SNPs were removed because they showed strong deviations from Hardy-Weinberg equilibrium (p-value $<1 \times 10^{-6}$).

After quality control, missing genotypes of the remaining individuals were filled in by imputation. We first tested the accuracy of imputation methods across a range of missing data (2% - 30%) by artificial removing SNPs from a fraction of our genotypes. Among the available family-based and population-based methods we assessed the following programs for imputation accuracy: BEAGLE [28], fastPHASE [29], MENDEL [30], random forest, SVD Impute, k-nearest neighbors [31], BLUP A matrix, Bayesian PCA, NIPALS, Probabilistic PCA [32]. BEAGLE provided the best accuracy for all missing data percentages, with accuracies exceeding 95% in all cases (Additional file 1). We therefore used BEAGLE to impute missing genotypes at 41,304 SNPs retained after the filtering steps discussed above, across all 168 G0 and 949 G1 individuals. The imputed genotype data was subsequently used in all genomic prediction analyses. LD between SNP pairs were measured using the squared correlation coefficient (r^2) for SNPs located on the same chromosome. Following Remington et al. [33], the decay of LD versus physical distance was then modelled using a nonlinear regression method.

We further estimated population structure and pairwise genomic relationships among the 1117 individuals by performing principal components analysis (PCA) [34] and by calculating genomic relationships among individuals [14] using 10,213 independent SNPs (LD-pruned) ($r^2 < 0.2$) calculated in PLINK [27]. Pedigree-based genetic relationship was estimated by using ABLUP in ASReml (see below for further information).

Statistical methods for genomic prediction

Four statistical methods were assessed for their ability to estimate the parameters in eq. (1) and for predicting GEBVs. These methods include *genomic best linear unbiased predictor* (GBLUP) [5], *ridge regression BLUP* (rrBLUP) [6], *Bayesian LASSO* (BL) [35], and *reproducing kernel Hilbert space* (RKHS) regression [17]. These methods were chosen to represent the variety of available approaches for genomic prediction. GBLUP represents a method which does not rely on marker effect estimation; rrBLUP estimates marker effects using linear and penalized parameters; BL represents a linear, parametric and Bayesian method for marker effect estimation; whereas RKHS represents a non-linear semi-parametric method. The performance of the four genomic prediction methods was compared with that of the commonly used pedigree-based BLUP (ABLUP) [36].

The GEBVs were estimated using the following mixed linear model:

$$y = 1\beta + Za + e \qquad (1)$$

where y is the vector of adjusted phenotypes of single trait, β is the vector of overall mean fitted as a fixed effect, a is the vector of random effects, and e is the vector of random residual effects. 1 and Z are incident matrix of β and a, respectively.

ABLUP

ABLUP is the standard method for predicting breeding values using the expected relatedness among individuals based on pedigree information [36]. For ABLUP, the vector of random additive effects (a) in Eq. (1) is assumed to follow a normal distribution $a \sim N(0, A\sigma_a^2)$, where A is the additive numerator relationship matrix estimated from pedigree information and the σ_a^2 is the additive genetic variance. The residual vector e is assumed as $e \sim N(0, I\sigma_e^2)$, where I is the identity matrix. Under these assumptions, Eq. (1) can be re-written as:

$$\begin{bmatrix} X^T X & X^T Z \\ Z^T X & Z^T Z + A^{-1} \frac{\sigma_e^2}{\sigma_a^2} \end{bmatrix} \begin{bmatrix} \hat{\beta} \\ \hat{a} \end{bmatrix} = \begin{bmatrix} X^T y \\ Z^T y \end{bmatrix} \qquad (2)$$

where σ_e^2 and σ_a^2 are estimated using a restricted maximum likelihood method. The estimated breeding values ($\hat{a}$) and fixed effects ($\hat{\beta}$) can be calculated directly from Eq. (2). ABLUP calculations were performed using ASReml 3.0 [37].

GBLUP

The GBLUP method is derived from ABLUP, but differs in that the matrix A in Eq. (2) is replaced with the genomic relationship matrix (G) that is calculated from genotypic data using $G = \frac{(M-P)(M-P)^T}{2\sum_{j=1}^{P} p_j(1-p_j)}$, where M is the matrix of samples with SNPs encoded as 0, 1, 2 (i.e. the number of minor alleles), P is the matrix of allele frequencies with the j-th column given by $2(p_j - 0.5)$, where p_j is the observed allele frequency of the samples [5]. In GBLUP, the random additive effects (a) in the Eq. (1) is assumed to follow $a \sim N(0, G\sigma_g^2)$, where σ_g^2 is the genomic-based genetic variance and GEBVs ($\hat{a}$) are again calculated from equation (2) but with A^{-1} replaced by G^{-1} and σ_a^2 replaced by σ_g^2. The GBLUP calculations were performed using ASReml 3.0 [37] and the G matrix was estimated using the "A.mat" function from the rrBLUP package in R [14].

rrBLUP

As opposed to the previous two methods, rrBLUP alters the notations of parameters a and Z in the Eq. (1), where Z now refers to a design matrix for SNP effects, rather than an incident matrix and a refers to SNP effects that are assumed to follow $a \sim N(0, I\sigma_m^2)$, where σ_m^2 denotes the proportion of the genetic variance contributed by each SNP [6]. With these alterations, Eq. (2) becomes:

$$\begin{bmatrix} X^T X & X^T Z \\ Z^T X & Z^T Z + I\lambda \end{bmatrix} \begin{bmatrix} \hat{\beta} \\ \hat{a} \end{bmatrix} = \begin{bmatrix} X^T y \\ Z^T y \end{bmatrix} \qquad (3)$$

where $\lambda = \sigma_e^2 / \sigma_u^2$ is the ratio between the residual and marker variances. A prediction for the GEBV for each individual is calculated as $\hat{g}_i = Z_i^T \hat{a}$ from equation (3), where Z_i^T is the SNP vector for individual i and $\hat{a}$ is the vector of estimated SNP effects. All calculations were performed using the "mixed.solve" function from the rrBLUP package in R [14].

Bayesian LASSO

The Bayesian LASSO (BL) method is the Bayesian treatment of LASSO regression as proposed by Legarra et al. [34]. In BL the vector of SNP effects, a in equation (1), is assumed to follow a hierarchical prior distribution with $a \sim N(0, \mathbf{T}\sigma_m^2)$, where $\mathbf{T} = \text{diag}\left(\tau_1^2, ..., \tau_p^2\right)$. τ_j^2 is assigned as $\tau_j^2 \sim Exp(\lambda^2)$, $j = 1, ..., p$. λ^2 is assigned as $\lambda^2 \sim Gamma(r, \delta)$. The residual variance σ_e^2 is assigned as $\sigma_e^2 \sim \chi^{-2}(df_e, S_e)$.

We implemented the BL method using the "BLR" function from the BLR package in R [38]. Here a Monte Carlo Markov Chains sampler was applied and prior parameters $(df_e, S_e, r, \delta,$ and $\lambda^2)$ were defined following the guidelines proposed by de los Campos et al. [39]. The chain length was 20,000 iterations, with the first 2000 excluded as burn-in and with a subsequent thinning interval of 100.

RKHS

RKHS assumes that the random additive effects in Eq. (1) are $a \sim N\left(0, \mathbf{K}\sigma_g^2\right)$, where $\mathbf{K}$ is computed by means of a Gaussian kernel that is given by $K_{ij} = \exp(-hd_{ij})$ [17]. h is a semi-parameter that controls how fast the prior covariance function declines as genetic distance increase and d_{ij} is the genetic distance between two samples computed as $d_{ij} = \sum_{k=1}^{p} \left(x_{ik} - x_{jk}\right)^2$, where x_{ik} and x_{jk} are kth SNPs ($k = 1, ..., p$) for the ith and jth samples, respectively. We implemented the RKHS method through the "BGLR" function from the BGLR package in R [40], which use a Gibbs sampler for the Bayesian framework and assigns the prior distribution of σ_g^2 and σ_e^2 as $\sigma_g^2 \sim \chi^{-2}\left(df_g, S_g\right)$ and $\sigma_e^2 \sim \chi^{-2}(df_e, S_e)$, respectively. Here we chose a multi-kernel model as suggested by Perez [40], where three h values were defined as $h_1 = 2/\left(5*\overline{d}\right)$, $h_2 = 2/\overline{d}$, $h_3 = 2*5/\overline{d}$, $\overline{d}$ was the median of d_{ij}. The Gibbs chain length was 20,000 iterations with the first 2000 iterations discarded as burn-in and a thinning interval set to 100.

Heritability estimation

We estimated the pedigree-based narrow-sense heritability (h_a^2) using the relationship matrix from the ABLUP method, and the narrow-sense genomic heritability (h_g^2) using the genomic relationship matrix from GBLUP (details in [41]). The respective heritabilities were calculated as:

$$h_a^2 = \frac{\sigma_a^2}{\sigma_y^2} \qquad h_g^2 = \frac{\sigma_g^2}{\sigma_y^2}$$

where σ_a^2 is the additive variance estimated from ABLUP, while σ_g^2 is the marker-based genetic variance estimated from GBLUP. σ_y^2 is the phenotypic variance of the population.

Size and genetic composition of the training and validation sets

We simultaneously assessed the impact of the size and genetic participation of G0 and G1 individuals in the training set (TS) and validation set (VS) of the genomic prediction models. Regarding relative TS/VS sizes, we divided all 1117 (G0 and G1) individuals into five different size groups with a TS:VS ratio of 1:1, 2:1, 3:1, 4:1 or 9:1. The corresponding sizes of the TS/VS were respectively 558/559, 743/374, 836/281, 892/225 and 1003/114 individuals. Within these pre-established size compositions, four scenarios were employed where the participation of G0 and G1 individuals were evaluated to assess the impact of varying the degrees of relationship and diversity between TS and VS. In the first scenario (CV$_1$) assignment of individuals to either TS or VS was random. For the second scenario (CV$_2$) all G0 parents were assigned to the TS and complemented with a random selection of G1 individuals up to the required number in the set, while the VS was composed exclusively of the remaining G1 individuals. The third (CV$_3$) and fourth (CV$_4$) scenarios were built based on minimizing and maximizing relatedness between TS and VS. The relatedness-based assignment of individuals was determined using the procedure described in Spindel et al. [9]. Briefly, 1117 individuals were assigned to 182 clusters based on their genotypes using a k-means clustering algorithm implemented in the "pamk" function from the fpc package in R. This method attempts to minimize the distance between individuals in a cluster and the centre of that cluster. Using the relatedness estimates, CV$_3$ was then built by assigning individuals to TS and VS based on dissimilarity, such that individuals from the same cluster were not allowed to be both in the same TS or VS. For CV$_4$ individuals from same cluster were forced to be either in the TS or VS to increase relatedness within TS and VS [9].

Genomic prediction models

We evaluated the effects of the five statistical methods (GBLUP, rrBLUP, BL, RKHS and ABLUP), five TS/VS sizes and four TS/VS composition scenarios (5*5*4 = 100 models in total) on the predictive ability (PA) of genomic prediction. For each of the 100 models, 200 replicate runs were carried out for each trait and the performance of the models were evaluated in terms of their PA ($r_{y, \hat{g}}$), which is defined as the Pearson correlation between the adjusted phenotypes and the GEBVs of the samples in the VS. ANOVA was performed with all effects declared as fixed on 80 out of the 100 models tested (the 20 ABLUP models were excluded) to partition the total variance into different sources (genomic prediction method, TS/VS size and genetic composition). The significant differences we found were further assessed by means of a paired t tests ($\alpha = 5\%$), adjusted by a Bonferroni correction. The 80 models, as described above, were used for assessing the impact of TS/VS composition and TS/VS size, while all 100 models were used to evaluate the statistical methods against ABLUP. All available SNPs were used in all the analyses of these models.

Numbers and genomic location of SNPs subsets

We finally assessed the impact of the number of SNPs and their locations (gene vs. intergenic region) on the PA of genomic prediction models. 12 subsets with different numbers of SNPs were generated by randomly selecting 10, 20, 50, 100, 200, 500, 1000, 2000, 5000, 10,000, 20,000 and 41,304 SNPs from all the available SNPs. For SNP location, SNPs subsets located in different regions of the genome were established by including SNPs located in four different regions: (i) coding sequences (CDS) only (11,786 SNPs); (ii) entire genic regions including CDS, UTRs, introns, and sequences 2 kb up and downstream of the gene (30,405 SNPs); (iii) intergenic regions (10,899 SNPs), and (iv) all 41,304 SNPs. The location and classification of each SNP was obtained by mapping SNPs onto the E.grandis genome using SnpEff [42]. Genomic prediction models were built for all four TS/VS compositions using only the two statistical methods (GBLUP and RKHS) that showed the best predictive performance in the previous analyses, using a TS/VS size ratio of 4:1 (892/224).

Results

Phenotypic trait correlations

Growth (height, volume, and CBH) and wood properties (basic density and pulp yield) were measured for all 168 G0 and 949 G1 individuals. The raw phenotypic data were adjusted using a mixed linear model to minimize the impacts of environment and age differences. The pairwise correlations between the adjusted traits were described by calculating Pearson correlation coefficients (Fig. 1). Growth traits were correlated with each other. Interestingly, however, while CBH and volume at age three and six years were highly correlated ($r = 0.92$ and 0.95 respectively), height at age three was only weakly correlated with

height at age 6 ($r = 0.36$). For wood properties traits, basic density was negatively correlated with pulp yield, although only weakly so ($r = -0.28$). Growth traits showed no correlations with wood traits ($r = - 0.1$ to 0.1).

Breeding population structure and relatedness

Population structure across G0 and G1 individuals was assessed by PCA based on 10,213 LD-pruned, independent SNPs ($r^2 < 0.2$). The first two PCs explained 6.07% and 3.8% of the total genetic variance (Fig. 2a) and clearly separated the G0 individuals of the two species, E.grandis and E.urophylla, with the E.grandis individuals further subdivided into two subgroups likely representing the two main provenances used in breeding programs in Brazil. The G1 individuals were generally projected into the space defined by their parents, but with a few outliers. The expected pedigree-based and realized genomic-based genomic relationships among G0 and G1 individuals were visualized using heatmaps (blue and red in Fig. 2b, respectively). The result of the genomic relationship analysis corroborated the PCA result, in which E. urophylla was clustered into a single group, whereas E. grandis formed two subgroups. The average values of the realized genomic relationships among what were considered to be full-sibs, half-sibs and unrelated individuals from the pedigree data were generally lower than the expected relationships values (0.309 vs. 0.5, 0.131 vs. 0.25 and 0.0056 vs. 0, respectively) (Table 1). This result suggests that pedigree errors were likely present in this population. These putative pedigree errors in turn negatively affected our ability to estimate the heritability of traits based on pedigree information, which were considerably lower than those estimated using genomic-based realized genetic relationships (Table 2).

Predictive abilities with different statistical methods

Estimates of PAs were obtained using different statistical methods, compositions and sizes of TS/VS for each trait (Additional file 2). An ANOVA showed that all these factors had a significant effect on the PA (P-value <0.005) (Additional file 3). Across the four genomic prediction methods used (GBLUP, rrBLUP, BL, and RKHS) the average PA varied from 0.27 to 0.274 (Additional file 4). All the four methods outperformed the pedigree-based ABLUP prediction (mean PA = 0.121) by an average of 80%–200% across the eight traits (Fig. 3). RKHS yielded a slightly better PAs for six out of eight traits and this method was particularly suitable for predicting traits that displayed lower heritabilities, such as CBH and height. The other three methods generally gave similar results across all traits, although with a slightly better performance than RKHS for pulp yield (Fig. 3).

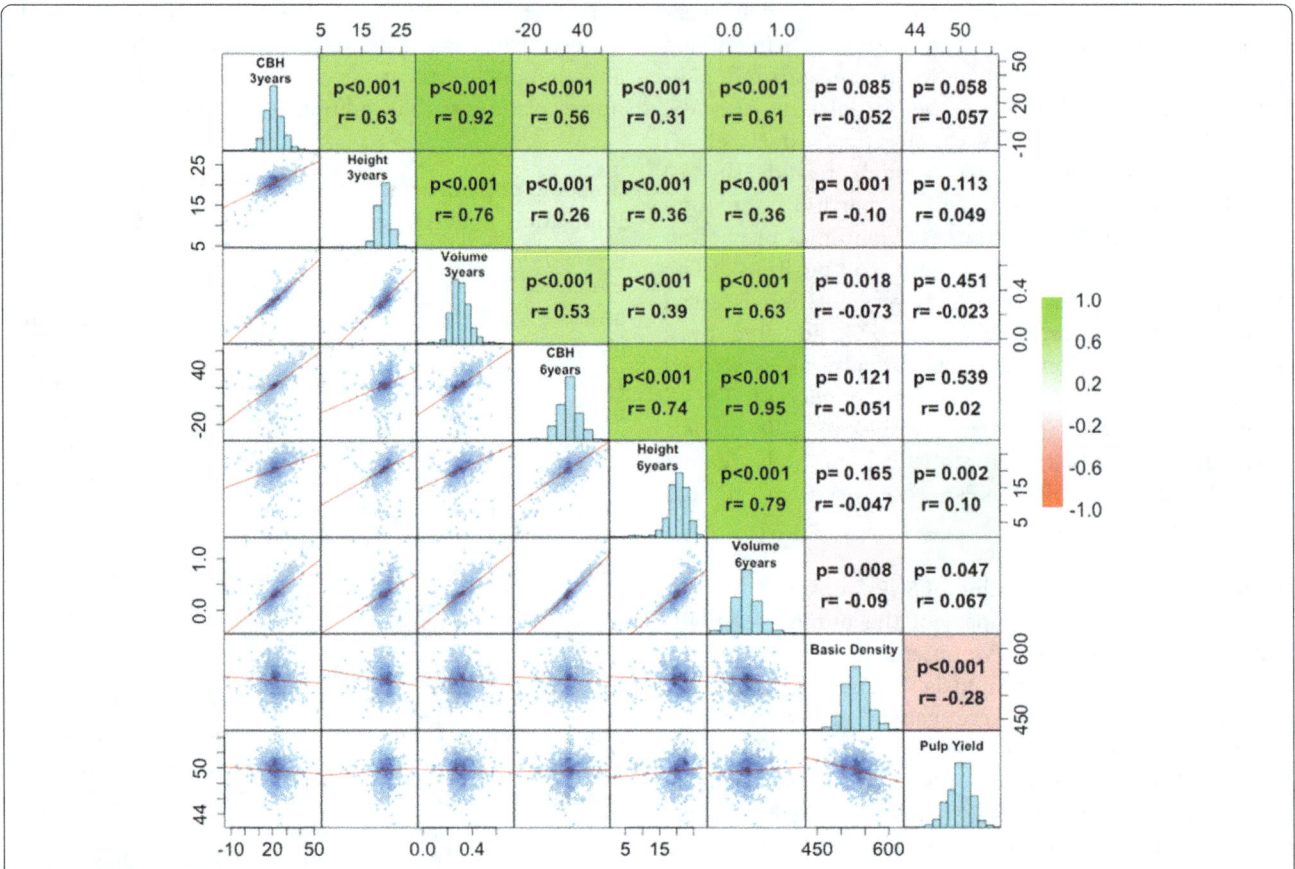

Fig. 1 Correlation and distribution of phenotypes. *Scatter plots* (*lower* off-diagonal) and correlations with probability values (*upper* off-diagonal; H_0: $r = 0$) for adjusted phenotypes between pairs of traits. *Color* key on the *right* indicates the strength of the correlations. Diagonal: histograms of the distribution of adjusted phenotypes values

Impact of TS/VS compositions and relative sizes on predictive ability

The average PAs differed significantly for the different TS/VS compositions tested and varied from 0.253 to 0.286 (Additional file 5). The genomic prediction model built with CV_2 (all G0 parents in the TS) showed the highest PAs for all traits except pulp yield, whereas models based on CV_3 (minimum relatedness between TS and VS) gave the worst predictions. The models based on CV_1 (random assignment) and CV_4 (maximum relatedness between TS and VS) showed no significant differences in PA (Fig. 4, Additional file 5). The average PA was significantly improved from 0.251 to 0.285, as the TS/VS ratio increased from 1:1 (558/559) to 9:1 (1003/113) (Additional file 6), irrespective of the prediction method (Fig. 3) or the genetic composition of TS/VS used (Fig. 4), clearly showing the importance of an adequate size of the training set to build prediction models. Furthermore, there was a steeper increase in PA when TS/VS ratio increased from 1:1 (558/559) to 2:1 (743/374) than from 2:1 (743/374) to 9:1 (1003/114) for all traits (Figs. 3 and 4).

Impact of the number of SNPs and their genomic location on predictive ability

Estimates of PA using different numbers of SNPs (Additional file 7) and subsets of SNPs in different genomic locations (Additional file 8) were obtained with two prediction methods, using a TS/VS ratio of 892/225 and using all the four different TS/VS compositions. An ANOVA showed that both the number of SNPs and their genomic location significantly affect the PA for both prediction methods (GBLUP and RKHS) (*P*-value <0.005), and that the number of SNPs has a larger impact than their genomic location (Additional file 9). The average PAs across all traits decreased from 0.278 to 0.113 when the number of SNPs used in the prediction models dropped from 41,304 to only 10, and the reduction was especially strong when the number of SNPs went below 5000 (Additional file 10). On the other hand, no significant improvement was generally seen in the average of PA when more than 5000 SNPs were used (Additional file 10, Fig. 5). The results obtained for the different traits suggest that traits with lower heritability are more sensitive to the reduction in the number of SNPs (Fig. 5). For instance, PA for basic density ($h^2 = 0.35$)

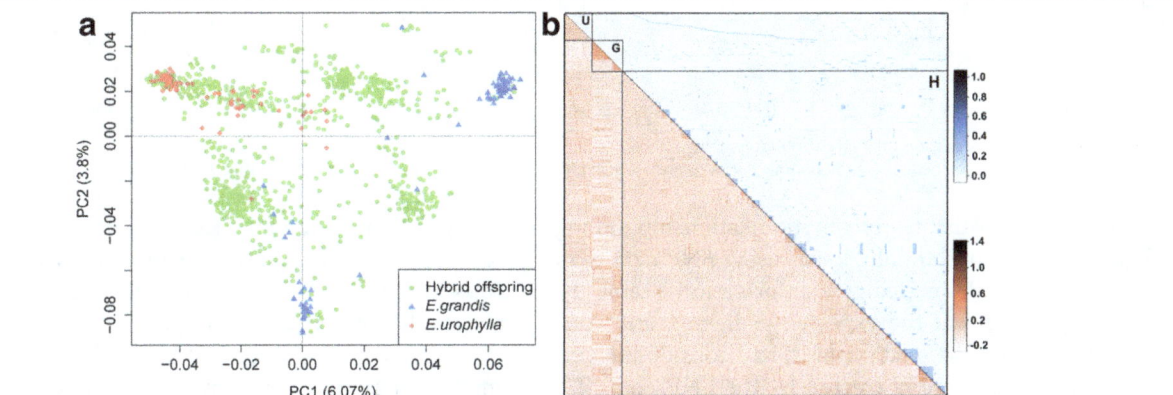

Fig. 2 Genetic structure and relatedness in the breeding population. (**a**) First two principal components of a PCA revealing population structure. *Dots* represent *E.grandis* (*blue*), *E.urophylla* (*red*) and their F$_1$ (*green*) individuals. (**b**) Heatmaps of the pairwise pedigree-expected relationships (*blue, upper* off-diagonal) and genomic-realized relationship (*red, lower* off-diagonal) of the 1117 individuals assigned to *E.grandis* (G), *E.urophylla* (U) and their hybrid progenies (H)

went from 0.47 to 0.24 (a 50% decrease) when the number of SNPs dropped from 40,000 to 10, whereas CBH of age three ($h^2 = 0.113$) decreased from 0.128 to 0.03 (a 77% decrease). Overall, slight significant differences were seen in PAs by using SNP sets located in different genomic regions (Fig. 6), the average PAs range from 0.270 to 0.284 (Additional file 11). Predictions using SNPs located in intergenic regions were marginally better than using SNPs in genic regions or all SNPs, except for pulp yield that could be better predicted based on models using SNPs from coding and gene regions (Fig. 6). When comparing the PA of models using SNPs in coding versus entire gene regions, the latter had a slightly better performance, most likely due to the larger number of SNPs used (30,504 vs. 11,786) and not due to any specific effect of genomic location. When we assessed the pairwise LD (r^2) among SNPs in the four regions tested, the extent of LD differed among them, with LD showing the most rapid decay in coding regions and the slowest decay in intergenic regions (Additional file 12).

Discussion

This study presents the results of an empirical evaluation of the accuracy of genomic prediction on growth and wood quality traits in *Eucalyptus* using data from a high-density SNP array. Our results are based on data from a two generations breeding population and provide additional encouraging results on the prospects of using

genomic prediction to accelerate breeding. We have assessed a range of factors, including the statistical methods used to estimate predictive ability, the size and composition of the training and validation sets as well as the number and genomic locations of SNPs used in the prediction model. Hereafter we will discuss how these factors influenced the prediction accuracy.

Genomic data corrected pedigree inconsistencies

All four genomic prediction methods performed significantly better than the pedigree-based evaluations for all complex traits assessed (Fig. 3). While similar results have been reported for animals [18, 43] and crop species [9, 36] across a number of traits, in forest trees prediction accuracies using genomic data have generally been similar or up to 10–30% lower than accuracies obtained using pedigree-estimated breeding values, including *Eucalyptus* [4], loblolly pine (*Pinus taeda*) [44], white spruce (*Picea glauca*) [45, 46], interior spruce (*Picea engelmannii × glauca*) [47, 48] and maritime pine (*Pinus pinaster*) [49]. Genomic predictions with lower accuracies than pedigree-based predictions could arise from insufficient marker density, such that not all casual variants are captured in the genomic estimate [41], or an overestimate of the pedigree-based prediction due to its inability of ascertaining the true genetic relationships in half-sib families [47]. Our result however differ from previous studies in forest trees due to the fact that the

Table 1 Pairwise expected pedigree-based and realized genomic-based relationships in the different family types

	Full-sib families (961)[a]			Half-sib families (12718)			Unrelated individuals (434252)		
	Min	Mean	Max	Min	Mean	Max	Min	Mean	Max
Pedigree-expected relationship	0.5	0.5	0.5	0.25	0.25	0.25	0	0	0
Genomic-realized relationship	−0.274	0.309	0.933	−0.464	0.131	0.908	−0.467	−0.056	0.891

[a]Number in parentheses indicate the number of pairwise estimates

Table 2 Pedigree-based and genomic heritabilities for each trait

	CBH (3)[a]	Height (3)	Volume (3)	CBH (6)	Height (6)	Volume (6)	Basic density	Pulp yield
h_a^{2b}	0.051(0.03)	0.074(0.04)	0.057(0.03)	0.085(0.04)	0.097(0.05)	0.068(0.04)	0.23(0.04)	0.27(0.05)
h_g^2	0.113(0.04)	0.171(0.05)	0.162(0.04)	0.184(0.04)	0.193(0.05)	0.196(0.04)	0.35(0.05)	0.46(0.05)

[a]Number in parentheses correspond the age at measurement;
[b]h_a^2 and h_g^2 correspond to the pedigree and genomic narrow-sense heritability, respectively, with their standard deviation in parenthesis

average pairwise estimates of genetic relationship among individuals were substantially lower using SNP data than expectations based on pedigree information (Table 1), clearly suggesting that the expected pedigrees, and consequently the pairwise relationships, had considerable inconsistencies that were corrected by the SNP data. We speculate that these inconsistencies likely derived from pollen contamination and/or mislabelling in the process of generating the full and half-sib families. Besides correcting potential pedigree errors, the relatively dense SNP data used in our study also was able to accurately capture the Mendelian sampling variation within families so that genetic variances estimates were based on the true proportion of the genome that is identical by descent (IBD) or state (IBS) among half- or full-sib individuals, resulting in improved estimates of trait heritability (Table 2).

Genomic predictions show that traits adequately fit the infinitesimal model

Overall, the different genomic prediction methods provided similar results for the all traits evaluated, with only a slight advantage for RKHS which showed better PAs for the low-heritability growth traits (Fig. 3). However, for pulp yield, RKHS was instead the worst performing method, and it is possible that the definition of a kernel

simply was not suitable for this particular trait [17]. Our results corroborate previous reports from both crops and animals [18, 50, 51], as well as forest trees. In loblolly pine, for example, the performance of rrBLUP and three Bayesian methods were only marginally different when compared across 17 traits with distinct heritabilities, with a small improvement seen for BayesA only for fusiform rust resistance where loci of relatively larger effect have been described [44]. Similar results were obtained for growth and wood traits in other forest trees showing no performance difference between rrBLUP and Bayesian methods [46, 48, 49]. This occurs despite simulation studies suggesting that Bayesian methods, like BL, should outperform univariate methods such as rrBLUP and GBLUP [6, 52, 53]. One possible reason for the apparent disagreement between simulations and empirical data sets could be that the true QTL effects for most of traits are relatively small and the distribution is less extreme than in simulated data [54]. Our results therefore support the proposal that either rrBLUP or GBLUP are effective methods in providing the best compromise between computational time and prediction efficiency [55] and that the quantitative traits assessed in our study adequately fit the assumption of the infinitesimal model.

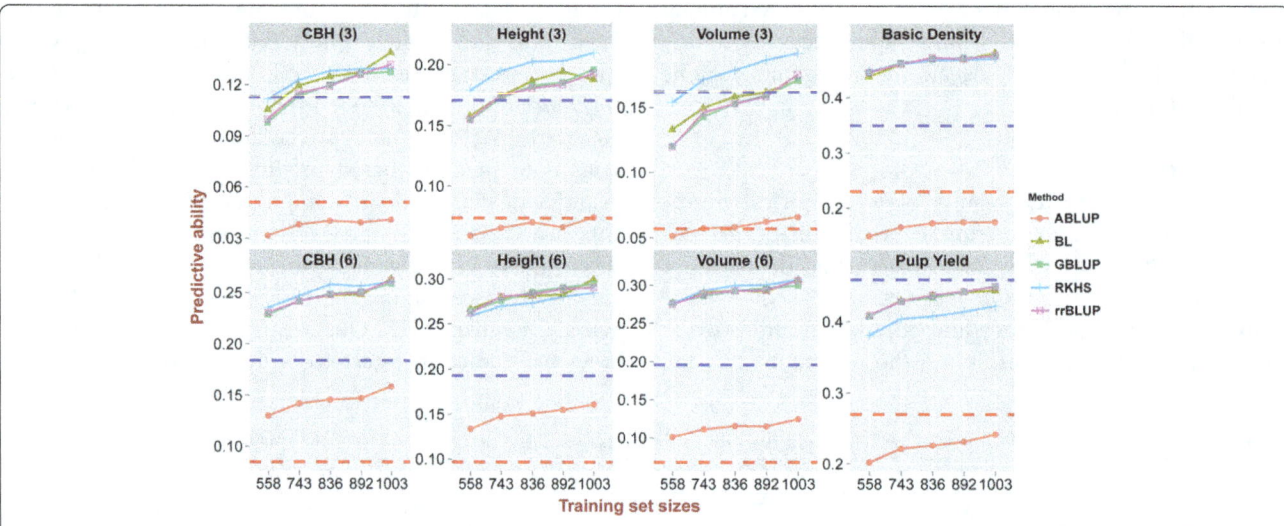

Fig. 3 Predictive abilities with different methods and increasing sizes of training sets. Predictive ability (y axis) estimated using five methods across five training set/validation set sizes in numbers of individuals (x axis) 558/559, 743/374, 836/281, 892/225 and 1003/114. *Red* and *blue dashed lines* indicate the pedigree-based (h_a^2) and genomic-realized (h_g^2) narrow-sense heritability respectively

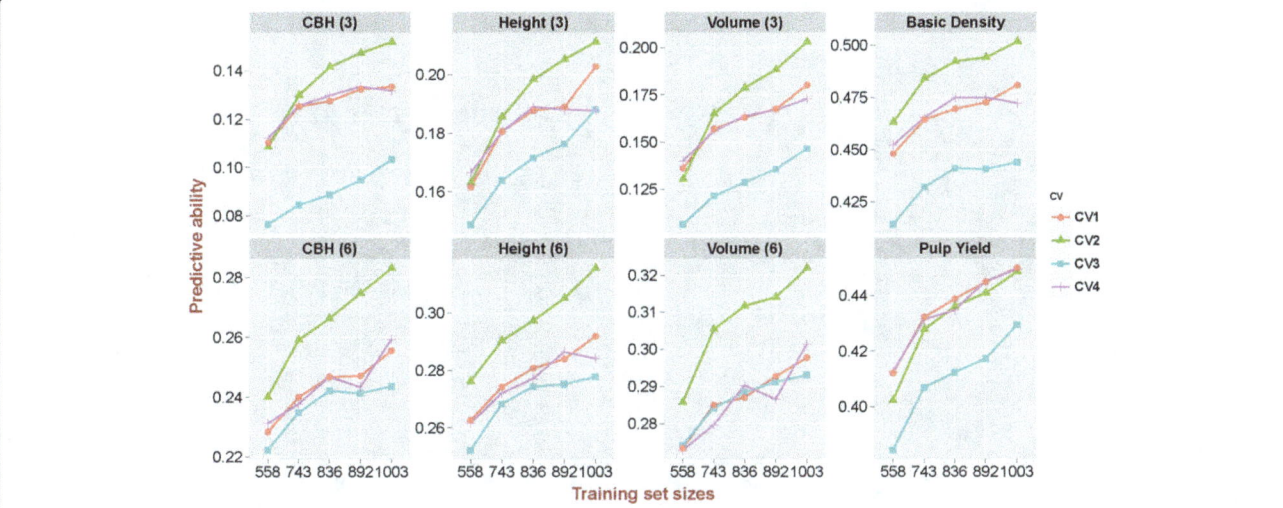

Fig. 4 Predictive abilities with variable levels of relatedness between training and validation sets. CV$_1$: random assignment of individuals to either training set (TS) or validation set (VS); CV$_2$: all the G0 pure species parents assigned to the TS; CV$_3$: minimum relatedness between TS and VS individuals; CV$_4$: maximum relatedness between TS and VS individuals. Estimates were obtained using GBLUP and RKHS across five TS/VS sizes in numbers of individuals (x axis): 558/559, 743/374, 836/281, 892/225 and 1003/114

Training set size, composition and relatedness strongly affect predictive ability

Our results show that the size and compositions of training and validation sets had the largest impact on the PA, irrespective of the analytical method used (Fig. 4). The average PA rapidly increased with increasing sizes of the TS and did not show any sign of plateauing. Earlier simulations of *Eucalyptus* breeding scenarios had in fact shown that with up to $N = 1000$ individuals in the TS, the accuracy would rapidly increase, and additional gains were seen up to $N = 2000$ individuals for traits with low heritabilities, for larger

numbers of QTLs involved in traits and for larger effective population size (N_e). After $N = 2000$ the predictive accuracy would tend to plateau irrespective of the N_e and genotyping density [22]. Simulations [19, 56] and proof-of-concept studies [57] in crop species also show improved PA with larger TS sizes. Larger training populations alleviate the probability of losing rare favourable alleles from the breeding population as generations of selection advance. Additionally, by sampling more individuals for training, a larger diversity is captured and better estimates of the marker effects are obtained which in turn positively

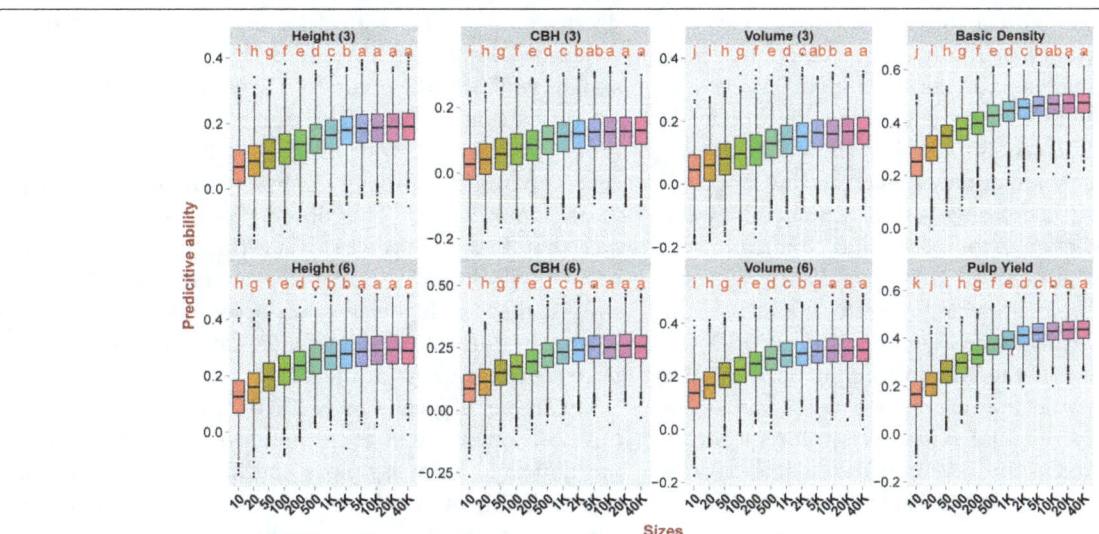

Fig. 5 Predictive abilities with increasing numbers of SNPs. Predictive ability estimated with GBLUP and RKHS with increasingly larger sets of SNP sampled at random from the total of 41,304 SNPs. Outliers are indicated by *black dots*. *Letters* indicate significant difference between the different models after Bonferroni adjustment ($P < 0.05$)

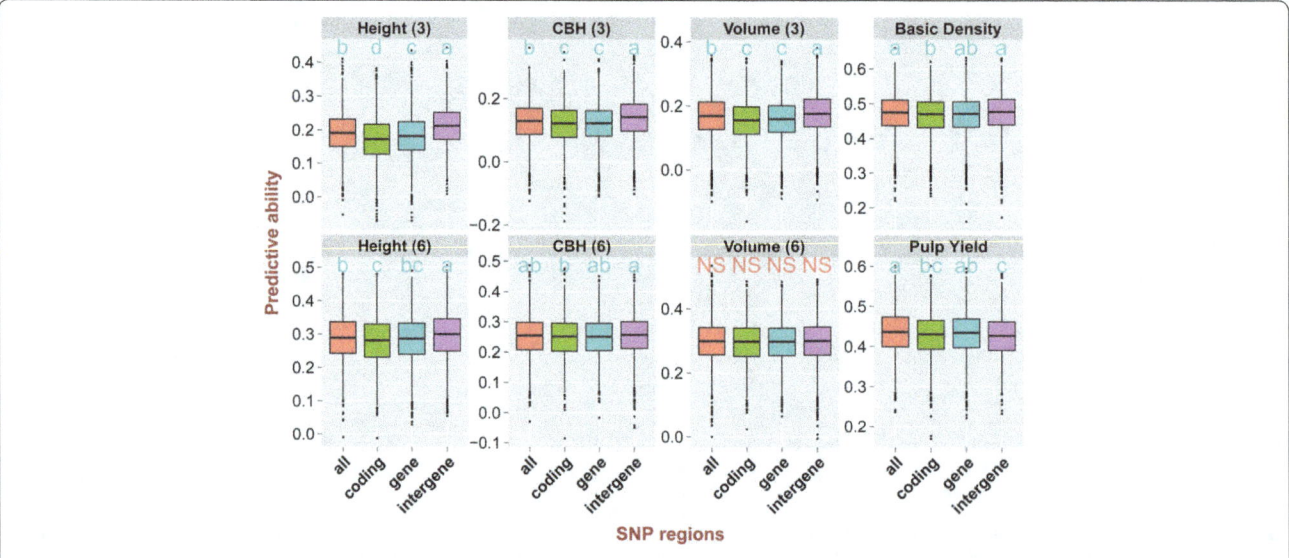

Fig. 6 Predictive abilities using SNPs located in different genomic regions. Predictive ability estimated with GBLUP and RKHS using 11,786 SNPs in coding DNA, 30,405 SNPs in genic regions (CDS, UTR, intron, and within 2 kb upstream and downstream of genes), 10,899 SNPs in intergenic regions and all 41,304 SNPs. *Letters* indicate significant difference between the different models after Bonferroni adjustment ($P < 0.05$)

impact predictions in cross-validations and future genomic selection candidates.

As expected, relatedness between TS and VS had a large impact on PAs for all traits. Prediction models built under scenario CV_3 (minimized relatedness between TS and VS) resulted in significantly worse predictions than in scenario CV_4 when relatedness was maximized. Our results are in line with previous reports in forest trees, such as white spruce [45, 46] and *Eucalyptus* [4], where models developed for one population had limited or no ability of predicting phenotypes in unrelated populations, suggesting that prediction models are largely population specific. With a lower relationship between TS and VS, the extent of LD is shorter and not stable across distantly related individuals in populations and the predictive ability of genomic prediction model is therefore reduced. Recent simulations show that the accuracy of genomic prediction models decline approximately linearly with increasing genetic distance between training and prediction populations [58]. Increased relatedness reduce the number of independently segregating chromosome segments and therefore increase the probability that chromosome segments that are IBD and which are sampled in the training population are also represented in the selection candidates. Our results provide additional experimental evidence that for successful implementation of GS the selection candidates have to show a close genetic relationship to the training population.

PAs were considerably higher when all the G0 parents were kept in the TS (scenario CV_2). This result could be due to two reasons. On one hand, by keeping all G0

parents in TS, we ensure that a large genetic diversity is available for model training, which could explain the positive impact of G0 inclusion on predictions. On the other hand, it is possible that by allocating all G0 individuals to the TS the positive effect we observe is strictly not due to increased predictive power but rather because we avoid the potentially negative impact of having pure species parents in the validation set in combination with G_1 progeny that were largely F_1 hybrids. In order to evaluate this, we estimated PA of genomic prediction models by using GBLUP and RKHS, having only the 168 G0 parents for TS and randomly selected 168 G1 individuals in VS. To control for the effect of the strongly reduced TS size, we compared this setup with random assignment of individuals to TS or VS but keeping the size of each at $N = 168$. The results showed considerably lower PAs (even zero or negative) when using only pure species parents to predict G1 hybrid progeny phenotypes (Additional file 13). This observation, together with the fact that PAs for scenario CV_4 (maximum relatedness between TS and VS) were also generally lower than CV_2, suggest that the higher PAs we observe for scenario CV_2 is mostly due to avoiding the negative effect of having pure species parents in the VS.

The issue of genomic prediction in hybrid breeding has been investigated so far only within species and only for domestic animals, more specifically for bovine and pig breeding in which selection is carried out in pure breeds but with the aim to improve crossbred performance [43, 59]. Results from simulations show that training on crossbred data provides good PAs by selecting purebred individuals for crossbred performance, although PAs drop with increasing distances between breeds [60]. When crossbred data is not available, separate purebred training populations can be used either separately or combined depending on the correlation of LD phase between the pure lines [61], which in turn is in part determined by the time of divergence between the populations. Compared to bovine breeds that belong to the same species and have diverged relatively recently (<300KYA) [62], the estimated divergence time between the two *Eucalyptus* species used in our study is much older, estimated at 2–5 MYA [63]. We therefore don't expect much correlation of LD phase between the two species and it is thus not surprising that training on the combined pure species sets with validation in F_1 hybrids resulted in poor PA. To the best of our knowledge, our results are the first ones to provide an initial look at the issue of genomic prediction from pure species to interspecific hybrids and our results indicate that, consistent with theoretical expectations, models have to be trained using hybrids if one is to predict phenotypes in hybrid selection candidates.

Number of SNPs is more important than SNP genomic location

Across all traits, no major improvement was detected in PA when more than 5000 SNPs were used (Additional file 10, Fig. 5), although a slight increase was observed for height of age three, basic density and pulp yield when using GBLUP based on 20,000 SNPs. Several studies have previously shown that considerably lower numbers of SNPs provided PAs equivalent to those observed using all SNPs available [24, 64]. The necessary number of SNPs needed for genomic prediction model depends on the extent of LD, which is strictly dependent on N_e. Our results, where we achieve equivalent PAs using either all or only 10–20% of the genotyped markers suggests that it represents a closed breeding population with a relatively modest N_e. This has been a common approach in domestic animals with the intent of developing low-density genotyping chips to reduce genotyping costs [8]. The main advantage of using reduced SNP panels is cost-effectiveness, although it is expected that using a higher density of markers will be necessary to mitigate the decay of PAs over generations due to the combined effect of recombination and selection on the patterns of LD [65]. It is also questionable whether it will be more cost effective to have targeted low-density SNP chips for specific populations or a full SNP chip that can be used across breeding populations of several organizations. By having a SNP chip that will accommodate several populations the cost-effectiveness and economy of scale of amassing many more samples to be genotyped with the same chip will likely be much larger

than the cost reduction observed by using a smaller number of SNPs on each specific population.

SNP location also contributed to the predictive ability of genomic prediction model although the effects were rather modest. PAs using SNPs in intergenic regions were slightly better than using SNPs in genic regions or using all SNPs, except for pulp yield that could be somewhat better predicted with SNPs in coding and gene regions (Fig. 6). This likely represents a random sampling effect and not any specific enrichment for functional variants for this trait. However, the decline of LD was slower for SNPs in intergenic regions than for SNPs in genic and/or coding regions (Additional file 12) and the slightly longer range of LD might help explain why using SNPs in intergenic regions provided better PAs. With slower LD decay, SNPs in intergenic regions might better capture QTLs across longer genomic segments than SNPs in coding regions where LD decays more rapidly.

Further issues affecting the accuracy of model prediction

Several issues remain to be investigated for successful adoption of genomic prediction in operational eucalypt breeding. First, how does the accuracy of genomic prediction decline over successive generations of selection due to the effects of recombination? Simulation studies illustrated that the prediction accuracy decline rapidly during early generations but this decline slows down in later generations [6, 16]. A GS model should therefore be updated after the phenotypes of next generation individuals become available. Second, how stable are genomic prediction models across multiple environments and how important is it to consider genotype by environment interactions in the models? The interaction between genomic prediction and environmental effects will essentially follow conventional G x E strategies. Prediction models are expected to be accurate across sites within the same breeding zone (an area within which a single population of improved trees can be planted without fear of maladaptation) but not necessarily across different breeding zones [12]. Furthermore, with genomic prediction, individuals are not evaluated on the basis of their own phenotypic performance, but on the basis of genomic information across other individuals, years and environments, which given an opportunity to evaluate the effect of particular genomic segments that are shared between individuals across multiple environments. Burgueno et al. [66] showed that models incorporating pedigree and marker data on wheat lines from multiple environments can substantially enhance prediction accuracy relative to only pedigree-based prediction or relative to genomics prediction models derived from single environments. Finally, we have only considered the additive genetic variance for building genomic prediction models in our eucalypt population, but it is possible, and perhaps even likely, that non-additive genetic effects play an important role in many breeding populations and specifically in populations consisting of early generation hybrids. A recent simulation study of genomic prediction in *Eucalyptus* breeding reported that genomic prediction including dominance effects performed better for clone selection where as non-additive effects did not improve the estimation of breeding value for parental selection [67]. To the best of our knowledge, no experimental data exist in forest trees regarding the ability of GS to predict the total genotypic value of individual trees, including both additive and non-additive effects.

Conclusions

Our experimental results provide further promising perspectives for the implementation of genomic prediction in *Eucalyptus* breeding programs. Genomic prediction largely outperformed pedigree-based prediction in our experiment, mainly due to the fact that our expected pedigree had major inconsistencies, resulting in gross underestimation of all pedigree-based estimates. This rather unexpected result illustrated an additional advantage of using SNP data and genomic prediction in breeding programs. While the main advantage of genomic prediction in eucalypt breeding will likely be the reduction of the breeding cycle length [4], the use of a genomic relationship matrix allowed us to obtain precise estimates of genetic relationship and heritabilities that we would otherwise not have had access to. Furthermore, our results corroborated the key role of relatedness as a driver of PA, the potential of using lower density SNP panels, and the fact that growth and wood traits adequately fit the infinitesimal model such that either GBLUP or rrBLUP represents a good compromise between computational time and prediction efficiency. In contrast to previous studies in *Eucalyptus*, we had accessed to both the pure species parents (*E. grandis* and *E. urophylla*) and their F$_1$ progeny. We show that models trained on pure species parents do not allow for accurate prediction in F$_1$ hybrids, likely due to the strong genetic divergence between the two species and lack of consistent patterns of LD between the two species and their hybrids.

Additional files

Additional file 1: Average accuracy of SNP imputation methods with increasing proportions of missing data. SNPs on chromosomes 6 and 8 were randomly removed from the dataset to generate specific missing data proportions. Accuracy between imputed and true SNP genotypes were subsequently calculated with the different methods.

Additional file 2: Predictive abilities on genomic selection model that comprises of statistical methods, genetic compositions and relative sizes of Training Set/Validation Set for each trait.

Additional file 3: ANOVA analysis of sources of variation affecting the predictive ability.

Additional file 4: Mean and standard deviation of predictive ability with the five prediction methods for the eight traits.

Additional file 5: Mean and standard deviation of predictive ability estimated with the four Training Set/Validation Set compositions.

Additional file 6: Mean and standard deviation of predictive ability estimated with the five relative sizes of Training Set/Validation Set expressed in proportions and numbers of individuals.

Additional file 7: Mean and standard deviation of predictive ability across increasing numbers of SNPs, statistical methods (RKHS and GBLUP), four Training Set/Validation Set compositions for each of eight traits.

Additional file 8: Mean and standard deviation of predictive ability estimated with SNPs in four genomic locations, with two statistical methods (RKHS and GBLUP), four Training Set/Validation Set compositions for each of eight traits.

Additional file 9: ANOVA of predictive ability with SNP genomic location and SNP number as sources of variation.

Additional file 10: Average predictive ability estimated with different numbers of SNPs fitted into the model.

Additional file 11: Average predictive abilities estimated using SNP sets located in different genomic regions.

Additional file 12: Decay of linkage disequilibrium (LD) with physical distance estimated with SNPs in different genomic locations. (a) A comparison of the decay of LD with physical distance in four classes of SNPs located with coding, genic, intergenic and all regions, respectively. Dots of pairwise LD versus physical distance and the LD decay for SNPs located in all regions (b), coding region (c), genic region (d) and intergenic region (e), respectively.

Additional file 13: Predictive abilities by training in pure species eucalypt parents and predicting in their F$_1$ hybrids. Predictive ability estimated under three training/validation sets (TS/VS) scenarios with two methods (GBLUP and RKHS) for each trait. PO168 (red boxes): all 168 *E. grandis* and *E. urophylla* pure species G0 parents used for training and 168 G1 random selected hybrid progeny for validation; random168 (green): randomly selected 168 individuals from all 1117 for TS and 168 randomly also for VS; random558 (blue): randomly divided all 1117 individuals into TS and VS of same size (558/558). Outlier estimates are indicated by black dots.

Abbreviations

BL: Bayesian LASSO; CBH: Circumference at breast height; CDS: Coding sequences; GBLUP: Genomic best linear unbiased predictor; GEBV: Genomic estimated breeding values; GRM: Genomic relationship matrix; GS: Genomic selection; IBD: Identity by descent; IBS: Identity by state; LD: Linkage disequilibrium; MAS: Marker-assisted selection; N_e: Effective population size; PA: Predictive ability; PCA: Principal components analysis; QTLs: Quantitative trait loci; RKHS: Reproducing kernel Hilbert space; rrBLUP: Ridge-regression best linear unbiased prediction; SNP: Single-nucleotide polymorphism; TS: Training set; VS: Validation set

Acknowledgements

We would like to thank Michelle Bayerl Fernandes for her contribution on phenotyping the breeding population. The computations were performed on resources provided by the Swedish National Infrastructure for Computing (SNIC) at UPPMAX and HPC2N.

Funding

The study has partly been funded through grants from Vetenskapsrådet and the Kempestiftelserna to PKI. BT gratefully acknowledges financial support from the Umeå Plant Science Centre (UPSC) "The Research School of Forest Genetics, Biotechnology and Breeding".

Authors' contributions

BT, BS and PKI conceived and designed the experiment; GSM phenotyped data; GSM and KZF collected samples for genotyping; DG was responsible for genotyping; BT analysed the data under DG and PKI's guidance; BT drafted the first version of the manuscript and BT, DG, BS and PKI critically contributed to the final version of the manuscript. All authors read and approved the final manuscript.

Competing interests

The authors declare that they have no competing interests.

Author details

[1]Umeå Plant Science Centre, Department of Ecology and Environmental Science, Umeå University, Umeå SE-90187, Sweden. [2]Biomaterials Division, Stora Enso AB, Nacka SE-13104, Sweden. [3]EMBRAPA Genetic Resources and Biotechnology – EPqB, Brasilia, DF 70770-910, Brazil. [4]Universidade Católica de Brasília- SGAN, 916 modulo B, Brasilia, DF 70790-160, Brazil. [5]Veracel Celulose S.A., Eunápolis, BA 45.820-970, Brazil. [6]Present address: Department of Plant Biology, Uppsala BioCenter, Swedish University of Agricultural Sciences, Uppsala SE-75007, Sweden.

References

1. Rezende GDSP, Resende MDV, Assis TF. *Eucalyptus* breeding for clonal forestry. In: Fenning T, editor. Challenges and opportunities for the world's forests in the 21st century. Dordrecht: Springer Netherlands; 2014. p. 393–424.
2. Myburg AA, Potts BM, Marques CM, Kirst M, Gion JM, Grattapaglia D, Grima-Pettenati J. Eucalyptus. Genome Mapping and Molecular Breeding in Plants. Volume 7. Edited by: Kole CR. New York: Springer, Forest trees; 2007. pp. 115-160.
3. Bison O, Ramalho M, Rezende G, Aguiar A, De Resende M. Comparison between open pollinated progenies and hybrids performance in *Eucalyptus grandis* and *Eucalyptus urophylla*. Silvae Genet. 2006;55(4–5):192–6.
4. Resende MD, Resende MF Jr, Sansaloni CP, Petroli CD, Missiaggia AA, Aguiar AM, et al. Genomic selection for growth and wood quality in *Eucalyptus*: capturing the missing heritability and accelerating breeding for complex traits in forest trees. New Phytol. 2012;194(1):116–28.
5. Goddard ME, Hayes BJ, Meuwissen THE. Using the genomic relationship matrix to predict the accuracy of genomic selection. J Anim Breed Genet. 2011;128(6):409–21.
6. Meuwissen THE, Hayes BJ, Goddard ME. Prediction of total genetic value using genome-wide dense marker maps. Genetics. 2001;157(4):1819–29.
7. Meuwissen T, Hayes B, Goddard M. Accelerating improvement of livestock with genomic selection. Annu Rev Anim Biosci. 2013;1:221–37.
8. Van Eenennaam AL, Weigel KA, Young AE, Cleveland MA, Dekkers JCM. Applied animal genomics: results from the field. Annu Rev Anim Biosci. 2014;2:105–39.
9. Spindel J, Begum H, Akdemir D, Virk P, Collard B, Redona E, et al. Genomic selection and association mapping in rice (*Oryza sativa*): effect of trait genetic architecture, training population composition, marker number and statistical model on accuracy of rice genomic selection in elite, tropical rice breeding lines. PLoS Genet. 2015;11(2):e1004982.
10. Windhausen VS, Atlin GN, Hickey JM, Crossa J, Jannink JL, Sorrells ME, et al. Effectiveness of genomic prediction of maize hybrid performance in different breeding populations and environments. G3-Genes Genom Genet. 2012;2(11):1427–36.
11. Isik F. Genomic selection in forest tree breeding: the concept and an outlook to the future. New Forest. 2014;45(3):379–401.
12. Grattapaglia D. Breeding Forest Trees by Genomic Selection: Current Progress and the Way Forward. In: Genomics of Plant Genetic Resources: Volume 1 Managing, sequencing and mining genetic resources. Edited by Tuberosa R, Graner A, Frison E. Dordrecht: Springer Netherlands; 2014. pp. 651–82.
13. de los Campos G, Hickey JM, Pong-Wong R, Daetwyler HD, MPL C. Whole-genome regression and prediction methods applied to plant and animal breeding. Genetics. 2013;193(2):327–45.
14. Endelman JB. Ridge regression and other kernels for genomic selection with R package rrBLUP. Plant Genome. 2011;4(3):250–5.
15. Silva FF E, Viana JM, Faria VR, de Resende MD. Bayesian inference of mixed models in quantitative genetics of crop species. Theor Appl Genet. 2013; 126(7):1749–61.
16. Habier D, Fernando RL, Dekkers JCM. The impact of genetic relationship information on genome-assisted breeding values. Genetics. 2007;177(4):2389–97.
17. De los Campos G, Gianola D, Rosa GJ, Weigel KA, Crossa J. Semi-parametric genomic-enabled prediction of genetic values using reproducing kernel Hilbert spaces methods. Genet Res. 2010;92(4):295–308.
18. Neves HH, Carvalheiro R, Queiroz SA. A comparison of statistical methods for genomic selection in a mice population. BMC Genet. 2012;13(1):100.
19. Hayes B, Daetwyler H, Bowman P, Moser G, Tier B, Crump R, Khatkar M, Raadsma H, Goddard M. Accuracy of genomic selection: comparing theory and results. In: Proceedings of the 18th Conference: Association for the Advancement of Animal Breeding and Genetics, Barossa Valley, Australia; 2009. pp. 34–37.
20. Wu X, Lund MS, Sun D, Zhang Q, Su G. Impact of relationships between test and training animals and among training animals on reliability of genomic prediction. J Anim Breed Genet. 2015;132(5):366–75.
21. Zhong S, Dekkers JC, Fernando RL, Jannink JL. Factors affecting accuracy from genomic selection in populations derived from multiple inbred lines: a barley case study. Genetics. 2009;182(1):355–64.
22. Grattapaglia D, Resende MDV. Genomic selection in forest tree breeding. Tree Genet Genomes. 2011;7(2):241–55.
23. Moser G, Khatkar MS, Hayes BJ, Raadsma HW. Accuracy of direct genomic values in Holstein bulls and cows using subsets of SNP markers. Genet Sel Evol. 2010;42
24. Su G, Brondum RF, Ma P, Guldbrandtsen B, Aamand GR, Lund MS. Comparison of genomic predictions using medium-density (similar to 54,000) and high-density (similar to 777,000) single nucleotide polymorphism marker panels in Nordic Holstein and red Dairy cattle populations. J Dairy Sci. 2012;95(8):4657–65.
25. MacLeod IM, Hayes BJ, Goddard ME. The effects of demography and long-term selection on the accuracy of genomic prediction with sequence data. Genetics. 2014;198(4):1671–84.
26. Silva-Junior OB, Faria DA, Grattapaglia D. A flexible multi-species genome-wide 60K SNP chip developed from pooled resequencing of 240 *Eucalyptus* tree genomes across 12 species. New Phytol. 2015;206(4): 1527–40.
27. Purcell S, Neale B, Todd-Brown K, Thomas L, Ferreira MA, Bender D, et al. PLINK: a tool set for whole-genome association and population-based linkage analyses. Am J Hum Genet. 2007;81(3):559–75.
28. Browning SR, Browning BL. Rapid and accurate haplotype phasing and missing-data inference for whole-genome association studies by use of localized haplotype clustering. Am J Hum Genet. 2007;81(5):1084–97.
29. Stephens M, Scheet P. Accounting for decay of linkage disequilibrium in haplotype inference and missing-data imputation. Am J Hum Genet. 2005;76(3):449–62.
30. Candes EJ, Recht B. Exact matrix completion via convex optimization. Found Comput Math. 2009;9(6):717–72.
31. Rutkoski JE, Poland J, Jannink JL, Sorrells ME. Imputation of unordered markers and the impact on genomic selection accuracy. G3-Genes Genom Genet. 2013;3(3):427–39.
32. Stacklies W, Redestig H, Scholz M, Walther D, Selbig J. pcaMethods - a bioconductor package providing PCA methods for incomplete data. Bioinformatics. 2007;23(9):1164–7.
33. Remington DL, Thornsberry JM, Matsuoka Y, Wilson LM, Whitt SR, Doeblay J, et al. Structure of linkage disequilibrium and phenotypic associations in the maize genome. P Natl Acad Sci USA. 2001;98(20): 11479–84.
34. Patterson N, Price AL, Reich D. Population structure and eigenanalysis. PLoS Genet. 2006;2(12):2074–93.
35. Legarra A, Robert-Granie C, Croiseau P, Guillaume F, Fritz S. Improved Lasso for genomic selection. Genet Res. 2011;93(1):77–87.
36. Crossa J, Campos Gde L, Perez P, Gianola D, Burgueno J, Araus JL, et al. Prediction of genetic values of quantitative traits in plant breeding using pedigree and molecular markers. Genetics. 2010;186(2):713–24.
37. Gilmour AR, Gogel B, Cullis B, Thompson R, Butler D. ASReml user guide release 3.0. UK https://www.vsni.co.uk/: VSN International Ltd, Hemel Hempstead; 2009.

38. Perez P. De los Campos G, Crossa J, Gianola D. Genomic-enabled prediction based on molecular markers and pedigree using the Bayesian linear regression package in R. Plant Genome. 2010;3(2):106–16.

39. los Campos G, Pérez P, Vazquez AI, Crossa J. Genome-enabled prediction using the BLR (Bayesian linear regression) R-package. In: Genome-wide association studies and genomic prediction. Edited by Gondro C, van der Werf J, Hayes B. Totowa, NJ: Humana Press; 2013: 299-320.

40. Perez P. De los Campos G. Genome-wide regression and prediction with the BGLR statistical package. Genetics. 2014;198(2):483–95.

41. de Los CG, Sorensen D, Gianola D. Genomic heritability: what is it? PLoS Genet. 2015;11(5):e1005048.

42. Cingolani P, Platts A, Wang LL, Coon M, Nguyen T, Wang L, et al. A program for annotating and predicting the effects of single nucleotide polymorphisms. SnpEff Fly. 2012;6(2):80–92.

43. Hidalgo AM, Bastiaansen JWM, Lopes MS, Harlizius B, Groenen MAM, de Koning DJ. Accuracy of predicted genomic breeding values in purebred and crossbred pigs. G3-Genes Genom Genet. 2015;5(8):1575–83.

44. Resende MF Jr, Munoz P, Resende MD, Garrick DJ, Fernando RL, Davis JM, et al. Accuracy of genomic selection methods in a standard data set of loblolly pine (*Pinus taeda* L.). Genetics. 2012;190(4):1503–10.

45. Beaulieu J, Doerksen T, Clement S, MacKay J, Bousquet J. Accuracy of genomic selection models in a large population of open-pollinated families in white spruce. Heredity. 2014;113(4):343–52.

46. Beaulieu J, Doerksen TK, MacKay J, Rainville A, Bousquet J. Genomic selection accuracies within and between environments and small breeding groups in white spruce. BMC Genomics. 2014;15:1048.

47. El-Dien OG, Ratcliffe B, Klapste J, Chen C, Porth I, El-Kassaby YA. Prediction accuracies for growth and wood attributes of interior spruce in space using genotyping-by-sequencing. BMC Genomics. 2015;16:370.

48. Ratcliffe B, El-Dien OG, Klapste J, Porth I, Chen C, Jaquish B, et al. A comparison of genomic selection models across time in interior spruce (*Picea engelmannii* x *glauca*) using unordered SNP imputation methods. Heredity. 2015;115(6):547–55.

49. Isik F, Bartholome J, Farjat A, Chancerel E, Raffin A, Sanchez L, et al. Genomic selection in maritime pine. Plant Sci. 2016;242:108–19.

50. Crossa J, Perez P, Hickey J, Burgueno J, Ornella L, Ceron-Rojas J, et al. Genomic prediction in CIMMYT maize and wheat breeding programs. Heredity. 2014;112(1):48–60.

51. Onogi A, Ideta O, Inoshita Y, Ebana K, Yoshioka T, Yamasaki M, et al. Exploring the areas of applicability of whole-genome prediction methods for Asian rice (*Oryza sativa* L.). Theor Appl Genet. 2015;128(1):41–53.

52. Clark SA, Hickey JM, van der Werf JHJ. Different models of genetic variation and their effect on genomic evaluation. Genet Sel Evol. 2011;43(1):1–9.

53. Honarvar M, Rostami M. Accuracy of genomic prediction using RR-BLUP and Bayesian LASSO. Eur J Exp Biol. 2013;3:42–7.

54. Daetwyler HD, Pong-Wong R, Villanueva B, Woolliams JA. The impact of genetic architecture on genome-wide evaluation methods. Genetics. 2010;185(3):1021–31.

55. Lorenz AJ, Chao S, Asoro FG, Heffner EL, Hayashi T, Iwata H, et al. Genomic selection in plant breeding: knowledge and prospects. Adv Agron. 2011;110

56. Lorenz AJ. Resource allocation for maximizing prediction accuracy and genetic gain of genomic selection in plant breeding: a simulation experiment. G3-Genes Genom Genet. 2013;3(3):481–91.

57. Riedelsheimer C, Endelman JB, Stange M, Sorrells ME, Jannink JL, Melchinger AE. Genomic predictability of interconnected biparental maize populations. Genetics. 2013;194(2):493–503.

58. Scutari M, Mackay I, Balding D. Using genetic distance to infer the accuracy of genomic prediction. PLoS Genet. 2016;12(9):e1006288.

59. Esfandyari H, Bijma P, Henryon M, Christensen OF, Sørensen AC. Genomic prediction of crossbred performance based on purebred landrace and Yorkshire data using a dominance model. Genet Sel Evol. 2016;48(1):1–9.

60. Ibáñez-Escriche N, Fernando RL, Toosi A, Dekkers JC. Genomic selection of purebreds for crossbred performance. Genet Sel Evol. 2009;41(1):1–10.

61. Esfandyari H, Sørensen AC, Bijma P. Maximizing crossbred performance through purebred genomic selection. Genet Sel Evol. 2015;47(1):1–16.

62. Murray C, Huerta-Sanchez E, Casey F, Bradley DG. Cattle demographic history modelled from autosomal sequence variation. Philos T R Soc B. 2010; 365(1552):2531–9.

63. Silva-Junior OB, Grattapaglia D. Genome-wide patterns of recombination, linkage disequilibrium and nucleotide diversity from pooled resequencing and single nucleotide polymorphism genotyping unlock the evolutionary history of *Eucalyptus grandis*. New Phytol. 2015;208(3):830–45.

64. Zhang Z, Ding X, Liu J, Zhang Q, de Koning DJ. Accuracy of genomic prediction using low-density marker panels. J Dairy Sci. 2011;94(7):3642–50.

65. Solberg TR, Sonesson AK, Woolliams JA, Meuwissen THE. Genomic selection using different marker types and densities. J Anim Sci. 2008;86(10):2447–54.

66. Burgueno J, de los Campos G, Weigel K, Crossa J. Genomic prediction of breeding values when modeling genotype x environment interaction using pedigree and dense molecular markers. Crop Sci. 2012;52(2):707–19.

67. Denis M, Bouvet J-M. Efficiency of genomic selection with models including dominance effect in the context of Eucalyptus breeding. Tree Genet Genomes. 2012;9(1):37–51.

15

Overexpression of *Arabidopsis P3B* increases heat and low temperature stress tolerance in transgenic sweetpotato

Chang Yoon Ji[1,2†], Rong Jin[1,2,3†], Zhen Xu[3†], Ho Soo Kim[1], Chan-Ju Lee[1,2], Le Kang[1,2], So-Eun Kim[1,2], Hyeong-Un Lee[4], Joon Seol Lee[4], Chang Ho Kang[5], Yong Hun Chi[5], Sang Yeol Lee[5], Yiping Xie[3], Hongmin Li[3], Daifu Ma[3] and Sang-Soo Kwak[1,2*]

Abstract

Background: Sweetpotato (*Ipomoea batatas* [L.] Lam) is suitable for growth on marginal lands due to its abiotic stress tolerance. However, severe environmental conditions including low temperature pose a serious threat to the productivity and expanded cultivation of this crop. In this study, we aimed to develop sweetpotato plants with enhanced tolerance to temperature stress.

Results: P3 proteins are plant-specific ribosomal P-proteins that act as both protein and RNA chaperones to increase heat and cold stress tolerance in *Arabidopsis*. Here, we generated transgenic sweetpotato plants expressing the *Arabidopsis* ribosomal P3 (*AtP3B*) gene under the control of the CaMV 35S promoter (referred to as OP plants). Three OP lines (OP1, OP30, and OP32) were selected based on *AtP3B* transcript levels. The OP plants displayed greater heat tolerance and higher photosynthesis efficiency than wild type (WT) plants. The OP plants also exhibited enhanced low temperature tolerance, with higher photosynthesis efficiency and less membrane permeability than WT plants. In addition, OP plants had lower levels of hydrogen peroxide and higher activities of antioxidant enzymes such as peroxidase and catalase than WT plants under low temperature stress. The yields of tuberous roots and aerial parts of plants did not significantly differ between OP and WT plants under field cultivation. However, the tuberous roots of OP transgenic sweetpotato showed improved storage ability under low temperature conditions.

Conclusions: The OP plants developed in this study exhibited increased tolerance to temperature stress and enhanced storage ability under low temperature compared to WT plants, suggesting that they could be used to enhance sustainable agriculture on marginal lands.

Keywords: Acidic ribosomal P-proteins, Heat stress, Low temperature stress, Protein chaperone, Sweetpotato

Background

To cope with climate change and environmental stresses such as drought, temperature variation, UV radiation, and salinity, plants have evolved sophisticated signaling and protective systems [1, 2]. Since plants are sessile organisms, they must utilize various mechanisms to respond and adapt to continuously changing environmental conditions [3–5]. Temperature variation is an especially important environmental factor that affects plant development and crop production [6, 7]. Understanding the molecular mechanisms underlying the plant response to various stresses has been a subject of great interest for many decades. Nevertheless, there is still a significant knowledge gap and, in general, we are unable to predict how well plants will cope with multiple environmental stress factors.

* Correspondence: sskwak@kribb.re.kr
†Equal contributors
[1]Plant Systems Engineering Research Center, Korea Research Institute of Bioscience and Biotechnology (KRIBB), 125 Gwahak-ro, Daejeon 34141, South Korea
[2]Department of Environmental Biotechnology, Korea University of Science and Technology (UST), 217 Gajeong-ro, Daejeon 34113, South Korea
Full list of author information is available at the end of the article

Sweetpotato (*Ipomoea batatas* [L.] Lam) is an important root crop worldwide [8, 9]. This crop is used as an alternative source of animal feed and industrial biomass for biomaterial and biofuel production, and it represents an abundant source of nutrients and natural antioxidant compounds for the human diet, such as anthocyanins, carotenoids, and Vitamin C and E [10–13]. Sweetpotato is well suited for growth on marginal lands due to its tolerance to abiotic stress [14]. Thus, sweetpotato represents an industrially valuable source of biomass to supplement grain-based bioenergy production and increase food security. Nevertheless, sweetpotato as a tropical crop is very sensitive to low temperature, posing critical threats to the productivity and geographical distribution. Opportunities for genetic improvement of sweetpotato using conventional breeding are limited due to its high male sterility, hexaploid nature, the lack of suitable germplasm, and issues with incompatibility [15, 16]. Thus, it is highly important to develop a sweetpotato cultivar with enhanced tolerance to severe abiotic stresses via genetic engineering.

To overcome temperature stress, plants must immediately recognize the outside temperature and communicate this information via signaling cascades, which activate distinct downstream proteins such as heat-shock proteins (HSPs) and cold-shock proteins (CSPs), initiating downstream temperature stress-related responses [17]. In sweetpotato, a number of genes have been identified that are involved in the responses to abiotic stresses, such as drought, oxidative, and salt stress [18–25]. However, little is known about the temperature stress response in sweetpotato.

A distinct lateral protuberance of the large ribosomal subunit in eukaryotic ribosomes called the "stalk", which contains highly conserved small ribosomal proteins with an acidic isoelectric point (pI 3–5) [26, 27]. Acidic ribosomal proteins (ARPs) are phosphorylated by several protein kinases to facilitate assembly into ribosomes [28] and, therefore, these proteins are also referred to as ribosomal P-proteins. P-proteins in various eukaryotes have three domains, including an α-helical N-terminal region and a central, flexible acidic hinge region, followed by a highly conserved C-terminus [27]. In all eukaryotes, these proteins are categorized into two groups, P1 and P2, based on primary sequence similarity [29], with the exception of plants, which contain an extra group, P3 [30]. The "stalk" is the active part of the ribosome structure and the center for interactions between mRNAs, tRNAs, and translation factors occur during protein synthesis [31]. The biological functions of eukaryotic ribosomal P-proteins are currently unclear. P-proteins are believed to function in the regulation of protein synthesis at the level of the protein elongation step. These proteins might also be involved in transcriptional processes and DNA repair [32]. In addition, phosphorylation is an important posttranslational step that regulates P-protein function [27]. Conserved phosphorylation sites have been identified in *Arabidopsis* [33] and maize [34, 35]. However, the biological significance of plant P-proteins has not been fully elucidated.

Recently, Kang et al. identified a plant-specific acidic ribosomal P3 protein (designated AtP3B) as a heat-shock protein from heat-treated *Arabidopsis* suspension cells [36]. The authors demonstrated that AtP3B has both protein and RNA chaperone activities. Overexpressing *AtP3B* increased tolerance to high- and low temperature stress in transgenic plants, whereas knockdown plants of *AtP3B* created by RNAi showed increased sensitivity to both stresses. Here, we developed transgenic sweetpotato plants that overexpressed *AtP3B* and evaluated their growth under heat and low temperature stress. Transgenic sweetpotato overexpressing *AtP3B* showed not only increased temperature stress tolerance, but also improved storage ability under low temperature stress conditions.

Results

Molecular characterization of transgenic sweetpotato overexpressing *AtP3B*

Transgenic sweetpotato plants overexpressing *AtP3B* under the control of the CaMV 35S promoter (Fig. 1a) were successfully generated via *Agrobacterium*-mediated transformation. We performed an initial screening of the putative transgenic sweetpotato plants using PCR analysis of genomic DNA with a portion of the 35S promoter and *AtP3B*-specific primers. The expected amplification profiles were acquired from eight transgenic lines, suggesting that the recombinant *AtP3B* gene had been integrated into the genomes of transgenic plants from eight independent lines, whereas no integration was detected in the wild type (WT) line (Fig. 1b). Transgenic plants harboring *AtP3B* under the control of the CaMV 35S promoter were designated "OP" plants. We propagated the eight independent OP lines in a growth chamber and subjected the plants to quantitative RT-PCR analysis using leaf discs to determine the transcription levels of *AtP3B* (Fig. 1c). *AtP3B* expression was strongly induced in lines OP1, OP30, and OP32 (Fig. 1c); we therefore selected these lines for further study. As shown in Fig. 1d, lines OP1, OP30, and OP32 contained single, double, and single copy insertions of AtP3B, respectively. When OP transgenic and WT plants were grown in the growth chamber, no visible phenotypic differences were detected under normal conditions (data not shown), indicating that *AtP3B* overexpression did not lead to phenotypic defects in transgenic sweetpotato plants.

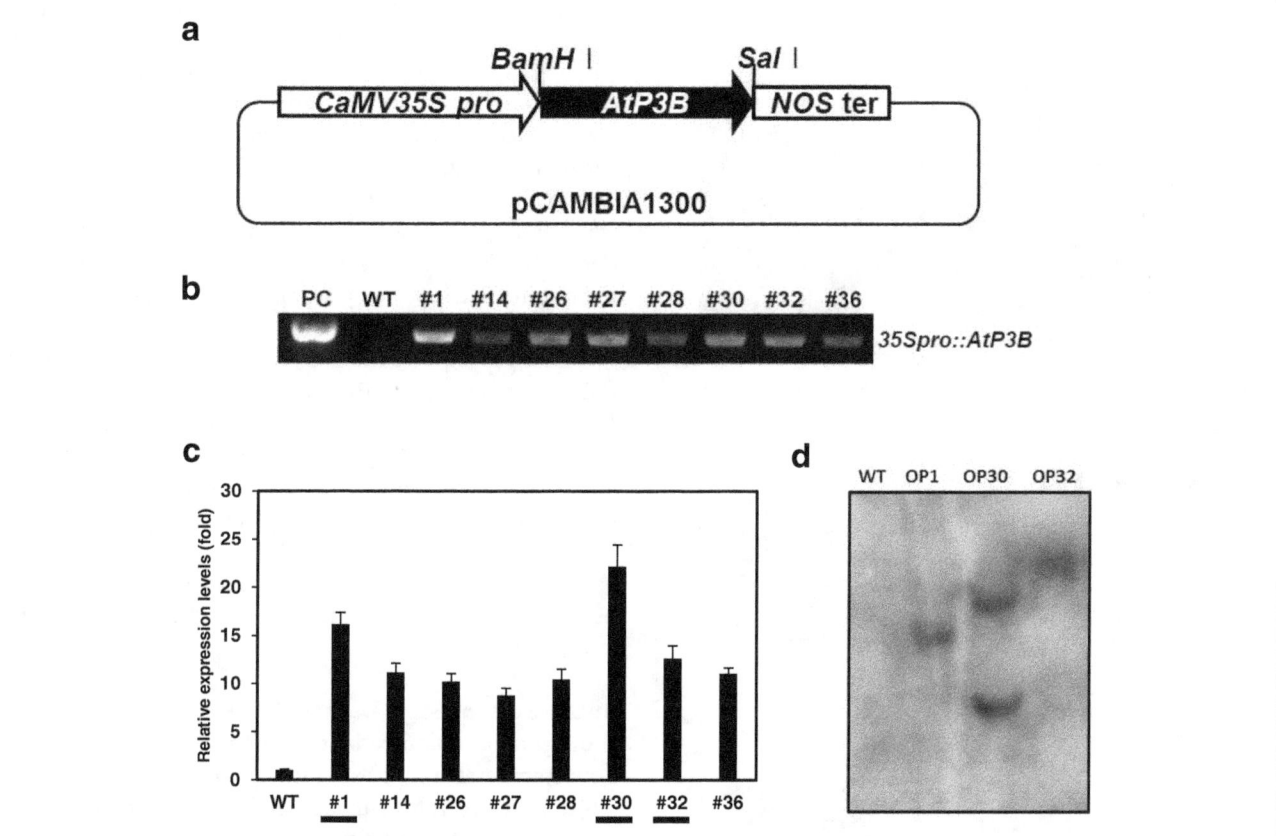

Fig. 1 Development and molecular characterization of transgenic sweetpotato plants overexpressing *AtP3B*. **a** Schematic diagram of the vector construct containing *AtP3B* under the control of the CaMV 35S promoter. **b** Genomic DNA PCR analysis using the *35Spro::AtP3B* primer set. PC, positive control. **c** qRT-PCR analysis of transgenic sweetpotato plants overexpressing *AtP3B*. Three independent transgenic lines (#1, #30, and #32) were selected for further characterization. **d** Southern blot analysis of OP plants; the integration and gene copy number of the construct in OP plants were confirmed using a ^{32}P–labeled probe designed based on the AtP3B cDNA fragment

Transgenic sweetpotato overexpressing *AtP3B* display enhanced tolerance to heat and low temperature stress

AtP3B-overexpressing transgenic *Arabidopsis* plants showed enhanced tolerance to heat and low temperature stress, whereas knockdown of *AtP3B* by RNAi led to increased sensitivity to both stresses [36]. In the current study, we evaluated the physiological functions of AtP3B using transgenic sweetpotato plants. First, we investigated the tolerance of WT and OP plants to heat stress conditions. Under normal conditions (25 °C), the phenotypes of OP plants did not differ from those of WT plants in terms of plant growth. However, when we treated 1-month-old WT or OP plants with heat stress (45 °C) for 12 h, the OP plants exhibited marked thermotolerance compared to WT plants (Fig. 2a). Following recovery via incubation at 25 °C, severe damage was observed in WT plants, whereas the OP plants exhibited only slight wilting (Fig. 2a). Under this heat stress condition, the photosynthetic efficiency (*Fv/fm*) of WT plants decreased by 35.3%, whereas the photosynthetic efficiency only decreased by 9.7%, 8.5%, and 15.0% in OP1, OP30, and OP32 plants, respectively (Fig. 2b). After a

24 h recovery, the photosynthetic efficiency of OP plants was maintained to approximately similar levels of normal conditions, whereas WT plants continued to show reduced photosynthetic efficiency. In addition, OP plants exhibited significantly lower levels of ion leakage than WT plants after heat stress treatment (Fig. 2c). These results suggest that OP sweetpotato plants are more tolerant to high temperature stress than WT plants due to overexpression of *AtP3B* in the transgenic plants.

AtP3B plays an essential role in cold stress tolerance in *Arabidopsis*, a process mediated by its RNA chaperone activity [36]. Therefore, we investigated whether the OP plants also exhibited increased tolerance to low temperatures due to *AtP3B* overexpression. We subjected soil-grown whole plants (1 month old) to low temperature conditions (4 °C) for 48 h, followed by recovery at 25 °C. After low temperature treatment, severe wilting and chilling injury were observed in the leaves of WT plants, whereas all OP plant lines (OP1, OP30, and OP32) showed only slight damage (Fig. 3a). Following a 24 h recovery period, the phenotypes of OP plants had returned to normal. However, WT plants still had slight

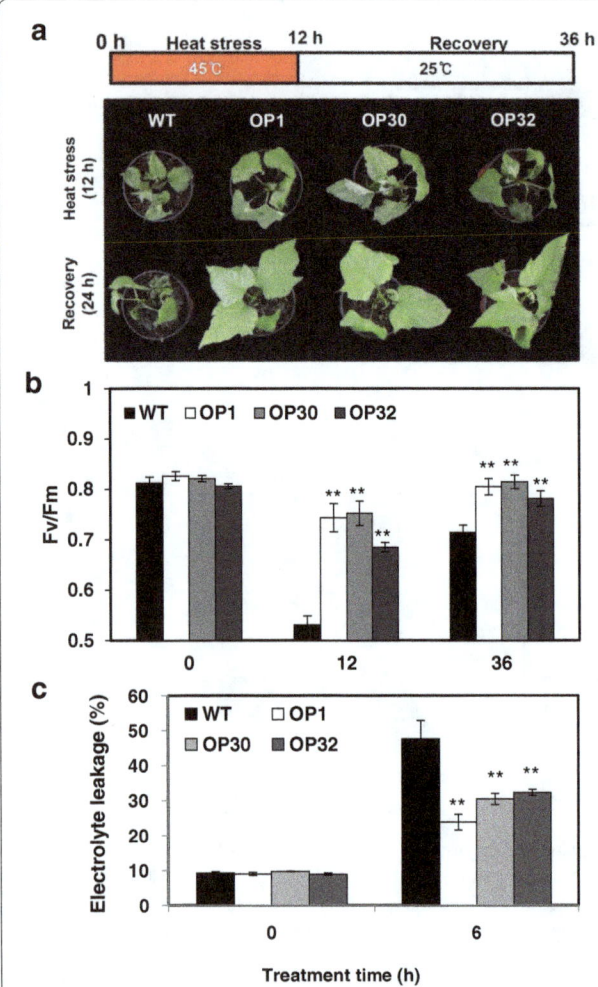

Fig. 2 Phenotypic and physiological analyses of OP plants under heat stress treatment (45 °C) and after recovery at 25 °C. **a** Visible damage in the leaves of sweetpotato plants after 12 h heat stress treatment and 24 h recovery. **b** PSII photosynthetic efficiency (Fv/fm) in the leaves of WT and OP plants after 12 h heat stress treatment and 24 h recovery. **c** Analysis of electrolyte leakage. Data are expressed as the mean ± SD of three replicates. Asterisks indicate significant differences between WT and OP plants by ANOVA at * $p < 0.05$ and ** $p < 0.01$

reactive oxygen species (ROS) production, is an important indicator of cell membrane damage under stress conditions [37]. Under normal conditions, the MDA contents of WT plants were similar to those of OP plants. However, after 4 °C treatment for 48 h, MDA levels were higher in WT plants than in OP plants. Following incubation at 25 °C for 24 h, the MDA contents were significantly higher in WT plants than in OP plants (Fig. 3d). These results indicate that the degree of cell membrane damage was greater in WT plants than in OP plants under low temperature stress.

We also examined the expression of *IbHSP*, *IbCBF* and *IbCOR* genes (*IbHSP17.6*, *IbHSP18.2*, *IbCBF3* and *IbCOR27*), which are candidate homologs of *Arabidopsis HSP*, *CBF* and *COR* genes. The transgenic sweetpotato plants exhibited higher expression levels of these target genes than did the WT plants (Additional file 1: Fig. S1). The altered expression levels of these *IbHSP*, *IbCBF* and *IbCOR* candidate genes might help explain the enhanced heat and cold stress tolerance of the transgenic sweetpotato plants.

AtP3B-overexpressing transgenic sweetpotato plants show enhanced antioxidant enzyme activity

Low temperature stress induces H_2O_2 accumulation, which can severely damage cells [38]. Thus, we investigated the H_2O_2 contents in WT versus OP plants after low temperature treatment. Under normal growth conditions, the H_2O_2 contents in OP and WT plants were similar (Fig. 4a). After low temperature treatment at 4 °C for 24 h and 48 h, the H_2O_2 levels in WT plants were 1.6- to 1.5-fold higher than those of OP (OP1, OP30, and OP32) plants (Fig. 4a). After 24 h recovery, the H_2O_2 contents in WT plants were still significantly higher than those of OP plants (Fig. 4a). These results indicate that *AtP3B* expression suppresses H_2O_2 accumulation under low temperature stress in transgenic sweetpotato plants.

Exposure of plants to unfavorable environmental conditions leads to the overproduction of ROS, which can cause significant oxidative damage to proteins, lipids, carbohydrates, and DNA [39]. ROS-scavenging enzymes such as superoxide dismutase (SOD), ascorbate peroxidase (APX), catalase (CAT), and peroxidase (POD) are very important for plants, as they help protect plants from toxic oxygen intermediates [6, 40, 41]. In particular, POD and CAT are major enzymes responsible for H_2O_2 scavenging during oxidative stress in plants. Under normal conditions, the POD activity levels in the three OP lines were similar to that of WT plants (Fig. 4b). However, the POD activity of OP plants significantly increased under low temperature treatment. After exposure to 4 °C for 24 h and 48 h, the OP plants exhibited an average of 2.2- and 2.3-fold higher POD activity

dehydration symptoms (Fig. 3a). The photosynthetic efficiency (Fv/fm) of both WT and OP plants decreased during the low temperature treatment. After exposure to 4 °C conditions for 48 h, the Fv/fm values of WT plants decreased by 13.5%, which were significantly lower than those of the three OP lines. The photosynthetic efficiency of all OP plants was restored to normal levels after a 24 h recovery at 25 °C, whereas the WT plants continued to show reduced Fv/fm values (Fig. 3b). In addition, OP plants exhibited significantly lower levels of ion leakage than WT plants after low temperature treatment (Fig. 3c). Malondialdehyde (MDA), a naturally occurring product of lipid peroxidation due to accelerated

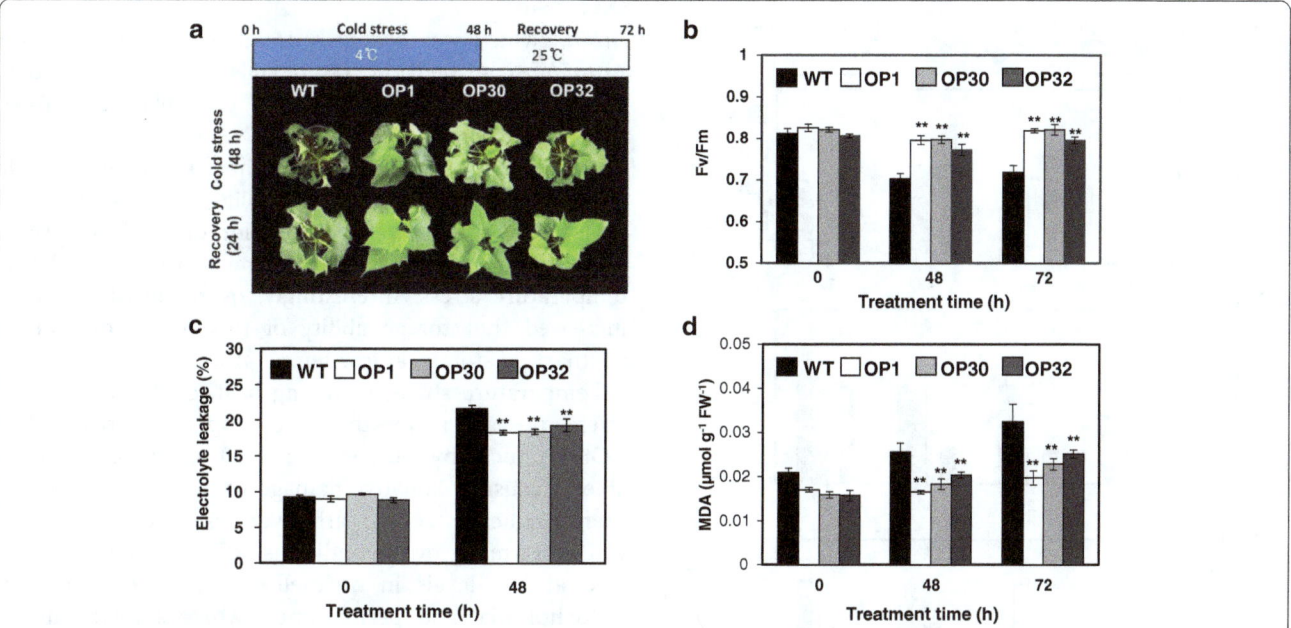

Fig. 3 Phenotypic and physiological analyses of OP plants under low temperature treatment (4 °C) and after recovery at 25 °C. **a** Visible damage in the leaves of sweetpotato plants after 48 h cold stress treatment and 24 h recovery. **b** PSII photosynthetic efficiency (*Fv/fm*), **c** Ion leakage in detached leaves treated with 4 °C for 48 h, and **d** MDA contents in the leaves of WT and OP plants after 48 h cold stress treatment and 24 h recovery. Data are expressed as the mean ± SD of three replicates. *Asterisks* indicate significant differences between WT and OP plants by ANOVA at * $p < 0.05$ and ** $p < 0.01$

than WT plants, respectively (Fig. 4b). Following a 24 h recovery, POD activity levels were similar among WT and OP plants, except for OP32. In addition, under normal conditions, CAT activity was lower in OP plants than in WT plants (Fig. 4c). CAT activity in WT plants was significantly reduced by low temperature treatment, whereas CAT activity in OP plants was not affected by cold stress. After 4 °C treatment for 48 h, the average CAT activity level was 3.7-fold higher in OP plants than in WT plants (Fig. 4c). After 24 h of recovery, the OP plants still exhibited higher CAT activity than WT plants (Fig. 4c). These results suggest that the enhanced low temperature stress tolerance of OP plants might be attributed to the increased activity of ROS-scavenging enzymes such as POD and CAT.

Yield of *AtP3B* transgenic sweetpotato plants under field conditions

To assess whether overexpressing *AtP3B* affects sweetpotato yields, we measured the yields of the aerial parts and tuberous roots of the transgenic lines. Under field conditions, the yields of aerial parts and tuberous roots were not significantly different between WT and OP plants, although one transgenic line (OP32) had slightly higher yields than WT plants (Fig. 5). The average shoot length of WT plants was approximately 118.6 cm, whereas that of OP plants was slightly higher (150.3 cm for OP1, 146.8 cm for OP30, and 158.4 cm for OP32)

(Fig. 5b). The average yield for the aerial parts of WT plants was 308 kg per are (a), while OP1, OP30, and OP32 plants produced 299.1, 375, and 361.6 kg a^{-1}, respectively (Fig. 5c). Moreover, we evaluated the average yields of tuberous roots, which varied among transgenic lines. The yields of tuberous roots in WT, OP1, OP30, and OP32 plants were similar (419.1 kg a^{-1} for NT, 413.8 kg a^{-1} for OP1, and 472.3 kg a^{-1} for OP30, and 541.5 kg a^{-1} for OP32; Fig. 5d).

The tuberous roots of OP transgenic sweetpotato show enhanced storage ability under low temperature conditions

Long-term exposure to low temperature causes a variety of chilling injuries in sweetpotato tuberous roots [42–46]. As mentioned, the OP transgenic lines exhibited increased tolerance to low temperature stress (Fig. 3). To further verify the low temperature stress resistance of OP transgenic sweetpotato, we investigated the physiological responses of tuberous roots from field-grown OP plants under low temperature storage. Sweetpotato tuberous roots from both WT and OP plants showed no symptoms of chilling injury or morphological changes when stored at 13 °C for 8 weeks (Fig. 6a). However, after storage for 8 weeks at 4 °C, severe morphological changes including surface wounding, darkening of internal tissues and susceptibility to decay were observed in tuberous roots of WT plants, whereas all three OP 77 lines (OP1, OP30,

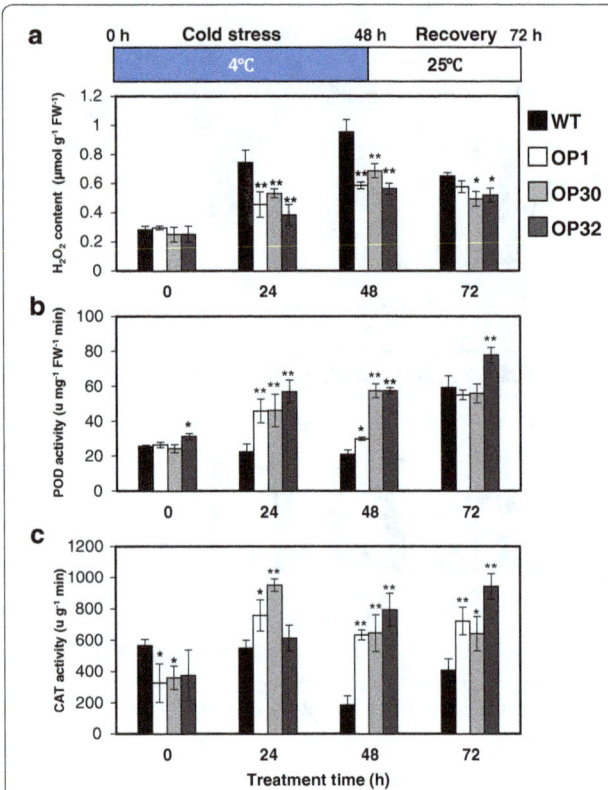

Fig. 4 Analysis of H_2O_2 contents and antioxidant enzyme activity in WT and OP plants under low temperature stress. **a** H_2O_2 contents in leaves of WT and OP plants after 48 h low temperature stress treatment and 24 h recovery. **b** and **c** Changes in antioxidant enzyme activity in WT and OP plants after 48 h low temperature stress treatment and 24 h recovery. POD activity (**b**) and CAT activity (**c**). Data are shown as mean ± SD of three independent measurements. *Asterisks* indicate significant differences between WT and OP plants by ANOVA at * $p < 0.05$ and ** $p < 0.01$

and OP32) showed only slight damage (Fig. 6a). In addition, we compared ion leakage levels between WT and OP tuberous roots under long-term exposure to low temperature. The tuberous roots of WT and OP plants exhibited similar levels of ion leakage when stored at 13 °C for 8 weeks (Fig. 6b). However, after incubation at 4 °C for 8 weeks, tuberous roots of OP plants had lower levels of ion leakage than those of WT plants (Fig. 6b). Chilling injuries caused by long-term exposure to low temperature include cellular membrane degradation [42]. Thus, we also measured the MDA contents of tuberous roots, which represent the degree of cell membrane damage, under chilling stress. Under storage at 13 °C for 8 weeks, the MDA contents of OP tuberous roots were similar to those of WT plants. After storage for 8 weeks at 4 °C, the MDA contents of WT tuberous roots were higher than those of OP plants overexpressing *AtP3B* (Fig. 6c). These results indicate that overexpressing *AtP3B* in sweetpotato increases storage ability under low temperature conditions.

Discussion

AtP3B, which was originally isolated from heat-treated *Arabidopsis* suspension culture cells, plays essential roles in both heat and cold tolerance. AtP3B plays dual roles as a protein chaperone and an RNA chaperone; it prevents protein aggregation under heat stress, whereas it supports RNA processing or stability under cold stress [36]. In this study, ectopic expression of *AtP3B* in sweetpotato resulted in enhanced tolerance to heat and low temperature stress. Interestingly, overexpressing *AtP3B* increased the storage ability of this root crop during postharvest storage at low temperature.

Temperature stress, including heat, cold, and freezing stress, poses a major threat to crop productivity [47]. ROS produced by these stresses are toxic molecules capable of causing oxidative damage to proteins, cell membranes, nucleic acids, carbohydrates, and lipids [48]. Under normal growth conditions, ROS are mainly generated at low levels in organelles such as chloroplasts, mitochondria, and peroxisomes, whereas their rate of production dramatically increases during temperature stress. The ROS, H_2O_2, plays dual roles in plants: at low concentrations, it acts as a signaling molecule involved in triggering defense responses to various biotic and abiotic stresses, whereas at high concentrations, it induces programmed cell death [49]. In the current study, while H_2O_2 levels significantly increased in WT sweetpotato under low temperature stress (Fig. 4a), these levels were lower in OP plants than in WT plants under these conditions. In addition, overexpressing *AtP3B* in sweetpotato increased POD and CAT activity under low temperature stress conditions (Fig. 4b and c). POD and CAT are the major enzymes responsible for H_2O_2 scavenging during oxidative stress in plants. In addition to H_2O_2 scavenging, plant PODs are also involved in plant growth and development, as well as the lignification, suberization, and cross-linking of cell wall compounds [50]. In addition, CAT eliminates H_2O_2 by breaking it down directly to form water and oxygen. Thus, CAT activity does not require reducing power and has a high reaction rate, but it has a low affinity for H_2O_2, thereby only removing high concentrations of H_2O_2 [51]. Interestingly, CAT activity was significantly higher in OP plants than in WT plants (Fig. 4c). These data are consistent with the notion that the increase in POD and CAT activity resulting from *AtP3B* expression in transgenic sweetpotato plants is correlated with low temperature stress tolerance via an H_2O_2-regulated stress response-signaling pathway.

Sweetpotato is a high-yielding, industrially valuable root crop. While sweetpotato is widely adapted to growth on marginal lands ranging from tropical to temperate zones, it is highly sensitive to low temperature stress. In addition, postharvest storage conditions for

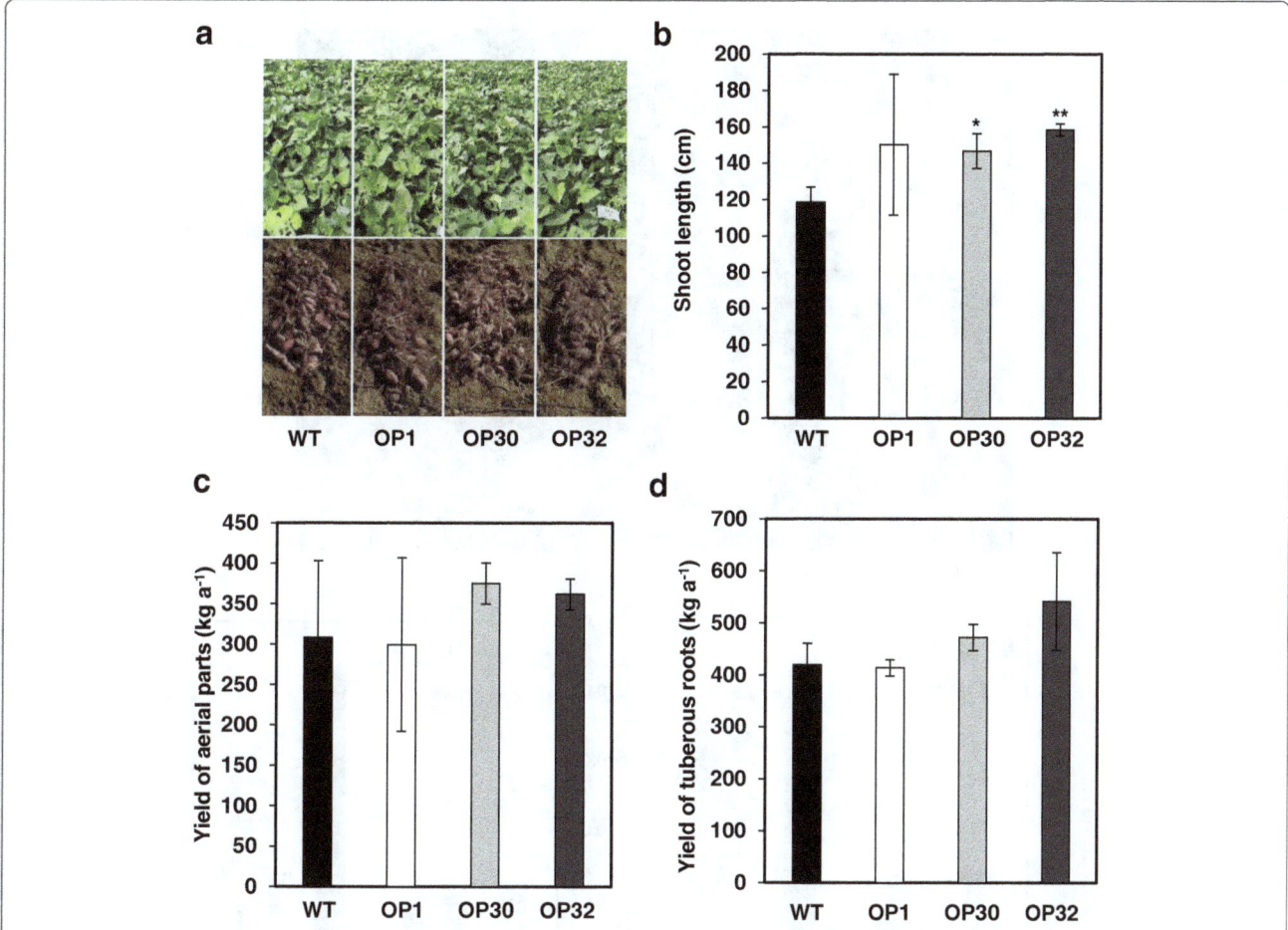

Fig. 5 Growth-related features of WT and OP plants under field conditions. **a** Photographs of aerial plant parts and tuberous roots. **b** Average shoot lengths of plants. **c** Average yields of aerial plant parts. **d** Average yields of tuberous roots. Data are means ± SD of three row replicates (40 individual plants were planted per line). *Asterisks* indicate significant differences between WT and OP plants by ANOVA at * $p < 0.05$ and ** $p < 0.01$

sweetpotato are a major issue affecting its use for industrial applications sweetpotato. Therefore, it is essential to genetically engineer sweetpotato with enhanced tolerance to low temperatures. We previously produced sweetpotato plants with enhanced tolerance to low temperature stress via genetic engineering. Transgenic sweetpotato plants overexpressing the soybean cold-inducible zinc finger protein gene *SCOF-1* under the control of an oxidative stress-inducible peroxidase (*SWPA2*) promoter exhibited enhanced tolerance to low temperature stress [52]. We also reported that transgenic sweetpotato plants expressing the *Arabidopsis* nucleoside diphosphate kinase 2 gene (*AtNDPK2*) exhibited enhanced tolerance to multiple environmental stresses, including cold, high salt, drought, and MV-mediated oxidative stress, due to increased H_2O_2-scavenging enzyme activity regulated by NDPK2 [53]. However, enhanced tolerance to low temperature stress in sweetpotato has not previously been correlated with increased postharvest storage ability under low temperatures. In the current study, we demonstrated that expressing *AtP3B* led

to increased H_2O_2 scavenging enzyme activity and enhanced tolerance to low and high temperature stress in transgenic sweetpotato. Furthermore, overexpressing *AtP3B* in sweetpotato increased its storage ability at low temperatures. We are currently focused on isolating and characterizing endogenous homologous genes of *AtP3B* in sweetpotato. Such work should pave the way for improving postharvest storage ability in sweetpotato without the loss of quality due to chilling injury.

Conclusion

In this study, we successfully developed transgenic sweetpotato plants overexpressing *AtP3B* under the control of the CaMV 35S promoter. As expected, the transgenic sweetpotato plants exhibited enhanced tolerance to heat and low temperature stress. After exposure to low temperature stress, the OP plants displayed less wilting and chilling symptoms than WT plants, which was associated with enhanced antioxidant enzyme activity. In addition, the OP plants exhibited a stronger ability to

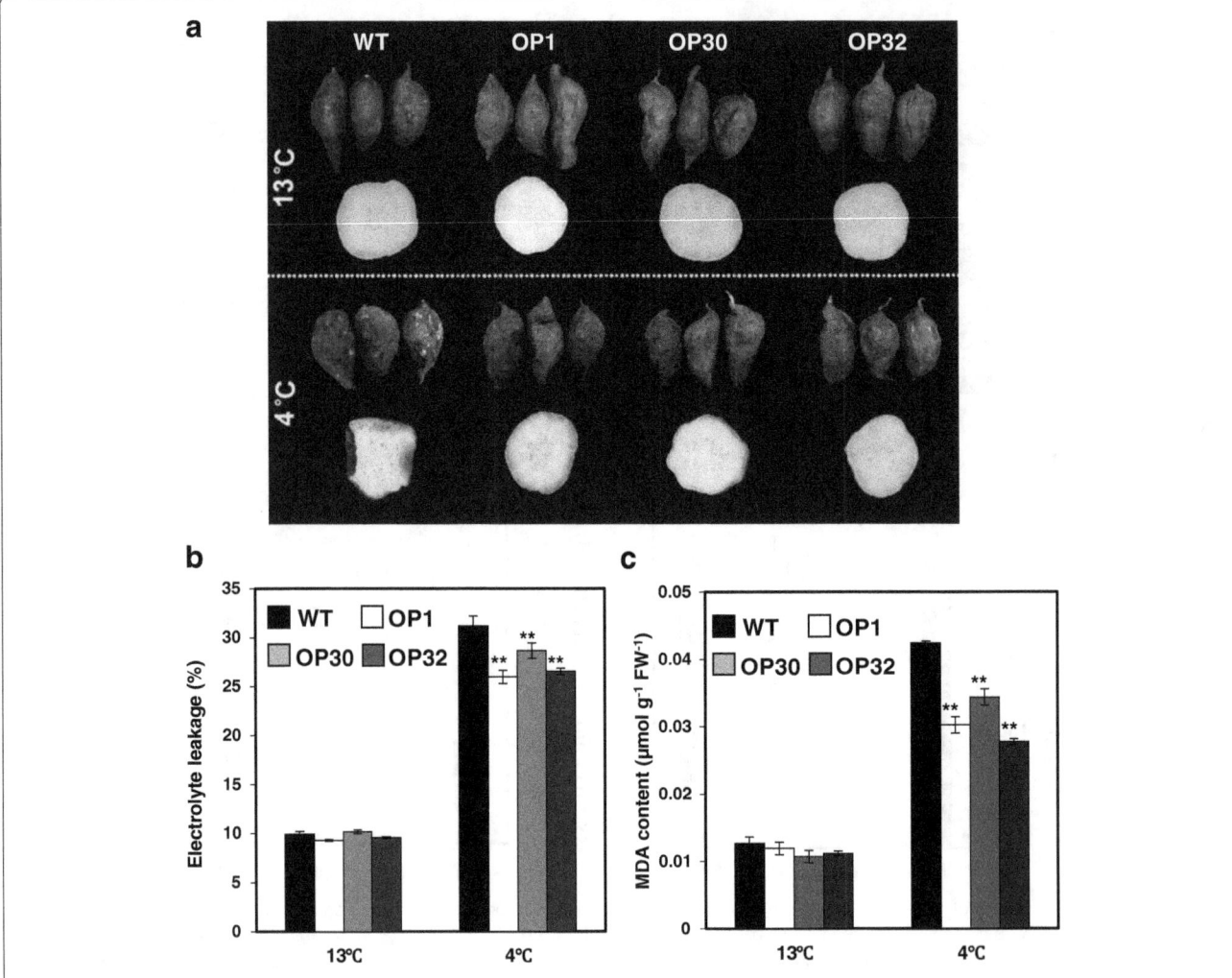

Fig. 6 Storage ability of transgenic sweetpotato tuberous roots during low temperature storage. **a** Photographs of WT and OP tuberous roots stored at 13 °C and 4 °C for 8 weeks. **b** Analysis of ion leakage and **c** MDA contents in tuberous roots of WT and OP plants stored in 13 °C and 4 °C for 8 weeks. Data are expressed as the mean ± SD of three replicates. *Asterisks* indicate significant differences between WT and OP plants by ANOVA at * $p < 0.05$ and ** $p < 0.01$

recover from low temperature stress than WT plants. Finally, the tuberous roots of OP transgenic sweetpotato showed enhanced storage ability at low temperatures compared to WT plants.

Methods
Plant materials
Sweetpotato (*Ipomoea batatas* [L.] Lam., cv. Xushu 29, one of the most widely grown varieties in Northwest China) plants were used in this study. The plants were cultivated in a growth chamber in soil at 25 °C under a 16 h/8 h (light/dark) photocycle. Embryogenic calli cultured on MS medium containing 1 mg l^{-1} 2,4-dichlorophenoxyacetic acid (2,4-D). The embryogenic calli were proliferated by subculture in fresh medium at 3 week intervals in the dark in a 25 °C incubator.

Vector construction and transformation
The *AtP3B* gene construct was generated in plant expression vector pCAMBIA1300 using CaMV 35S promoter and a NOS terminator sequence. CaMV 35Spro::AtP3B plasmids were transformed into embryogenic calli from sweetpotato via *Agrobacterium*-mediated transformation as described by Lim et al. [54]. The transformed embryogenic calli were selected on MS medium containing 400 mg l^{-1} cefotaxime, and 25 mg l^{-1} hygromycin, and subcultured in fresh medium at 3 week intervals. After transgenic sweetpotato plants were generated, vine cuttings from the transgenic sweetpotato and WT plants were used for propagation. Regenerated plants were transplanted into pots and grown in a greenhouse for further analysis.

Southern blot analysis

Up to 30 μg genomic DNA of indicated sweetpotato plants was digested with *Eco*RI (Roche, Manheim, Germany) and separated on a 0.8% agarose gel. The separated DNA was hybridized with an [α-^{32}P]dCTP labeled probe after transferred to a positively charged nylon membrane (Bio-Rad, CA, USA). The probe was designed by the coding sequence of 35S and *AtP3B* gene (5′-CCGGAAAC-CTCCTCGGATTC-3′, 5′-ATCTCCAGCG-CAAGCTTGTT-3′). Autoradiography was used to detect the hybridization signals.

Gene expression analysis

Genomic DNA was extracted from putative WT (cv. Xushu 29) and transgenic plant using purified genomic DNA in premix (Bioneer, Korea). The specific primer set based on the sequence of part of the 35S promoter and *AtP3B* gene (5′-CTACAAATGCCATCATTGCG-3′, 5′-CTCTTCCTCCTT-TGTGGCTG-3′) was used in PCR analysis. Total RNA was isolated from the third leaf, of sweetpotato shoot tips, using TRIzol reagent (Invitrogen, USA), and reverse transcribed using TOPscript™ RT DryMIX (dT18) (Enzynomics, Korea) according to the manufacturer's instructions. The gene-specific primers used for PCR analysis were as follows: the AtP3B primer set (5′-GATGATTGAGCCTGCGATTC-3′, 5′-TTACC CCTTTTCACCAGCAC-3′) was used to amplify a cDNA encoding AtP3B. The total synthesized cDNA was also used to amplify the ubiquitin extension gene (*UBI*) as a reference gene using *UBI* gene-specific primers (5′-TCGACAATGTGAAGGCAAAG-3′, 5′-CT TGATCTTCTTCGGCTTGG-3′) [55]. All quantitative RT-PCR analysis was conducted with Ever-Green 20 Fluorescent Dye (BioFACT, Korea) in a CFX96 Touch Real-time PCR Detection System (MJ Research, USA). All reactions were repeated at least three times.

Stress treatment of whole plants

WT and OP plants were grown at 25 °C (60% relative humidity, 16 h/8 h [light/dark] photoperiod) in a growth chamber for 1 month and utilized for the temperature stress tolerance assay. For the heat stress experiments, the sweetpotato plants were transferred to a growth chamber maintained at 45 °C for 12 h and then returned to 25 °C for 24 h recovery. For low temperature stress experiments, the plants were transferred to a growth chamber maintained at 4 °C for 48 h and then returned to 25 °C for 24 h recovery.

Low temperature storage of sweetpotato tuberous roots

WT and OP sweetpotato tuberous roots were harvested, followed by curing. The storage temperature was then changed to 13 °C (optimal storage condition) at relative humidity >80%. Roots were sampled for the control time point (0 week) and at specified internals during an 8 week exposure to storage at optimal storage condition (13 °C) and low temperature storage condition (4 °C), respectively.

Analysis of photosynthetic activity

The photosynthetic activity was estimated based on the maximal yield of the photochemical reaction in PSII (Fv/ fm). The Fv/fm values in the 3rd-4th mature leaves from top of 1-month-old WT and OP plants were measured using a portable Chl Fluorescence Meter (Handy PEA, England) after 30 min of dark adaption

Analysis of lipid peroxidation

Malondialdehyde (MDA) content, a marker of lipid peroxidation, was determined according to a modified thiobarbituric acid (TBA) method [56]. The MDA content was determined spectrophotometrically at A_{532} and A_{600}. The experiments were repeated three times.

Ion leakage measurements

Ion leakage was measured in root tissues according to Lieberman et al. with minor modifications [57]. The leakage from tissue slices was determined using conductivity measurements of solutions surrounding 40 discs per replication, which were approximately 2 mm thick and 1 cm in diameter. The discs were washed ten times in deionized water and drained. The samples were placed in test tubes containing 20 ml of deionized water at 25 °C for 1 h at a shaking speed of 60 rpm. Conductivity was measured using an ion conductivity meter (S230 SevenCompact™, Mettler Toledo, Switzerland). The values were compared with the total conductivity of the solution after autoclaving at 121 °C for 15 min. The experiments were repeated three times.

Quantification of hydrogen peroxide (H_2O_2)

Hydrogen peroxide (H_2O_2) levels were determined according to Velikova et al. with minor modifications [58]. Leaf tissues (100 mg) were ground in liquid nitrogen and extracted with 1 ml of 0.1% (*w/v*) TCA. The homogenate was centrifuged at 12,000 rpm for 15 min, and 0.5 ml of the supernatant was combined with 0.5 ml 1 M KI. After a 1 h reaction in darkness, the absorbency of the supernatant was read at 390 nm. The H_2O_2 content was calculated using a standard curve. The experiments were repeated three times.

Antioxidant enzyme activity analysis

The third leaves of sweetpotato plants were homogenized in cold condition with 0.1 M potassium phosphate buffer (pH = 7). The homogenate was centrifuged at 12,000 g for 15 min at 4 °C. The supernatant was used immediately for enzyme assays. POD activity was assayed as described by Kwak et al. [59] using pyrogallol as a substrate. CAT activity was assayed as described in Aebi et al. [60]. The experiments were repeated three times.

Field cultivation of transgenic sweetpotato plants

Sweetpotato field cultivation was conducted in 2015 at the Living Modified Organism (LMO) field of National Institute of Crop Science, Muan, South Korea. Stems were subsequently cut to 15 cm in length, and 40 stems per line were transplanted in the field with three replications. The field cultivation method was performed according to Park et al. [61]. At harvest time, shoot lengths and the fresh weights of the aerial parts and tuberous roots were recorded.

Additional file

Additional file 1: Figure S1. Transcript levels of *IbHSP*, *IbCBF*, and *IbCOR* genes in WT and OP plants. qRT-PCR analysis of *IbHSP*, *IbCBF*, and *IbCOR* genes including *IbHSP17.6*, *IbHSP18.2*, *IbCBF3* and *IbCOR27* was performed using mRNA isolated from sweetpotato leaves. Under normal growth conditions (25 °C), WT and OP plant samples were taken, and the transcript levels of these genes were normalized to tubulin gene (*TUB*) expression. The error bars represent the mean ± SD of three biological replicates. Asterisks indicate significant differences between WT and OP plants by ANOVA at * $p < 0.05$ and ** $p < 0.01$. **Table S1.** Primers used for PCR analysis.

Abbreviations

APX: Ascorbate peroxidase; CaMV: Cauliflower mosaic virus; CAT: Catalase; CBF: C-repeat-binding factor; COR: Cold-responsive; HSP: Heat-shock protein; MDA: Malondialdehyde; MV: Methyl viologen; POD: Peroxidase; qRT-PCR: Quantitative reverse-transcription PCR; ROS: Reactive oxygen species; WT: Wild type

Acknowledgments

This work was supported by grants from the National Research Foundation of Korea (NRF) funded by the Korean Government (2015053321), and the KRIBB initiative program.

Authors' contributions

CYJ, RJ, ZX, and S-SK conceived and designed the experiments. CYJ, CHK, YHC, HSK, C-JL, LK, S-EK, H-UL, and JSL performed the experiments. CYJ, CHK, YHC, SYL, YX, HL, DM, and S-SK analyzed the data. CYJ, HSK, and S-SK wrote the paper. All authors read and approved the final manuscript.

Competing interests

The authors declare that they have no competing interests.

Author details

¹Plant Systems Engineering Research Center, Korea Research Institute of Bioscience and Biotechnology (KRIBB), 125 Gwahak-ro, Daejeon 34141, South Korea. ²Department of Environmental Biotechnology, Korea University of Science and Technology (UST), 217 Gajeong-ro, Daejeon 34113, South Korea. ³Sweetpotato Research Center, Jiangsu Academy of Agricultural Science, Xuhuai Road, Xuzhou, Jiangsu 221131, China. ⁴Bioenergy Crop Research Center, National Institute of Crop Science, Rural Development Administration, Muan 58545, South Korea. ⁵Division of Applied Life Science (BK21 Plus program) and Plant Molecular Biology and Biotechnology Research Center, Gyeongsang National University, 501 Jinjudae-ro, Jinju 52828, South Korea.

References

1. Smirnoff N. Plant resistance to environmental stress. Curr Opin Biotechnol. 1998;9:214–9.
2. Tuteja N. Abscisic acid and abiotic stress signaling. Plant Signal Behav. 2007;2:135–58.
3. Chinnusamy V, Schumaker K, Zhu JK. Molecular genetic perspectives on cross-talk and specificity in abiotic stress signaling in plants. J Exp Bot. 2004;55:225–36.
4. Yamaguchi-Shinozaki K, Shinozaki K. Transcriptional regulatory networks in cellular responses and tolerance to dehydration and cold stresses. Annu Rev Plant Biol. 2006;57:781–803.
5. Das K, Roychoudhury A. Reactive oxygen species (ROS) and response of antioxidants as ROS-scavengers during environmental stress in plants. Front Environ Sci. 2014;2:53.
6. Iba K. Acclimative response to temperature stress in higher plants: approaches of gene engineering for temperature tolerance. Annu Rev Plant Biol. 2002;53:225–45.
7. Ahuja I, de Vos RC, Bones AM, Hall RD. Plant molecular stress responses face climate change. Trends Plant Sci. 2010;15:664–74.
8. Bovell-Benjamin A. Sweetpotato, a review of its past, present, and future role in human nutrition. Adv Food Nutr Res. 2007;52:1–59.
9. FAO. 2013. www.fao.org/giews/english/fo/index.htm.
10. Yoshinaga M, Yamakawa O, Nakatani M. Genotypic diversity of anthocyanin content and composition in purple-fleshed sweet potato [*Ipomoea batatas* (L.) Lam]. Breeding Sci. 1999;49:43–7.
11. Teow CC, Truong V-D, McFeeters RF, Thompson RL, Pecota KV, Yencho GC. Antioxidant activities, phenolic and β-carotene contents of sweet potato genotypes with varying flesh colours. Food Chem. 2007;103:829–38.
12. Zhang LA, Zhao H, Gan MZ, Jin YL, Gao XF, Chen QA, Guan JF, Wang ZY. Application of simultaneous saccharification and fermentation (SSF) from viscosity reducing of raw sweetpotato for bioethanol production at laboratory, pilot and industrial scales. Bioresour Technol. 2011;102:4573–9.
13. Ji CY, Kim YH, Kim HS, Ke Q, Kim GW, Park SC, Lee HS, Jeong JC, Kwak SS. Molecular characterization of tocopherol biosynthetic genes in sweetpotato that respond to stress and activate the tocopherol production in tobacco. Plant Physiol Biochem. 2016;106:118–28.
14. Ziska LH, Runion GB, Tomecek M, Prior SA, Torbet HA, Sicher R. An evaluation of cassava, sweet potato and field corn as potential carbohydrate sources for bioethanol production in Alabama and Maryland. Biomass Bioenergy. 2009;33:1503–8.
15. Dhir SK, Oglesby J, Bhagsari AS. Plant regeneration via somatic embryogenesis, and transient gene expression in sweet potato protoplasts. Plant Cell Rep. 1998;17:665–9.
16. Lebot V. Tropical root and tuber crops: cassava, sweet potato, yams and aroids. Oxfordshire, UK: CABI; 2009.
17. Bita CE, Gerats T. Plant tolerance to high temperature in a changing environment: scientific fundamentals and production of heat stress-tolerant crops. Front Plant Sci. 2013;4:273.
18. Kim SH, Ahn YO, Ahn MJ, Jeong JC, Lee HS, Kwak SS. Cloning and characterization of an orange gene that increases carotenoid accumulation and salt stress tolerance in transgenic sweetpotato cultures. Plant Physiol Biochem. 2013;70:445–54.
19. Kim SH, Kim YH, Ahn YO, Ahn MJ, Jeong JC, Lee HS, Kwak SS. Downregulation of the lycopene e-cyclase gene increases carotenoid synthesis via the b-branch-specific pathway and enhances salt-stress tolerance in sweetpotato calli. Physiol Plant. 2013;147:432–42.
20. Wang LJ, He SZ, Zhai H, Liu DG, Wang YN, Liu QC. Molecular cloning and functional characterization of a salt tolerance associated gene *IbNFU1* from sweetpotato. J Integr Agric. 2013;12:27–35.
21. Liu DG, Wang LJ, Liu CL, Song XJ, He SZ, Zhai H, Liu QC. An *Ipomoea batatas* Iron-sulfur cluster scaffold protein gene, *IbNFU1*, is involved in salt tolerance. PLoS One. 2014;9:e93935.

22. Liu DG, Wang LJ, Zhai H, Song XJ, He SZ, Liu QC. A novel a/b-hydrolase gene *IbMas* enhances salt tolerance in transgenic sweetpotato. PLoS One. 2014;9:e115128.
23. Liu DG, He SZ, Song XJ, Zhai H, Liu N, Zhang DD, Ren ZT, Liu QC. *IbSIMT1*, a novel salt-induced methyltransferase gene from *Ipomoea batatas*, is involved in salt tolerance. Plant Cell Tissue Organ Cult. 2015;120:701–15.
24. Park S, Kim HS, Jung YJ, Kim SH, Ji CY, Wang Z, Jeong JC, Lee H-S, Lee SY, Kwak SS. Orange protein has a role in phytoene synthase stabilization in sweetpotato. Sci Rep. 2016;6:33563.
25. Wang B, Zhai H, He S, Zhang H, Ren Z, Zhang D, Liu Q. A vacuolar Na+/H+ antiporter gene, *IbNHX2*, enhances salt and drought tolerance in transgenic sweetpotato. Scientia Hort. 2016;201:153–66.
26. Liljas A. Comparative biochemistry and biophysics of ribosomal proteins. Int Rev Cytol. 1991;124:103–36.
27. Tchórzewski M. The acidic ribosomal P proteins. Int J Biochem Cell Biol. 2002;34:911–5.
28. Zinker S, Warner J. The ribosomal proteins of Saccharomyces Cerevisiae phosphorylated and exchangeable proteins. J Biol Chem. 1976;251:1799–807.
29. Wool IG, Chan YL, Glück A, Suzuki K. The primary structure of rat ribosomal proteins P0, P1 and P2 and a proposal for a uniform nomenclature for mammalian and yeast ribosomal proteins. Biochimie. 1991;73:861–70.
30. Bailey-Serres J, Vangala S, Szick K, Lee CHK. Acidic phosphoprotein complex of 60S ribosomal subunit of maize seedling roots. Plant Physiol. 1997;114:1293–305.
31. Wittmann-Liebold B. Ribosomal proteins: their structure and evolution. In: Hardesty B, Kramer G, editors. Structure function and genetics of ribosomes. New York: Springer-Verlag; 1986. p. 326–61.
32. Tchórzewski M, Boldyreff B, Grankowski N. Extraribosomal function of acidic ribosomal P1-protein YP1alpha from Saccharomyces Cerevisiae. Acta Biochim Polon. 1999;46:901–10.
33. Carroll A, Heazlewood J, Ito J, Millar A. Analysis of the Arabidopsis cytosolic ribosome proteome provides detailed insights into its components and their post-translational modification. Mol Cell Proteomics. 2008;7:347.
34. Aguilar R, Montoya L. Sanchez de Jimenez E. Synthesis and phosphorylation of maize acidic ribosomal proteins implications in translational regulation. Plant Physiol. 1998;116:379–85.
35. Szick K, Springer M, Bailey-Serres J. Evolutionary analyses of the 12-kDa acidic ribosomal P-proteins reveal a distinct protein of higher plant ribosomes. Proc Natl Acad Sci U S A. 1998;95:2378–83.
36. Kang CH, Lee YM, Park JH, Nawkar GM, Oh HT, Kim MG, Lee SI, Kim WY, Yun DJ, Lee SY. Ribosomal P3 protein AtP3B of Arabidopsis acts as both protein and RNA chaperone to increase tolerance of heat and cold stresses. Plant Cell Environ. 2016;39:1631–42.
37. Hodges DM, DeLong JM, Forney CF, Prange RK. Improving the thiobarbituric acid-reactive-substances assay for estimating lipid peroxidation in plant tissues containing anthocyanin and other interfering compounds. Planta. 1999;207:604e611.
38. ÓKane D, Gill V, Boyd P, Burdon R. Chilling, oxidative stress and antioxidant responses in Arabidopsis Thaliana Callus. Planta. 1996;198:371–7.
39. Apel K, Hirt H. Reactive oxygen species: metabolism, oxidative stress, and signal transduction. Annu Rev Plant Biol. 2004;55:373–99.
40. Asada K, Takahashi M. Production and scavenging of active oxygen in photosynthesis. In: Kyle DJ, Osmond CB, Arntzen CJ, editors. Photoinhibition. Amsterdam: Elsevier; 1987. p. 227–87.
41. Mittler R, Vanderauwera S, Gollery M, Van Breusegem F. Reactive oxygen gene network of plants. Trends Plant Sci. 2004;9:490–8.
42. Yamaki S, Uritani I. Mechanism of chilling injury in sweet potato VII. Changes in mitochondrial structure during chilling storage Plant Cell Physiol. 1972;13:795e805.
43. Lieberman M, Craft CC, Audia W, Wilcox M. Biochemical studies of chilling injury in sweetpotatoes. Plant Physiol. 1958;33:307e311.
44. Minamikawa T, Akazawa T, Uritani I. Mechanism of cold injury in sweetpotatoes II. Biochemical mechanism of cold injury with special reference to mitochondrial activities Plant Cell Physiol. 1961;2:301e309.
45. Picha D. Chilling injury, respiration, and sugar changes in sweet potatoes stored at low temperature. J Am Soc Hortic Sci. 1987;112:497e502.
46. Uritani I. Biochemistry on postharvest metabolism and deterioration of some tropical tuberous crops. Bot Bull Acad Sin. 1999;40:177e183.
47. Boyer JS. Plant productivity and environment. Science. 1982;218:443–8.
48. Foyer CH, Noctor G. Redox homeostasis and antioxidant signaling: a metabolic interface between stress perception and physiological responses. Plant Cell. 2005;17:1866–75.
49. Quan LJ, Zhang B, Shi WW, Li HY. Hydrogen peroxide in plants: a versatile molecule of the reactive oxygen species network. J Integr Plant Biol. 2008;50:2–18.
50. Passardi F, Cosio C, Penel C, Dunand C. Peroxidases have more functions than a Swiss army knife. Plant Cell Rep. 2005;24:255–65.
51. Willekens H, Chamnongpol S, Davey M, Schraudner M, Langebartels C, Van Montagu M, Inze D, Van Camp W. Catalase is a sink for H2O2 and is indispensable for stress defence in C3 plants. EMBO J. 1997;16:4806–16.
52. Kim YH, Kim MD, Park SC, Yang KS, Jeong JC, Lee HS, Kwak SS. *SCOF-1*-expressing transgenic sweetpotato plants show enhanced tolerance to low-temperature stress. Plant Physiol Biochem. 2011;49:1436–41.
53. Kim YH, Lim S, Yang KS, Kim CY, Kwon SY, Lee HS, Wang X, Zhou Z, Ma D, Yun DJ, et al. Expression of Arabidopsis *NDPK2* increases antioxidant enzyme activities and enhances tolerance to multiple environmental stresses in transgenic sweetpotato plants. Mol Breed. 2009;24:233–44.
54. Lim S, Yang KS, Kwon SY, Paek KY, Kwak SS, Lee HS. Agrobacterium-mediated genetic transformation and plant regeneration of sweetpotato (*Ipomoea batatas*). J Plant Biotechnol. 2011;31:267–71.
55. Park SC, Kim YH, Ji CY, Park S, Jeong JC, Lee HS, Kwak SS. Stable internal reference genes for the normalization of real-time PCR in different sweetpotato cultivars subjected to abiotic stress conditions. PLoS One. 2012;7:e51502.
56. Horie T, Motoda J, Kubo M, Yang H, Yoda K, Horie R, Chan WY, Leung HY, Hattori K, Konomi M. Enhanced salt tolerance mediated by AtHKT1 transporter-induced Na+ unloading from xylem vessels to xylem parenchyma cells. Plant J. 2005;44:928–38.
57. Lieberman M, Craft CC, Audia W, Wilcox M. Biochemical studies of chilling injury in Sweetpotatoes. Plant Physiol. 1958;33:307–11.
58. Velikova V, Yordanov I, Edreva A. Oxidative stress and some antioxidant systems in acid rain-treated bean plants: protective role of exogenous polyamines. Plant Sci. 2000;151:59–66.
59. Kwak SS, Kim SK, Lee MS, Jung KH, Park IH, Liu JR. Acidic peroxidases from suspension-cultures of sweet potato. Phytochemistry. 1995;39:981–4.
60. Aebi H. Catalase in vitro. Meth Enzymol. 1984;105:J21–6.
61. Park SC, Kim YH, Kim SH, Jeong YJ, Kim CY, Lee JS, Bae JY, Ahn MJ, Jeong JC, Lee HS, Kwak SS. Overexpression of the *IbMYB1* gene in an orange-fleshed sweet potato cultivar produces a dual-pigmented transgenic sweet potato with improved antioxidant activity. Physiol Plant. 2015;153:525–37.

PERMISSIONS

LIST OF CONTRIBUTORS

Margarita Hadjipieri, Egli C. Georgiadou, Vlasios Goulas, Vasileios Fotopoulos and George A. Manganaris
Department of Agricultural Sciences, Biotechnology and Food Science, Cyprus University of Technology, 3603 Lemesos, Cyprus

Alicia Marin, Huertas M. Diaz-Mula and Francisco A. Tomás-Barberán
Quality, Safety, and Bioactivity of Plant Foods, CEBAS-CSIC, P.O. Box 164, Espinardo, Murcia, Spain

Wei Hu, Yan Yan, Juhua Liu, Hongxia Miao, Weiwei Tie, Zehong Ding, XuPo Ding, Chunlai Wu, Yang Liu and Biyu Xu
Key Laboratory of Biology and Genetic Resources of Tropical Crops, Institute of Tropical Bioscience and Biotechnology, Chinese Academy of Tropical Agricultural Sciences, Haikou, Hainan, China

Zhiqiang Jin
Key Laboratory of Biology and Genetic Resources of Tropical Crops, Institute of Tropical Bioscience and Biotechnology, Chinese Academy of Tropical Agricultural Sciences, Haikou, Hainan, China
Key Laboratory of Genetic Improvement of Bananas, Hainan province, Haikou Experimental Station, China Academy of Tropical Agricultural Sciences, Haikou, Hainan, China

Jiashui Wang
Key Laboratory of Genetic Improvement of Bananas, Hainan province, Haikou Experimental Station, China Academy of Tropical Agricultural Sciences, Haikou, Hainan, China

Haitao Shi
Hainan Key Laboratory for Sustainable Utilization of Tropical Bioresources, College of Agriculture, Hainan University, Haikou, China

Carla Ibañez, Julia Bellstädt, Kathrin Denk, Marcel Quint and Carolin Delker
Institute of Agricultural and Nutritional Sciences, Martin Luther University Halle-Wittenberg, Betty-Heimann-Str. 5, 06120 Halle (Saale), Germany

Department of Molecular Signal Processing, Leibniz Institute of Plant Biochemistry, Weinberg 3, 06120 Halle (Saale), Germany.

Tom Peterson
Department of Molecular Signal Processing, Leibniz Institute of Plant Biochemistry, Weinberg 3, 06120 Halle (Saale), Germany

Yvonne Poeschl and Andreas Gogol-Döring
German Centre for Integrative Biodiversity Research (iDiv) Halle-Jena-Leipzig, Deutscher Platz 5e, 04103 Leipzig, Germany
Institute of Computer Science, Martin Luther University Halle-Wittenberg, Von-Seckendorff-Platz 1, 06099 Halle (Saale), Germany

Jian Xu, Ji Li, Li Cui, Ting Zhang, Zhe Wu, Pin-Yu Zhu, Yong-Jiao Meng, Kai-Jing Zhang, Xia-Qing Yu, Qun-Feng Lou and Jin-Feng Chen
State Key Laboratory of Crop Genetics and Germplasm Enhancement, Nanjing Agricultural University, Nanjing 210095, China

Margaret A. Carpenter, Martin Shaw, Rebecca D. Cooper, Tonya J. Frew, Ruth C. Butler, Sarah R. Murray, Leire Moya, and Gail M. Timmerman-Vaughan
The New Zealand Institute for Plant & Food Research Limited, PO Box 4704, Christchurch, New Zealand

Clarice J. Coyne
USDA-ARS Western Regional Plant Introduction Station, 59 Johnson Hall, WSU Pullman, Pullman, Washington WA 99164-6402, USA

Guo-qing Song
Plant Biotechnology Resource and Outreach Center, Department of Horticulture, Michigan State University, East Lansing, MI 48824, USA

Xuan Gao
Plant Biotechnology Resource and Outreach Center, Department of Horticulture, Michigan State University, East Lansing, MI 48824, USA

Key Laboratory for the Conservation and Utilization of Important Biological Resources, College of Life Sciences, Anhui Normal University, Wuhu 241000, China

Melanie Pajon and Vicente J. Febres
Horticultural Sciences Department, Institute of Food and Agricultural Sciences, University of Florida, 2550 Hull Road, Gainesville, FL 32611, USA

Gloria A. Moore
Horticultural Sciences Department, Institute of Food and Agricultural Sciences, University of Florida, 2550 Hull Road, Gainesville, FL 32611, USA
Plant Molecular and Cellular Biology Program, University of Florida, Gainesville, FL 32611, USA

Hamama Islam Butt, Zhaoen Yang, Qian Gong, Eryong Chen, Xioaqian Wang, Ge Zhao, Xiaoyang Ge, Xueyan Zhang and Fuguang Li
State Key Laboratory of Cotton Biology, Institute of Cotton Research of Chinese Academy of Agricultural Science (ICR, CAAS), Anyang 455000, China

Farzaneh Yazdanpanah, Henk W.M. Hilhorst and Leónie Bentsink
Wageningen Seed Laboratory, Laboratory of Plant Physiology, Wageningen University, Droevendaalsesteeg 1, 6708, PB, Wageningen, The Netherlands

Johannes Hanson
Umeå Plant Science Center, Department of Plant Physiology, Umeå University, SE-901 87 Umeå, Sweden
Department of Molecular Plant Physiology, Utrecht University, Padualaan 8, 3584, CH, Utrecht, The Netherlands

L. Fattorini, A. Veloccia, F. Della Rovere, S. D'Angeli, G. Falasca and M. M. Altamura
Dipartimento di Biologia Ambientale, Sapienza Università di Roma, Roma, Italy

Baodi Bi, Jingliang Tang, Shuang Han, Jinggong Guo and Yuchen Miao
Institute of Plant Stress Biology, State Key Laboratory of Cotton Biology, Department of Biology, Henan University, 85 Minglun Street, Kaifeng 475001, China

Zhaoen Yang, Wenqiang Qin, Fuguang Li
Xinjiang Research Base, State Key Laboratory of Cotton Biology, Xinjiang Agricultural University, Urumqi 830052, China

Institute of Cotton Research, Chinese Academy of Agricultural Sciences, Anyang 455000, China

Qian Gong, Zuoren Yang, Yuan Cheng, Lili Lu, Xiaoyang Ge, Chaojun Zhang and Zhixia Wu
Institute of Cotton Research, Chinese Academy of Agricultural Sciences, Anyang 455000, China

Christelle Bonnet Steve Lassueur and Philippe Reymond
Department of Plant Molecular Biology, University of Lausanne, Biophore Building, 1015 Lausanne, Switzerland

Camille Ponzio, Rieta Gols and Marcel Dicke
Laboratory of Entomology, Wageningen University, P.O. Box 16, 6700 AA Wageningen, The Netherlands

Biyue Tan
Umeå Plant Science Centre, Department of Ecology and Environmental Science, Umeå University, Umeå SE-90187, Sweden
Biomaterials Division, Stora Enso AB, Nacka SE-13104, Sweden

Pär K. Ingvarsson
Umeå Plant Science Centre, Department of Ecology and Environmental Science, Umeå University, Umeå SE-90187, Sweden
Present address: Department of Plant Biology, Uppsala BioCenter, Swedish University of Agricultural Sciences, Uppsala SE-75007, Sweden

Björn Sundberg
Biomaterials Division, Stora Enso AB, Nacka SE-13104, Sweden

Dario Grattapaglia
EMBRAPA Genetic Resources and Biotechnology – EPqB, Brasilia, DF 70770-910, Brazil
Universidade Católica de Brasília- SGAN, 916 modulo B, Brasilia, DF 70790-160, Brazil

Gustavo Salgado Martins and Karina Zamprogno Ferreira
Veracel Celulose S.A., Eunápolis, BA 45.820-970, Brazil

Ho Soo Kim
Plant Systems Engineering Research Center, Korea Research Institute of Bioscience and Biotechnology (KRIBB), 125 Gwahak-ro, Daejeon 34141, South Korea

Chang Yoon Ji, Chan-Ju Lee, Le Kang, So-Eun Kim and Sang-Soo Kwak
Plant Systems Engineering Research Center, Korea Research Institute of Bioscience and Biotechnology (KRIBB), 125 Gwahak-ro, Daejeon 34141, South Korea
Department of Environmental Biotechnology, Korea University of Science and Technology (UST), 217 Gajeong-ro, Daejeon 34113, South Korea

Rong Jin
Plant Systems Engineering Research Center, Korea Research Institute of Bioscience and Biotechnology (KRIBB), 125 Gwahak-ro, Daejeon 34141, South Korea
Department of Environmental Biotechnology, Korea University of Science and Technology (UST), 217 Gajeong-ro, Daejeon 34113, South Korea
Sweetpotato Research Center, Jiangsu Academy of Agricultural Science, Xuhuai Road, Xuzhou, Jiangsu 221131, China

Zhen Xu, Yiping Xie, Hongmin Li and Daifu Ma
Sweetpotato Research Center, Jiangsu Academy of Agricultural Science, Xuhuai Road, Xuzhou, Jiangsu 221131, China

Hyeong-Un Lee and Joon Seol Lee
Bioenergy Crop Research Center, National Institute of Crop Science, Rural Development Administration, Muan 58545, South Korea

Chang Ho Kang, Yong Hun Chi and Sang Yeol Lee
Division of Applied Life Science (BK21 Plus program) and Plant Molecular Biology and Biotechnology Research Center, Gyeongsang National University, 501 Jinjudae-ro, Jinju 52828, South Korea

Index